Biology and Biochemistry of Nitrogen Fixation

Studies in Plant Science, 1

Biology and Biochemistry of Nitrogen Fixation

Edited by

M.J. Dilworth and A.R. Glenn

School of Biological & Environmental Sciences, Murdoch University, Murdoch, Western Australia 6150

ELSEVIER

Amsterdam — New York — Oxford — Tokyo 1991

ELSEVIER SCIENCE PUBLISHERS B.V.
Sara Burgerhartstraat 25
P.O. Box 211, 1000 AE Amsterdam, The Netherlands

Distributors for the United States and Canada:

ELSEVIER SCIENCE PUBLISHING COMPANY INC.
655, Avenue of the Americas
New York, NY 10010, U.S.A.

ISBN 0-444-88960-4

This book is printed on acid-free paper.

Printed in The Netherlands

TABLE OF CONTENTS

Acknowledgements

The genesis of every book has a beginning and an end. Upon its completion the work will hopefully appear on bookshelves and be much used by appreciative purchasers. If successful, the authors of the volume will have the satisfaction of having contributed to the dissemination of ideas and knowledge. However, we have become much more aware of the contributions of those whose names do not grace the list of authors, but whose work nevertheless has made this volume possible. We would like to thank the unknown typists and graphic artists who got text and diagrams into a suitable form. Our understanding of the power, limitations and incompatibilities of computer systems has increased significantly during the preparation of this book. To Kaye Brotherwood our ever-willing fixer of disk reading problems - our eternal thanks. To our long-suffering secretary Marika Gurney - our thanks and the best present of all - the knowledge that the book is finished. Our research group staff learnt the art of treading softly when all was not going well and provided encouragement when the task before us seemed too formidable. Lastly, we would like to thank our families especially Mary and Odile for sharing the struggle and the celebration with us. Now that the book is completed we hope to see more of all of them.

Andrew Glenn
Mike Dilworth
September, 1990.

CHAPTER 1

THE BIOLOGY AND BIOCHEMISTRY OF NITROGEN FIXATION: A LOOK FORWARD

ANDREW R. GLENN and MICHAEL J. DILWORTH
School of Biological and Environmental Sciences, Murdoch University, Murdoch, Western Australia 6150

The biology and biochemistry of nitrogen fixation continue to exercise a potent fascination for researchers from a wide variety of disciplines. Some of these seek to more fully exploit the practical benefits of prokaryotic N_2 fixation, in the field or forest; for others it is the intrinsic interest in biological systems that carry out such a complex process under the mildest of conditions. Both the practical and curiosity-driven research have their place; in truth they are difficult to totally separate since they overlap and feed each other. The widespread application of N_2 fixation in agriculture and forestry will happen only as and when basic and strategic research are carried out in concert. Because of the potential practical benefits from N_2 fixation, it has generated a significant research activity. However, like all rapidly changing human activities, there is a need to pause at intervals and to critically evaluate where we are going, and to separate rigorously-established fact from oft-repeated myth. This book arises out of such a questioning; all the contributors were asked to critically review one area of nitrogen fixation research and to evaluate progress. This we believe they have done.

We have chosen to begin the book with what is known about the chemistry, biochemistry, genetics and physiology of the various nitrogenase systems (Chapters 2 to 7). The occurrence of free-living diazotrophs and their function in associative symbioses follow (Chapters 8 and 9), with the legumes and their root nodule bacteria taking a prominent position (Chapters 10 to 14). The actinorhizal and cyanobacterial systems are considered in Chapters 15 and 16, leading on to an assessment of the prospects for genetic enhancement of nitrogen fixation (Chapter 17).

Diazotrophic bacteria can fix N_2 at atmospheric temperatures and pressures, whereas the Haber process for the industrial production of ammonia requires pressures of 200 atm and 800°C - data that provide eloquent estimony to the extraordinary nature of nitrogenase. Electrochemical reduction of N_2 has now been achieved at atmospheric temperature and pressure, giving promise of industrial application and possibly to models for nitrogenase action. It is, however, more accurate now to talk of the family of nitrogenases. After many years of study of the Mo-nitrogenase system (nif), there was initial scepticism about the work of Paul Bishop and collaborators on the possibility of Mo-independent nitrogenases. Genetic analysis and the purification of the proteins, however, have unequivocally established the existence not only of a V-nitrogenase system (vnf), but also of a nitrogenase-3 system (anf). These discoveries raise some significant questions about the mechanism of N_2 reduction in the various systems - especially after it had been assumed for many years that Mo was an obligate requirement for enzymatic N_2 reduction. Old paradigms have had to make way for an expanded view of the nitrogenases. However, the chemistry of metal-N_2 compounds is now on a sound footing (see Richards, Chapter 4) and this should contribute to the work of the biochemist struggling to understand the structure, biosynthesis and especially the mode of action of the various cofactors such as FeMoco and FeVaco in the respective nitrogenases. The nature of the cofactor in nitrogenase-3, if there is one, is unknown. Understanding the chemistry of V and Fe in the alternative nitrogenases may provide greater insight into the enzymatic requirements for N_2 reduction. The nitrogenase biochemist's prayer "Lord, let me have some crystals of purified nitrogenase - bigger ones would be better, but smaller will do!" seems to have been at least partially answered. Although there is general acceptance of the Thorneley-Lowe model for the mechanism of Mo-nitrogenase and site-directed mutagenesis of the nif genes can identify amino acid residues which are important to catalytic activity, resolution of the 3-dimensional X-ray crystallographic structure of the protein will be immensely valuable. The discovery of the alternative nitrogenases has not only given fresh impetus to nitrogenase biochemistry, but also raised some intriguing questions about the origins, distribution and rationale for the existence of multiple systems.

Currently the properties of the proteins and the organization of the nif, vnf and anf nitrogenases genes are known for only a small number of organisms (see Pau, Chapter 3 and Robson, Chapter 7). Little is known about how widespread such systems are, or their ecological significance. Investigations with free-living diazotrophs (see Silvester and Musgrave, Chapter 8) only establish the fact of $^{15}N_2$ incorporation and have not to date identified if this occurs via the Mo, V or nitrogenase-3 systems. Attempts to find acetylene-dependent ethane formation (an indicator of the V-nitrogenase and of nitrogenase-3) in Mo-deficient legumes have been unsuccessful (M.J. Dilworth and J.F. Loneragan, unpublished), suggesting that the V-nitrogenase or nitrogenase-3, even if they exist in root nodule bacteria, may have little or no agronomic significance. No doubt there will be further attempts to explore the ecological significance of the alternative nitrogenases, though such studies will need a level of biochemical expertise that has not always been present in ecological studies.

The presence of three different nitrogenase systems in Azotobacter offers the evolutionary biologist a real challenge. Are the systems related? Have they diverged from one another and if so when in time did this occur? At the moment the evidence from one nitrogenase component contradicts that from another (Pau, Cahpter 3). However, we will undoubtedly see much more on the evolution of nitrogenase genes in the future - perhaps providing the opportunity for those inclined to construct evolutionary trees to overlay the data with preconceived ideas, an argument pursued by Silvester and Musgrave in their chapter on diazotrophs (Chapter 8).

The genetic regulation of the various nitrogenase genes (nif, vnf and anf) is likely to prove a fertile research ground (see Elmerich, Chapter 6, and Robson, Chapter 7). The large number of papers on the nif genes in a range of bacteria is likely to be followed by an equal volume on vnf and anf. The nif systems in the various organisms are usually considered separately, but we asked Claudine Elmerich to attempt the daunting task of comparing and contrasting the various nif systems. It is clear that ntr or analogous systems are generally involved in the regulation of nif, but there are differences in the details of the relationship between N-supply and nif in different organisms. The extent to which there are common regulatory elements controlling nif, vnf and anf is not yet clear; some genes may function in the assembly of both Mo and V-nitrogenases. The involvement of metals in the regulation of nitrogenases and hydrogenases leads to important questions about how such apparently low concentrations are recognised by the cell. Relatively little is known about the mechanisms of uptake and metabolism of Mo, V, Fe and Ni in diazotrophic bacteria.

In addition to genetic regulation, the activity of nitrogenases is also affected by physiological factors, the supply of electrons and ATP and especially pO_2 (see Miller, Chapter 2, and Bergersen, Chapter 5). The energy supply in some systems seems to differ with the degree of anaerobiosis. In the bacteroids of legume root nodules, the energy supply appears to be C_4-dicarboxylates, though the rationale for the apparently obligate requirement for such compounds to be provided by the plant is still not clear (See Rosendahl and the editors, Chapter 12). The roles of electron transport and membrane energization in the N_2-fixing systems are gradually becoming clearer (see Bergersen, Chapter 5).

Hydrogen evolution is a consequence of the activity of nitrogenase. Some organisms possess genes for the uptake hydrogenase (Hup), which can recycle H_2 with a saving of energy and an increase in overall efficiency. There has been considerable debate about the field significance of the uptake hydrogenase, some reports suggesting that the hup system is agronomically important and others that it is not. The overall value of hup in the field remains to be proven. Since the soil is an assembly of heterogeneous environments, it is conceivable that in some habitats the hup system may be valuable, but not in all.

The number of known diazotrophs continues to expand as some long-standing myths about diazotrophic physiology are exposed and isolation techniques are modified (see Chapter 8). Although there are some pseudomonads that can fix N_2, it remains surprising to us that this extremely versatile bacterial group is so poorly represented among the free-living diazotrophs. Classical approaches to pure culture may well overlook a diazotroph which is dependent on a community of microbes for its nutrition. The contributions of the diazotrophs to N inputs into particular environments have ranged, in the words of Silvester and Musgrave, from "the insigificant to the improbable". There is clearly more to be done on the ecological importance of the free-living diazotrophs, particularly in ways that check the claimed rates of N_2 fixation with reasoned estimates of the likely carbon inputs required. Similarly, the vexed question of the benefits of associative N_2 fixation is still far from resolved (see Klucas, Chapter 9). Despite claims for benefits from inoculation it is not known whether such effects are due to N_2 fixation. Indeed, short-lived plant growth and yield responses to inoculation with particular strains of non-diazotrophic bacteria have beeen known since the 1920's. The significance of particular associations between plants and diazotrophs requires further attention and a more critical evaluation of the data.

The cyanobacteria are a major group of diazotrophs and progress in understanding the biology of the group has been steady (see Rowell and Kerby, Chapter 16). The formation of symbiotic associations between cyanobacteria and plants appears to be mainly restricted to lower plants, though it is not apparent why there are so few associations with angiosperms. Comparison of the morphological and biochemical processes involved in the cyanobacteria and rhizobial symbioses has not very often been attempted, but such comparisons may yield useful information. The development of appropriate genetic systems should allow even better dissection of the control mechanisms leading to heterocyst differentiation and its connection with nitrogenase induction. The elegant genetic rearrangements that have been reported to lead to re-formation of nifHDK in some strains of cyanobacteria seem to be the exception rather than the rule. Indeed, one of the consistent refrains throughout this book is that generalizations based on detailed studies of a few organisms or N_2-fixing systems can be very misleading. For good practical reasons, experimental systems are usually chosen on the basis of ease of culture and physiological manipulation and availability of genetic systems, rather than whether they are typical or atypical. It is only when a number of different systems have been studied that patterns begin to emerge. Learning more and more about one system, however elegant, may be less useful than casting a wider net. Paul Bishop might well have said:

"Two roads diverged in a wood, and I -
I took the one less well travelled by,
And that has made all the difference." (Robert Frost: The Road Not Taken)

Our advances in understanding the ecology of root nodule bacteria has not kept pace with developments in physiology and genetics. As Barnet (see Chapter 10) points out, we have often assumed that studies about inoculation success are synonymous with studies in ecology. Our attempts to study these organisms in the soil have been limited by the techniques that are available. Even the much-used technique of antibiotic resistance marked strains is fraught with danger and has in-built limitations that are not always realised. There is now increasing evidence that the competitiveness of strains can be modified by relatively simple changes to the genome. In our laboratory, a whole series of spontaneous chloramphenicol-resistant mutants of Rhizobium leguminosarum bv. trifolii were able to nodulate clover and fix N_2 as well as the wild type when studied in pure culture. When mixed with the wild type, however, they were markedly reduced in competitiveness and produced substantially fewer nodules than expected (F.J. Rynne, unpublished). The use of such strains in the field would produce totally misleading results. In the light of these observations, any mutants that are used in the field as control strains should be shown to be as competitive as the wild type. The development of specific DNA probes, hybridization and PCR techniques may help in the unequivocal identification of root nodule bacteria in the soil.

The genetics of nodulation has been one of the most competitive and rapidly expanding areas of research in nitrogen fixation (see Sobral and colleagues, Chapter 11). Many genes have been identified and sequenced though many nucleotide sequences still lack physiological functions to go with them. Perhaps when sequence libraries are considerably expanded the function of more of the genes may be established. The role of the flavones in the induction of the nod genes has been pursued by a number of groups and provides a clear example of the complex interactions between the legume host and the root nodule bacteria. The process of recognition between the bacteria and the legume host has attracted a great deal of research; the role of lectins still finds proponents and opponents, and the particular role of exopolysaccharides and lipopolysaccharides are still not clear, though a number of mutants lacking the exopolysaccharides are defective in nodulation.

Once the root nodule bacteria have differentiated into bacteroids, they are totally dependent on the host for mineral and organic nutrients. Although rhizobia are physiologically diverse, they probably receive significant concentrations of only a very limited range of carbon sources, even though small quantities of a wide range of materials can pass into the bacteroid (see Rosendahl and the editors, Chapter 12). Indirect evidence from studies on the physiology of the bacteroid and direct measurements of the permeability of the peribacteroid membrane both point to this membrane having a profound controlling influence on the flow of materials from the host cytosol into the bacteroids. The development of shuttle models (e.g., a malate-aspartate shuttle) has considerably expanded ways of thinking about how exchanges between rhizobia and legumes might take place, though there is a need for data from a wider range of systems before the role of shuttle mechanisms can be accepted. The consequences of these transfers of material into, and out of, the bacteroids for pH changes in the peribacteroid space remain to be determined, preferably by non-destructive methods.

Ammonia assimilation in the legume root nodule takes place in the host cytoplasm (see Atkins, Chapter 13). The dichotomy between ureide and amide producing plants is now well-established, but the detailed reasons underlying the dichotomy are not understood. The enzymology of the assimilation pathways is now the subject of detailed examination. Inhibitors of various steps in the assimilation process have been used quite extensively in whole plant studies, but they appear to be considerably less specific than was originally thought. Although the biochemistry of assimilation is now reasonably well understood, the high activities of glutamate dehydrogenase in most legume nodules are still unexplained. Molecular biology is contributing significantly to our understanding of which enzymes or isozymes are specifically induced in the nodule and are therefore primarily concerned with nitrogen fixation. The mechanisms for induction of nodule-specific enzymes required for ammonia assimilation and the inducers themselves are yet to be identified.

Understanding of the actinorhizal non-legume symbioses has accelerated with the _in vitro_ cultivation of at least some of the endophytes (see Schwintzer and Tjepkema, Chapter 15). The taxonomy of the genus _Frankia_ is still far from adequate. The ecology of various frankiae is gradually being clarified, but the complications of the spore-positive and spore-negative strains and the change in strain distribution over time have perhaps not been widely appreciated. However, these bacteria still lack detailed genetic systems, hampering investigation of this group. Reference to the genetic systems found in some of the antibiotic-producing actinomycetes may help in the development of such systems in the frankiae. The role of haemoglobin in actinorhizal nodules remains to be determined; in some actinorhizal nodules the haemoglobin content is very low and its capacity for facilitating oxygen transport at low dissolved O_2 concentration would seem to be slight. It is also somewhat surprising to find that its concentrations in some other nodules is as high as those in legume nodules when the nitrogenase in the endophyte is less sensitive to oxygen than it is in legume bacteroids.

Limitations to the productivity of the legume root nodule symbioses come from various environmental stresses (see Robson and Bottomley, Chapter 14). It is becoming clear that too little attention has been paid to identifying the factors limiting productivity, or to which of the partners is more sensitive to the stress. There is some evidence that naturally occurring gene pools, especially for acid tolerance, can be used to select bacterial and/or plant types with increased productivity in areas where the symbiosis is under environmental stress.

The world will continue to look to leguminous and actinorhizal plants and their associated bacteria to provide food, pastures and firewood, and to increase soil fertility. The potential for manipulating these systems is one of immense importance (see Phillips, Chapter 17). The use of the appropriate host legume or cultivar, and the appropriate root nodule bacteria, can increase productivity. The immense gene pool among the legumes has not yet been fully explored for agronomically useful species. In addition, there are still possibilities for new rhizobial isolates. The role that genetic engineering will ultimately play in the continued exploitation of N_2 fixation by legumes remains unclear. While the influence of molecular biology on our understanding of the process of nodulation and fixation has been profound, it has yet to put a single additional nodule on field legumes, whereas employing the variations in the natural gene pool still contributes substantially to agronomy. It will not be straightforward to construct rhizobial strains with desirable characteristics that will perform well in the field against the indigenous strains. Although the genetic conversion of relatively ineffective inoculant strains to a level of performance approaching that of regular inoculant strains may be achievable, it may have absolutely no practical value in the field.

The rapid developments in plant molecular biology may soon allow the development of genetically modified legumes with characteristics suitable for use in a range of different environments. In addition, the "holy grail" of the nitrogen-fixing wheat plant continues to exercise the hopes of some researchers. Recent claims for the hormone-mediated infection of wheat plants by root nodule bacteria may indeed represent the first step in this direction, or they may merely be an interesting but trivial artifact of no practical value. Only time will tell which of these is correct - but the construction of a new symbiosis will not be a simple task. The Rhizobium-legume symbiosis is the product of a complex set of interactions and depends on carefully-defined environmental conditions for the bacteria to fix N_2 at any significant rate. Even if genetically engineered associations of any agronomic significance can be constructed, it will be important that appropriate regulations are in place before they are released for commercial use. Molecular biology is an important tool in the hands of biologists seeking to understand and appropriately exploit the biology and biochemistry of nitrogen-fixing systems - but it needs to be used to answer important biological questions rather than as an end in itself. The study of N_2 fixation requires a genuinely interdisciplinary approach so that the ecologist benefits from the techniques developed by the molecular biologist, the nitrogenase biochemist from the skills of the chemist, and the rhizobial geneticist and physiologist from the observations and insights of the agricultural scientist. If this book contributes to increasing interdisciplinary approaches and an even more critical evaluation of the literature on the biology and biochemistry of nitrogen fixation it will have served its purpose.

CHAPTER 2

MOLYBDENUM NITROGENASE

R.W. MILLER
Plant Research Centre, Agriculture Canada, Ottawa, Canada K1A 0C6

1 THE MOLYBDENUM NITROGENASE PROTEINS

The properties, structure and function of isolated molybdenum nitrogenase have been reviewed recently (Burgess, 1985; Lowe et al., 1985; Eady, 1986; Smith et al., 1987). This chapter will attempt to use the physical, genetic and biochemical information obtained from studies of purified molybdenum nitrogenases to consider the function of the enzyme *in vivo*.

Only procaryotes can reduce dinitrogen to a form that supports cellular growth. Cell-free preparations of the proteins responsible for this reaction were only obtained when anaerobic techniques of preparation were developed (Schneider et al., 1960; McNary and Burris, 1962; Bulen et al., 1964). A cell-free extract of *Clostridium pasteurianum* (Mortenson et al., 1967) that reduced dinitrogen was separated into two iron-sulphur proteins which, with ATP, Mg^{2+} ions and a reduced ferredoxin, were required for catalytic activity under anaerobic conditions (Mortenson, 1964). Dithionite replaced ferredoxin as low potential reductant in cell free systems (Bulen et al., 1965; Mortenson, 1966). The nitrogenase components were designated 1 and 2 according to the order that they were eluted from DEAE-cellulose. The purified components are now known as the molybdenum-iron protein (MoFe protein) and the iron protein (Fe protein). The MoFe protein was renamed dinitrogenase and the iron protein dinitrogenase reductase based on early kinetic studies of the dilute purified proteins (Hageman and Burris, 1978). These designations suggest that the Fe protein functions only as a reductant for the MoFe protein. More recent kinetic data indicate that the Fe protein participates directly in an enzyme-substrate complex during substrate reduction and hence the mechanistic implications of this nomenclature are not valid (Thorneley and Lowe, 1985).

Since many families of heterotrophic, phototrophic and cyanobacteria fix N_2, a shorthand nomenclature has developed to indicate the organism producing the nitrogenase components (Eady et al., 1972). Thus, the MoFe protein of *C. pasteurianum* is designated Cp1, the Fe protein of *Klebsiella pneumoniae* is Kp2, etc. Since catalytically active nitrogenase proteins expressed under conditions of molybdenum deficiency may contain vanadium or possibly other metal ions in place of molybdenum (Chapters 3 & 7), an additional indication of the metal serving the role of molybdenum must be indicated as in $Ac1^{Mo}$ and $Ac1^{V}$ to distinguish the proteins from Azotobacter chroococcum (Robson, 1986; Robson et al., 1986). Since A. vinelandii has three different nitrogenase systems (Bishop et al., 1988; Pau et al., 1988), this more complicated nomenclature has become necessary. Because Fe proteins and MoFe or VFe proteins isolated from different sources may form catalytically active or inactive complexes (Emerich et al., 1978; Eady et al., 1982; Eady et al., 1987), this nomenclature is convenient for comparing the various nitrogenase systems.

1.1 Structure of the MoFe protein

The MoFe protein isolated from all sources is a tetramer composed of 2 different subunits giving a classical tetrahydral $\alpha_2\beta_2$ structure (Fig. 1). Dissociation of the tetramer into smaller units may occur at high dilution resulting in complete loss of activity. Further $MgCl_2$ concentration-dependent aggregation of the tetrameric units of Av1 can also occur under the conditions of varying protein concentration encountered during chromatography or analytical ultracentrifugation. These effects result in distorted sedimentation patterns from which consistent molecular weights are difficult to calculate (Voordouw et al., 1983). Amino acid sequences of the α subunit from eight species and of the β subunit from six species have been reported (Eady, 1986; Smith et al., 1987). The subunit peptides have molecular weights ranging from 50 to 60 kD and exhibit slightly different mobilities on SDS acrylamide gel electrophoresis depending on the commercial preparation of SDS used (Kennedy et al., 1976). Amino acid sequences from either protein or nucleotide sequencing of the α and β subunits bear some homology, suggesting their evolution from a common ancestral peptide. The amino acid residues of Kp1 which are highly conserved include cysteines at positions 63, 89, 155, 184 and 275 of the α subunit and at positions 69, 94 and 152 of the β subunit. Some of these cysteine residues are essential for metal cluster binding and catalytic activity (Kent et al., 1990). The MoFe protein β subunit has conserved sequences resembling nucleotide-binding sites of nucleotide-binding proteins (Robson, 1984).

1.2 The molybdenum-iron cofactor and iron sulphur clusters

The metal content of highly purified MoFe protein varies somewhat according to source and isolation procedures, but concensus values for molybdenum, iron and acid labile sulphide are 2 Mo, 30-32 Fe and 24-34 S^{2-} atoms per tetramer (Lowe et al., 1985). Four discrete, cubane Fe_4S_4 clusters, designated as P clusters and containing about half of the total iron, can be extruded from the tetramer (Kurtz et al., 1979). Low temperature Mossbauer studies of a reconstituted MoFe protein from a mutant of K. pneumoniae containing P clusters enriched in ^{57}Fe have provided evidence for the presence of 2 distinct types of P clusters per half Kp1 molecule (McLean et al., 1987). Alternative arrangements of P cluster iron into 2 Fe_8S_8 clusters have also been considered (Hagen et al., 1987). Adventitiously bound iron in some nitrogenase preparations can partly account for variations in iron content.

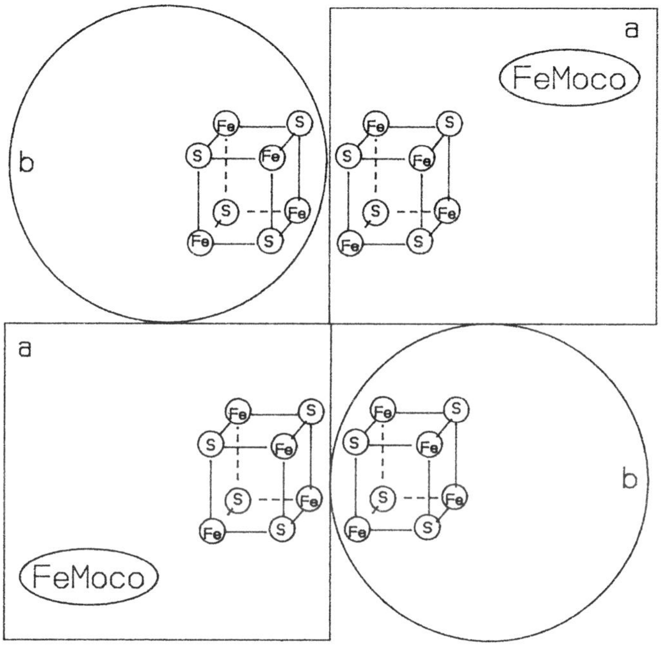

Fig. 1. Schematic diagram of the MoFe protein

Extraction of the MoFe protein with N-methyl formamide or other solvents (Wink et al., 1989) removes 2 moles of a metal cluster, designated FeMoco or the M cluster, containing roughly half of the total iron and labile sulphide of the tetrameric protein (see Stiefel and Cramer, 1985). Although FeMoco is very labile in the presence of oxygen, the isolated cofactor complements the apoprotein of nitrogenase accumulated by certain mutants or tungstate-grown wild type cells, all of which are defective in the synthesis of functional cofactor (Shah et al., 1977; Hawkes and Smith, 1983). The isolated cofactor contains 1 Mo, 6-8 Fe and 4-9 acid labile sulphide atoms and displays an EPR signal originating from a Kramer's doublet of a spin 3/2 iron system located in a distorted rhombic ligand field (Palmer et al., 1972). This signal indicates more rhombic distortion than that of the dithionite-reduced MoFe protein (True et al., 1988).

In the absence of crystals suitable for X-ray crystallography, the 3 dimensional structure of FeMoco has been investigated with x-ray absorption spectroscopy (XAS). A partial structure has been proposed for isolated FeMoco based on calculated inter-atomic distances for Fe-Fe, Fe-S and Fe-Mo (Garner et al., 1989), with molybdenum at the vertex of a cubane-like Fe_3MoS_4 structure which forms part of a larger octahedal cluster with 6 metal atoms arranged around a central iron atom (Fig. 2). Eight sulphur atoms are coordinated with the metal atoms in this model with sulphur being the nearest neighbour of all the iron atoms. Since molybdenum is located at an exterior vertex of this model and each metal atom is coordinated to 4 sulphur atoms, octahedral coordination of molybdenum could be achieved by 2 ligands (possibly O or N) external to the cluster and provided by the MoFe apoprotein or by extraction solvents. Figure 2 shows some of the metal-sulphur bonds of isolated FeMoco, but spatially locates

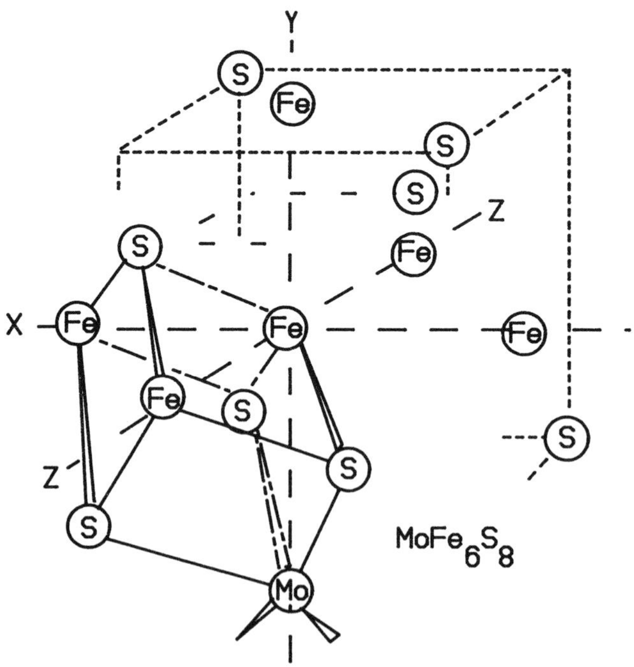

Fig. 2. Partial FeMoco structure (after Garner et al., 1989).

only some (dashed lines) of the iron and sulphur atoms of the complete molecule. FeMoco is considered to contain the site of binding and reduction of nitrogenase substrates (see Chapter 4; Stiefel and Cramer, 1985).

Oxidation of the MoFe protein with excess thionine immediately abolishes the FeMoco S=3/2 EPR signal and removes up to 8 electrons, slowly producing an EPR spectrum attributed to a transition within an excited state of a spin 7/2 system of the P clusters (Hagen et al., 1987). Integrated intensities of nitrogenase EPR signals are given in Table 1. Although a role for oxidized P clusters during catalytic turnover has not been established, the reduction of dinitrogen and other substrates requires the accumulation and transfer of 8 reducing equivalents within the MoFe protein-substrate complex. However, the P clusters do not become oxidised to EPR active states during normal catalytic turnover unless reductant supply becomes limiting.

TABLE 1. Redox states of nitrogenase proteins

Protein	Redox State	Function	EPR signals		
			Spin system	g values	Intensity (e^-)
MoFe protein					
	Super-reduced FeMoco	Reduces substrates in protein complex	-	-	-
	Dithionite-reduced FeMoco	Resting state	3/2	4.3,3.7,2.0	1.0
	Dye-oxidized	Binds ADP tightly	-	-	-
	Reduced P clusters	e^- source for substrate reduction?			
	Dye-oxidized P-clusters	?	7/2	10.4,5.6,1.8	0.87[b]
Fe protein	Reduced, $2e^-$	-	-	-	-
	Reduced, $1e^-$	Reduces MoFe protein-substrate complex	3/2 1/2	3.4,4.8,5.9 1.94	0.6[a] 0.4[a]
	Oxidized	Dissociation from MoFe protein is rate limiting	-	-	-

a. Intensity of signals varies with bound ADP, ATP.
b. Determined only for Kp1, Ac1 (Hagen et al., 1987)

The P and M clusters account for all functional iron and molybdenum atoms, and are responsible for the redox properties of the MoFe protein. The structure of the iron sulphur clusters differs in several respects from that of Fe_4S_4 clusters. of bacterial ferredoxins. The MoFe protein is usually purified in the dithionite-reduced form in which the P clusters are EPR silent. Further reduction of the MoFe protein (E_m < -430 mV) results in an EPR-silent super-reduced state capable of reducing substrates bound to the molybdenum centre when complexed with the Fe protein and MgATP.

1.3 Structure of the Fe protein

The Fe protein in all diazotrophs is a dimer of 2 identical polypeptides. A single Fe_4S_4 cluster bound to the dimer is the low potential redox centre responsible for reducing FeMoco during catalytic turnover. Molecular weights in the range 24-36 kD and amino acid sequences for 9 bacteria have been reported (see Eady, 1986). Fe proteins from various sources show 60-70% homology in primary structure and contain 5 conserved cysteine residues and 3 sets of possible nucleotide binding sites (Robson, 1984). The high degree of homology between different Fe proteins is responsible for the *in vitro* cross complementation between the Fe and MoFe proteins, although variations in redox potential and binding affinity then alter the kinetic parameters of nitrogenase catalysis.

In the presence of ammonia, the Fe protein of *Rhodospirillum rubrum* is inactivated *in vivo* by the covalent linkage of adenosine diphosphoribose through its terminal ribosyl group to a guanyl nitrogen of an arginine residue (Pope et al., 1985). This regulatory modification can be reversed by a MnATP-requiring activating enzyme from the chromatophores. Activation requires removal of at least part of the modifying group. Oxygen induces a reversible modification of the Fe protein of heterocystous cyanobacteria (Tabita et al., 1988).

1.4 Iron-sulphur cluster of the Fe protein

Reduction of the Fe_4S_4 cluster of isolated Fe protein by dithionite to an EPR active state requires a single electron. The reduced cluster gives evidence of both spin 1/2 and spin 3/2 systems, yielding EPR signals with a total integrated intensity near 1 per mole Fe protein (Hagen et al., 1985; Lindahl et al., 1985). Reduction by dithionite is biphasic, with an initial rapid process followed by slower processes which, in the case of Ac2, sum to an additional electron per mole (Lowe et al., 1985). This result may indicate that part of the isolated Fe protein has been damaged, or that the native protein is, in fact, a 2 electron carrier. Activities measured *in vitro* may therefore be only half those expected for the undamaged protein functioning *in vivo*. The concentration of active Fe protein sites remains unclear, since the maximum activity of a labile protein can only be inferred from indirect studies.

Both ADP and ATP bind tightly to the reduced and oxidised forms of the Fe protein, but the Fe protein alone is incapable of hydrolyzing ATP. In the absence of nucleotides, the redox potentials of Fe proteins isolated from various sources range from -240 mV for Cp2 to -393 mV for Av2. In the presence of ATP, redox potentials are between -280 and -430 mV, making the Fe protein a better reductant (see Lowe et al., 1985). Binding MgATP also increases the reactivity of the Fe_4S_4 cluster with chelating agents (Ljones and Burris, 1978). Oxygen, or cold treatment in the case of Cp2, reduce the ability of chelating agents to remove iron. These effects indicate a major nucleotide-induced change in conformation of the protein and allow calculation of a difference in absorbance between oxidized and reduced forms of the isolated, catalytically active Fe protein at 430 nm of $\Delta E = 6.6\ mM^{-1}cm^{-1}$.

2 BIOSYNTHESIS OF MOLYBDENUM NITROGENASE

The genes coding for the regulation and synthesis of the nitrogenase enzyme system have been characterized in K. pneumoniae (see Dixon et al., 1987a; Merrick, 1988), where twenty one complementing loci (nifJCHDKTYENXUSVWZMFLABQ) have been identified within a contiguous cluster (Chapter 6). The α and ß subunits are coded, respectively, by the nifD and nifK structural genes of the nifHDK operon. Although the nif operons of aerobic diazotrophs are not contiguous, the K. pneumoniae system has provided a pattern against which the the equivalent genes of other procaryotes can be compared.

Derepression of the nif transcriptional units requires low fixed nitrogen and oxygen levels and is coordinate as shown by the expression of lac gene fusions within the nif cluster of K. pneumoniae (Dixon et al., 1980). Molybdate is required for optimal expression of the autogenously regulated nifHDK operon. The products of the nifLA genes of K. pneumoniae (see Dixon, 1987), or functionally similar genes of other diazotrophs (Hill, 1988), regulate nif gene expression in response to intracellular ammonia and oxygen concentrations.

2.1 Structural genes of Mo nitrogenase

Site directed mutagenesis in the MoFe protein structural nifDK genes of K. pneumoniae and A. vinelandii has provided insight into the roles of several conserved amino acid residues. In mutant proteins of K. pneumoniae, the substitution of alanine for cysteine at α subunit positions 63, 89, 155 and 275 and at positions 69, 94 and 152 of the ß subunit result in the loss of catalytic activity and of the EPR signal from bound FeMoco (Kent et al., 1990). Serine activity and of the EPR signal from bound FeMoco (Kent et al., 1990). Serine substitutions in any of the equivalent positions except 153 (Av1) or 152 (Kp1) of the ß subunit also resulted in a loss of activity, indicating that these conserved sulfhydryl groups are required for the synthesis of the normal MoFe holoprotein (Brigle et al., 1987a; Dean et al., 1988). Extracts of a mutant with an alanine substitution at residue 275 containing unbound FeMoco complemented another mutant Kp1 lacking FeMoco, suggesting that α subunit cysteine-275 may provide one of the sulphur ligands of FeMoco. Substitution of cysteine-184 of Kp1 (equivalent to cysteine-183 in Av1) with alanine diminished component 1 activity but did not eliminate the FeMoco EPR signal. Substitution of serine for cysteine-152 (conserved) or alanine for cysteine-112 (non-conserved) of the ß subunit did not seriously affect the activity of Kp1.

Each of the four P clusters would require 4 external ligands and each FeMoco cluster would require 2 or more external ligands. The number of conserved cysteine mercaptide ligands available is therefore insufficient to provide all of the external ligands required for the MoFe protein. Other α-subunit ligands of FeMoco have been identified as histidine-195 and glutamine-191 (Scott et al., 1990). Because the P clusters are spectroscopically atypical compared to bacterial ferredoxins, and at least one conserved cysteine (152) of Kp1 can be substituted by serine without activity loss, oxygen or nitrogen ligands may also be involved in binding the metal clusters of the MoFe protein. The further application of site directed mutagenesis and analysis of the component 1 gene products should contribute to an understanding of the structure and function of the MoFe protein.

2.2 Cofactor synthesis and processing

Synthesis of FeMoco is ATP-dependent and requires the expression of *nifQ*, *nifB*, *nifV*, *nifN*, *nifE* and *nifH* (Imperial et al., 1984; Filler et al., 1986; Robinson et al., 1987), while the maturation of catalytically active Fe protein requires *nifM* (Howard et al., 1986; Kennedy et al., 1986). The putative roles of the *nif* operons involved in FeMoco synthesis have been indicated by experiments in which mutants were examined for replacement or complementation by specific factors (see Hoover et al., 1988a). Mutants of *K. pneumoniae* in *nifQ* lack a 19.7 kd peptide (deduced from the DNA sequence, Buikema et al., 1987), which appears to be required for efficient processing of molybdate. *NifQ* mutants can be partially restored by high molybdate concentrations or by replacement of sulphate by cysteine or cystine in growth media.

Mutants in *nifB* accumulate an apoprotein lacking FeMoco but containing a full complement of P clusters. The apoprotein can be activated *in vitro* with molecular size of 50.9 kd (Buikema et al., 1987), is membrane bound in *K. pneumoniae* (soluble in *A. vinelandii*) and is oxygen labile. The *nifB* protein is required in stoichiometric amounts for FeMoco synthesis *in vitro* and *in vivo* and may donate a an iron-sulphur FeMoco precursor early in the process (Hoover et al., 1988a). Mutants in *nifNE*, like those in *nifB*, produce a nitrogenase that can be activated with FeMoco.

Homocitrate biosynthesis, which is required for FeMoco synthesis in a cell free system, requires the *nifV* product (Shah et al., 1986; Hoover et al., 1987). Addition of homocitrate to growth media of *nifV* mutants results in reversion to the wild type MoFe protein (McLean et al., 1981; Hoover et al., 1988a). Citrate can substitute for homocitrate, resulting in the production of an altered nitrogenase which reduces nitrogen poorly and shows CO inhibition of H^+ reduction while retaining acetylene reduction activity (Hoover et al., 1988b). Although homocitrate is apparently essential for assembly of active MoFe protein, its intact carbon skeleton has not been identified in extracted FeMoco. However, since at least part of the molecule is incorporated into protein-bound FeMoco or FeMoco synthesized by a cell-free system (Hoover et al., 1988a), a derivative of homocitrate may provide all or part of the oxygen ligands of FeMoco.

2.3 Assembly of nitrogenase proteins

Biosynthesis of fully active MoFe protein requires insertion of the MoFe cofactor into both halves of the nitrogenase tetramer. The *nifN* and *nifE* products form a tetrameric protein with marked homology to the dinitrogenase tetramer, including conserved cysteine residues in the α subunit (Brigle et al., 1987b), suggesting that it is involved in preassembly of FeMoco. The *nifH* product is also required for the synthesis of FeMoco (Filler et al., 1986; Robinson et al., 1987) and for its insertion into the MoFe apoprotein (Robinson et al., 1989). The catalytically active form of the Fe protein is not required for these functions, since *nifM* mutants that express an inactive form of the *nifH* product still synthesize active MoFe protein.

The *nifM* and *nifH* proteins direct the synthesis of the Fe protein (Howard et al., 1989; Wyatt et al., 1989), but little additional information is available as to the synthesis and assembly of either its Fe_4S_4 cluster or of the P clusters of component 1. The Fe_4S_4 cluster is bound by cysteine residues 97 and 132 of each subunit (Hausinger and Howard, 1983). An inactive 2Fe species of the Fe protein, generated by treating dye-oxidized protein with ATP and an iron chelator, can be restored to an active form by a process requiring a sulphurtransferase (rhodanese), thiosulphate, ferric citrate and dihydrolipoate (Pagani et al., 1987). Iron-sulphur and iron-molybdenum-sulphur structures not bound to protein can be generated in a similar manner (Bonomi et al., 1988). The relationship between these *in vitro* cluster syntheses and the processes required for the *de novo* synthesis of the Fe_4S_4 clusters of nitrogenase is unknown.

2.4 Stability of the nitrogenase proteins

The Fe protein ($t_{1/2}$ = 30 sec.) and the MoFe protein ($t_{1/2}$ = 10 min.) cannot tolerate excess oxygen yet both can be oxidized anaerobically by dyes without loss of activity. The Fe protein can react with sub-stoichiometric amounts of oxygen without loss of activity producing hydrogen peroxide (Thorneley and Ashby, 1989). Indeed, they suggested that the reduction of oxygen by component 2 might account for a part of the high oxygen uptake rates of some aerobic diazotrophs. The inactivation of the Fe protein by excess oxygen is then due to destructive alteration of the Fe_4S_4 cluster or other group(s) in the Fe protein. The reaction of excess Fe protein with dioxygen may therefore have a protective role. ATP causes a significant increase in the susceptibility of the Fe protein to destructive oxidation.

Cp2 is unique among Fe proteins in being cold labile near 0°C (Moustafa and Mortenson, 1969). Wild type Mo nitrogenase components isolated from all species are stable for months when frozen rapidly in liquid nitrogen, or for periods of 24 hrs or more when held at room temperature in the presence of dithionite. As mentioned previously, the MoFe protein may dissociate or aggregate with activity loss under certain conditions. In general, the highly purified and concentrated native tetramer is stable in buffered (pH 8) 200-400 mM NaCl solutions, although species do differ in this regard.

3 CATALYTIC REACTIONS AND MECHANISM OF Mo NITROGENASE

All substrate reduction reactions require active Fe protein, MoFe protein, MgATP, protons and a low potential reductant. Data relating to substrate specificity, steady state and pre-steady state reaction kinetics and reaction mechanism have generally been obtained with highly purified Kp, Ac, Cp and Av proteins. The chemical reductant, sodium dithionite (Bulen et al., 1965), has been substituted for physiological reductants in most *in vitro* studies. While the utilization of dithionite in nitrogenase-catalyzed reactions can be quantified, the Fe protein catalyzes the dismutation of dithionite under anaerobic conditions (McKenna et al., 1988).

The presence of inactive forms of the nitrogenase proteins in highly purified preparations can be demonstrated and accounted for in kinetic calculations (Thorneley and Lowe, 1985). This is important in any interpretation of kinetic data, since it is the concentration of catalytically active sites that is of importance both in vitro and in vivo. In conformity with the literature, the MoFe protein will be assumed here to operate as a dimer of catalytically independent units composed of a single molecule of the α and β subunits, one FeMoco centre and 2 non-identical Fe_4S_4 clusters (McLean et al., 1988).

The electron transport chain necessary for the reduction of dinitrogen by the Kp nif system in vivo (Chapter 6) requires the passage of reducing equivalents from a third, low-potential, reduced protein to the oxidized Fe protein complexed with 2 MgADP molecules (Deistung et al., 1985). Flavodoxin hydroquinone, produced by membrane-bound pyruvate:flavodoxin oxidoreductase (Hill and Kavanagh, 1980), is the physiological reductant for nitrogenase in K. pneumoniae. Formation of a binary complex between flavodoxin hydroquinone and Kp2(2MgADP) results in the rapid one electron reduction of Kp2 (Thorneley and Deistung, 1988). Ferredoxins may act in place of flavodoxin as the immediate reductant of component 2 in anaerobic diazotrophs (Mortenson, 1964; Yates, 1980).

3.1 Proton and dinitrogen reduction

The binding and reduction of several classes of nitrogenase substrates has been comprehensively reviewed (Burgess, 1985). Since nitrogen fixation takes place in an aqueous environment at physiological pH, protons are always available to the catalytic site of nitrogenase. Hydrogen production under argon has often been used as an assay for nitrogenase activity since in the absence of other reducible substrates all electron flux is directed to protons. Typical specific activities for the MoFe protein with saturating Fe protein and MgATP fall in the range 1400-2000 nmoles H_2 min^{-1} mg^{-1} protein. While reducing equivalents enter the system in single electron steps, electrons appear to move in pairs in the reduction of substrates, indicating that reduced intermediates of substrates other than protons normally remain bound. Hydrogen production accounts for a measurable portion of the electron flux through the system under all practical conditions, although some substrates should theoretically suppress all H_2 production at high concentrations. With dinitrogen as substrate, the minimum amount of hydrogen production accounts for 25% of the total electron flux and this ratio cannot be improved upon even at very high pN_2 (Simpson and Burris, 1984). Evidence for a bound hydrazido intermediate in the reduction of dinitrogen has been reported (Thorneley et al., 1978).

Since 6 electrons are required for the production of 2 moles of ammonia, the minimum stoichiometric requirements for nitrogen fixation are 8 electrons and 16 ATP hydrolyzed per dinitrogen molecule reduced. In practice, this stoichiometry is achieved only under optimal conditions in the presence of excess reductant and an ATP-generating system. Since the distribution of electrons between dinitrogen and proton reduction is strongly influenced by ADP concentration and by the Fe protein/ MoFe protein ratio, the optimum ATP/2e ratio of 4 is probably seldom achieved in vivo (Upchurch et al., 1980; Miller et al., 1986). An electron allocation coefficient can be calculated by dividing the electron flux to a given substrate by the total electron flux as measured by hydrogen production under argon. The root nodules of legumes infected with appropriate Rhizobium strains seldom exceed 60% of electron flux to dinitrogen reduction.

Hydrogen is a competitive inhibitor (K_i = 3kPa) only of the reduction of N_2 and this inhibition may be of practical importance where H_2 accumulates in the vicinity of nitrogen-fixing cells. In addition, nitrogenase catalyzes the reductant- and ATP-dependent exchange of D_2 with enzyme-bound H_2 resulting in the release of HD, but does not promote the exchange of the isotope with solvent protons. Dinitrogen, unlike any of the alternative substrates, stimulates HD formation. Studies of this reaction suggest that H_2 and N_2 bind at a common site and that the prior formation of at least 2 metal-hydride bonds at this site is necessary for the binding and subsequent reduction of N_2 (Guth and Burris, 1983; Thorneley and Lowe, 1985).

3.2 Reduction of alternative substrates

In addition to protons and dinitrogen, molybdenum nitrogenase reduces a wide range of alternative substrates including isonitriles, nitriles, alkynes and alkenes. Gaseous substrates in order of decreasing apparent K_m values are: allene (40 kPa), N_2O (40 kPa), N_2 (10-20 kPa), acetylene (0.3-2 kPa), cyclopropene (1.2 kPa) and diazurine (0.12 kPa). Solutes reduced in order of decreasing apparent K_m (solution concentration) are: C_2H_2 (20-30 mM), C_2H_2NC (10-25 mM), HCN (4.5 mM), N_3^- (1-3 mM), N_2O (1 mM), cyanamide (H_3NCN, 0.8 mM), CH_2=CHNC (0.8 mM) and CH_3NC (0.7 mM). Apparent K_m values are generally determined over a range of nitrogenase component ratios. The values given are estimates for neutral buffers at temperatures of 25-30°C (Burgess, 1985). Nitrogenase substrates can be classified according to electron allocation to various products, inhibition of total electron flux, suppression of proton reduction at high substrate concentration and inhibition of reduction by CO or NO. The reduction of all substrates except protons by wild-type molybdenum nitrogenase is inhibited by CO. A $nifB^-$ mutant of K. pneumoniae lacking the ability to process molybdate synthesizes a MoFe protein that cannot reduce N_2 but does catalyze proton reduction which is atypical in being CO-sensitive (Hawkes and Smith, 1983). This provides strong evidence for the binding and reduction of N_2 at the FeMoco centre.

The wide substrate range of nitrogenase has inspired reflection on the reasons for the evolution of the enzyme. Detoxification of HCN (Postgate, 1982) and organic isonitriles might have been one of the original functions of nitrogenase. Since the major products of HCN reduction are NH_3, CH_4 and CH_3NH_2, HCN can serve as a nitrogen source for the growth of aerobic diazotrophs (Stouthamer et al., 1988). Cyanide ion, a minor component of HCN solutions at pH 7, is an inhibitor of nitrogenase (Li et al., 1982). Recent evidence indicates that cyanide may induce a modification of the MoFe protein prior to its reduction (Lowe et al., 1990).

The isonitriles and cyanamide (Miller and Eady, 1988b) also represent substrates having more than a single pathway of multiple electron reduction. Generally, 6 and 8 electron pathways give rise to a mixture of major products including methane, methylamines and ammonia (from cyanamide). The involvement of free radicals as intermediates in the reduction of isonitriles leads to the formation of C_2 and C_3 products through an insertion mechanism (Burgess, 1985). Electron balance calculations for these reactions can be complicated by the formation of products not accounted for in the analysis of reaction mixtures.

Azide, like N_2 does not appear to suppress more than 75% of proton reduction at physiological pH (Dilworth and Thorneley, 1981). In this case the 1:1 molar stoichiometry between hydrogen produced and substrate reduced is not maintained since azide is a 2, 6 or 8 electron acceptor (Rubinson et al., 1985). This leaves N_2 as the sole substrate where stoichiometric equivalence between hydrogen production and substrate reduction is obligatory.

The reduction of alkynes, specifically the 2 electron reduction of acetylene to ethylene (Dilworth, 1966), has provided researchers with a very useful nitrogenase assay. Ethylene is not normally reduced further in this assay, but can act as a substrate and inhibit hydrogen evolution under certain circumstances (Ashby et al., 1987). The usefulness of the method lies in its relative simplicity and high sensitivity and the solubility and ready diffusion of the substrate. Acetylene effectively suppresses dinitrogen reduction by preventing binding of the natural substrate to the MoFe protein. The method is suitable for estimating the total nitrogenase activity of intact nodules, whole cells and isolated enzyme preparations. Standard nitrogenase assays employing measurements of either ethylene or hydrogen under argon are generally carried out at 30°C. However, the temperature dependence of the system is complicated by several factors (see Section 4).

Sources of error in the application of the acetylene reduction assay method to plant tissues include the induction of variable diffusion barriers in legume root nodules (Witty et al., 1986; Hunt et al., 1988), substrate inhibition at high concentrations (Burgess, 1985), stimulation of total electron flow at assay temperatures below 20°C (Thorneley and Eady, 1977) and residual reduction of protons. Apparent K_m values range from 0.3 to 2.1 kPa acetylene (Hardy, 1979) and hence, at partial pressures of 10 to 20 kPa, virtually all (>98%) available electron flux through nitrogenase can be measured as ethylene. The predominant product from the reduction of acetylene in D_2O is cis-$C_2H_2D_2$ (Dilworth, 1966; Kelly, 1968).

Nitrous oxide (N_2O) and nitrite ion (Vaughn and Burgess, 1988) are unique nitrogenase substrates in which N-O bonds are reduced to give N_2 and ammonia, respectively. Nitrite also irreversibly inactivates the Fe protein (Meyer, 1981). The reduction of nitrite to ammonia and the irreversible inhibition of electron transfer to the MoFe protein represent possible sources of error in measurements of nitrogenase activity in legume root nodules and in free living diazotrophs where nitrite may be formed from nitrate.

3.3 Electron accumulation and allocation

The reduction of substrates by multiple electron pairs requires that reducing equivalents be stored in some way by the enzyme substrate complex before the release of reaction products. Chemical quenching of reaction mixtures during turnover has allowed the detection of hydrazine resulting from the destruction of a hydrazido intermediate of N_2 reduction (Thorneley et al., 1978). The many possible sites for electron storage within the enzyme-substrate complex have made it difficult to specify the exact means of electron accumulation. Although the FeMoco centre cycles between super-reduced states and the dithionite-reduced state, the role of the P centres during turnover has not been defined. Evidence for the existence of oxidation states of S=7/2 (Hagen et al., 1987) and for transient EPR signals originating from the P clusters during turnover suggests that these redox centres may themselves function as electron donors for reduction of substrates under conditions of limiting reductant.

The allocation of electrons to protons and other substrates varies with nitrogenase component ratio since this parameter controls the rate of introduction of electrons into the system. Component ratio thus exerts a profound influence on the proportion of products formed. Optimal allocation of reducing equivalents to N_2 requires high rates of electron flux and high ratios of Fe protein to MoFe protein with saturation at about 10:1 (Burgess, 1985). There are small but reproducible differences in the effect of component ratio on electron allocation with nitrogenases from different species. However, the isolated nitrogenase proteins may not be identical in this respect with the native, intracellular proteins. Maximum donation of electrons to HCN, cyanamide (Miller and Eady, 1988) and isonitriles is favoured by molar equivalence between the nitrogenase components (Burgess, 1985). Since some alternative substrates also inhibit total electron flux through the system, simple hyperbolic substrate saturation curves are not observed in these cases.

3.4 Adenine nucleotides and energy transduction

Although the initial steps in the reduction of N_2 are highly endothermic, the overall reduction to ammonia is exothermic (Leigh, 1977). Even though the ATP requirement for reduction of nitrogenase substrates has been established since 1969 (Hadfield and Bulen, 1969), the means by which the free energy of hydrolysis of ATP is utilized in the process remains unclear. While the Fe protein binds 2 moles of ATP relatively tightly with reported K_d values ranging from 15 to 70 μM (see Thorneley and Lowe, 1985), ATP is not hydrolyzed by the purified protein alone. Kp2(2MgATP) must bind to the MoFe protein before hydrolysis to ADP and orthophosphate can occur.

Despite the presence of amino acid sequences suggesting nucleotide-binding sites (Robson, 1984), the purified, myokinase-free MoFe protein appears to bind ATP quite weakly, making it extremely difficult to measure binding constants and on/off rates under equilibrium conditions (Miller et al., 1980). High protein concentrations are required and species or experimental differences affecting binding rates, binding affinities and reductant-independent hydrolysis of ATP have led to differing interpretations of binding studies (Cordewener et al., 1987). NMR studies strongly indicate the existence of potential nucleotide-binding sites on the MoFe protein (Kimber et al., 1982). ADP has recently been shown to bind much more tightly than ATP to both thionine-oxidized and dithionite-reduced Ac1. That reaction is followed by such a slow dissociation of the nucleotide from at least one of the 2 observed binding sites that equilibrium binding constants for the Ac1-ADP complex could not be calculated (Miller & Eady, 1989b).

The best interpretation of reported binding data involves ATP in a bridging role between the components of nitrogenase. Binding measurements made with the isolated proteins cannot mimic this reaction and ATP undergoes reductant-independent hydrolysis in the protein complex. Myokinase-free preparations of Kp1 catalyze the slow exchange of ^{32}P-orthophosphate with both the α and β phosphate groups of ATP, forming radiolabeled ADP and ATP (Miller et al., 1980). Catalytic activity and exchange of inorganic phosphate with ADP are strongly inhibited by o-phthalaldehyde (R.W. Miller, unpublished). Gvozdev et al. (1984) have suggested that this reagent blocks ATPase activity of the nitrogenase complex by covalently linking an essential -SH group of the Fe protein to a nearby $-NH_2$ group of the MoFe protein. While adenine nucleotides clearly play an important role in inducing a major conformational change in the Fe protein and lower the redox potential, the actual site of ATP hydrolysis probably resides on the MoFe protein. Energy transduction by the nitrogenase complex must directly involve bound ATP since it is absolutely required for electron transfer between the nitrogenase components. The tight binding of ADP by the MoFe protein may play a role in the rate-limiting process during catalytic turnover.

3.5 Catalytic cycles and mechanism

Detailed studies of the reduction of alternative substrates have provided insights into the nature and function of the substrate binding sites of nitrogenase. A "2 site" model seems to accommodate most of the available data with at least one site being the reduced FeMoco centre. Another binding site might involve P clusters or more probably a different redox state of FeMoco. Chemical modeling of molybdenum complexes suggests that the molybdenum atom is the site of bonding of N_2 and other nitrogenous substrates (Chapter 4).

Two sets of catalytic cycles have been proposed (Thorneley and Lowe, 1985) to account for the pre-steady state and steady state data that have been accumulated with the Kp1 plus Kp2 system at 23°C. The first describes the transfer of electrons from the Fe protein to the MoFe protein (large arrow in Fig. 3). The Fe protein oxidation-reduction cycle requires the assembly of a Kp2(2MgATP) complex that can be reduced by the SO_2^- free radical from dithionite. Binding between the Fe protein-MgATP complex and Kp1 is rapid (k_1=5x10^7 M^{-1} s^{-1}), while the next step is much slower (k_2= 200 s^{-1}) and involves a single electron being passed to the FeMoco centre and ATP being hydrolyzed. Pre-steady state observations (Cordewener et al., 1987; Thorneley et al., 1989) suggest that the transfer of electrons to the FeMoco centre is preceded by the hydrolysis of ATP to give bound ADP plus orthophosphate. This step may therefore result from 2 discrete processes, although orthophosphate is not released immediately. At 6°C, the ATP cleavage reaction and subsequent electron transfer may be reversible because of the temperature dependence of the equilibrium position of this reaction (Thorneley et al., 1989). When both Kp2 and substrate concentrations are saturating, the slowest step in the reduction of substrates is the dissociation of the nitrogenase component proteins (k_3=6.4 sec^{-1}).

The Fe protein cycle is basic to the MoFe protein cycle that leads to the reduction of substrates. In the model, avoidance of wasteful hydrogen production requires minimization of the concentration of H_2 evolving species such as E_3H_3 (Thorneley and Lowe, 1985). High component protein or ADP concentrations inhibit the dissociation of the protein complex by mass action effects. The slow dissociation of an isolated Ac1$(MgADP)_2$ complex reported for the thionine-oxidized MoFe protein (Miller and Eady, 1989b) suggests that nucleotide binding sites located on the MoFe protein could retard the dissociation of the Ac2(2MgADP)Ac1 complex. The relatively weak interaction of ATP observed with the isolated MoFe protein may minimize non-productive hydrolysis of ATP and require that nucleotides be introduced into the enzyme complex by first binding to the Fe protein. Each turn of the Fe protein cycle requires the interaction of the two nitrogenase components, the Fe protein carrying 2 moles of adenine nucleotide during association (2MgATP) and dissociation (2MgADP). After replacement of ADP with ATP in the free, oxidized Kp2(2MgADP) complex, and its reduction by dithonite, the cycle continues until sufficient electrons have accumulated on the MoFe protein-substrate complex to allow release of the final products of substrate reduction (Thorneley and Lowe, 1985).

The MoFe protein cycle shown in Fig. 3 can account for the stoichiometry and steady state kinetics of nitrogenase with a minimum of assumptions (Thorneley and Lowe, 1985). E_0 represents one of 2 halves of the tetrameric Kp1 molecule. Dinitrogen reduction in this scheme requires that the MoFe protein first undergo a minimum of 3 single electron reduction steps each requiring completion of one

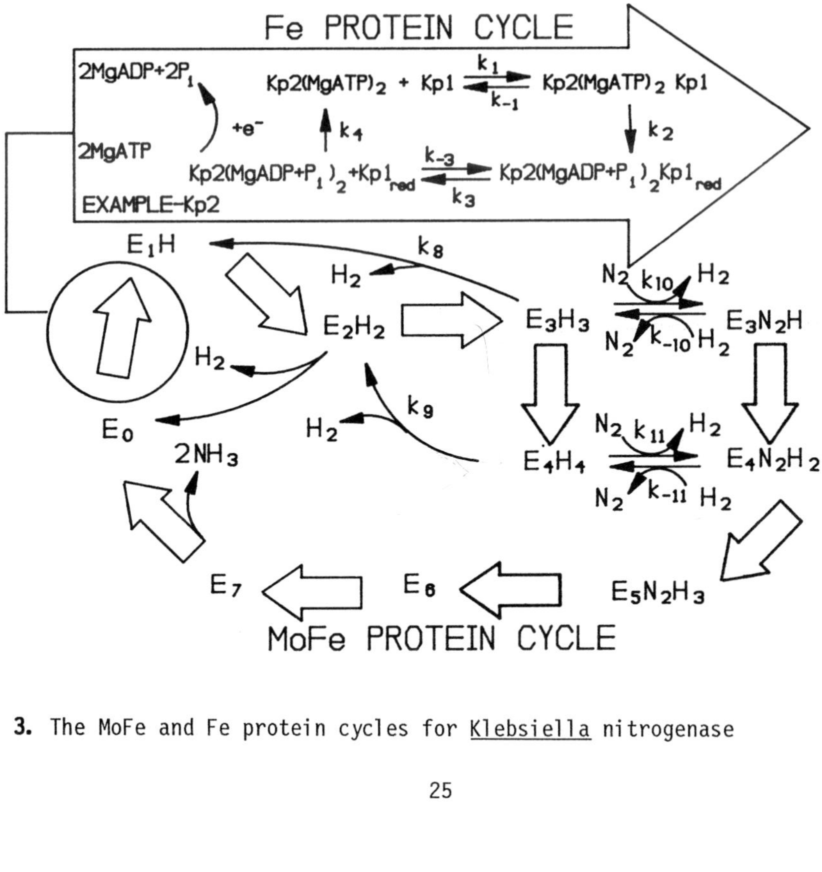

Fig. 3. The MoFe and Fe protein cycles for Klebsiella nitrogenase

Fe protein cycle (indicated inside large arrow in Fig. 3). N_2 can bind only to E_3H_3 or E_4H_4 with displacement of a dihydrogen molecule, possibly from a molybdenum hydride as suggested by Chatt (1980). Additional ATP-dependent reductions then lead to the release of 2 moles of ammonia under ideal conditions.

The dissociation reaction that follows ATP hydrolysis and electron transfer in the Fe protein cycle is the rate-limiting process at protein concentrations approaching the physiological. Under conditions where the MoFe protein concentration greatly exceeds that of the Fe protein, the second order association reaction (k_1) can become rate limiting (Hageman and Burris, 1978). Hydrogen can only be evolved from free Kp1 following slow dissociation of the Fe protein, so that association between the nitrogenase components at any stage in the cycle prevents release of H_2. Any conditions that retard the dissociation of the complex therefore inhibit hydrogen evolution.

D_2 can also displace H_2 from E_3H_3 leading to the formation of HD by randomization of the hydride hydrogen atoms. N_2 may stimulate this isotope exchange by displacing HD. Hydrogen can displace dinitrogen from either the E_3 or E_4 states, thus explaining hydrogen inhibition of N_2 reduction. Acid or base treatment after binding of N_2 and further accumulation of electrons leads to the release of hydrazine from the $E_4N_2H_4$ state.

Substrates other than N_2 are postulated to bind to the MoFe protein in the E_1 or E_2 states. These states represent more oxidised forms of the protein complex than those required for N_2 binding. If binding of alternative substrates occurs prior to the accumulation of 2 electrons (E_1H), hydrogen need not be released at all as in the case which is approached at saturating concentrations of acetylene. Binding of alternative substrates at sites representing these initial redox states would competitively exclude protons and explain the ability of high concentrations of such substrates to suppress hydrogen production to less than one mole per mole of substrate reduced.

4 LIMITATIONS TO PHYSIOLOGICAL FUNCTION OF Mo NITROGENASE

Diazotrophs may synthesize up to 10% of cellular protein as nitrogenase polypeptides, attaining calculated intracellular MoFe protein concentrations above 30 μM when nitrogenase is fully derepressed. Nitrogenase component concentrations can be estimated either by complementation assays of cellular extracts using prepurified nitrogenase components or by immunoassay. The concentration of catalytically active sites is the preferred parameter, but even then extrapolation to the intracellular situation assumes a minimum of damage or modification to the nitrogenase proteins on extraction. In the intact cell, estimated molar ratios of Fe protein to MoFe protein generally lie in the range 2-4:1.

Under physiological conditions, ATP and reductant supply are probably inadequate to support maximum turnover rates of the concentrated nitrogenase proteins, despite the slow, rate-limiting process of the reaction (Upchurch & Mortenson, 1980; Miller et al., 1988). The presence of ADP (an inhibitor of nitrogenase activity) at concentrations near those of ATP in diazotrophic cells must further retard the overall process. High intracellular protein concentrations aid in maximizing the allocation of electrons to N_2 reduction but the measured proportion of total electron flux used to produce ammonia seldom exceeds 60%, except where uptake hydrogenase systems are present.

4.1 Oxygen tolerance of Mo nitrogenase *in vivo*

Protection of nitrogenase against inactivation by oxygen in aerobes is generally considered to require high cellular respiratory rates combined with low extracellular oxygen concentrations, conditions which relieve oxygen repression of nitrogenase synthesis (Hill, 1988). The appreciable fluxes of dioxygen required to support nitrogenase activity in terms of respiration-dependent ATP and reductant supply are provided to the symbiotic rhizobia by facilitated diffusion via oxyleghaemoglobin at very low dissolved O_2 concentrations (Chapter 12).

In azotobacters, the nitrogenase proteins may also be protected from oxygen inactivation by forming a complex with an Fe_2S_2 protein (Haaker and Veeger, 1977; Robson, 1979) isolated prior to the discovery of the nitrogenase complex (Shethna et al., 1968). When growing diazotrophically, these organisms respond to increased solution rates of O_2 by switching off nitrogenase activity. This process is readily reversible and may involve conformational protection of the proteins or diversion of electrons from nitrogenase by oxidation of flavodoxin (see Postgate, 1982). Reactions between dioxygen and the MoFe protein and complexes of the nitrogenase proteins require more investigation.

4.2 Temperature dependence

Nitrogenase activities *in vivo* and *in vitro* have often been measured at temperatures well above those in soil, 30°C being the routine assay temperature for the isolated proteins. Shifting from 30°C to 21°C more than doubles the activation energy for acetylene reduction (Burns, 1969). The activation energy for formation of a 1:1 complex between the nitrogenase components (Thorneley et al., 1975) was high (418 kJ/mol) below 17°C, while above this temperature the association reaction was temperature independent ($\Delta H^*=0$). While the association reaction is not rate limiting above 17°C, the process below this temperature may be significantly slower. Hence, nitrogenase displays different kinetic characteristics in the temperature range applicable to symbiotic associations in soil. In fact, between 5°C and 20°C, the Q_{10} for proton or acetylene reduction by purified nitrogenase is nearly 5. In vitro, N_2 reduction by Mo nitrogenase is even more markedly affected in this temperature range (Thorneley and Eady, 1977). Measurement of Mo nitrogenase activity below 5°C requires high protein concentrations, long incubation times and sensitive detection methods.

In contrast, the same activities of nitrogenase in excised alfalfa nodules show far less drastic temperature dependence, with Q_{10} between 1 and 2 from 5°C to 20°C. Nitrogen-fixing legume root nodules and isolated R. meliloti bacteroids maintain nitrogenase activity to much lower temperatures than do the isolated proteins (Miller et al., 1986). The discrepancy between how nitrogenase responds to temperature in vitro and its activity in vivo implies the operation of a temperature compensation mechanism that presumably depends on other regulatory systems rather than on nitrogenase biochemistry. This conclusion assumes that drastic alteration of the catalytic properties of the enzyme does not occur on isolation.

4.3 Reductant and adenine nucleotide supply

In aerobes, cellular respiration provides the ATP and low potential reductant for the reduction of N_2. Dependence of rates of nitrogen fixation on cellular membrane potential and cellular energy charge or ATP/ADP ratios in the R. leguminosarum bacteroid system have been thoroughly investigated (Laane et al. , 1978; Laane et al., 1979). Flavodoxin-NADH oxidoreductase is membrane bound in this organism. The effect of ATP/ADP ratio on nitrogenase activity was small as long as energy charge exceeded 0.75, a situation which may not apply under conditions of respiratory limitation. Membrane potentials exceeding -250 mV were, however, an absolute requirement for nitrogenase activity (Laane et al., 1979).

In Bradyrhizobium and in several Rhizobium species, specialized respiratory terminal oxidases having increased O_2 affinity are utilized in place of the normal cytochrome a-a_3 complex to support N_2 fixation at low oxygen tension (Appleby et al., 1980). A similar system is expressed in R. meliloti bacteroids that lack the terminal oxidase predominating in free-living cells (Miller and Tremblay, 1983). Succinate, malate or fumarate oxidation by a specific respiratory system appears to be required to support nitrogenase activity in isolated R. meliloti bacteroids that retain coupling between nitrogenase and respiration (Miller et al., 1988).

It is possible to retain active nitrogenase proteins inside isolated R. meliloti bacteroids when they are made permeable to ATP and dithionite by treatment with chlorpromazine. When the nitrogenase is assayed with non-limiting reductant and an ATP-generating system that avoids ADP inhibition, activities of proton or acetylene reduction are roughly 4 times those for untreated bacteroids (Miller et al., 1988). The nitrogenase activity normally measured in intact bacteroids may therefore be only a fraction of the total possible, and may explain the apparent ability of bacteroids to maintain activity to lower temperatures than with the isolated enzyme.

Energy charge and ATP/ADP ratio within the unmodified bacteroids have been shown to increase with declining temperature in the range 25°C-5°C (Miller et al., 1980). By analogy with the Klebsiella system, the expected accompanying decline in ADP inhibition of the dissociation of the $Rm2(MgADP)_2Rm1$ may allow an increase in turnover rates that compensates for the adverse effects of the larger activation energy needed (Miller et al., 1980).

4.4 Potential for improvement

Nitrogenase has evolved as a system having low turnover rates and requiring high protein concentrations to regulate the energy cost. While improvements in the rates of dissociation of the nitrogenase components might result from site-directed mutagenesis of key amino acid residues in the polypeptides, this approach may initially only produce further insights into mechanism. Increasing the electron flux through the nitrogenase system could improve the energy efficiency (Lowe et al., 1990). Increasing the rates of reduction of the Fe protein could, for example, be achieved by increasing the rate of flavodoxin reduction, which in turn would require greater supplies of respiratory substrates and oxygen. A strategy aimed at increasing succinate transport in bacteroids has been initiated (Birkenhead et al., 1988; Watson et al., 1988), but the rate-limiting supply of oxygen in many symbiotic and non-symbiotic situations (Miller et al., 1986; Hunt et al., 1987; Bergersen and Turner, 1988; Vessey et al., 1988) may prevent any significant gains from this strategy alone.

The concepts of excess nitrogenase capacity and limiting ATP supply which allow the R. meliloti bacteroid system to operate sub-optimally at higher temperatures and become more efficient at lower temperatures suggests that higher than normal levels of the nitrogenase components, especially component 2, might increase symbiotic nitrogen fixation. Gene dosage effects of the nif structural genes or of positive regulators (e.g. nifA or equivalent genes) might amplify protein production. Alternative nitrogenase systems employing metals other than molybdenum might provide an advantage at growth temperatures below 10°C (Miller and Eady, 1989a). Nitrogenase systems that have evolved in "arctic" rhizobia that normally operate below 5°C (Prevost et al., 1987) might also improve the performance of symbiotic diazotrophs in temperate zone agriculture.

5 REFERENCES

Appleby, C.A., Bergersen, F.J., Ching, T.M., Gibson, A.H., Gresshoff, P.M. and Trinick, M.J., 1980. Cytochromes of rhizobia from parasponia, pea and soybean nodules, and from nitrogen fixing continuous cultures. In: A.H. Gibson and W.E. Newton (Editors), Current Perspectives in Nitrogen Fixation. Griffin Press, Netley, S. Australia, p. 369.

Ashby, G.A., Dilworth, M.J. and Thorneley, R.N.F., 1987. Klebsiella pneumoniae nitrogenase. Inhibition of hydrogen evolution by ethylene and the reduction of ethylene to ethane. Biochem. J., 247: 547-554.

Bergersen, F.J. and Turner, G.L., 1988. Steady state activities of soybean bacteroids with nanomolar oxygen. In: H. Bothe, F.J. de Bruijn and W.E. Newton, (Editors), Nitrogen Fixation: Hundred Years After. Gustav Fischer, Stuttgart & New York, p. 650.

Birkenhead, K., Manian, S.S. and O'Gara, F., 1988. Dicarboxylic acid transport in Bradyrhizobium japonicum: use of Rhizobium meliloti gene(s) to enhance nitrogen fixation. J. Bacteriol., 170: 184-189.

Bishop, P.E., Premakumar, R., Joerger, R.D., Jacobson, M.R., Dalton, D.A., Chisnell, J.R. and Wolfinger, E.D., 1988. Alternative nitrogen fixation systems in Azotobacter vinelandii. In: H. Bothe, F.J. de Bruijn and W.E. Newton (Editors), Nitrogen Fixation: Hundred Years After. Gustav Fischer, Stuttgart & New York, pp. 71-79.

Bonomi, F., Iamatti, S., Kurtz, D.M. and Pagani, S., 1988. Enzymic synthesis of nitrogenase-related iron-sulphur-molybdenum structures. In: H. Bothe, F.J. de Bruijn and W.E. Newton (Editors), Nitrogen Fixation: Hundred Years After. Gustav Fischer, Stuttgart & New York, p. 63.

Brigle, K.E., Setterquist, R.A., Dean, D.R., Cantwell, J.S., Weiss, M.C. and Newton, W.E., 1987a. Site-directed mutagenesis of the nitrogenase MoFe protein of Azotobacter vinelandii. Proc. Natl. Acad. Sci. USA, 84: 7066-7069.

Brigle, K.E., Weiss, M.C., Newton, W.E. and Dean, D.R., 1987b. Products of the iron-molybdenum cofactor-specific biosynthetic genes, nifE and nifN, are structurally homologous to the products of the nitrogenase molybdenum-iron protein genes, nifD and nifK. J. Bacteriol., 169: 1547-1553.

Buikema, W.J., Klingensmith, J.A., Gibbons, S.L. and Ausubel, F.M., 1987. Conservation of structure and location of Rhizobium meliloti and Klebsiella pneumoniae nifB genes. J. Bacteriol., 169: 1120-1126.

Bulen, W.A., Burns, R.C. and Lecomte, J.R., 1964. Nitrogen fixation: cell-free system with extracts of Azotobacter. Biochem. Biophys. Res. Commun., 17: 265-271.

Bulen, W.A., Burns, R.C. and Lecomte, J.R., 1965. Nitrogen fixation: hydrosulfite as electron donor with cell-free preparations of Azotobacter vinelandii and Rhodospirillum rubrum. Proc. Natl. Acad. Sci. USA, 53: 532-539.

Burgess, B.K., 1985. Substrate reactions of nitrogenase. In: T.G. Spiro (Editor), Molybdenum Enzymes. John Wiley and Sons, Inc., New York, pp. 161-220.

Burns, R.C., 1969. The nitrogenase system from Azotobacter. Activation energy and divalent cation requirement. Biochim. Biophys. Acta, 171: 253-259.

Chatt, J., 1980. Chemistry relevant to the biological fixation of nitrogen. In: W.D.P. Stewart and J.R. Gallon (Editors), Nitrogen Fixation. Ann. Proc. Phyt. Soc., Academic Press, London, p. 1-18.

Cordewener, J., Asbroek, A.T., Wassink, H., Eady, R., Haaker, H. and Veeger, C., 1987. Binding of ADP and orthophosphate during ATPase reaction of nitrogenase. Eur. J. Biochem., 162: 265-270.

Dean, D.R., Brigle, K.E., May, H.D. and Newton, W.E., 1988. Site-directed mutagenesis of the nitrogenase MoFe protein. In: H. Bothe, F.J. de Bruijn and W.E. Newton (Editors), Nitrogen Fixation: Hundred Years After. Gustav Fischer, New York, pp. 107-113.
Deistung, J., Cannon, F.C., Cannon, M.C., Hill, S. and Thorneley, R.N.F., 1985. Electron transfer to nitrogenase in Klebsiella pneumoniae. NifF gene cloned and the gene product, a flavodoxin, purified. Biochem. J., 231: 743-753.
Dilworth, M.J., 1966. Acetylene reduction by nitrogen-fixing preparations from Clostridium pasteurianum. Biochim. Biophys. Acta, 127: 285-294.
Dilworth, M.J. and Thorneley, R.N.F., 1981. Nitrogenase of Klebsiella pneumoniae. Hydrazine is a product of azide reduction. Biochem. J. 193: 971-983.
Dixon, R., Eady, R.R., Espin, G., Hill, S., Iaccarino, M., Kahn, D. and Merrick, M., 1980. Analysis of regulation of Klebsiella pneumoniae nitrogen fixation nif gene cluster with gene fusions. Nature, 286: 128-132.
Dixon, R.A., Austin, S., Buck, M., Drummond, M., Hill, S., Holtel, A., MacFarlane, S., Merrick, M. and Minchin, S., 1987. Genetics and regulation of nif and related genes in Klebsiella pneumoniae. Phil. Trans. Roy. Soc. Lond. B, 317: 147-158.
Dixon, R.A., 1987. Genetic regulation of nitrogen fixation. In: J.A. Cole and S. Ferguson (Editors), The Nitrogen and Sulphur Cycles. Cambridge University Press, Cambridge, pp. 417-438.
Eady, R.R., Smith, B.E., Cook, K.A. and Postgate, J.R., 1972. Nitrogenase of Klebsiella pneumoniae. Purification and properties of the component proteins. Biochem. J., 128: 655-675.
Eady, R.R., Kahn, D. and Buchanan-Wollaston, V., 1982. The molecular enzymology of nitrogen fixation. Israel J. Bot., 31: 45-60.
Eady, R.R., 1986. Enzymology of free living diazotrophs. In: W.J. Broughton and A. Pühler (Editors), Nitrogen Fixation, Vol. 4, Molecular biology. Clarendon Press, Oxford, pp. 1-49.
Eady, R.R., Robson, R.L., Richardson, T.H., Miller, R.W. and Hawkins, M., 1987. The vanadium nitrogenase of Azotobacter chroococcum: Purification and properties of VFe protein. Biochem. J., 244: 197-207.
Emerich, D.W., Ljones, T. and Burris, R.H., 1978. Nitrogenase: Properties of the catalytically inactive complex between the Azotobacter vinelandii MoFe protein and the Clostridium pasteurianum Fe protein. Biochim. Biophys. Acta, 527: 359-369.
Filler, W.A., Kemp, R.M, Ng, J.C., Hawkes, T.R., Dixon, R.A. and Smith, B.E., 1986. The nifH gene product is required for the synthesis or stability of the iron-molybdenum cofactor of nitrogenase from Klebsiella pneumoniae. Eur. J. Biochem., 160: 371-377.
Garner, C.D., Arber, J.M., Harvey, I., Hasnain, S.S., Eady, R.R., Smith, B.E., de Boer, E. and Wever, R., 12989. Characterization of molybdenum and vanadium centres in enzymes by X-ray absorption spectroscopy. Polyhedron, 8: 1649-1652.
Gvozdev, R., Mitsova, I.Z. and Blazhchuk, I.S., 1984. Reciprocal distribution of the catalytically active amino-groups of the molybdenum iron site and catalytically active sulfhydryl-group of nitrogenase iron protein. Biokhimiya 49: 1134-1137.
Guth, J.H. and Burris, R.H., 1983. Inhibition of nitrogenase-catalyzed NH_3 formation by H_2. Biochemistry, 22: 5111-5122.
Haaker, H. and Veeger, C., 1977. Involvement of the cytoplasmic membrane in nitrogen fixation by Azotobacter vinelandii. Eur. J. Biochem., 77: 1-10.
Hadfield, K.L. and Bulen, W.A., 1969. Adenosine triphosphate requirement of nitrogenase from Azotobacter vinelandii. Biochemistry, 8: 5103-5108.
Hageman, R.V. and Burris, R.H., 1978. Nitrogenase and nitrogenase reductase associate and dissociate with each catalytic cycle. Proc. Natl. Acad. Sci. USA, 75: 2699-2702.

Hagen, W.R., Eady, R.R., Dunham, W.R. and Haaker, H., 1985. A novel s=3/2 signal associated with native Fe-proteins of nitrogenase. FEBS Lett., 189: 250-254.
Hagen, W.R., Wassink, H., Eady, R.R., Smith, B.E. and Haaker, H., 1987. Quantita-tive EPR of s=7/2 system in thionine-oxidized MoFe proteins of nitrogenase. A redefinition of the P-cluster concept. Eur. J. Biochem., 169: 457-465.
Hardy, R.W.F., 1979. Reducible substrates of nitrogenase. In: R.W.F. Hardy, F. Bottomley and R.C. Burns (Editors), A Treatise on Dinitrogen Fixation. John Wiley and Sons, New York, pp. 515-568.
Hausinger, R.P. and Howard, J., 1983. Thiol reactivity of the nitrogenase Fe protein from Azotobacter vinelandii. J. Biol. Chem., 258: 13846-13492.
Hawkes, T.R. and Smith, B.E., 1983. Purification and characterization of the inactive MoFe protein (NifB-Kpl) of the nitrogenase from nifB mutants of Klebsiella pneumoniae. Biochem. J., 209: 43-50.
Hill, S., 1988. How is nitrogenase regulated by oxygen? FEMS Microbiol. Rev., 54: 111-130.
Hill, S. and Kavanagh, E.P., 1980. Roles of nifF and nifJ gene products in electron transport to nitrogenase in Klebsiella pneumoniae. J. Bacteriol., 141: 470-475.
Hoover, T.R., Robertson, A.D., Cerny, R.L., Hayes, R.N., Imperial, J., Shah, V.K. and Ludden, P.W., 1987. Identification of the V factor needed for synthesis of the iron-molybdenum cofactor of nitrogenase as homocitrate. Nature, 329: 855-857.
Hoover, T.R., Imperial, J., Ludden, P.W. and Shah, V.K., 1988a. Biosynthesis of the iron-molybdenum cofactor of nitrogenase. Biofactors, 1: 199-205.
Hoover, T.R., Imperial, J., Liang, J., Ludden, P.W. and Shah, V.K., 1988b. Dinitrogenase with altered substrate specificity results from the use of homocitrate analogues for in vitro synthesis of the iron-molybdenum cofactor. Biochemistry, 27: 3647-3652.
Howard, K., Collet, T., White, T. and Orme-Johnson, W.H., 1989. Klebsiella pneumoniae nifH protein produced in E. coli contains no iron in the absence of nifM. J. Inorg. Biochem., 36: 172.
Howard, K.S., McLean, P.A., Hansen, F.B., Lemley, P.V., Koblan, K.S. and Orme-Johnson, W.H., 1986. Klebsiella pneumoniae nifM gene product is required for stabilization and activation of nitrogenase iron protein in Escherichia coli. J. Biol. Chem., 261: 772-778.
Hunt, S., King, B., Canvin, D. and Layzell, D., 1987. Steady and nonsteady state gas exchange characteristics of soybean nodules in relation to the oxygen diffusion barrier. Plant Physiol., 84: 164-172.
Hunt, S., Gaito, S.T. and Layzell, D.B., 1988. Model of gas exchange and diffusion in legume nodules. ii. Characterization of the diffusion barrier and estimation of the concentrations of CO_2, H_2 and N_2 in the infected cells. Planta, 173: 128-141.
Imperial, J., Ugalde, R.A., Shah, V.K. and Brill, W.J., 1984. Role of the nifQ gene product in the incorporation of molybdenum into nitrogenase in Klebsiella pneumoniae. J. Bacteriol., 158: 187-194.
Kelly, M., 1969. Comparisons and cross reactions of nitrogenase from Klebsiella pneumoniae, Azotobacter chroococcum and Bacillus polymyxa. Biochim. Biophys. Acta, 191: 527-540.
Kennedy, C., Eady, R.R., Kondorosi, E. and Rekosh, D.K., 1976. The molybdenum-iron protein of Klebsiella pneumoniae nitrogenase: Evidence for non-identical subunits from peptide 'mapping'. Biochem. J., 155: 383-389.
Kennedy, C., Gamal, R., Humphrey, R., Ramos, J., Brigle, K. and Dean, D., 1986. The nifH, nifM and nifN genes of Azotobacter vinelandii: Characterisation by Tn5 mutagenesis and isolation from pLAFR1 gene banks. Mol. Gen. Genet., 205: 318-325.

Kent, H.M., Ioannidis, I., Gormal, C., Smith, B.E. and Buck, M., 1989. Site-directed mutagenesis of the Klebsiella pneumoniae nitrogenase-Effects of modifying conserved cysteine residues in the α- and ß-subunits. Biochem. J., 264: 257-264.

Kimber, S.J., Bishop, E.O. and Smith, B.E., 1982. Evidence on the role(s) of ATP in the mechanism of nitrogenase, from proton NMR relaxation studies on metal and nucleotide binding to the molybdenum-iron protein. Biochim. Biophys. Acta, 705: 385-395.

Kurtz, D.M., McMillan, R.S., Burgess, B.K., Mortenson, L.E. and Holm, R.H., 1979. Identification of iron-sulphur centers in the iron-molybdenum proteins of nitrogenase. Proc. Nat. Acad. Sci. USA, 76: 4986-4989.

Laane, C., Haaker, H. and Veeger, C., 1978. Involvement of the cytoplasmic membrane in nitrogen fixation by Rhizobium leguminosarum bacteroids. Eur. J. Biochem., 87: 147-153.

Laane, C., Krone, W., Konings, W.N., Haaker, H. and Veeger, C., 1979. The involvement of the membrane potential in nitrogen fixation by bacteroids of Rhizobium leguminosarum. FEBS Lett., 103: 328-332.

Leigh, G.J., 1977. Chemistry of dinitrogen. In: W.E. Newton, J.R. Postgate and C. Rodriguez-Barrueco (Editors), Recent Developments in Nitrogen Fixation. Academic Press, London, pp. 1-24.

Li, J.-G., Burgess, B.K. and Corbin, J.L., 1982. Nitrogenase reactivity: cyanide as substrate and inhibitor. Biochemistry, 21: 4393-4402.

Lindahl, P.A., Day, E.P., Kent, T.A., Orme-Johnson, W.H. and Munck, E., 1985. Mossbauer, EPR and magnetization studies of the Azotobacter vinelandii Fe protein. Evidence for a [4Fe-4S]$^{+1}$ cluster with spin s = 3/2. J. Biol. Chem., 260: 11160-11173.

Ljones, T. and Burris, R.H., 1978. Nitrogenase: The reaction between the Fe protein and bathophenanthrolinedisulfonate as a probe for interactions with MgATP. Biochemistry, 17: 1866-1872.

Lowe, D.J., Fisher, K., Thorneley, R.N.F., Vaughn, S.A. and Burgess, B.K., 1990. The kinetics and mechanism of the reaction of cyanide with molybdenum nitrogenase from Azotobacter vinelandii. Biochemistry, 28: 8460-8466.

Lowe, D.J., Thorneley, R.N.F. and Smith, B.E., 1985. Nitrogenase. In: P.M. Harrison (Editor), Metalloproteins, Part 1:Metal Proteins with Redox Roles. Verlag Chemie, Basel, pp. 207-249.

McKenna, C.E., Menard, R., Dao, C.J., Stephens, P.J. and McKenna, M.C., 1988. Self-oxidation of dithionite-Av2 solutions: The phenomena defined and explained. In: H. Bothe, F.J. de Bruijn and W.E. Newton (Editors), Nitrogen Fixation: Hundred Years After. Gustav Fischer, New York, p. 131.

McLean, P.A. and Dixon, R.A., 1981. Requirement of nifV gene for production of wild-type nitrogenase enzyme in Klebsiella pneumoniae. Nature, 292: 655-656.

McLean, P.A., Papaefthymiou, V., Orme-Johnson, W.H. and Munck, E., 1987. Isotopic hybrids of nitrogenase. Mossbauer study of MoFe protein with selective ^{57}Fe enrichment of the P cluster. J. Biol. Chem., 262: 12900-12903.

McLean, P.A., Papaefthymiou, V., Munck, E. and Orme-Johnson, W.H., 1988. Use of isototpic hybrids of the MoFe protein to study the mechanism of nitrogenase catalysis. In: H. Bothe, F.J. de Bruijn and W.E. Newton (Editors), Nitrogen Fixation: Hundred Years After. Gustav Fischer, Stuttgart, pp. 101-106.

McNary, J.E. and Burris, R.H., 1962. Energy requirements for nitrogen fixation by cell-free preparations from Clostridium pasteurianum. J. Bacteriol., 84: 598-599.

Merrick, M.J., 1988. Organisation and regulation of nitrogen fixation genes in Klebsiella pneumoniae. In: H. Bothe, F.J. de Bruijn and W.E. Newton (Editors) Nitrogen Fixation: Hundred Years After. Gustav Fischer, Stuttgart and New York, pp. 293-302.

Meyer, J., 1981. Comparison of carbon monoxide, nitric oxide, and nitrite as inhibitors of the nitrogenase from Clostridium pasteurianum. Arch. Biochem. Biophys., 210: 246-256.
Miller, R.W. and Tremblay, P.A., 1983. Cytoplasmic membrane of Rhizobium meliloti bacteroids. I. Alterations in lipid composition, physical properties and respiratory proteins. Canad. J. Biochem. Cell Biol., 61: 1334-1340.
Miller, R.W., Al-Jobore, A. and Berndt, W.B., 1986. Properties of the alfalfa-Rhizobium meliloti symbiotic nitrogenase system in vivo and in vitro. Can. J. Biochem. Cell Biol., 64: 556-564.
Miller, R.W. and Eady, R.R., 1988. Cyanamide:a new substrate for nitrogenase. Biochim. Biopys. Acta, 952: 290-296.
Miller, R.W., McRae, D.G., Al-Jobore, A. and Berndt, W.B., 1988. Respiration supported nitrogenase activity of isolated Rhizobium meliloti bacteroids. J. Cell. Biochem., 38: 35-49.
Miller, R.W., Robson, R.L., Yates, M.G. & Eady, R.R., 1980. Catalysis of exchange of terminal phosphate groups of ATP and ADP by purified nitrogenase proteins. Can. J. Biochem., 58: 542-548.
Miller, R.W. and Eady, R.R., 1989a. Molybdenum and vanadium nitrogenases of Azotobacter chroococcum. Low temperature favours N_2 reduction. Biochem. J., 256: 429-432.
Miller, R.W. and Eady, R.R., 1989b. Molybdenum nitrogenase of Azotobacter chroococcum. Tight binding of MgADP to the MoFe protein. Biochem. J., 263:725-729.
Mortenson, L.E., 1964. Ferredoxin requirement for nitrogen fixation by extracts of Clostridium pasteurianum. Biochim. Biophys. Acta, 81: 473-478.
Mortenson, L.E., 1966. Components of cell-free extracts of Clostridium pasteurianum required for ATP-dependent H_2 evolution from dithionite and for N_2 fixation. Biochim. Biophys. Acta, 127: 18-25.
Mortenson, L.E., Morris, J.A. and Jeng, D.Y., 1967. Purification, metal composition and properties of molybdo-ferredoxin, and azo-ferredoxin, two of the components of the nitrogen-fixing system of Clostridium pasteurianum. Biochim. Biophys. Acta, 141: 516-522.
Moustafa, E. and Mortenson, L.E., 1969. Properties of azo-ferredoxin purified from nitrogen-fixing extracts of Clostridium pasteurianum. Biochim. Biophys. Acta, 172: 106-115.
Pagani, S., Eldridge, M. and Eady, R.R., 1987. Nitrogenase of Klebsiella pneumoniae: Rhodanese-catalyzed restoration of activity of the inactive 2Fe species of the 2Fe protein. Biochem. J., 244: 485-488.
Palmer, G., Multani, J.S., Cretney, W.C., Zumft, W.G. and Mortenson, L.E., 1972. Electron paramagnetic resonance studies on nitrogenase. 1. The properties of molybdoferredoxin and azoferredoxin. Arch. Biochem. Biophys., 153: 325-332.
Pau, R.N., Mitchenall, L.A., Eady, R.R. and Robson, R.L., 1988. Genetics of Mo-independent nitrogenases. In: H. Bothe, F.J. de Bruijn and W.E. Newton (Editors), Nitrogen Fixation: Hundred Years After. Gustav Fischer, Stuttgart & New York, p. 322.
Pope, M.R., Murrell, S.A. and Ludden, P.W., 1985. Covalent modification of the iron protein of nitrogenase from Rhodospirillum rubrum by adenosine diphosphorylation of a specific arginyl residue. Proc. Natl. Acad. Sci. USA, 82: 3173-3177.
Postgate, J.R., 1982. The Fundamentals of Nitrogen Fixation. Cambridge University Press, Cambridge (UK).
Prevost, D., Antoun, H. and Bordeleau, L.M., 1987. Effects of low temperature on nitrogenase activity in sainfoin (Onobrychis viciifolia) nodulated by arctic rhizobia. FEMS Microbiol. Ecol., 45: 205-210.

Robinson, A.C., Dean, D.R. and Burgess, B.K., 1987. Iron-molybdenum cofactor biosynthesis in Azotobacter vinelandii requires the iron protein of nitrogenase. J. Biol. Chem., 262: 14327-14332.
Robinson, A.C., Chun, T.W., Li, J.-G. and Burgess, B.K., 1989. Iron-molybdenum cofactor insertion into the apo-MoFe protein of nitrogenase involves the iron protein-MgATP complex. J. Biol. Chem., 264: 10088-10095.
Robson, R.L., 1979. Characterization of an oxygen-stable nitrogenase complex isolated from Azotobacter chroococcum. Biochem. J., 181: 569-575.
Robson, R.L., 1984. Identification of possible adenine nucleotide-binding sites in nitrogenase Fe- and MoFe-proteins by amino acid sequence comparison. FEBS Lett., 173: 394-398.
Robson, R.L., 1986. Nitrogen fixation in strains of Azotobacter chroococcum bearing deletions of a cluster of genes coding for nitrogenase. Arch. Microbiol., 146: 74-79.
Robson, R.L., Eady, R.R., Richardson, T.H., Miller, R.W., Hawkins, M. and Postgate, J.R., 1986. The alternative nitrogenase of Azotobacter chroococcum is a vanadium enzyme. Nature, 322: 388-390.
Rubinson, J.F., Burgess, B.K., Corbin, J.L. and Dilworth, M.J., 1985. Nitrogenase reactivity: azide reduction. Biochemistry, 24: 273-283.
Scott, D.J., May,, H.D., Newton, W.E., Brigle, K.E. and Dean, D.R., 1990. Role for the nitrogenase MoFe protein α-subunit in FeMo-cofactor binding and catalysis. Nature, 343: 188-190.
Schneider, K.C., Bradbeer, C., Singh, R.N., Wang, L.C., Wilson, P.W. and Burris, R.H., 1960. Nitrogen fixation by cell-free preparations from microorganisms. Proc. Natl. Acad. Sci. USA, 46: 726-733.
Shah, V.K. and Brill, W.J., 1977. Isolation of an iron-molybdenum cofactor from nitrogenase. Proc. Natl. Acad. Sci. USA, 74: 3249-3253.
Shah, V.K., Imperial, J., Ugalde, R.A., Ludden, P.W. and Brill, W.J., 1986. In vitro synthesis of the iron-molybdenum cofactor of nitrogenase. Proc. Natl. Acad. Sci. USA, 83: 1636-1640.
Shethna, Y.I., Dervartanian, D.V. and Beinert, H., 1968. Non heme (iron-sulphur) proteins of Azotobacter vinelandii. Biochem. Biophys. Res. Commun., 31: 862-868.
Simpson, F.B. and Burris, R.H., 1984. A nitrogen pressure of 50 atmospheres does not prevent evolution of hydrogen by nitrogenase. Science, 224: 1095-1097.
Smith, B.E., Campbell, F., Eady, R.R., Eldridge, M., Ford, C.M., Hill, S., Kavanagh, E.P., Lowe, D.J., Miller, R.W., Richardson, T.H., Robson, R.L., Thorneley, R.N.F. and Yates, M.G., 1987. Biochemistry of nitrogenase and the physiology of related metabolism. Phil. Trans. Roy. Soc. Lond., B 317: 131-146.
Stiefel, E.I. and Cramer, S.P., 1985. Chemistry and biology of the iron-molybdenum cofactor of nitrogenase. In: T.G. Spiro (Editor), Molybdenum Enzymes. John Wiley and Sons, pp. 89-118.
Stouthamer, A.H., Stam, H., de Vries, W. and van Klerken, M., 1988. Some aspects of nitrogen fixation in free living cultures of Rhizobium. In: H. Bothe, F.J. de Bruijn and W.E. Newton (Editors), Nitrogen Fixation: Hundred Years After. Gustav Fischer, Stuttgart & New York, pp. 257-261.
Tabita, F.R., Van Baalen, C.,Smith, R.L. and Kumar, D., 1988. In vivo protection and recovery of nitrogenase from oxygen in heterocystous cyanobacteria. In: H. Bothe, F.J. de Bruijn and W.E. Newton (Editors), Nitrogen Fixation: Hundred Years After. Gustav Fischer, Stuttgart & New York, pp. 189-194.
Thorneley, R.N.F., Eady, R.R. and Yates, M.G., 1975. Nitrogenases of Klebsiella pneumoniae and Azotobacter chroococcum. Biochim. Biophys. Acta, 403: 269-284.
Thorneley, R.N.F. and Eady, R.R., 1977. Distinction between proton-reducing and acetylene-reducing forms of the enzyme: Effect of temperature and component protein ratio on substrate-reduction kinetics. Biochem. J., 196: 457-461.

Thorneley, R.N.F., Eady, R.R. and Lowe, D.J., 1978. Biological nitrogen fixation by way of an enzyme bound dinitrogen-hydride intermediate. Nature, 272: 557-558.
Thorneley, R.N.F. and Lowe, D.J., 1985. Kinetics and mechanism of the nitrogenase enzyme system. In: T.G. Spiro (Editor), Molybdenum Enzymes. John Wiley and Sons, Inc., New York, pp. 221-284.
Thorneley, R.N.F. and Deistung, J., 1988. Electron-transfer studies involving flavodoxin and a natural redox partner, the iron protein of nitrogenase. Biochem. J., 253: 587-595.
Thorneley, R.N.F. and Ashby, G.A., 1989. Oxidation of nitrogenase iron protein by dioxygen without inactivation could contribute to high respiration rates of Azotobacter species and facilitate nitrogen fixation in aerobic environments. Biochem. J., 261: 181-187.
Thorneley, R.N.F., Ashby, G., Howarth, J.V,, Miller, N.C. and Gutfreund, H., 1989. A transient kinetic study of the nitrogenase of Klebsiella pneumoniae by stopped-flow calorimetry. Biochem. J., 264: 657-661.
True, A.E., Nelson, M.J., Venters, R.A., Orme-Johnson, W.H. and Hoffman, B.M., 1988. ^{57}Fe hyperfine coupling tensors of the FeMo cluster in Azotobacter vinelandii MoFe protein: determination by polycrystalline endor spectroscopy. J. Am. Chem. Soc., 110: 1935-1943.
Upchurch, R.G. and Mortenson, L.E., 1980. In vivo energetics and control of nitrogen fixation: changes in the adenylate energy charge and adenosine 5'-diphosphate/adenosine 5'-triphosphate ratio of cells during growth on dinitrogen versus growth on ammonia. J. Bacteriol., 143: 274-284.
Vaughn, S. and Burgess, B.K., 1988. Nitrogenase reactivity towards nitrate and nitrite. In: H. Bothe, F.J. de Bruijn and W.E. Newton (Editors), Nitrogen Fixation: Hundred Years After. Gustav Fischer, Stuttgart & New York, p. 138.
Voordouw, G., Haaker, H., Van Breemen, J.F.L., Van Bruggen, E.F.J. and Eady, R.R., 1983. Structure of the Mo-Fe protein component of Azotobacter vinelandii nitrogenase. Eur. J. Biochem., 136: 397-401.
Vessey, J.K., Walsh, K.B., and Layzell, D.B., 1988. Oxygen limitation of N_2 fixation in soybean. Plant Physiol., 86: Supp. 123.
Watson, R.J., Chan, Y.-K., Wheatcroft, R., Yang, A.-F. and Han, S., 1988. Rhizobium meliloti genes required for C_4 dicarboxylate transport and symbiotic nitrogen fixation are located on a megaplasmid. J. Bacteriol., 170: 927-934.
Wink, D.A., McLean, P.A., Hickman, A.B. and Orme-Johnson, W.H., 1989. A new method for extraction of iron-molybdenum cofactor (FeMoco) from nitrogenase adsorbed to DEAE-cellulose. 2. Solubilization of FeMoco in a wide range of organic solvents. Biochemistry, 28: 9407-9412.
Witty, J.F., Minchin, F.R., Skot, L. and Sheehy, J.E., 1986. Nitrogen fixation and oxygen in legume root nodules. In: B.J. Miflin (Editor), Oxford Surveys of Plant Molecular and Cell Biology, Vol.3. Oxford University Press, Oxford, pp. 275-314.
Wyatt, P. and Merrick, M., 1989. The roles of the nifW, nifZ, and nifM genes of Klebsiella pneumoniae in nitrogenase biosynthesis. Eur. J. Biochem., 178: 675-682.
Yates, M.G., 1980. Biochemistry of Nitrogen Fixation. In: B.J. Miflin (Editor), Biochemistry of Plants, Vol. 5. Academic Press, New York, pp. 1-64.

CHAPTER 3

THE ALTERNATIVE NITROGENASES

RICHARD N. PAU
AFRC Institute of Plant Science Research, Nitrogen Fixation Laboratory, University of Sussex, Brighton BN1 9RQ, U.K.

1 INTRODUCTION

After demonstrating that molybdenum was required for nitrogen fixation by Azotobacter chroococcum, Bortels (1936) showed that Mo could be replaced by vanadium (V). He also noted that, unlike A. chroococcum, Azotobacter vinelandii appeared to fix nitrogen in the absence of Mo and V. Later studies confirmed that the growth rate of A. vinelandii strain OP in the presence of V is about 70% the rate in the presence of Mo, and in the absence of both Mo and V about 30% the rate with Mo (Becking, 1962; McKenna et al., 1970; Nagatani and Brill, 1974). After purified nitrogenases from diverse organisms were shown to contain Mo, diazotrophic growth in the absence of Mo was perplexing. Did Mo have unique properties essential for binding and reduction of N_2, or could it be replaced by V?

The generally accepted view at the time was that N_2 reduction required Mo. Nitrogenase isolated from cells grown in medium containing V exhibited several altered catalytic properties, and contained V (McKenna et al., 1970). It was studied in an attempt to elucidate the role of Mo, a small quantity of which was present in the V-containing enzyme, though not enough to account for the observed activity. It was then argued that V might have structurally altered the enzyme to allow more effective utilization of Mo (Benemann et al., 1972). This was disputed when it was shown that the electron transfer activity of the V-enzyme was greatly underestimated when measured by acetylene reduction. The activity and altered kinetic properties of the V-enzyme were taken to mean that V had a catalytic role, functionally replacing Mo (Burns et al., 1971; Burns et al., 1976).

Later, the extraction of a Mo-containing cofactor, the characterization of the nif gene cluster, and the demonstration of a high degree of conservation among nitrogenase genes, showed the similarity of the enzymatic apparatus for nitrogen fixation in a wide variety of organisms and reinforced the conviction that Mo is essential for activity (Ruvkun and Ausubel, 1980; Shah et al., 1984). Interest in V waned. Resolution of the conflicting views concerning the role of V eventually came from genetic analysis of the biochemically distinct nitrogenases in Azotobacter (see Chapter 7, and Bishop et al., 1980, 1982; Joerger and Bishop, 1988a; Pau, 1989; Bishop and Joerger, 1990). At present A. chroococcum and A. vinelandii are the only organisms in which alternative nitrogenase(s) have been extensively characterized both biochemically and genetically. This chapter will describe the biochemistry of nitrogenases from these two species.

2 OCCURRENCE

The decisive evidence for an 'alternative' nitrogenase system in A. vinelandii was the demonstration that a strain (CA11) lacking structural genes for the conventional (Mo) nitrogenase could fix nitrogen provided Mo was not added to the medium (Bishop et al., 1986a; Bishop et al., 1986b). An analogous deletion strain of A. chroococcum (MCD1155) was unable to grow under the same conditions, but its growth was stimulated by low concentrations of V (Robson et al., 1986a). Isolation of a V nitrogenase from this deletion strain, and then from an analogous strain of A. vinelandii grown in the presence of V, established the reality of a V nitrogenase in both species (Hales et al., 1986; Robson et al., 1986). Deletions in structural genes for both the Mo and V nitrogenases of A. vinelandii produced a strain which can fix nitrogen provided both Mo and V are absent, indicating that growth in their absence, originally observed in mutant strains of this species, is due to a third, genetically distinct, enzyme: nitrogenase-3 (Pau et al., 1989). This enzyme, isolated from strain CA11.6 (a tungstate-resistant derivative of strain CA11) grown in the absence of added Mo and V, lacks significant quantities of Mo and V (Chisnell et al., 1988). Strains with deletions in genes for both Mo and V nitrogenases in A. chroococcum (Robson et al., 1989), and with deletions in genes for the Mo, V, and nitrogenase-3 in A. vinelandii (R.N. Pau and L.A. Mitchenall, unpublished), do not fix nitrogen. In A. chroococcum, therefore, the V enzyme is the only alternative nitrogenase, whereas A. vinelandii has two alternative nitrogenases: the V nitrogenase and nitrogenase-3. As the V nitrogenases of A. chroococcum and A. vinelandii show only very few differences in amino acid sequence (see Section 8.1.ii), they will not be distinguished except where necessary. The abbreviations Ac, Av, and Kp will be used to identify the origins of the nitrogenase components from A. chroococcum, A. vinelandii, and Klebsiella pneumoniae, respectively.

3 GENERAL STRUCTURAL AND BIOCHEMICAL RELATIONSHIPS

The alternative nitrogenases are broadly similar to Mo nitrogenases in most physico-chemical properties (Table 1), and in their requirements for activity. All are inactivated by oxygen, though both alternative nitrogenases are more sensitive than Mo nitrogenase (Eady et al., 1987; Chisnell et al., 1988). The

TABLE 1. Physicochemical properties and genes of Azotobacter nitrogenases

1. Genes and polypeptides

Mo NITROGENASE *A. vinelandii*	Component 1			Component 2
Gene	nifD		nifK	nifH
Subunit structure	α2	--	β2	γ2
No of amino acids	492	--	523	290
Molecular weight	55288	--	59459	31516
Isoelectric point	6.32	--	6.42	4.50
V NITROGENASES				
A. vinelandii				
Gene	vnfD	vnfG	vnfK	vnfH
Subunit structure	α2	δ2	β2	γ2
No of amino acids	474	113	475	290
Molecular weight	53877	13372	52775	31042
Isoelectric point	6.37	4.87	5.69	4.60
A. chroococcum				
Gene	vnfD	vnfG	vnfK	vnfH
Subunit structure	α2	δ2	β2	γ2
No of amino acids	474	113	475	290
Molecular weight	53834	13405	52854	31073
Isoelectric point	6.37	5.03	5.69	4.64
NITROGENASE-3				
A. vinelandii				
Gene	anfD	anfG	anfK	anfH
Subunit strucure	α1 or α2	nd	β2	γ2
No of amino acids	518	132	462	275
Molecular weight	58326	15369	51153	29883
Isoelectric point	6.98	5.08	5.15	4.72

2. Metal contents (g-atoms.mol^{-1})

	Component 1				Component 2		
	Ac-Mo[a,c]	Av-V[b]	Ac-V[c]	Av-3[d]	Av-Mo[e]	Av-V[e]	Av-3[e]
S^{2-}	20.0	nd	19.4	17.6	4.0	4.4	3.8
Fe	23.0	9.3	20.8	24.2	3.5	3.6	3.7
V	0.06	0.62	2.0	<0.01	-	-	-
Mo	1.9	<0.05	0.06	0.13	-	-	-

Molecular weights and isoelectric points obtained from sequence data (Jacobson et al., 1989, Robson et al., 1986b, Robson et al., 1989, Joerger et al., 1990) using the program PeptideSort (Devereux et al., 1984).

Metal content data from: a, Yates and Planqué 1975; b, Hales et al., 1986; c, Eady et al., 1987; d, Chisnell et al., 1988; e, Premakumar et al., 1989.

Av-Mo, Ac-Mo, Av-V, Ac-V and Av-3 refer to Mo, V and nitrogenase-3 of A. vinelandii and A. chroococcum. nd = not determined.

half-life of A. chroococcum V nitrogenase in air is about 40 s compared with 10 min for Mo nitrogenase. All are composed of a smaller (M_r = about 60000) homodimeric protein, component 2 (Fe protein), which transfers electrons in an ATP-dependent reaction from a low potential reductant to a larger (M_r = about 230000) heteromeric protein, component 1 (MoFe and VFe proteins for the Mo and V nitrogenases, respectively). The latter contains the site for reduction of dinitrogen to ammonia. The principal physical features which distinguish them from the Mo nitrogenase are their metal contents (Table 1) and subunit structure (Section 8.1.i).

4 ENZYME ACTIVITY

4.1 Introduction

In the 1970's, the first biochemical studies of partially purified nitrogenases isolated from Azotobacter grown in the presence of V showed that replacement of Mo by V changed the activity of the enzyme towards various substrates. When deletion strains of A. chroococcum and A. vinelandii lacking structural genes for Mo nitrogenase became available in 1986, more highly purified V nitrogenases followed.

Less is known about nitrogenase-3. Although there has long been evidence for its existence, albeit generally unaccepted, the first study of the purified enzyme has only recently appeared (Chisnell et al., 1988).

Besides the reduction of nitrogen to ammonia, all nitrogenases catalyze the reduction of protons to dihydrogen, as well as the reduction of a number of other substrates. Extensive analysis of the mechanism of Mo nitrogenase has shown that catalysis entails formation of a complex of the two component proteins, transfer of an electron from component 2 to one of two independent and identical halves of the component 1, and dissociation of the two components. Substrate reduction requires transfer of more than two electrons, and so requires sequential electron transfer cycles (Thorneley and Lowe, 1985; Chapter 2). The similarities of the alternative nitrogenases to Mo nitrogenase imply a similar mechanism of substrate reduction.

4.2 Nitrogen and proton reduction

The Mo and V nitrogenases have similar apparent Km values (14-16 kPa) with N_2 as substrate (Burns et al., 1971; Eady et al., 1987). They also transfer electrons equally efficiently for each mole of ATP hydrolysed (Eady et al., 1987). Replacement of Mo by V therefore does not appear to affect N_2 binding to nitrogenase, or the evolution of H_2, which competes with N_2 reduction. However, the distribution of electrons between N_2 reduction and H_2 evolution under optimal conditions for N_2 reduction is different for the Mo and V enzymes (Burns et al., 1971; Eady et al., 1987). Mo nitrogenase allocates 75% of the total electron flow through the enzyme to N_2 reduction, the remainder going to proton reduction. With V nitrogenases only about 50% of the total electron flow is allocated to N_2 reduction under the assay conditions tested.

This difference in electron allocation to N_2 and proton reduction by Mo and V nitrogenases is differentially affected by temperature (Miller and Eady, 1988). As the temperature is lowered, the specific activities for N_2 reduction decrease for both enzymes, but more markedly for Mo than for V nitrogenase. The allocation of electrons during enzyme turnover is considered to be determined by competition between H_2 evolution and binding of reduced component 2 to component 1 to form a more reduced complex. This complex binds N_2 with release of H_2 (Thorneley and Lowe, 1985). The more effective reduction of N_2 by V nitrogenase at low temperatures is a function of the component 2 protein of the V nitrogenase, and so may be associated with electron transfer, and be determined by the rate of dissociation of the two nitrogenase components (Miller and Eady, 1988).

The specific activity for N_2 reduction by nitrogenase-3 is only about 4% of that of Mo nitrogenase (Table 2). The reason for this low activity is not known. It is unlikely to be an intrinsic property of the enzyme because the growth rate of strains which only synthesize nitrogenase-3 is much higher than would be expected were nitrogenase-3 to have the very low activity measured <u>in vitro</u>, and optimal purification or assay conditions have probably not been attained. While nitrogenase-3 apparently allocates an even smaller proportion (20-30%) of electrons to N_2 reduction than V nitrogenase (Chisnell et al., 1988), some caution is necessary until the reasons for low in vitro activities are better understood. Both alternative nitrogenases are therefore probably less efficient at reducing N_2 than Mo nitrogenase.

4.3 Acetylene reduction to ethylene

Reduction of acetylene to ethylene by both Mo and V nitrogenases appears to be effected by similar highly stereospecific mechanisms, for isotopic analyses showed that identical products (predominantly <u>cis</u>-$^2H_2$-ethylene) were formed by both enzymes in the presence of ^2H_2O (Dilworth et al., 1988). This has been interpreted as indicating that acetylene binds sideways to a metal in the active site (Hardy,1979). The apparent K_m value (2 kPa) for acetylene reduction by Mo nitrogenase is about a third of that for V nitrogenase (Dilworth et al., 1988).

TABLE 2. Specific activities of Azotobacter nitrogenases at 30°C.

		Product ($nmol.min^{-1}.mg^{-1}$ protein)			
		NH_4^+	H_2	C_2H_4	Reference
K. pneumoniae	Mo nitrogenase	990	2100	1693	Eady, 1989
A. vinelandii	Mo nitrogenase	1140	2200	2000	Hales et al., 1986
A. vinelandii	V nitrogenase	660	1400	220	Hales et al., 1986
A. chroococcum	V nitrogenase	350	1348	608	Eady et al., 1987
A. vinelandii	Nitrogenase-3	38	253	28	Chisnall et al., 1988

At saturating concentrations of acetylene, H_2 evolution by Mo nitrogenase is almost completely eliminated, with almost all of the electrons utilized in acetylene reduction. By contrast, H_2 evolution by V nitrogenase in the presence of high concentrations acetylene (5 times K_m) is still 33% of the rate in the absence of acetylene (Dilworth et al., 1988). The specific activity for acetylene reduction by nitrogenase-3 is lower than that of V nitrogenase (Chisnell et al., 1988). While acetylene is a good substrate for Mo nitrogenases, it is a relatively poor substrate for both alternative nitrogenases. The high allocation of electrons to proton reduction by both alternative nitrogenases makes H_2 evolution a better measure of the relative activity of different nitrogenases.

4.4 Acetylene reduction to ethane

Whereas ethylene is the only product detected when acetylene is reduced by Mo nitrogenases, the fully reduced product, ethane, is a minor product of the reaction catalyzed by V nitrogenase (Dilworth et al., 1987). Under standard assay conditions (15 kPa acetylene, 30°C) the ethane/ethylene ratio is about 0.01-0.03. The relative amount of ethane produced increases with increasing temperature and electron flux. Ethane, which was at first considered diagnostic for V nitrogenase, is also produced when acetylene is reduced by nitrogenase-3 (Pau et al., 1989). In this case, slightly higher amounts (ethane/ethylene ratio = 0.04-0.05) are formed. It is therefore an easily identifiable product of acetylene reduction by both alternative nitrogenases.

4.5 Ethylene reduction

Although Mo nitrogenase does not reduce acetylene to ethane, both Mo (Kp and

Av) and V nitrogenase (Ac) reduce ethylene to ethane (Ashby et al., 1987; Dilworth et al., 1988). Reduction of ethylene to ethane requires moderately high electron flux and substrate concentration. Ethylene reversibly inhibits proton reduction by Mo nitrogenase, and is a more potent inhibitor at low electron flux, when the enzyme is on average in a less reduced state. Addition of ethylene to Mo nitrogenase while it is reducing protons results in the appearance of a novel EPR signal (g_{av} = 2.04), suggesting that ethylene binds to a transient reduced state of the enzyme, possibly to a metal monohydride (Lowe et al., 1978). A common scheme for the paths leading to the formation of ethylene from acetylene, and ethane from ethylene has been suggested (Ashby et al., 1987). It is consistent with kinetic and spectroscopic data for acetylene and ethylene reduction by the Mo nitrogenase of K. pneumoniae, and based on the chemistry of model Mo-alkynyl complexes. A transient metal hydride is considered to combine with acetylene to give a metal-alkyne complex which is reduced to form a metal-alkene complex. This common intermediate can either be further reduced to yield ethane, or converted to a hydride-ethylene complex, which dissociates to form ethylene. In the presence of ethylene alone the hydride-ethylene complex is formed directly, and yields ethane via conversion to the common intermediate. When acetylene is reduced by Mo nitrogenase the concentrations of ethylene produced would yield extremely low amounts of ethane. Moreover, reduction of ethylene to ethane by Mo nitrogenase is strongly inhibited by acetylene (Ashby et al., 1987). Ethane is therefore not observed as a product of acetylene reduction by Mo nitrogenase.

V nitrogenase also reduces ethylene to ethane with an apparent Km value (172 kPa) similar to that for Mo nitrogenase (Dilworth et al., 1988). When acetylene is reduced by V nitrogenase, the ratio of ethane to ethylene does not change with increasing acetylene concentrations, and Km values for acetylene reduction to ethylene and to ethane are identical. A single site is therefore probably involved in the reduction of acetylene to the two products. However, a presteady state lag has been observed for ethane, but not for ethylene formation. The lag is not eliminated by preincubation of the enzyme with ethylene, suggesting that an intermediate is slowly formed on the route from acetylene to ethane, but not to ethylene (Dilworth et al., 1988). For V nitrogenase, there does not appear to be a common intermediate during reduction of acetylene to ethylene and ethane as in the scheme proposed for ethylene and ethane formation by Mo nitrogenase, and ethylene and ethane may be formed via different routes. Alternatively, if there is a common intermediate as postulated for Mo nitrogenase, then there is an additional step which generates the presteady-state lag.

4.6 Inhibition by carbon monoxide

CO does not inhibit H_2 evolution by Mo or V nitrogenases (Dilworth et al., 1988), except in nifV mutants of K. pneumoniae (Hawkes et al., 1984). CO does, however, inhibit reduction of both N_2 and acetylene by V nitrogenase. The K_i values for CO inhibition of N_2 reduction and of acetylene reduction to ethylene by V nitrogenase (Dilworth et al., 1988) are much higher than those observed for Mo nitrogenase (Burns et al., 1971). Less severe inhibition of V nitrogenase may be due to the known weaker binding of CO to V than to Mo. However, CO strongly inhibits reduction of acetylene to ethane by V nitrogenase while only weakly inhibiting reduction of ethylene to ethane (Dilworth et al., 1988). This again suggests that at least some of the steps to ethane formation from the two substrates are distinct.

4.7 Determinants of ethane formation

Ethane formation from acetylene by V nitrogenase is associated with the VFe protein, for the hybrid enzyme formed by combination of the VFe protein with Fe protein of Mo nitrogenase produces ethane (Dilworth et al., 1987). A V-containing cofactor (called FeVaco or FeVco), analogous to the iron-molybdenum co-factor (FeMoco) of Mo nitrogenase, can be extracted into N-methylformamide from the VFe protein of V nitrogenase (Smith et al., 1988). FeVaco activates the cofactor-less Mo nitrogenase component 1 protein purified from a $nifB^-$ strain of K. pneumoniae which is otherwise inactive. The FeVaco-activated component 1 protein reduces acetylene to yield both ethylene and ethane (ethane/ethylene ratio = 0.08), but does not reduce N_2. Since the same protein from the $nifB^-$ strain produces only C_2H_4 from C_2H_2 when activated with FeMoco, the chemical properties of the cofactor would appear to determine the route(s) of C_2H_2 reduction.

Other evidence, however, points to the ligand environment of the cofactor being able to influence the product pattern from acetylene reduction. Thus, when a strain which can only synthesize nitrogenase-3 is grown in the absence of Mo and V, and then transferred to medium containing Mo, reduction of acetylene to ethane rapidly increases about 10-fold, so that the amount of ethane produced almost equals that of ethylene. A species of nitrogenase-3 with a high Mo content has been purified from this strain, and the Mo appears to be incorporated into a cofactor species R.R. Eady, M.E. Eldridge, D.J. Lowe and R.N. Pau, unpublished). In this case a Mo-containing cofactor in an heterologous environment provided by nitrogenase-3 polypeptides results in an enzyme which reduces acetylene to give ethane as a major product. However, it is also possible that Mo in a cofactor environment for nitrogenase-3 may be behaving chemically very differently from Mo in FeMoco.

Single amino acid changes at a specific site, probably near the cofactor, can dramatically change the acetylene-reducing activity of A. vinelandii Mo nitrogenase (Scott et al., 1990). Replacing glutamine-191 and histidine-195 in the α-chain of component 1 by lysine and glutamine respectively results in enzymes with decreased activity, but which reduce acetylene to a mixture of ethylene and ethane and N_2 not all. Mutant strains synthesizing these enzymes do not appear to reduce nitrogen.

4.8 Acetonitrile, propene

Both Mo and V nitrogenase reduce acetonitrile. The allocation of electrons to acetonitrile reduction is greater for reduction by partially purified V nitrogenase than for Mo nitrogenase (Burns et al., 1971). Propene, which is poorly reduced by Mo nitrogenase (apparent K_m value = 50kPa), is not reduced by V nitrogenase (McKenna et al., 1979; Dilworth et al., 1988). Thus, all substrates examined so far compete with proton reduction less efficiently when reduced by the alternative nitrogenases compared with Mo nitrogenase. Unlike the reduction of acetylene, both possible reduction products of acetonitrile reduction, propene and propane, are produced by both Mo and V nitrogenases. Reduction of acetonitrile by V nitrogenase is analogous to reduction of acetylene in that a greater amount of the more reduced product, propane, is produced (Burns et al., 1971).

5 ELECTRON TRANSFER TO COMPONENT 1

Although many of the differences in enzyme activity described above are probably related to substrate binding at the active site, some are due to other aspects of nitrogenase activity. The ability to transfer electrons from component 2 to other component 1 proteins has been tested by making heterologous crosses, combining component proteins from different nitrogenases (Chisnell et al., 1988; Premakumar et al., 1989). The hybrid enzymes obtained by complementing A. vinelandii Mo nitrogenase component 1 with component 2 proteins of the V nitrogenase and nitrogenase-3 have about 50% and 10% of the specific activity of Mo nitrogenase for acetylene reduction, respectively. Further, component 2 of A. vinelandii nitrogenase-3 does not prevent complex formation between $Av1^{Mo}$ and $Av2^{Mo}$ (Premakumar et al., 1989). Component 1 of nitrogenase-3 combined with component 2 of either Mo or V nitrogenase has negligible activity.

6 ELECTRON DONORS TO COMPONENT 2

Oxidised component 2 proteins of A. chroococcum Mo and V nitrogenases are both reduced at similar rates by sodium dithionite, indicating that they have similar electron transfer sites (Bergström et al., 1988). In A. vinelandii the potential electron donors in vivo are flavodoxin, and three ferredoxins (Klugkist et al., 1986; Martin et al., 1989). How many of these are involved, and with which nitrogen-fixing system, has not been established. The mid-point potentials of A. chroococcum Mo and V nitrogenase component 2 proteins are consistent with flavodoxin being a competent donor to both nitrogenases (Bergström et al., 1988). A strain of A. vinelandii lacking flavodoxin and ferredoxin I grows diazotrophically in the presence of Mo (Martin et al., 1989). A gene encoding a ferredoxin-like protein is in the same operon as vnfH, the gene for V nitrogenase component 2 (Robson et al., 1989; Joerger et al., 1990). Since vnfH is expressed in the absence of Mo, whether V is present or not (Bishop and Joerger, 1990), the product of this ferredoxin-like gene may be involved in electron transfer to both alternative nitrogenases. A similar gene is located in the same operon as nifB, which is required for synthesis of all nitrogenase cofactors (Joerger and Bishop, 1988b) and may be expressed when any of the three A. vinelandii nitrogenases is synthesized.

7 METAL CENTRES

7.1 V nitrogenase

The similar Fe and S^{2-} contents of the alternative and Mo nitrogenases (Table 1) point to the presence of comparable metal centres in all of them. All components 2 probably have a single $[Fe_4S_4]$ centre. Component 1 of Mo nitrogenase is considered to have two types of redox centres: four 'P' clusters, which are probably Fe_4S_4 clusters of unusual properties, and a pair of cofactor centres. Low temperature magnetic circular dichroism (MCD) spectra and MCD magnetization behaviour of component 1 of A. vinelandii V nitrogenase show that it has clusters with electronic and magnetic properties very like those of the 'P' clusters of the Mo enzyme (Morningstar et al., 1987).

The principal differences in metal content are the presence or absence of Mo, and the replacement of Mo by V in V nitrogenases. Results from both EPR and Mossbauer spectra indicate that the biologically unique 3/2 spin state EPR signal of Mo nitrogenase component 1 protein originates in the two cofactor centres in each molecule. The major features of the EPR spectrum of dithionite-reduced Mo nitrogenase component 1 proteins are g = 4.3, 3.6. 2.0 signals associated with the ground state doublet of an S = 3/2 spin system. The EPR spectra of dithionite-reduced V nitrogenase component 1 proteins are different. They have weak signals with g = 5.58, 5.40, 4.34, 2.04 and 1.93 (A. vinelandii) and 5.6, 4.35, 3.77 and 1.93 (A. chroococcum) (Eady et al., 1987; Morningstar and Hales, 1987). The feature at about g = 5.5 is given by a formally spin forbidden transition from the ground state. The EPR spectra from V nitrogenases are nevertheless analogous to those of Mo nitrogenase component 1 protein in that they possess a paramagnetism which, in part, can be interpreted as arising from an S = 3/2 spin system. It differs from the Mo nitrogenase component 1 protein in possessing much higher rhombicity. The g = 3.77 signal in A. chroococcum VFe protein, which is absent from A. vinelandii VFe protein, can be interpreted as originating from an additional species. The g = 1.93 feature, which is absent in Mo nitrogenase, arises from an S = 1/2 spin system, probably from a reduced Fe-S cluster.

The EPR spectrum of FeVaco extracted from A. chroococcum V nitrogenase shows a very weak signal with g values of 4.5, 3.6 and 2.0, arising from a S = 3/2 spin system. Absence of the feature at g = 5.6 suggests that its presence depends on the ligand environment of cofactor in the enzyme (Smith et al., 1988). The VFe protein of A. vinelandii exhibits a third signal which has not been detected in MoFe protein. This is a broad absorption signal (g = 2.8-5.5) with a peak near g = 3.5. It is only observed in dispersion spectra with high power and low temperature, and may originate from a paramagnetic site coupled to a metal cluster (Hales et al., 1989).

The environment surrounding the unique Mo and V in Mo and V nitrogenases and their extruded cofactors has been probed by X-ray absorption spectroscopy.

TABLE 3. Structural data for the coordination shells around Mo and V

	M-Fe		M-S		M-O, N, C		
	N	nm	N	nm	N	nm	Reference
Kp MoFe protein	3 (1)	0.267	4 (1)	0.237	2 (1)	0.210	Eidsness et al., 1986
Av FeMoco	2.6	0.270	3.1	0.238	3.1	0.210	Conradson et al., 1987
Ac VFe protein	3 (1)	0.275	3 (1)	0.231	3 (1)	0.215	Arber et al., 1989
Ac FeVaco	3 (1)	0.274	2 (1)	0.232	4 (1)	0.214	Smith et al., 1988
Av VFe protein	3 (1)	0.276	3 (1)	0.234	3 (1)	0.214	George et al., 1988
$[Me_4N]VFe_3S_4Cl_3(dmf)_3]$	3	0.277-0.278	3	0.233-0.234	3	0.211-0.245	Kovacs et al., 1986

The data were deduced from X-ray absorption spectroscopy of Mo nitrogenase, V nitrogenase and the extruded cofactors FeMoco and FeVaco, and from X-ray crystallographic analysis of $[Me_4N]VFe_3S_4Cl_3(dmf)_3]$.
N = number of atoms; M = Mo or V

Extended X-ray absorption fine structure (EXAFS) data can be simulated by three coordination shells around Mo and V atoms, indicating that Mo and V are part of Mo- or V-Fe-S clusters with similar local radial environments (Table 3). The small differences between the distances of the Fe coordination shells of the Mo nitrogenase and those of V nitrogenases may reflect slight differences in cluster geometry (Eidsness et al., 1986; Conradson et al., 1987; Arber et al., 1987a, b; George et al., 1988; Smith et al., 1988). The distances for the shells of V nitrogenase and FeVaco are in good agreement with X-ray crystallographically-determined distances for the synthetic compound $[Me_4N][VFe_3S_4Cl_3(DMF)_3]$ (Arber et al., 1987a, b). This compound contains a VFe_3S_4 cubane-like cluster in which V is octahedrally coordinated.

The vanadium K-edge X-ray absorption spectra of the VFe proteins of A. chroococcum and A. vinelandii closely resemble that of the synthetic compound $[Me_4N][VFe_3S_4Cl_3(DMF)_3$, again indicating that they both have similar electronic structures and oxidation states. Small differences in the extended absorption near edge regions of the spectra of the V nitrogenase and the model V cubane may reflect differences in the detailed geometry of the two clusters. Iron K-edge EXAFS data for both FeMoco and FeVaco indicate the presence of Fe_2S_2M (M = Mo,V,Fe) rhombs with Fe..Fe separations of about 0.261 nm. A longer range Fe..Fe separation at about 0.365 nm indicates that these rhombs, in both cofactors, are part of a more extended structure (Arber et al.1988, Garner et al., 1989).

7.2 Nitrogenase-3

The nature of the metal fulfilling the role of Mo in the cofactor centre of nitrogenase-3 is not yet known. Since strain CA11, lacking genes for the Mo nitrogenase, grows in 'spent' medium from which Mo (and presumably V) have been removed both by chemical extraction and previous growth of Azotobacter (Bishop et al., 1986), A. vinelandii must be able to fix N_2 in the absence of Mo and V. Nanomolar concentrations of Mo inhibit growth. Extracts of strains grown in the absence of Mo and V do not exhibit the g = 3.6 EPR signal characteristic of Mo nitrogenase (Bishop et al., 1986). Purified nitrogenase-3 contains very little Mo or V (Table 1; Chisnell et al., 1988) and no significant quantities of W, Cr, or Re. Although some Zn (0.4 g-atom.mol^{-1}) is found, similar amounts are present in Mo nitrogenase preparations. Other than Fe, none of the transition metals (Co, Cr, Cu, Mn, Ni, Re, W, Ti, Zn) stimulates diazotrophic growth when added to batch cultures of a strain of A. vinelandii only able to synthesize nitrogenase-3 (Pau et al., 1989). However, nitrogenase-3 does form enzymes containing either Mo or V. A Mo-containing species can be extracted from a strain which can only synthesize nitrogenase-3, grown in medium containing non-repressive levels of Mo. The purified Mo-containing nitrogenase-3 reduces acetylene with the production of a high proportion of ethane, but apparently not N_2. (R.R. Eady, M.E. Eldridge, D.J. Lowe and R.N. Pau, unpublished). Whole cell studies suggest that nitrogenase-3 can also incorporate added V, though the activity of the enzyme is then less than that of the enzyme synthesized in the absence of Mo and V (Pau et al., 1989). These species of nitrogenase-3 containing Mo and V do not therefore appear to explain diazotrophic growth by strains synthesizing only nitrogenase-3. Together, these findings suggest that in nitrogenase-3 Fe may have assumed the role which Mo plays in Mo nitrogenase. Conclusive proof of this is difficult, for Fe is a major constituent of all the metal centres, and it is therefore premature to conclude that nitrogenase-3 is an Fe nitrogenase. The chemistry of Fe is restricted predominantly to that of Fe(II) and Fe(III) (and perhaps Fe(IV)) within a narrow range of oxidation levels, while Mo and V have a wide range of oxidation states and form multiply bonded ligands in high oxidation state compounds. If indeed Fe has the same functions in nitrogenase-3 that Mo and V serve in Mo and V nitrogenases, we will need to know more about the Fe-N_2 chemistry involved and about what features, if any, are shared with the Mo and V systems.

7.3 Tungstate

Tungstate is a long-established inhibitor of nitrogenase activity. An *A. vinelandii* mutant capable of growth in N-free Mo-deficient medium containing 1-10 mM W has been shown to contain an active nitrogenase with W and Mo in a 1:1 ratio (Hales and Case, 1987). The specific activity of the enzyme is lowered, though the fraction of electrons allocated to nitrogen reduction is similar to that characteristic of the conventional Mo enzyme. The EPR spectrum of the component 1 protein had two signals associated with S = 3/2 EPR spin states. One is characteristic of FeMoco, while the other has been assigned to a W-containing cofactor. Unlike the FeMoco signal, the latter does not decrease in amplitude during turnover, suggesting that the W-containing centre is not reduced, which is consistent with the chemistry of complexes of tungsten.

8 AMINO ACID SEQUENCES

8.1 Component 1 proteins

(i) Subunit structure. The organization of the nitrogenase structural genes of *A. vinelandii* and the relationships between their gene products are shown in Fig. 1. The most striking difference between the polypeptides of the component 1 proteins of the alternative nitrogenases and those of Mo nitrogenases is the presence of a small (M_r = 13000-15000) additional subunit (δ), which has no counterpart in Mo nitrogenases (Table 1, Fig. 1). It is the product of a gene (G) located between genes D and K, which encode the α- and ß-subunits of the component 1 proteins (Joerger et al., 1989; Robson et al., 1989; Joerger et al., 1990). The genes for the component 1 proteins of the alternative nitrogenases therefore have a similar organization, which is distinct from that of Mo nitrogenases. Both V nitrogenases and nitrogenase-3 have been shown to have an

$\alpha_2\beta_2\delta_2$ structure, homologous to the $\alpha_2\beta_2$ structure of Mo nitrogenases, but with the additional δ subunits (Chisnell et al., 1988; Robson et al., 1989). However, nitrogenase-3 also has a form with an $\alpha_1\beta_2$ structure (the stoichiometry with the δ subunit was not determined in this species). The δ subunits of the V nitrogenase and nitrogenase-3 show 39% sequence identity (Joerger et al., 1990). The anfG product lacks a segment corresponding to residues 1-19 at the NH_2-terminus of the vnfG product. The most conserved segment (of 5 identical amino acids) contains a cysteine residue. The amino acid sequences of the δ subunits do not show close homology with any known protein. Their function is not yet known, though they are likely to play an important role in the alternative nitrogenases.

(ii) Potential metal cluster ligands in component 1. Comparison of the sequences of Mo nitrogenase proteins, and those of the nifE product, has provided the primary data for experiments aimed at establishing the potential ligands to metal clusters (Brigle et al., 1987; Howard et al., 1989; Kent et al., 1989; Scott et al., 1990). The structural genes for the V nitrogenases of A.

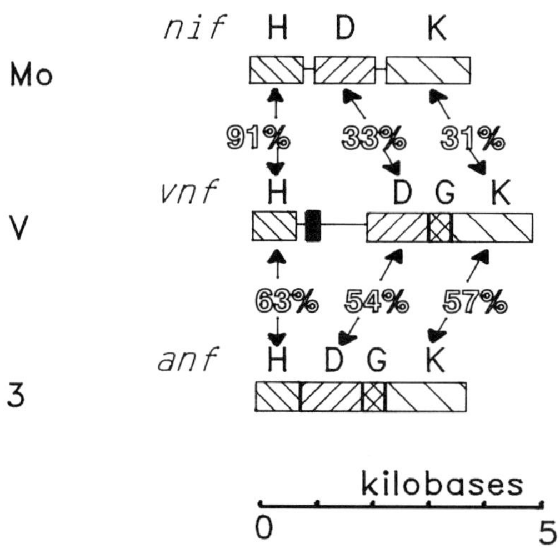

Fig. 1. Organization of the genes encoding the Mo and V nitrogenases, and nitrogenase-3, of A. vinelandii, and the percentage identical sequence matches between their polypeptide products. The Mo nitrogenase, V nitrogenase and nitrogenase-3 genes are distinguished by the names nif, vnf, and anf respectively.

chroococcum and A. vinelandii have been sequenced (Robson et al., 1986b; Raina et al., 1988; Robson et al., 1989; Joerger et al., 1990); the deduced amino-acid sequences show few differences - 9 in the vnfH gene products, 5 in the vnfD gene products and 25 in the vnfK gene products (Joerger et al., 1990). Comparison of the sequences of the V nitrogenases and nitrogenase-3 component 1 proteins with those of Mo nitrogenases is of interest because it provides information on potential metallocluster binding domains from enzymes with similar cofactors. The most likely metal cluster ligands are conserved cysteine residues. Five conserved cysteine residues in the α-subunit and three in the β-subunit of all Mo nitrogenases are also present in the corresponding subunits of the alternative nitrogenases (Fig. 2a). The first three in the α-subunit have approximately the same inter-cysteinyl spacing as those in the β-subunit. Substitution of the conserved cysteines in <u>K. pneumoniae</u> and <u>A. vinelandii</u> Mo nitrogenase component 1

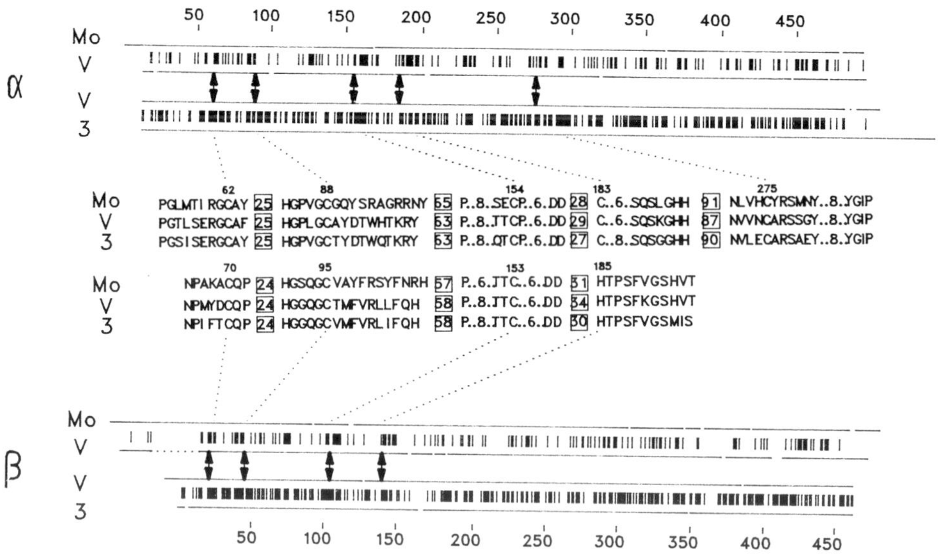

Fig. 2. Comparison of the positions of identical amino acids in aligned sequences of the α and β polypeptide chains of <u>A. vinelandii</u> Mo and V nitrogenase, and between V nitrogenase and nitrogenase-3. The sequences are represented by horizontal lines with spaces or dotted lines where gaps have been inserted to align the sequences. Solid vertical lines between the sequences indicate the positions of identical amino acids. The alignments and graphical plots were made with the programs Gap and Gapshow (Devereux et al., 1984). The positions of the invariant cysteines in the α and β chains and histidine-185 in the β chain are shown by double headed arrows. Homologous sequences adjacent to the invariant cysteines and His-β185 are shown between the schematic representations of the chains. The Cys and His residues are numbered according to the Mo nitrogenase sequence. Inter-cysteine and cysteine-histidine distances are given within squares.

proteins has shown that they are all required for maximum nitrogenase activity. A variety of data indicates that the conserved regions adjoining Cys-α275 and Cys-α184 may be involved with FeMoco-binding (Kent et al., 1989; Scott et al., 1990). However, the number of cysteines is insufficient to provide thiol ligands to the putative 'P' and cofactor centres. Other residues may provide N or O ligands. A histidine residue in the ß-subunit is in a highly conserved segment which has the relative position corresponding to the Cys-184 in the α-subunit. Four other conserved histidine residues (His-α83, His-α191, His-α192, His-ß90) are adjacent to conserved cysteine residues in all the nitrogenases (Fig 2b). Substitution of histidine-α195 by asparagine and glutamate-α191 by lysine in Mo nitrogenase enables it to reduce acetylene to ethane (Scott et al., 1990), a characteristic property of the alternative nitrogenases. All of the amino acid substitutions investigated so far in Mo nitrogenaes involve residues conserved in the altenative nitrogenases. Dissimilarities between the polypeptide regions associated with distinct metal cluster centres in the alternative and Mo nitrogenases must involve other residues, or relative spacing between residues, which affect the geometry and chemistry of the centres. Comparison of the segments associated with conserved cysteines shows several changes common to the alternative nitrogenases (Fig. 2).

8.2 Component 2 proteins

Compared with the component 1 subunits, the nifH, vnfH and anfH products show a high degree of sequence conservation (Fig 3). All have a conserved N-terminal segment terminated by lysine, which is homologous to the ATP-binding region of adenylate kinase and other nucleotide-binding phosphotransferases

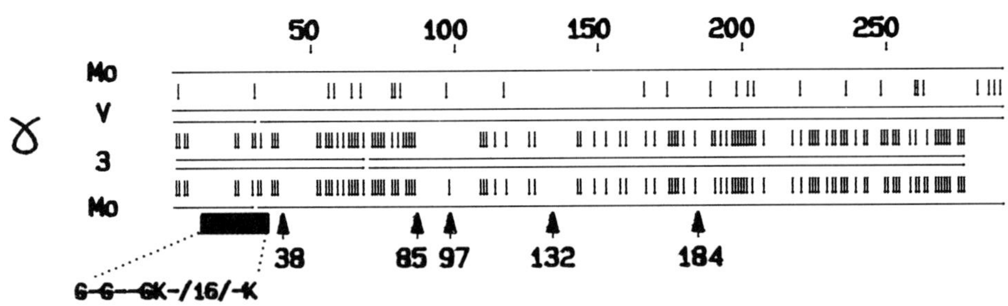

Fig. 3. Comparison of the distribution of non-identical amino acids between the γ polypeptide chains of Mo and V nitrogenases and nitrogenase-3 of A. vinelandii. The sequences are plotted graphically as in Fig 2., except that the vertical lines represent sequence differences. Arrows indicate the positions of conserved cysteines, which are numbered according to the Mo nitrogenase sequence. The position of the segment homologous to ATP binding proteins is shown by the solid rectangle, and the conserved residues are shown below it.

(Robson, 1984, Fry et al., 1986). Five conserved cysteine residues, two of which are implicated in Fe_4S_4 cluster binding, are associated with regions which show few amino acid substitutions (Howard et al., 1989). The products of nifH and vnfH are very similar and show significant differences from the product of anfH. This accords with the results with heterologous crosses (Section 5), which show that the component 2 proteins of the Mo and V nitrogenases do not transfer electrons to component 1 of nitrogenase-3. The dissimilarities between the nifH and vnfH products and the anfH product are located in the same segments of the polypeptide (Fig. 3). The most variable segments are 35 residues C-terminal to Cys-85 and the C-terminal half of the molecule. In addition, the C-terminal segments of the nifH and vnfH products are both 14 amino acids longer than the anfH product. The products of the five *Clostridium pasteurianum* nif-like genes also lack the C-terminal amino acids present in *A. vinelandii* nifH and vnfH products (Wang et al., 1988). One of these shows significant homology (82% identity) with the product of *A. vinelandii* anfH.

9 CONCLUSIONS

Nitrogenases constitute a family of enzymes which utilize more than one metal in catalysis. The evolution of this family seems to have involved gene duplication with the development of the ability to use different metals (Mo, V, Fe) in catalysis as well as in the regulation of the various genes involved in N_2 fixation (Chapters 6, 7). As a result, azotobacters may be able to fix nitrogen in environments with low concentrations of Mo, albeit less efficiently than in its presence. Biochemical differences between the Mo and alternative nitrogenases appear to be predominantly associated with changes in the cofactor centres, though it is clear that their evolution has also involved numerous other changes. The nature of the cofactor in nitrogenase-3 remains to be to be unequivocally determined. As envisaged in the first biochemical studies of the V-containing nitrogenase, the existence of nitrogenases utilizing different metals provides valuable opportunities for comparative biochemical studies aimed at understanding the mechanism of nitrogen fixation.

10 REFERENCES

Arber, J.M., Dobson, B.R., Eady, R.R., Stevens, P., Hasnain, S.S., Garner, C.D. and Smith, B.E., 1987a. Vanadium K-edge X-ray absorption spectra of the VFe protein of the vanadium nitrogenase of *Azotobacter chroococcum*. Nature, 325: 372-374.

Arber, J.M., Dobson, B.R., Eady, R.R., Flood, A.C., Garner, C.D., Gormal, C.A., Hasnain, S.S., Smith, B.E. and Stevens, P., 1987b. X-ray absorbtion spectroscopy of the VFe protein of *Azotobacter chroococcum* and the iron-molybdenum cofactor of *Klebsiella pneumoniae*. Rec. Trav. Chim. Pays-Bas, 106: 237.

Arber, J.M., Flood, A.C., Garner, C.D., Gormal, C.A., Hasnain, S.S.and Smith, B.E., 1988. Iron K-edge X-ray absorbtion spectroscopy of the iron-molybdenum cofactor of nitrogenase from Klebsiella pneumoniae. Biochem. J., 252: 421-425.

Arber, J.M., Dobson, B.R., Eady, R.R., Hasnain, S.S., Garner, C.D., Matsushita, T., Nomura, M. and Smith, B.E., 1989. Vanadium K-edge X-ray absorption spectroscopy of the functioning and thionine-oxidized forms of the VFe-protein of the vanadium nitrogenase from Azotobacter chroococcum. Biochem. J., 258: 733-737.

Ashby, G.A., Dilworth, M.J. and Thorneley, R.N.F., 1987. Klebsiella pneumoniae nitrogenase. Inhibition of hydrogen evolution by ethylene and the reduction of ethylene to ethane. Biochem. J.,247: 547-554.

Becking, J.H., 1962. Species differences in molybdenum and vanadium requirements and combined nitrogen utilization by Azotobacteriaceae. Plant Soil, 16: 171-201.

Benemann, J.R., McKenna, C.E., Lie, R.F., Traylor, T.G. and Kamen,M.D., 1972. The vanadium effect in nitrogen fixation by Azotobacter. Biochim. Biophys. Acta, 264: 25-38.

Bergström, J., Eady, R.R. and Thorneley, R.N.F., 1988. The vanadium-and molybdenum-containing nitrogenases of Azotobacter chroococcum. Comparison of mid-point potentials and kinetics of reduction by sodium dithionite of the iron proteins with bound magnesium adenosine 5'-diphosphate. Biochem. J., 251: 165-169.

Bishop, P.E., Hawkins, M.E. and Eady, R.R., 1986a. Nitrogen fixation in molybdenum-deficient continuous culture by a strain of Azotobacter vinelandii carrying a deletion of the structural genes for nitrogenase (nifHDK). Biochem. J., 238: 437-447.

Bishop, P.E., Jarlenski, D.M.L. and Hetherington, D.R., 1980. Evidence for an alternative nitrogen fixation system in Azotobacter vinelandii. Proc. Natl. Acad. Sci. USA, 77: 7342-7346.

Bishop, P.E., Jarlenski, D.M.L. and Hetherington, D.R., 1982. Expression of an alternative nitrogen fixation system in Azotobacter vinelandii. J. Bacteriol., 150: 1244-1251.

Bishop, P.E. and Joerger, R.D., 1990. Genetics and molecular biology of alternative nitrogen fixation systems. Ann. Rev. Pl. Physiol. Pl. Mol. Biol., in press.

Bishop, P.E., Premakumar, R., Dean, D.R., Jacobson, M.R., Chisnell, J.R., Rizzo, T.M. and Kopczynski, J.B., 1986b. Nitrogen fixation by Azotobacter vinelandii strains carrying deletions in structural genes for nitrogenase., Science 232: 92-94.

Bortels, H., 1936. Weitere Untersuchungen über die Bedeutung von Molybdän, Vanadium, Wolfram und anderen Erdschenstoffen für stickstoffbindende und andere Mikroorganismen. Zentr. Bakteriol. Parasitenk., 95: 193-218.

Brigle, K.E., Setterquist, R.A., Dean, D.R., Cantwell, J.S., Weiss, M.C. and Newton, W.E., 1987. Site-directed mutagenesis of the nitrogenase MoFe protein of Azotobacter vinelandii. Proc. Natl. Acad Sci. USA., 84: 7066-7069.

Burns, R.C., Fuchsman, W.H. and Hardy, R.W.F., 1971. Nitrogenase from vanadium-grown Azotobacter: isolation, characteristics, and mechanistic implications. Biochem. Biophys. Res. Commun., 42: 353-358.

Burns, R.C., Stasny, J.T. and Hardy, R.W.F., 1976. Isolation and characteristics of a modified nitrogenase from Azotobacter vinelandii including 'vanadium-Fe' protein from cells grown on medium enriched in vanadium. In: W.E. Newton and C.J. Nyman (Editors), Proc. 1st International Symposium on Nitrogen Fixation, Washington State Univ. Press, Pullman, pp. 196-207.

Chisnell, J.R., Premakumar, R. and Bishop, P.E., 1988. Purification of a second alternative nitrogenase from a nifHDK deletion strain of Azotobacter vinelandii. J. Bacteriol., 170: 27-33.

Conradson, S.D., Burgess, B.K., Newton, W.E., Mortenson, L.E. and Hodgson, K.O., 1987. Structural studies of the molybdenum site in the MoFe protein and its FeMo cofactor by EXAFS. J. Am. Chem. Soc., 109: 7507-7515.
Devereux, J., Haeberli, P. and Smithies, O., 1984. A comprehensive set of sequence analysis programs for the VAX. Nucleic Acids Res., 12: 387-395.
Dilworth, M.J., Eady, R.R. and Eldridge, M.E., 1988. The vanadium nitrogenase of Azotobacter chroococcum. Reduction of acetylene and ethylene to ethane. Biochem. J., 249: 745-751.
Dilworth, M.J., Eady, R.R., Robson, R.L. and Miller, R.W., 1987. Ethane formation from acetylene as a potential test for vanadium nitrogenase in vivo. Nature, 327: 167-168.
Eady, R.R., 1989. The vanadium nitrogenase of Azotobacter. Polyhedron, 8: 1695-1700.
Eady, R.R., Robson, R.L., Richardson, T.H., Miller, R.W. and Hawkins, M., 1987. The vanadium nitrogenase of Azotobacter chroococcum. Biochem. J., 244: 197-207.
Eidsness, M.K., Flank, A.M., Smith, B.E., Flood, A.C., Garner, C.D. and Cramer, S.P., 1986. EXAFS of Klebsiella pneumoniaenitrogenase MoFe protein from wild-type and nifV mutant strains. J. Am. Chem. Soc., 108: 2746-2747.
Fry, D.C., Kuby, S.A. and Mildvan, A.S., 1986. ATP-binding site of adenylate kinase: Mechanistic implications of its homology with ras-encoded p21, F_1-ATPase, and other nucleotide-binding proteins. Proc. Natl. Acad. Sci. USA., 83: 907-911.
George, G.N., Coyle, C.L. Hales, B.J. and Cramer, S.P., 1988. X-ray absorbtion of Azotobacter vinelandii vanadium nitrogenase. J. Am. Chem. Soc., 110: 4057-4059.
Garner, C.D., Arber, J.M., Harvey, I., Hasnain, S.S., Eady, R.R., Smith, B.E., de Boer, E. and Wever, R., 1989. Characterization of molybdenum and vanadium centres in enzymes by X-ray absorption spectroscopy. Polyhedron, 8: 1649-1652.
Hales, B.J. and Case, E.E., 1987. Nitrogen fixation by Azotobacter vinelandii in tungsten-containing medium. J. Biol. Chem., 262: 16205-16211.
Hales, B.J., Case, E.E., Morningstar, J.E., Dzeda, M.F. and Mauterer, L.A., 1986. Isolation of a new vanadium-containing nitrogenase from Azotobacter vinelandii. Biochemistry, 25: 7251-7255.
Hales, B.J., True, A.E. and Hoffman, B.M., 1989. Detection of a new signal in the EPR spectrum of vanadium nitrogenase from Azotobacter vinelandii. J. Am. Chem. Soc., 111: 8519-8520.
Hardy, R.W.F., 1979. Reducible substrates of nitrogenases. In: R.W.F. Hardy, F. Bottomley and R.C. Burns (Editors), A Treatise on Dinitrogen Fixation. J. Wiley, New York, pp. 515-568.
Hawkes, T.R., McLean, P.A. and Smith, B.E., 1984. Nitrogenase from nifV mutants of Klebsiella pneumoniae contains an altered form of the iron-molybdenum cofactor. Biochem. J., 217: 317-321.
Howard, J.B., Davis, R., Moldenhauer, B. Cash, V.L. and Dean, D., 1989. Fe:S cluster ligands are the only cysteines required for nitrogenase Fe-protein activities. J. Biol. Chem., 264: 11270-11274.
Jacobson, M.R., Brigle, K.T., Bennett, L.T., Setterquist, R.A., Wilson, M.S., Cash, V.L., Beynon, J., Newton, W.E. and Dean, D.R., 1989. Physical and genetic map of the major nif gene cluster from Azotobacter vinelandii. J. Bacteriol., 171: 1071-1027.
Joerger, R.D. and Bishop, P.E., 1988a. Bacterial alternative nitrogen fixing systems. CRC Critical Rev., 16: 1-14.
Joerger, R.D. and Bishop, P.E., 1988b. Nucleotide sequence of the nifB-nifQ region of Azotobacter vinelandii. J. Bacteriol., 170: 1475-1487.

Joerger, R.D., Jacobson, M.R., Premakumar, R., Wolfinger, E.D. and Bishop, P.E., 1989. Nucleotide sequence and mutational analysis of the structural genes (anfHDK) for the second alternative nitrogenase from Azotobacter vinelandii. J. Bacteriol., 171: 1075-1086.
Joerger, R.D., Loveless, T.M., Pau, R.N., Mitchenall, L.A., Simon, B.H. and Bishop, P.E., 1990. Nucleotide sequence and mutational analysis of the structural genes for nitrogenase 2 of Azotobacter vinelandii. J. Bacteriol., in press.
Kent, H.M., Ionnidis, I., Gormal, C., Smith, B.E. and Buck, M., 1989. Site-directed mutagenesis of the Klebsiella pneumoniae nitrogenase. Biochem. J., 264: 257-264.
Klugkist, J., Haaker, H. and Veeger, C., 1986. Studies on the mechanism of electron transport to nitrogenase in Azotobacter vinelandii. Eur. J. Biochem., 155: 41-46.
Kovacs, J.A. and Holm, R.H., 1986. Assembly of vanadium-iron- sulphur cubane clusters from mononuclear and linear trinuclear reactants. J. Am. Chem. Soc., 108: 340-341.
Lowe, D.J., Eady, R.R. and Thorneley, R.N.F., 1978. Electron-paramagnetic-resonance studies on nitrogenase of Klebsiella pneumoniae: evidence for acetylene- and ethylene-nitrogenase transient complexes. Biochem. J., 173: 277-290.
Martin, A.E., Burgess, B.K., Iismaa, S.E., Smart, C.T., Jacobson, M.R. and Dean, D.R., 1989. Construction and characterization of an Azotobacter vinelandii strain with mutations in the genes encoding flavodoxin and ferredoxin I. J. Bacteriol., 171: 3162-3167.
McKenna, C.E., Benemann, J.R. and Traylor, T.G., 1970. A vanadium containing nitrogenase preparation: implications for the role of molybdenum in nitrogen fixation. Biochem. Biophys. Res. Commun., 41: 1501-1508.
McKenna, C.E., McKenna, M.C. and Huang, C.W., 1979. Low stereoselectivity in methylacetylene and cyclopropene reductions by nitrogenase. Proc. Natl. Acad. Sci. USA., 76: 4773-4777.
Miller, R.W. and Eady, R.R., 1988. Molybdenum and vanadium nitrogenases of Azotobacter chroococcum: Low temperature favours N_2 reduction by vanadium nitrogenase. Biochem. J., 256: 429-432.
Morningstar, J.E. and Hales, B.J., 1987. Electron paramagnetic resonance study of the vanadium-iron protein of nitrogenase from Azotobacter vinelandii. J. Am. Chem. Soc., 109: 6854-6855.
Morningstar, J.E., Johnson, M.K., Case, E.E. and Hales, B.J., 1987. Characterization of the metal clusters in the nitrogenase molybdenum-iron and vanadium-iron proteins of Azotobacter vinelandii using circular dichroism spectroscopy. Biochemistry., 26: 1785-1800.
Nagatani, H.H. and Brill, W.J., 1974. Nitrogenase. V. The effect of Mo, W and V on the synthesis of nitrogenase components in Azotobacter vinelandii. Biochim. Biophys. Acta, 362: 160-166.
Pau, R.N., 1989. Nitrogenases without molybdenum. Trends in Biochem. Sci., 14: 183-186.
Pau, R.N., Mitchenall, L. and Robson, R.L., 1989. Genetic evidence for an Azotobacter vinelandii nitrogenase lacking molybdenum and vanadium. J. Bacteriol., 171: 124-129.
Premakumar, R., Chisnell, J.R. and Bishop, P.E., 1989. A comparison of the three dinitrogen reductases expressed by Azotobacter vinelandii. Can. J. Microbiol., 35: 344-348.
Raina, R., Reddy, M.A., Ghosal, D. and Das, H.K., 1988. Characterization of the gene for the Fe-protein of the vanadium dependent alternative nitrogenase of Azotobacter vinelandii and construction of a Tn5 mutant. Mol. Gen. Genet., 214: 121-127.

Robson, R.L., 1984. Identification of possible adenine nucleotide-binding sites in nitrogenase Fe- and MoFe- proteins by amino acid comparison. FEBS Lett., 173: 394-198.
Robson, R.L., Eady, R.R., Richardson, T.H., Miller, R.W., Hawkins, M. and Postgate, J.R., 1986a. The alternative nitrogenase of Azotobacter chroococcum is a vanadium enzyme. Nature, 322: 388-390.
Robson, R.L., Woodley, P.R. and Jones, R., 1986b. Second gene (nifH*) coding for a nitrogenase iron-protein in Azotobacter chroococcum is adjacent to a gene coding for a ferrodoxin-like protein. EMBO J., 5: 1159-1163.
Robson, R.L., Woodley, P.R., Pau, R.N. and Eady, R.R., 1989. Structural genes for the vanadium nitrogenase from Azotobacter chroococcum. EMBO J., 8: 1217-1224.
Ruvkun, G.B. and Ausubel, F.M., 1980. Interspecies homology of nitrogenase genes. Proc. Natl. Acad. Sci. USA., 77: 191-195.
Scott, D.J., May, H.D., Newton, W.E., Brigle, K.E. and Dean, D.R., 1990. Role of the nitrogenase MoFe protein α-subunit in FeMo-cofactor binding and catalysis. Nature., 343: 188-190.
Shah, V.K., Ugalde, R.A., Imperial, J. and Brill, W.J., 1984. Molybdenum in nitrogenase. Ann. Rev. Biochem., 53: 231-257.
Smith, B.E., Eady, R.R., Lowe, D.J. and Gormal, C., 1988. The VFe protein of V-nitrogenase of Azotobacter chroococcum contains an iron-vanadium cofactor. Biochem. J., 250: 299-302.
Thorneley, R.N.F. and Lowe, D.J., 1985. Kinetics and mechanism of the nitrogenase enzyme system. In: T.G. Spiro (Editor) Molybenum Enzymes. J. Wiley, New York, pp. 222-284.
Wang, S-Z., Chen, J-S. and Johnson, J.L., 1988. The presence of five nifH-like sequences in Clostridium pasteurianum: sequence divergence and transcriptional properties. Nucl. Acids Res., 16: 439-454.
Yates, M.G. and Planqué, K., 1975. Nitrogenase from Azotobacter chroococcum. Eur. J. Biochem., 60: 467-476.

CHAPTER 4

THE CHEMISTRY OF DINITROGEN REDUCTION

R.L. RICHARDS
AFRC IPSR Nitrogen Fixation Laboratory, University of Sussex, Brighton, BN1 9RQ, East Sussex, U.K.

1 INTRODUCTION

The activation and reduction of dinitrogen by molybdenum compounds has been studied by chemists since the initial discovery by Bortels (1930) of the requirement for molybdenum by nitrogen-fixing bacteria.

These studies became focused into two areas as a result of the biochemical characterisation of the nitrogenase proteins described in Chapters 2 and 3. The first is the investigation of the structure of the metallic centre of nitrogenase i.e. what is the environment of the metal centre, or centres, which are involved in substrate binding and reduction?

The second area, upon which this chapter concentrates, concerns the mechanism of metal-mediated reduction of substrates to the observed products. Substrates of interest are not only dinitrogen, but also the proton, isocyanides, cyanides and alkynes. The interaction of carbon monoxide with nitrogenase is also important, since this is an ubiquitous ligand to metals of the type found in nitrogenase, particularly in their lower oxidation states.

Although it is convenient to use the above classification to describe the types of study being undertaken, the overriding aim is to unify the two areas and produce an accurate, clearly understood model of the active metallic centre of nitrogenase.

2 THE METALS IN NITROGENASE

Nitrogenases from a range of bacteria have been shown to contain as their essential metals iron and molybdenum or iron and vanadium while a third possible nitrogenase may contain only iron (Chapter 3). The most commonly found and most thoroughly studied nitrogenase is that containing iron and molybdenum, and this will be the nitrogenase discussed from now on, unless otherwise stated.

Evidence has accumulated (Chapter 2; Hawkes et al., 1984) to implicate molybdenum as the metal centre at which substrate binding and activation occurs, although there is still no direct evidence for interaction of molybdenum in functioning nitrogenase with any substrate.

There are two molybdenum atoms in the large nitrogenase protein, together with about thirty iron atoms and an equal number of sulphide ions. The molybdenum appears to be bound in a still imprecisely defined, extractable cluster molecule called the iron-molybdenum cofactor (FeMoco) (Rawlings et al., 1978). This anionic species has the approximate empirical composition $MoFe_{6-8}S_{6-8}$ (Nelson et al., 1983). The vanadium nitrogenase appears to have a similar cofactor (FeVaco) in which Mo is replaced by V (Eady, 1989).

The two FeMoco units appear to be the site of binding and reduction of substrate by nitrogenase and a major research target is to define the structure of the cofactor. A definitive structure awaits suitable crystals for X-ray analysis. The main spectroscopic tool of investigation has been X-ray absorption spectroscopy (EXAFS and EXANES), which indicates that in the immediate environment of molybdenum (and vanadium) there are three sulphur atoms, together with three oxygen or nitrogen atoms (N and O are indistinguishable by this technique) (Hedman et al., 1988).

As a guide to the likely structure of such a molecule, (assuming that the cofactor is a single metal-cluster species), chemists have synthesised a number of cluster complexes which attempt to reproduce the stoichiometry and spectroscopic properties of FeMoco or FeVaco. Examples of such complexes are shown in Fig. 1. Although the required environment of ligands to Mo (or V) is approached in some cluster molecules, and the Mo/Fe stoichiometry in others, none shows an exact correspondence. Moreover, although in some compounds of this type a nitrogenase substrate such as cyanide, or inhibitor (CO) is bound to Mo or to V, no reduction of substrates has been achieved, nor has any interaction with N_2 been demonstrated. Nevertheless, elegant chemistry has been developed to achieve the synthesis of these molecules and further modification of them should lead to a close structural model for the cofactors.

3 MODELS FOR THE BINDING OF DINITROGEN AND OTHER SUBSTRATES AT METAL CENTRES OF NITROGENASE

In parallel with the above attempts to define the metal structure of nitrogenase, are studies of the binding of the various nitrogenase substrates at metal centres and their activation towards reduction.

A number of the important properties of nitrogenase should be stressed at this point. The first is the ability of the enzyme to reduce N_2 to NH_3 under ambient conditions of temperature and pressure. Although the reduction of N_2 to NH_3 is well known in the industrial (Haber) process, the conditions required at temperatures of the order of 300°C and elevated pressure (around 100 atmospheres). The second is the ability of nitrogenase to reductively cleave multiple bonds in alternative substrates such as $C{\equiv}N^-$ and $RN{\equiv}C$, but not that in $HC{\equiv}CH$ (Burgess, 1985). It is also of additional interest that the reduction of $HC{\equiv}CH$ by molybdenum nitrogenase is almost exclusively to $CH_2{=}CH_2$ and essentially stereospecifically to cis-CHD=CHD when the reaction is carried out in D_2O. Although C_2H_6 is not a significant product of reduction of $HC{\equiv}CH$ by molylxenum nitrogenase, C_2H_6 is a significant and apparently diagnostic product of its reduction by vanadium nitrogenase (Dilworth et al., 1987; Eady, 1989).

A final major point is that N_2 reduction is always accompanied by reduction of protons, as that the overall stoichiometry of the nitrogenase reaction is as in Reaction (1).

$$N_2 + 8H^+ + 8e^- = 2NH_3 + H_2 \qquad \text{. . . . Reaction (1)}$$

These features must be taken into account when designing model systems to aid understanding of the mechanism of action of the metallic centres of nitrogenase.

4 BINDING OF DINITROGEN AND RELATED SUBSTRATES TO METALS

The first step in the conversion of N_2 or alternative substrate to product is to bind it to a metal centre and thereby inctease its chemical reactivity. Since the initial discovery of such binding of N_2, to ruthenium in $[Ru(NH_3)_5(N_2)]^{2+}$ (Allen and Senoff, 1965), most transition metals have been shown to form stable compounds in which dinitrogen is bound to a single metal centre as a terminal end-on ligand. There also exists a large number of examples where dinitrogen binds between two or more metals in a bridging mode (Chatt et al., 1978; Henderson et al., 1983); Richards, 1983). This chapter will concentrate only on those compounds which are well-studied and in which reactivity of dinitrogen has been demonstrated.

4.1 Dinitrogen as a terminal ligand.

A metal site at which dinitrogen binds in a terminal end-on fashion is electron rich; characterized by a relatively low formal oxidation state and relative ease of oxidation.

A robust co-ligand framework such as is provided by tertiary phosphines aids the stability of these compounds, binding such metals as molybdenum, tungsten and rhenium. The maximum number of dinitrogen molecules known to bind to a single metal is three, and a typical preparation of such a compound at ambient temperature and pressure is shown in Reaction (2) (Anderson et al., 1986a).

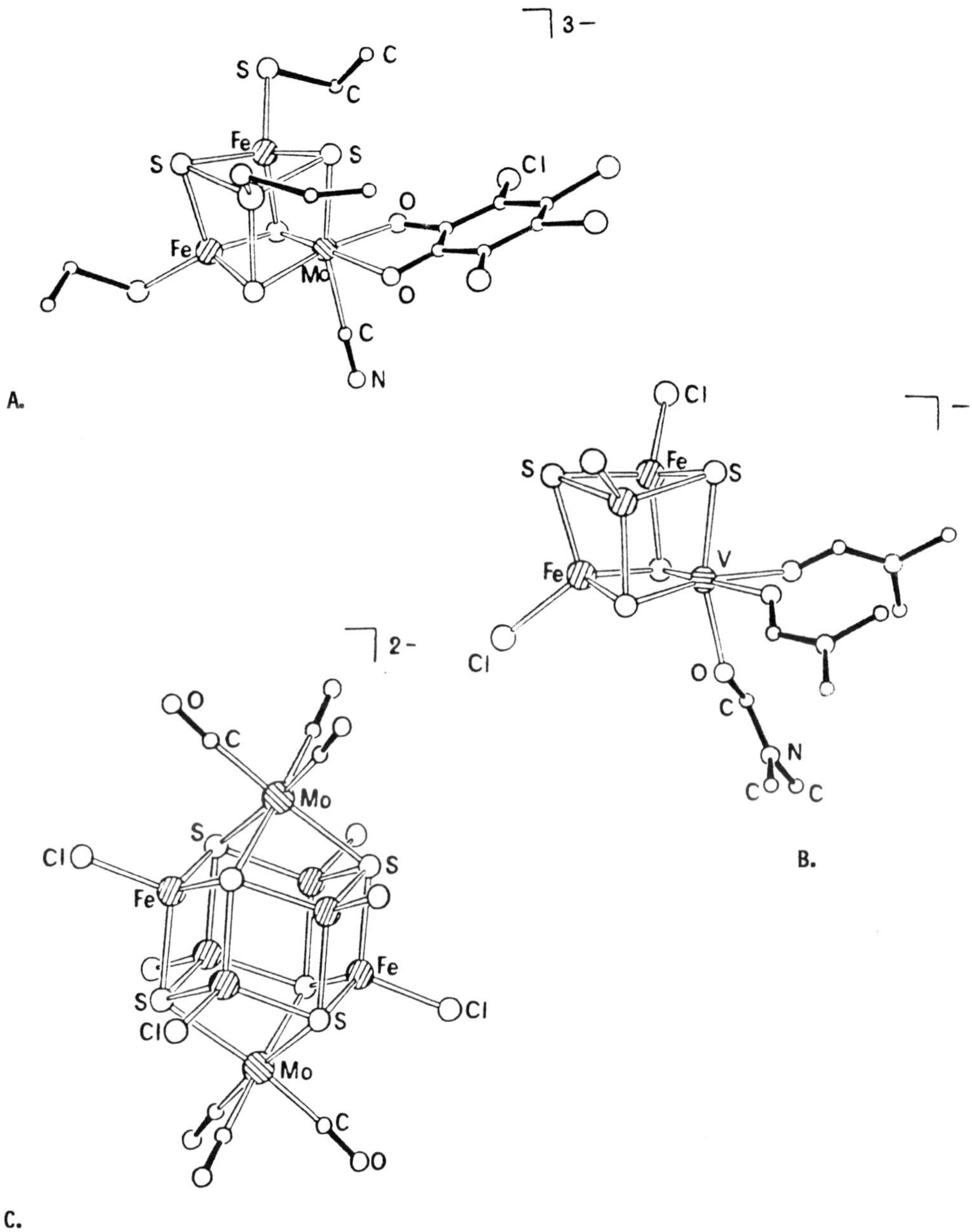

Fig. 1. Structures of some cluster complexes relevant to nitrogenase cofactors. A. Palermo et al., (1984). B. Kovacs and Holm (1987). C. Coucouvanis and Kanatzidis (1985).

$$[MoCl_3(thf)_3] \xrightarrow[N_2/P]{Mg/thf} \underline{cis}\text{-}[Mo(N_2)_2P_4] \xrightarrow[-P]{N_2} \underline{mer}\text{-}[Mo(N_2)_3P_3]$$

. . . . Reaction (2)

thf = tetrahydrofuran
P = PPr_2Ph

The dinitrogen ligands in these molybdenum complexes and in their analogues are reactive (see Section 5). It is the nature of the binding of dinitrogen, involving synergic 'back-donation' from the metal to stabilise the metal-N_2 interaction, which confers reactivity on the N_2 ligand (Chatt et al., 1978). This perturbation of N_2 on binding is demonstrated, inter alia, by an increase of N-N distance on binding, from 1.098Å in free N_2 to about 1.12Å, although the extent of increase in a particular complex is not a criterion of reactivity.

Although the phosphine co-ligands do not reproduce the sulphur environment found in nitrogenase, they do allow a definitive chemistry of N_2 at metal centres to be developed. However, in the molybdenum complexes of dinitrogen with entirely sulphur-donor co-ligands which have been synthesized recently (Fig. 2), the reactivity of dinitrogen is very similar to that shown in the phosphorus-ligated environment (Richards, 1988; Adachi et al., 1989).

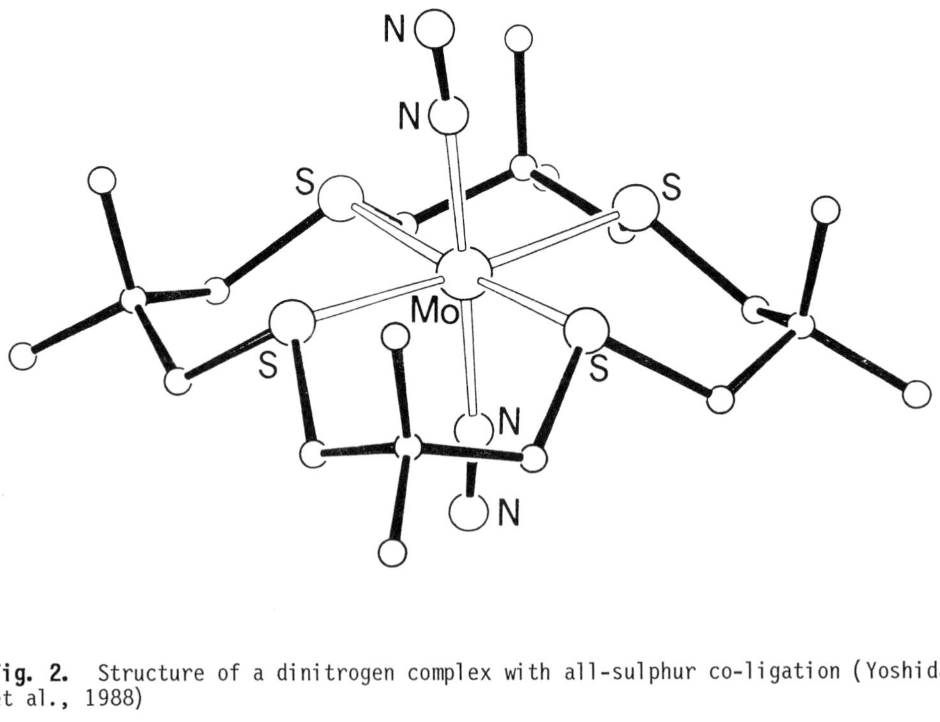

Fig. 2. Structure of a dinitrogen complex with all-sulphur co-ligation (Yoshida et al., 1988)

Other sites at which N_2 binds include WP_4 (P = tertiary phosphine) and $ReClP_4$. These sites are often used to parallel those of molybdenum because not only do they show similar behaviour, as expected from the relationship of their constituent metals to Mo in the periodic table, but they tend to form kinetically stable compounds which allow reaction intermediates to be isolated.

Alternative substrates of nitrogenase also bind at these centres, allowing their reductive chemistry to be explored in parallel with that of N_2. Synthetic routes to these compounds generally involve displacement of N_2 by the alternative substrate e.g. in Reaction (3)(Chatt et al., 1980).

trans-$[Mo(N_2)_2(dppe)_2]$ + 2RNC = trans-$[Mo(RNC)_2(dppe)_2]$ + $2N_2$

dppe = $Ph_2PCH_2CH_2PPh_2$ . . . . Reaction (3)

Importantly, hydride ligands can also occupy these N_2-binding sites and can be displaced by incoming N_2 from such a complex as $[MoH_4(dppe)_2]$. This displacement reaction is greatly accelerated in the presence of protons. A proposed mechanism for this reaction (Scheme 1) suggests that the substitution rate is increased because of the formation, under the influence of a proton, of a dihydrogen ligand which is more easily displaced by N_2. Thus, for every N_2 bound to Mo, one H_2 is displaced, a point to which we will return (Henderson, 1987).

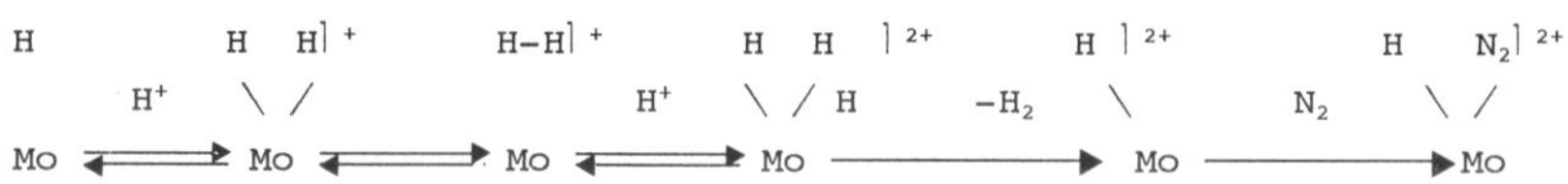

Scheme 1. Mechanism of displacement of H_2 from Mo by N_2

It is notable that at these metal centres N_2 is generally more strongly bound that its reduction product NH_3 or those ligands such as H_2O which might be competitive with N_2 in a biological environment.

The bound dinitrogen ligand can itself act to displace ligands from a second metal, producing compounds with dinitrogen bridging two metal centres. This is one route to this class of complex which is discussed in the next section.

4.2 Dinitrogen as a bridging ligand.

These compounds may be prepared by displacement reactions (Reaction 4, Cradwick et al., 1975; Reaction 5, Rocklage et al., 1982) or directly (Reaction 6, Anderson et al., 1986b). (For other examples, see Chatt et al., 1978, Henderson et al., 1983 and Murray and Schrock, 1985).

$2\underline{Re}N_2$ + $[MoCl_4(thf)_2]$ = $[\underline{Re}(\mu\text{-}N_2)MoCl_4(\mu\text{-}N_2)\underline{Re}\}]$ + 2thf

$\underline{Re}$ = $ReCl(PMe_2Ph)_4$ . . . . Reaction (4)

$2[Ta(CHMe_3)Cl(PMe_3)_4]$ + N_2 = $[\{Ta(CHMe_3)Cl(PMe_3)_2\}_2(\mu\text{-}N_2)]$ + $4PMe_3$

. . . . Reaction (5)

$$[WCl_4(PEt_2Ph)_2] \xrightarrow{Mg/N_2} [\{W(N_2)_2(PEt_2Ph)_3\}_2(\mu\text{-}N_2)]$$

. . . . Reaction (6)

The nature of the binding of bridging dinitrogen varies widely depending upon the metals which are bridged. In general the M-N-N-M (M=metal) system is linear, with the N-N bond distance varying over a considerable range, from ca 1.12Å, (essentially similar to the distances in terminal N_2 complexes) to ca 1.3Å.

In the latter case, the bridging ligand has been likened to a hydrazide(4-) ion (Rocklage et al., 1982), and as might be anticipated, such a perturbed N_2 species is highly reactive (section 5.2).

Important amongst binuclear complexes is $[V\{(2\text{-}Me_2NCH_2C_6H_4)_2(C_5H_5N)\}_2(\mu\text{-}N_2)]$ (Fig. 3a), the first stable dinitrogen complex of vanadium to be characterised (Edema et al., 1989); its reactivity is yet to be investigated.

The environment of the metal in binuclear complexes can also vary widely; nitrogen or phosphorus donor co-ligands are common, with oxygen or sulphur co-ligands in the case of tantalum complexes e.g. $[\{Ta(XAr)_3(thf)_2\}_2(\mu\text{-}N_2)]$ (X=S or O; Ar= $2,6\text{-}C_6H_3Pr^i_2$) (Schrock et al., 1988).

Bridging dinitrogen has also been postulated to occur in intermediates in systems which reduce dinitrogen, particularly those including molybdenum, vanadium (e.g. Fig. 3b) and iron (Shilov, 1980, 1988). Although such intermediates have not yet been isolated, the reactions in which they might be involved are discussed in Section 5.2.

5 REACTIONS OF LIGATING DINITROGEN

If the binding of dinitrogen to a metal centre is regarded as the first step in the reaction sequence of nitrogenase, then the next stage is conversion of the bound N_2 to ammonia. This has been achieved and studied in detail for mononuclear complexes, particularly of molybdenum and tungsten. Reduction of dinitrogen in well-characterised, bridging-dinitrogen complexes gives hydrazine (Section 5.2).

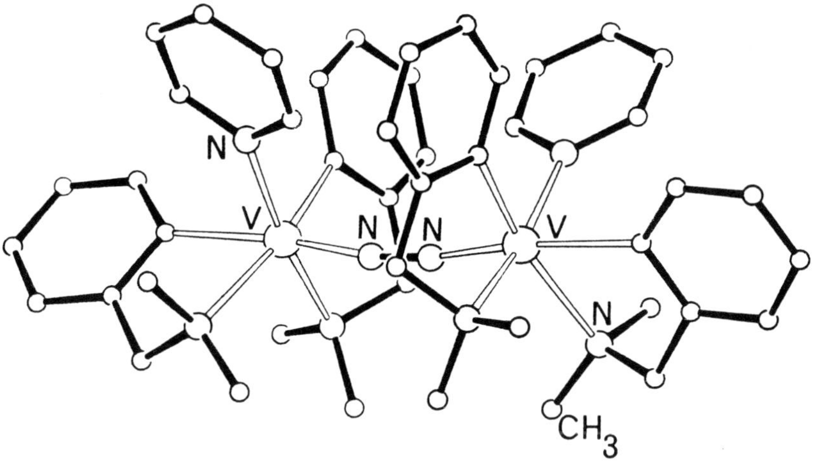

Fig. 3(a). Structure of a μ-N_2 complex of vanadium (Edema et al., 1989)

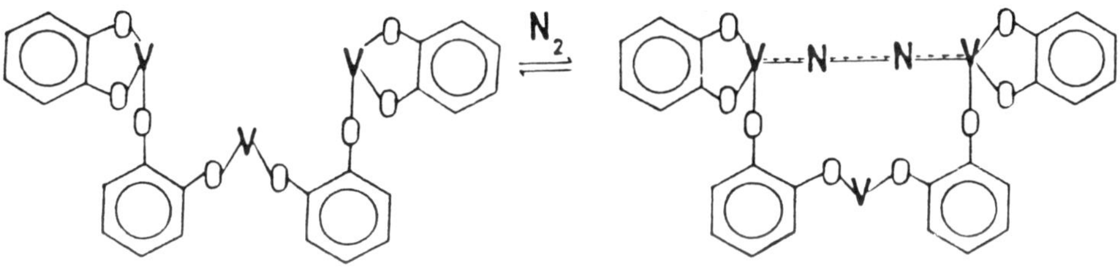

Fig. 3(b). μ-N_2 intermediate proposed by Shilov (1980).

5.1 <u>Terminal dinitrogen.</u>

The best-characterized system in which dinitrogen is reduced to ammonia involves complexes of the type $[M(N_2)_xP_{6-x}]$(M=Mo or W; x=1, 2 or 3; P= tertiary phosphine). The dinitrogen in these compounds, as a consequence of its binding to the electron-rich metal centre, is much more nucleophilic than is free N_2 (Section 4). It is now sufficiently basic to be attacked by protons from a variety of mineral acids, typically H_2SO_4, HBr or HTosyl (Tosyl= 4-$CH_3C_6H_4SO^{2-}$) (Dilworth and Richards, 1982; Pickett et al., 1988).

Protic attack occurs initially at the terminal (β) nitrogen atom and continues stepwise as shown in Scheme 2a. The steps in this Scheme are based upon isolation and structural characterisation of the intermediates shown and/or their observation in solution, particularly by ^{15}N NMR spectroscopy (Anderson et al., 1981; George et al., 1986; Galindo et al., 1987; Pickett et al., 1988).

Of these intermediates, the hydrazido(2-) (=N-NH_2) stage gives particularly stable compounds, which have been studied in detail and inter alia shown to give hydrazine when treated with acid or base under certain conditions (Anderson et al., 1981; Baumann et al., 1985). This behaviour mimics that shown by nitrogenase, which gives hydrazine rather than ammonia when quenched by acid or base at an early stage in its reduction cycle (Thorneley et al., 1978).

The hydrazidium (=N-N^+H_3) ligand, which represents the next stage of reduction, has only been structurally characteriseda in one case, where stabilisation of the ligand is aided in the solid by extensive hydrogen bonding between the =N-N^+H_3 ligand protons and chloride ions (Galindo et al., 1987). Extensive hydrogen bonding of this type also occurs in solid complexes of the =N-NH_2 ligand, and raises the question of whether such interactions could play a role in the enzymic mechanism by aiding the formation of intermediate steps in the reduction process (Sellmann, 1989).

The cycle of Scheme 2a includes as a first step the displacement of H_2 to form the initial dinitrogen complex, an established reaction mentioned in Section 4. Thus the overall stoichiometry is that of the enzyme process (Reaction 1).

Scheme 2a has been incorporated in a satisfying way into a model of the overall enzymatic process based upon observation of the rates of reaction of the nitrogenase proteins during the reduction of N_2. It is shown in outline in Scheme 3, detailed discussion of which can be found in Thorneley and Lowe (1985).

It therefore seems reasonable to propose that a reduction cycle like that in Scheme 2a occurs in nitrogenase, assuming that the molybdenum atoms act independently, an assumption based on the evidence of Hawkes and Smith (1984). In the chemical reactions which form the basis of Scheme 2a, the electrons necessary for N_2 reduction are supplied by the molybdenum; these electrons must be replaced for the cycle to recommence. It is envisaged that this would occur in nitrogenase by stepwise supply of electrons from the iron protein to the molybdenum in the MoFe protein, via the iron-sulphur cluster electron-transfer system.

It has been established in the model metal-complex system that electrons can be supplied to the molybdenum to continue the reduction process, and this is the basis of an electrochemical synthesis of ammonia at ambient temperature and pressure (Pickett et al., 1986, 1988)(Scheme 2b).

In principle, the electrochemical process is the same as that shown in Scheme 2a, but without the H_2 displacement step. The tungsten (or less effectively molybdenum) dinitrogen complex chosen as catalyst is trans-[$W(N_2)_2(dppe)_2$] because in it the chelating dppe ligands form a robust binding site, which resists degradation as the reaction cycle proceeds. The source of acid in the system (thf solvent) is HTosyl. This choice is important because it is necessary for the acid anion group to act as a ligand only during part of the cycle and Tosyl is particularly effective as a leaving group.

The detailed mechanism of the electrochemical process is more complicated that the outline shown in Scheme 2b; it involves intra-molecular proton transfer from a hydrazide intermediate and the turnover number is as yet low. A problem in the system is the competing reduction of protons to dihydrogen, which increases if the proton concentration is raised. Reduction of protons is probably inescapable in such a system and is, of course, a feature of nitrogenase.

This type of cyclic system is likely to be modified towards use of a robust sulphur environment for the metal, following the direction taken in the synthesis of dinitrogen complexes.

Some other systems have been developed which are claimed to reduce dinitrogen via mononuclear dinitrogen complexes of molybdenum. Typically, these systems are prepared by reducing a molybdate salt with borohydride in a borate buffer, under dinitrogen and in the presence of potential co-ligands such as cysteine, cyanide ion or bovine serum albumin. Yields of ammonia in the last case, although low in absolute terms, are apparently as high as 70 mol NH_3 per Mo atom. It has been proposed that these reduction proceeds through an intermediate having dinitrogen bound side-on to Mo. This mode of binding is rare, however, and no direct evidence for such an intermediate, or any other, has been obtained (Schrauzer, 1980).

The cycle of reduction shown in Scheme 2 is currently the best-characterised description of the reduction of N_2 to ammonia at a metal site.

5.2 Bridging dinitrogen.

As might be expected, bridging-dinitrogen complexes in which the N-N distance is very long are very reactive and protonate easily to give hydrazine. Although several such compounds give hydrazine on protonation (Rocklage et al., 1982; Dilworth and Richards, 1982), only a few have received detailed study. The compounds $[M\{(S_2CNEt_2)\}_2(\mu\text{-}N_2)]$ (M=Nb or Ta) react sufficiently slowly with HBr to allow a kinetic study of the reaction, which appears to follow the mechanism of Scheme 4 (Henderson et al., 1990). That study showed the rate of protonation of bridging N_2 to be several orders of magnitude slower than that of terminal N_2. Although this comparison must be treated with caution, since the conditions of metal, co-ligands and solvent all vary, there nevertheless seems to be little kinetic advantage for reduction of dinitrogen between metals as opposed to bonding at a single metal.

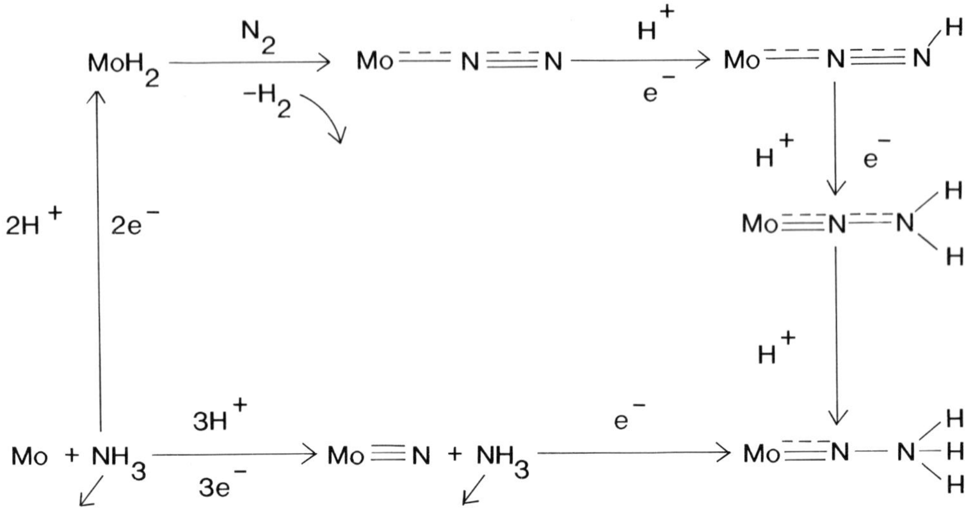

Scheme 2a. The mechanism of protonation and reductive cleavage of ligating N_2

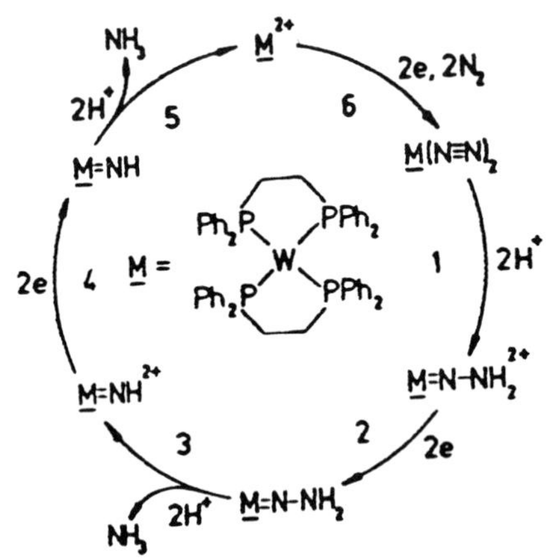

Scheme 2b. Electrochemical synthesis of ammonia

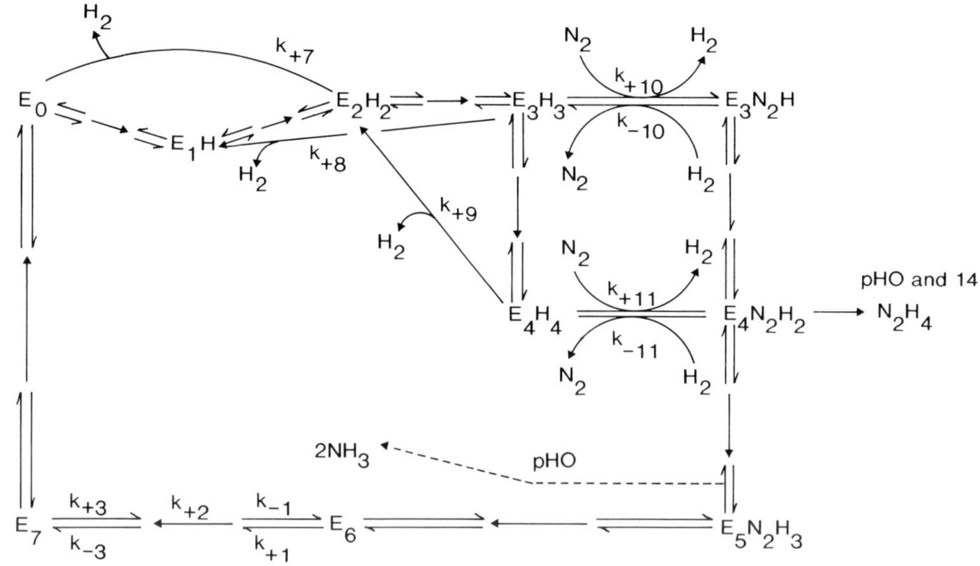

Scheme 3. General scheme for nitrogenase function (Thorneley and Lowe, 1985)

In general, only hydrazine is produced from a bridging-dinitrogen complex unless a further reductant, such as a Grignard reagent, is added to the system (Dilworth and Richards, 1982). In most cases, this is not too surprising, because the metals in the most reactive complexes can only supply a maximum of four electrons for the reduction process. However, under the conditions studied, hydrazine was the only product from the tungsten complex $[\{W(N_2)_2(PEt_2Ph)_3\}_2(\mu\text{-}N_2)]$ which has twelve electrons available from the two metals (Anderson et al.,

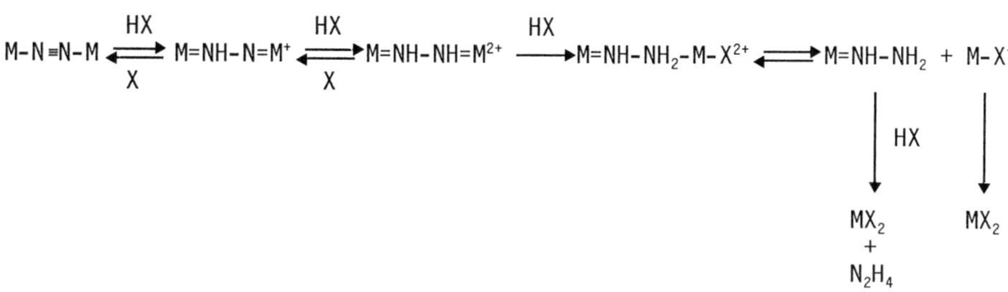

Scheme 4. Mechanism of protonation of bridging N_2 in $[M\{(S_2CNEt_2)\}_2(\mu\text{-}N_2)]$ (M=Nb or Ta) (Henderson et al., 1990)

1986b). In the absence of labelling experiments, it should not be assumed that the hydrazine produced from this compound originated solely from the μ-N_2 ligand. Where such labelling has been carried out, as with the complex $[\{ZrCp^1(^{15}N_2)_2\}_2(\mu$-$N_2)]$ (Cp^1=C_5Me_5), protonation gives N_2H_4 which has 50% ^{15}N label, implying that the terminal N_2 ligands are involved in the reaction mechanism (Manriquez et al., 1976).

As noted for terminal N_2 complexes, there are some N_2-reducing systems based on Mo, V, Nb or Ta, for which μ-N_2 intermediates have been proposed. Probably the most important, because of their apparent simplicity, are those obtained using divalent vanadium, either in alkaline suspension with magnesium hydroxide, or at a controlled pH around 10 in aqueous methanol together with catechol (Shilov, 1980, 1988). Hydrazine or ammonia can be obtained depending upon the conditions of temperature and pressure and although μ-N_2 intermediates such as that shown in Fig. 3b have been proposed for these reactions, none has yet been observed.

6 REDUCTION OF ALTERNATIVE SUBSTRATES

If the models discussed above for the reduction of N_2 at metal centres are reasonable, then they should reproduce the general pattern of substrate reduction by nitrogenase. In the main this is true for both terminal N_2 and some bridging N_2 systems.

The technique used in these studies is to bind the alternative substrate to a metal site in place of N_2 (Section 4), and then attempt to carry out a reduction of the ligand using the same conditions as were used to reduce N_2. Occasionally, several different metal complexes have been used in the investigation of one substrate, to enable intermediate stages of reduction to be trapped, and thus allow mechanistic conclusions to be drawn.

The most-studied alternative substrates are cyanide ion, isocyanides, alkynes and to a lesser extent organic cyanides. In general, the pattern of substrate reaction follows that of nitrogenase reasonably well in terms of products. Moreover, for isocyanide, cyanide ion and organic cyanides, the dominant feature in their reactions is that bonding to the metal centre induces susceptibility to attack by protons at the atom β to the metal, as is observed for N_2 (Pombeiro et al., 1986; Pombeiro, 1989).

This reaction pattern is illustrated in Scheme 5a for cyanide ion and isocyanides and carries the implication for cyanide reduction by nitrogenase that this ion bonds to the metal centre through carbon. Recent mechanistic studies of cyanide reduction by nitrogenase support this view (Lowe et al., 1989).

$$^*M{-}C{\equiv}NR \xrightarrow[e^-]{H^+} {}^*M{\equiv}C{-}NHR \xrightarrow{H^+} {}^*M{\equiv}C{-}N^+H_2R \xrightarrow[5e^-]{4H^+} M + RNH_2 + CH_4$$

$$^*M{-}C{\equiv}N^- \xrightarrow{H^+} {}^*M{=}C{=}N{-}H \xrightarrow[e^-]{H^+} {}^*M{\equiv}C{-}NH_2 \xrightarrow[5e^-]{5H^+} M + CH_4 + NH_3$$

$$^*M{-}N{\equiv}CR \xrightarrow[e^-]{H^+} {}^*M{-}N{=}CHR$$

Scheme 5a. Protonation of RNC, CN^- and RCN at N_2-binding metal sites

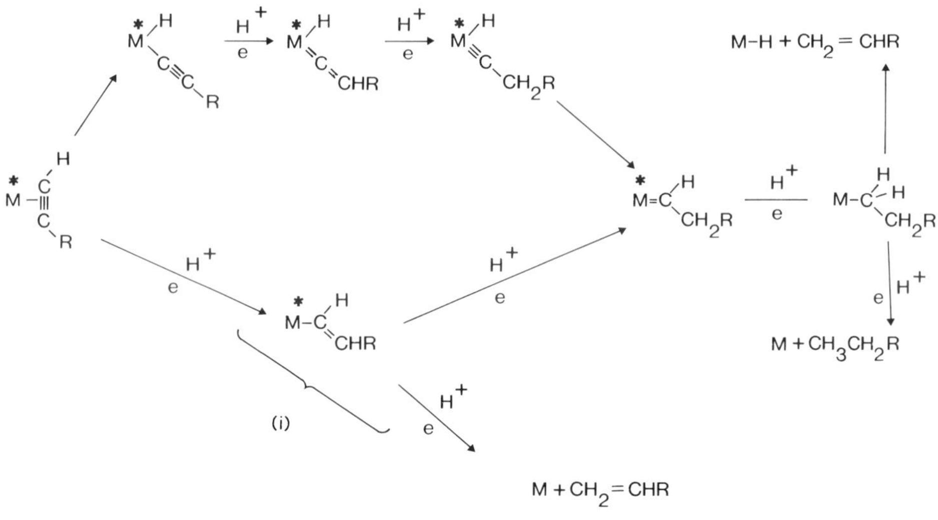

Scheme 5b. Reduction of alkynes at N_2-binding metal sites

In both 5(a) and 5(b), * denotes characterised example of this ligand at an N_2-binding site.

Protic attack at the β-N atom of bound organic cyanide also occurs at these metal sites, to give the -N=CH(R) ligand. However, cleavage of the C-N bond has not been observed and organic cyanides bind relatively weakly (through nitrogen) to the metal. Recent studies of the reactions of these substrates with pure nitrogenase have thrown doubts upon the early reports of reductive cleavage of organic cyanides by the enzyme (D. J. Lowe and K. Fisher, unpublished observations).

The reduction of alkynes at metal centres is very complicated since a variety of pathways can be followed, including polymerisation. The outline of behaviour shown in Scheme 5b summarises observations for a range of systems and also suggests how divergent pathways for Mo- and V-nitrogenases (to C_2H_4 and C_2H_6, respectively) might arise (Hughes et al., 1986; Pombeiro, 1989). C_2H_2 reduction has also been studied in the borohydride-molybdate systems of Schrauzer (1980)(Section 5). The suggested mechanism for these reductions to C_2H_4 is similar to route (i) of Scheme 5b. On balance, the well-established route via alkenyl complexes (Kashef and Richards, 1989) seems the most likely, inter alia on grounds of simplicity. The stereospecific production of cis-CHD=CHD has also been observed for some of these systems. All of the above substrates are also reduced by the molybdenum and vanadium systems for which μ-N_2 intermediates have been proposed in dinitrogen fixation (Shilov, 1980). Again the observed products are similar to those obtained from nitrogenase reductions, including stereospecific reduction of C_2H_2.

7 CONCLUSIONS

During the past two decades, stimulated by the biochemical studies of nitrogenase, chemists have demonstrated that the reduction of dinitrogen under ambient conditions of temperature and pressure can be achieved by binding dinitrogen at certain metal centres.

There is still much to be learned about the types of metal-ligand environment which can be used to activate N_2 and other substrates and of mechanisms of reduction of dinitrogen in terminal- and bridging-modes of bonding. Moreover, knowledge of reduction of N_2 at vanadium and at iron must be further developed to aid our understanding of the function of the alternative nitrogenases.

Nevertheless, a pattern of reduction of substrates at metal centres is emerging which is consistent with that of nitrogenase action. At the present time, the detailed behaviour of these substrates at mononuclear sites is the most thoroughly studied and on balance a similar site for nitrogenase is favoured by this author. However, the possibility that dinitrogen (or some reduced derivative) bridges two metals at some stage in the nitrogenase reaction cannot be excluded (Sellmann, 1989) and indeed, a bimetallic system might be a characteristic of the vanadium and iron nitrogenases. Detailed X-ray structural information of the nitrogenase centre is necessary before firmer conclusions can be drawn.

8 REFERENCES

Adachi, T., Nabyoshi, S., Tatsuo, U., Manabu, K. and Yoshida, T., 1989. Reactions of a molybdenum(0) dinitrogen complex containing a crown thioether with isocyanides. J. Chem. Soc. Chem. Commun., 1320-1322.

Allen, A.D. and Senoff, C.V., 1965. Nitrogenopentammine-ruthenium(II) complexes. Chem. Commun. 621

Anderson, S.N., Chatt, J., Fakley, M.E. and Richards, R.L., 1981. Hydrazido(2-)-complexes as intermediates in the conversion of ligating dinitrogen into ammonia and hydrazine. J. Chem. Soc. Dalton Trans., 1973-1980.

Anderson, S.N., Hughes, D.L. and Richards, R.L., 1986a. Preparation of mer-$[Mo(N_2)_3(PPr_2Ph)_3]$ from trans-$[Mo(N_2)_2(PPr_2Ph)_4]$. J. Chem. Soc. Dalton Trans., 1591-1594.

Anderson, S.N., Hughes, D.L. and Richards, R.L., 1986b. Dinitrogen complexes of tungsten with bulky phosphine co-ligands. J. Chem. Soc. Dalton Trans., 245-252.

Baumann,, J.A., Bossard, G.E., George, T.A., Howell, D.B., Koczon, L.M., Lester, R.K. and Noddings, C.M., 1985. Formation of ammonia and hydrazine from reactions of acids with bis(dinitrogen) complexes of molybdenum. Inorg. Chem., 24: 3568-3578.

Bortels, H., 1930. Molybdenum as a catalyst in nitrogen fixation. Arch. Mikrobiol., 1: 333-342.

Burgess, B.K., 1985. Substrate reactions of nitrogenase. In: T.G. Spiro (Editor), Molybdenum Enzymes, Wiley, Chichester, pp. 161-217.

Chatt, J., Dilworth, J.R. and Richards, R.L., 1978. Recent advances in the chemistry of nitrogen fixation. Chem. Rev., 78: 589-625.

Chatt, J., Pombeiro, A.J.L. and Richards, R.L., 1980. Protonation of isonitriles ligating molybdenum (0) and tungsten (0). J. Chem. Soc. Dalton Trans., 492-498.

Coucouvanis, D. and Kanatzidis, M.G., 1985. Synthesis and characterization of a new Fe/Mo/S cluster containing the $[Fe_6Mo_2S_6]^{2+}$ core. J. Amer. Chem. Soc., 107: 5005-5006.

Cradwick, P.D., Chatt, J., Crabtree, R.H. and Richards, R.L., 1975. Preparation and X-ray structure of a trinuclear dinitrogen-bridged complex. J. Chem. Soc. Chem. Commun., 351-352.

Dilworth, J.R. and Richards, R.L., 1982. Reactions of dinitrogen promoted by transition metal compounds. In: G. Wilkinson, F.G.A. Stone and E.W. Abel (Editors), Comprehensive Organometallic Chemistry. Pergamon Press, Oxford, Vol. 5, pp. 1073-1106.

Dilworth, M.J., Eady, R.R., Robson, R.L. and Miller, R.W., 1987. Ethane formation from acetylene as a potential test for vanadium nitrogenase in vivo. Nature, 327: 167-168.

Eady, R.R., 1989. The vanadium nitrogenase of Azotobacter. Polyhedron, 8: 1695-1700.

Edema, J.J.H., Meetsma, A. and Gambarotta, S., 1989. Divalent vanadium and nitrogen fixation. J. Amer. Chem. Soc., 111: 6878-6880.

Galindo, A., Hills, A., Hughes, D.L. and Richards, R.L., 1987. A complex of the hydrazidium($=N-N^+H_3$) ligand. J. Chem. Soc. Chem. Commun., 1815-1816.

George, T.A., Kocson, C.M., Tisdale, R.C., Gebrayes, K. and Zubieta, J., 1986. Structural characterisation of two isomeric hydrazido(2-) intermediates in the conversion of molybdenum-bound dinitrogen to ammonia and hydrazine. Inorg. Chem., 25: 405-407.

Hawkes, T.R., McLean, P.A. and Smith, B.E., 1984. Nitrogenase from nif mutants of Klebsiella pneumoniae contains an altered form of the iron-molybdenum cofactor. Biochem. J., 217: 317-321.

Hawkes, T.R. and Smith, B.E., 1984. The inactive MoFe protein ($NifB^-$ Kp1) of the nitrogenase from nifB mutants of Klebsiella pneumoniae. Biochem. J., 223: 783-792.

Hedman, B., Frank, P., Gheller, S.F., Roe, A.L., Newton, W.E. and Hodgson, K.O., 1988. New structural insights into the iron-molybdenum cofactor from Azotobacter vinelandii nitrogenase. J. Amer. Chem. Soc., 110: 3798-3805.

Henderson, R.A., 1987. The detection and reactivity of $[MoH_4(H_2)(PPh_2CH_2CH_2PPh_2)_2]^{2+}$. J. Chem. Soc. Chem. Commun., 1670-1672.

Henderson, R.A., Leigh, G.J. and Pickett, C.J., 1983. The chemistry of nitrogen fixation and models for the reactions of nitrogenase. Advances in Inorg. Chem. and Radiochem., 27: 197-292.

Henderson, R.A., Morgan, S.H. and Stephens, A.N., 1990. The chemistry of niobium and tantalum dithiocarbamato-complexes. Part 5. The kinetics and mechanism of the hydrazine-forming reactions between $[\{M(S_2CNEt_2)_3\}_2(\mu-N_2)]$(M=Nb or Ta) and acid. J. Chem. Soc. Dalton Trans., 1101.

Hughes, D.L., Pombeiro, A.J.L. and Richards, R.L., 1988. Reductive cleavage of N_2, RNC and RCN at dinitrogen-binding centres. In: H.Bothe, F.J. de Bruijn and W.E. Newton (Editors), Nitrogen Fixation: Hundred Years After. G.Fischer, New York, p. 66.

Kashef, N. and Richards, R.L., 1989. The reactions of trans-$[MoH_4(Ph_2PCH_2CH_2PPh_2)_2]$ with $RCO_2C{\equiv}CH$(R = Me or Et) in the presence of acids. Preparation of hydrido-σ-alkenyl complexes of molybdenum(IV). J. Organometallic Chem., 365: 309-315.

Kovacs, J.A. and Holm, R.H., 1987. Structural chemistry of vanadium-iron-sulphur clusters containing the cubane-type $[VFe_3S_4]^{2+}$ core. Inorg. Chem., 26: 711-718.

Lowe, D.J., Fisher, K., Thorneley, R.N.F., Vaughn, S.A. and Burgess, B.K., 1989. Kinetics and mechanism of the reaction of cyanide with molybdenum nitrogenase from Azotobacter vinelandii. Biochemistry, 28: 8460-8466.

Manriquez, J.M., Sanner, R.D., March, R.E. and Bercaw, J.E., 1976. Reduction of molecular nitrogen to hydrazine. J. Amer. Chem. Soc., 98: 3042-3044.

Murray, R.C. and Schrock, R.R., 1985. Preparation of an unsubstituted hydrazido(1-) complex and an authentic high oxidation state ditungsten dinitrogen complex. J. Amer. Chem. Soc., 107: 4557-4558.

Nelson, M.J., Levy, M.A. and Orme-Johnson, W.J., 1983. Metal and sulphur composition of the iron-molybdenum cofactor of nitrogenase. Proc. Natl. Acad. Sci. USA, 80: 147-150.

Palermo, R.E., Singh, R., Baskin, J.K. and Holm, R.H., 1984. Molybdenum atom ligand substitution reactions of $MoFe_3S_4$ cubane-type clusters: synthesis and structures of clusters containing Mo-bound pseudo-substrates of nitrogenase. J. Amer. Chem. Soc., 106: 2600-2612.

Pickett, C.J., Ryder, K.S. and Talarmin, J., 1986. Electron-transfer reactions in nitrogen fixation. J. Chem. Soc. Dalton Trans., 1453-1457.

Pickett, C.J., Cate, K., McDonald, C.J., Mohammed, M.Y., Ryder, K.S. and Talarmin, J., 1988. Electron transfer reactions in nitrogen fixation. In: H. Bothe, F.J. de Bruijn and W.E. Newton (Editors), Nitrogen Fixation: Hundred Years After. G.Fisher, New York, pp. 51-55.

Pombeiro, A.J.L., 1989. Electrophilic β-addition to isocyanides, cyanide and alkyne-derived ligands. Polyhedron, 8: 1595-1600.

Rawlings, J., Shah, V.K., Chisnell, J.R., Brill, W.J., Zimmermann, R., Munck, E. and Orme-Johnson, W.H., 1978. Novel metal cluster in the iron-molybdenum cofactor of nitrogenase. J. Biol. Chem., 253: 1001-1004.

Richards, R.L., 1983. Dinitrogen complexes and their reactions. In: A. Muller and W.E. Newton (Editors): Nitrogen Fixation; The Chemical-Biochemical-Genetic Interface, Plenum Press, London, pp. 275-300.

Richards, R.L., 1988. Biological nitrogen fixation. Chemistry in Britain, 133-136.

Rocklage, S.M., Turner, H.W., Fellmann, J.D. and Schrock, R.R. 1982. Preparation of tantalum μ-dinitrogen complexes from molecular nitrogen and reduced tantalum complexes. Organometallics, 1: 703-707.

Schrauzer, G.N., 1980. Studies of the mechanism of biological nitrogen fixation with functional model systems. In: J. Chatt, L. Pina, R.L. Richards (Editors), New Trends in the Chemistry of Nitrogen Fixation. Academic Press, London, pp. 103-119.

Schrock, R.R., Wesolek, M., Liu, A.H., Wallace, K.C. and Dewan, J.C., 1988. Thiolato dinitrogen [or hydrazido(4-)] complexes, $[\{Ta(SR_3)_3(thf)_2\}_2(\mu\text{-}N_2)]$(Ar= 2,6-$C_6H_3Pr^i_2$, 2,4,6-$C_6H_2Pr^i_2$) and phenoxide analogues. Inorg. Chem., 27: 2050-2054.

Shilov, A.E., 1980. Comparison of biological nitrogen fixation with chemical analogues. In: J. Chatt., L. Pina and R.L. Richards (Editors), New Trends in the Chemistry of Nitrogen Fixation. Academic Press, London, pp. 121-150.

Shilov, A.E., 1988. Chemical models of biological nitrogen fixation. Recent advances. In: H. Bothe., F.J. de Bruijn and W.E. Newton (Editors), Nitrogen Fixation: Hundred Years After. G.Fischer, New York. pp.57-61.

Thorneley, R.N.F., Eady, R.R. and Lowe, D.J., 1978. Biological nitrogen fixation by way of an enzyme-bound dinitrogen-hydride intermediate. Nature, 272: 557-558.

Thorneley, R.N.F. and Lowe, D.J., 1985. Kinetics and mechanism of the nitrogenase enzyme system. In: T.G. Spiro (Editor), Molybdenum Enzymes. Wiley, London, pp. 221-284.

Yoshida, T., Adachi, T., Kaminaka, M. and Ueda, T., 1988. A novel molybdenum(0) dinitrogen complex containing crown thioether as the sole auxiliary ligand. J. Amer. Chem. Soc., 110: 4872-4873.

CHAPTER 5

PHYSIOLOGICAL CONTROL OF NITROGENASE AND UPTAKE HYDROGENASE

F.J. BERGERSEN
CSIRO Division of Plant Industry, GPO Box 1600, Canberra, A.C.T., Australia

1 INTRODUCTION

The regulation of nitrogenases and uptake hydrogenases is closely related. In this chapter, attention is concentrated on the physiology of nitrogenase and uptake hydrogenase activities and is a selective rather than an exhaustive coverage.

1.1 Nitrogenase

The intrinsic properties of the nitrogenase proteins (Chapters 2,3) are basic to the regulation of its activity. The extreme sensitivity of nitrogenase to inactivation by O_2 and the energetic advantages of aerobic metabolism generate a tension between N_2-fixing and energy-generating metabolism in aerobic and microaerophilic diazotrophs. Consequently, regulation of nitrogenase by O_2 is extremely important (see Hill, 1988; Yates, 1988). The absolute and relative concentrations of the nitrogenase proteins and the cellular concentrations of MgATP and MgADP are major determinants of nitrogenase kinetics (see Thorneley and Lowe, 1985) (Fig. 1) and are thus important in physiological control, affecting the allocation of electrons to proton reduction (H_2 evolution) and N_2 reduction (NH_3 production), as well as affecting rates of reaction.

Other controls arise from competition between interacting pathways of metabolism. Allocation of resources to carbon storage, for example, may regulate N_2 fixation by diverting reducing power from catabolic to anabolic pathways (Bergersen and Turner, 1990a,b). The energetic advantage to an organism of using combined nitrogen rather than N_2 gives rise to complex regulation of nitrogenase activity (Drozd et al., 1972) and gene expression (Chapters 6 and 7). Consequently, control of activity by combined nitrogen (chiefly NH_4^+) and by energy supply are important in nitrogenase regulation.

1.2 Uptake hydrogenases

Hydrogen is intimately involved with N_2 fixation. Evolution of H_2 is part of the chemistry of the binding and reduction of N_2 by nitrogenase (Fig. 1). Also,

H_2 is a competitive inhibitor of N_2 fixation (Hwang et al., 1973; Thorneley and Lowe, 1985). Evolution of H_2 from nitrogenase, inevitable though it is, represents a net loss of energy, unless the H_2 can be reprocessed to recover it. Hydrogenases are widespread in microorganisms, the most common using protons as terminal electron acceptor in pathways of energy production, and showing little H_2 uptake capacity. In contrast, the uptake hydrogenases of aerobic diazotrophs have about 100-fold greater capacity for H_2 uptake than for H_2 evolution and oxidise H_2, leading to ATP synthesis. Their role is an important adjunct to N_2 fixation, especially in plant symbioses, where an energy drain through loss of H_2 may be reflected in lower yields.

2 NITROGENASE: FACTORS DETERMINING ACTIVITY AND REACTION RATES

Until recently, it was generally accepted that nitrogenases from all sources were very similar indeed (Chapter 2), an assumption which the discovery of alternative nitrogenases in *Azotobacter vinelandii* and *A. chroococcum* (Bishop et al., 1988; Chapter 3) renders invalid. Although the presence or absence of Mo is important in regulation of the Mo- and V-nitrogenases, and of nitrogenase-3, it seems unlikely that overcoming Mo deficiency was a sufficient basis for the evolution of the alternatives. The finding that synthesis and activity of the Mo- and V-nitrogenases are differentially affected by temperature may have wide ecological significance (Eady et al., 1988a,b). In *Azotobacter*, V-nitrogenase is favoured over Mo-nitrogenase at low temperature (5-10°C).

However, the Mo-nitrogenases are the most widely distributed and this chapter will therefore only consider Mo-nitrogenases and their physiological control.

2.1 Reaction rates and electron allocation to N_2 reduction and H_2 evolution (consequences of the reaction sequence)

The reduction of N_2 to NH_3 by nitrogenase, at its most efficient, has the following stoichiometry:

$$N_2 + 8e^- + 8H^+ + 16MgATP = 2NH_3 + H_2 + 16MgADP + 16P_i \quad \ldots\ldots (1)$$

The Lowe-Thorneley model (Fig. 1) of nitrogenase action (Thorneley and Lowe, 1985) describes the reduction of N_2 to NH_3 in terms of enzyme kinetics. The model takes account of most of the observations made in experiments with isolated nitrogenase proteins (Lowe and Thorneley, 1984; Thorneley and Lowe, 1984a,b), and predicts accurately many of the observed features of N_2 fixation *in vivo*. The reaction sequence (Fig. 1b) requires eight sequential one electron reductions of the active centres of the MoFe protein, in each of which (RC, Fig. 1) a reduced

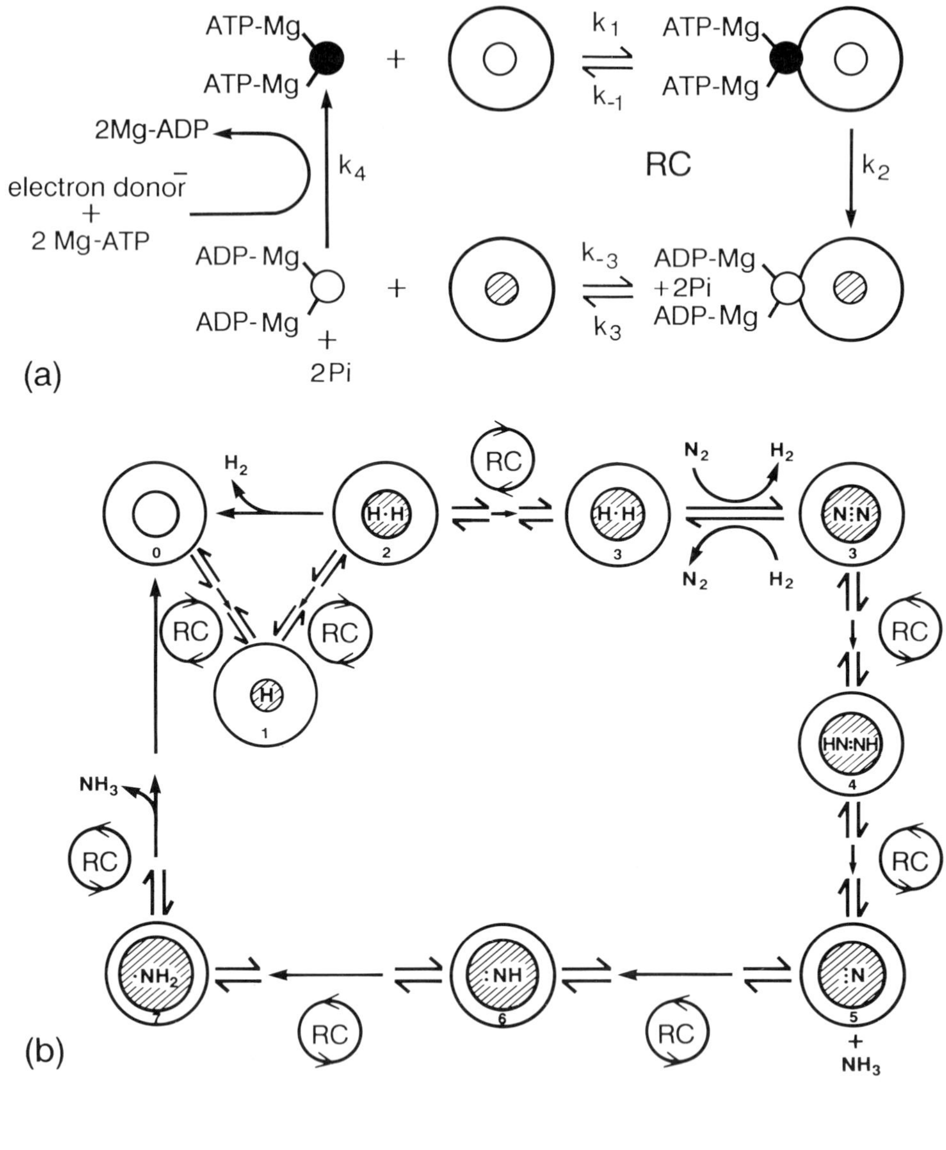

Fig 1. A diagrammatic representation of the reaction sequence in the production of 2 molecules of NH_3 from one of N_2. (Adapted and simplified from Lowe and Thorneley, 1984 and Eady et al., 1988). For further details of legend, see over.

(a) The reduction cycle (RC) in which the Fe protein (small circle) is reduced (shaded small circle) and binds MgATP, before binding to the MoFe protein at one of two specific sites and then reducing one of the two independent reactive centres of the MoFe protein. The hachured inner circle represents a reactive centre which is more reduced than that represented by the open inner circle. The overall reaction rate is limited by k_3.

(b) The reaction model for turnover of MoFe protein and reduction of N_2 to NH_3. Diagrammatically, the reactive centre is first hachured and then increased in size to indicate each successive one-electron reduction by the R.C. (a). With reduction, a metal in the reactive centre is first hydrided (1 & 2) and then either H_2 is evolved, returning the site to the ground (oxidized) state, or after a further reductive step, H_2 is displaced from the reactive centre by N_2 and successive reductions of bound N intermediates follow. The formation of a nitride at 5 and the subsequent steps (6,7) have not been experimentally determined. The model assumes that k_1 and k_{-1} in (a) are independent of the reduction level of the MoFe protein.

Fe protein molecule carrying 2MgATP complexes with the MoFe protein and then reduces an active centre thereon, with concomitant hydrolysis of ATP. The MoFe protein/Fe protein-MgADP complex dissociates in the rate-determining step of the cycle, which is completed when the Fe protein is reduced by a nitrogenase electron donor, either physiological or artificial, and MgATP replaces MgADP.

In the model, neither of the two independent active centres of the MoFe protein interacts with substrates (H^+, N_2, C_2H_2) or releases reduced products (H_2, NH_3, C_2H_4) until dissociated from the oxidised Fe protein-MgADP complex. Three reductions of the active centre (Fig. 1b) must occur before N_2 can be bound. If the third reduction is limited by low electron flux, or by low ratios of Fe protein to MoFe protein, wasteful evolution of H_2 ensues (Reaction 1). After the third reduction, N_2 may bind, leading to the unavoidable evolution of H_2. It is at this step that H_2 competitively inhibits N_2 fixation (Hwang et al., 1973). Stepwise reduction of the bound N_2 then proceeds with release of NH_3 (Fig. 1b). Wasteful evolution of H_2 is minimised when the intracellular concentration of nitrogenase is high (Klugkist et al., 1984; Thorneley et al., 1985), where N_2 concentration is high relative to H_2, and when the concentration of Fe protein is high relative to that of MoFe protein (Thorneley et al., 1985). These favourable conditions can be deduced from the nature of the equilibria shown in Figure 1. Differences between _in vitro_ and _in vivo_ electron allocation to H_2 evolution have been discussed by Yates (1988).

2.2 Generation of ATP

The requirement for at least 16 moles of ATP per mole of N_2 reduced, the importance of ATP concentration, and of ATP/ADP ratios in the nitrogenase reaction sequence, show the importance of ATP in regulating nitrogenase activity. It is therefore important to understand the mechanisms of ATP generation in the different types of diazotrophs.

(i) Anaerobic diazotrophs. The fermentative anaerobic bacteria appear to depend entirely on substrate-level phosphorylation for production of ATP. Given the fermentation pathway and the ATP yield per mole of substrate, the growth yield of cells per mole of ATP can be determined and used to estimate ATP costs in N_2 fixation from the difference between the growth yields of cells grown on N_2 and NH_4^+. Thus, Daesch and Mortenson (1967) found about 20 moles ATP/mole N_2 for sucrose-limited continuous cultures of Clostridium pasteurianum and Hill (1976) reported 29 moles ATP/mole N_2 for glucose-limited, anaerobic, continuous cultures of Klebsiella pneumoniae (cf. 16 moles/mole N_2 in Reaction 1). It will be of interest to see how ATP is produced in a group of strictly anaerobic bacteria which use cellulose as a sole source of carbon for diazotrophic growth (Leschine et al., 1988).

(ii) Aerobic and microaerobic diazotrophs. Some aerobic diazotrophs have branched respiratory chains (e.g. A. vinelandii, Downs and Jones, 1975; Bradyrhizobium japonicum, Appleby, 1969), probably differing in P/O ratios, and the relative electron transport capabilities of the branches is affected by growth conditions, especially the availability of O_2 and growth rate in chemostat culture (Stouthamer, 1983). Surprisingly, little attention has been paid to the special properties of the high-affinity terminal oxidases in diazotrophs, which produce ATP for nitrogenase activity when the concentration of dissolved O_2 is only of the order of 10-60 nM [e.g. in continuous N_2-fixing cultures of Rhizobium sp., Azospirillum brasilense (Bergersen and Turner, 1980) and soybean nodule bacteroids (Bergersen and Turner, 1980, 1990a)].

In bacteroids from soybean nodules, rates of O_2 consumption at low concentrations of O_2, rates of nitrogenase activity and ATP/ADP ratios were related (Appleby et al., 1975; Bergersen and Turner, 1975), but such relationships were not found with bacteroids from pea root nodules (Laane et al., 1978). However, oxidative phosphorylation in membrane vesicles prepared from A. vinelandii and from pea nodule bacteroids (Laane et al., 1979) was most efficient when dissolved O_2 was undetectable by an O_2 electrode in the medium (i.e. when the dissolved O_2 was probably less than 0.1 μM). These examples indicate that respiratory chains operating at very low concentrations of O_2 produce ATP efficiently. Under O_2-limiting conditions (5 to 8 nM free dissolved O_2 in the presence of oxyleghaemoglobin) we have found that, in soybean nodule bacteroids utilizing only endogenous reserves of ß-hydroxybutyrate (Bergersen and Turner, 1990a,b), N_2 fixation is directly correlated with O_2 consumption. The slope of the regression line indicates that these bacteroids fix N_2 with an efficiency of 0.39 moles NH_3 per mole of O_2 consumed (Bergersen and Turner 1990a); i.e. 5.12

moles of O_2 per mole of N_2. If the P/O ratio were 2, this would represent 20.5 moles of ATP per mole of N_2 fixed, a value only 20% above the minimum (Reaction 1) and similar to the value obtained for anaerobic C. pasteurianum (Daesch and Mortensen, 1967).

Although the activity of nitrogenase (Fig. 1a) is clearly regulated by ATP concentration and the ATP/ADP ratio, direct measurement of ATP and ADP may be misleading. During steady state N_2-fixation in whole cells, the rates of ATP/ADP turnover are substantial and concentrations of ATP and ADP are technically difficult to establish. While N_2-fixing continuous cultures of Rhizobium sp. strain CB756 showed greatest nitrogenase rates (3.3 nmol C_2H_4 min^{-1} mg^{-1}) with ATP/ADP of 2 and energy charge of 0.66 (Ching et al., 1981), and Laane et al. (1978) found ATP/ADP of 2.5 to 4 during maximum nitrogenase activity in pea nodule bacteroids, we (F.J. Bergersen and G.L. Turner, unpublished) have sometimes measured ATP/ADP ratios in soybean bacteroids of only about 0.5. These values are clearly inconsistent with the substantial nitrogenase activity at the time the reactions were quenched.

(iii) Facultatively aerobic diazotrophs. Diazotrophic members of the Enterobacteriaceae (e.g. Citrobacter freundii and K. pneumoniae) fix N_2 both anaerobically and microaerobically. For K. pneumoniae, microaerobic conditions seem to confer an energetic advantage, $Y_{glucose}$ (g dry wt. bacteria produced per mole of glucose consumed) being 1.5 to 1.9 times greater for cultures fixing N_2 under microaerobic (O_2-limited) rather than anaerobic conditions (Hill, 1976). The difference appeared to be in greater efficiency of aerobic ATP production rather than in more ATP-efficient N_2 fixation.

(iv) Photosynthetic diazotrophs. In principle, there are five sources of ATP available for N_2 fixation in photosynthetic diazotrophs, viz. (1) photophosphorylation, (2) oxidative phosphorylation, (3) anaerobic phosphorylation via electron transport chains terminating in inorganic sulphur or nitrogen compounds, (4) substrate-level phosphorylation and (5) an uptake hydrogenase system.

In the light, photophosphorylation via PSI is likely to be the major source of ATP for N_2 fixation by anoxyphotobacteria (Vignais et al., 1985) and may be important also in the heterocysts of filamentous cyanobacteria (see Gallon, 1980). Maryan et al. (1986) found it impossible to show unequivocally that photophosphorylation provided adequate ATP for N_2 fixation in continuously illuminated cultures of Gloeothece, a non-heterocystous cyanobacterium. However, for Gloeothece spp. growing under alternating light and darkness, N_2 fixation occurred only in the dark, when it was supported by glucan catabolism coupled to respiratory ATP production (Gallon and Chaplin, 1988). Oxidation of H_2 via uptake hydrogenase is a major mechanism for ATP production in photosynthetic diazotrophs (Bothe et al., 1984; Vignais et al., 1985).

(v) Symbiotic systems. In all N_2 fixing symbioses, ATP must almost certainly be provided within the microsymbiont, since ATP is unlikely to cross the membranes between the host and the symbiont. Therefore, since most symbiotic systems are aerobic, modulation of nitrogenase activities by host metabolism occurs indirectly by alteration of the supply of energy-yielding substrates or of O_2. In soybean nodules on intact plants treated for 5d with 5, 20 and 30% O_2, nitrogenase increased with pO_2, as did ATP/ADP, but the increases were not proportional (Heckmann et al., 1989). Changes to the gaseous diffusion resistance of the inner cortex of legume root nodules in response to applied external stresses (see Witty et al., 1986), which lead to restricted supply of O_2, should be regarded as modulating supply of ATP, as well as 'protecting' nitrogenase against damage by O_2.

2.3 Reductant supply for nitrogenase

(i) Ferredoxins and flavodoxins. Yates (1977) concluded that the only known biological electron carriers capable of transferring electrons to nitrogenases were ferredoxins and flavodoxins and summarized the properties of these carriers. This view has prevailed until challenged recently by the results of Haaker et al. 1988) and Häger et al. (1988). In anaerobic heterotrophs, reduction of the electron carriers is effected by pyruvate:ferredoxin- or pyruvate: flavodoxin-oxidoreductases (Shah et al., 1983).

(ii) Pyridine nucleotides. How pyridine nucleotides act as intermediates for the generation of reductants for nitrogenase has always been a problem, because the redox potential of the NAD(P)H/NAD(P) couple is insufficient to generate adequate concentrations of reduced low potential carriers (Table 1). The sources proposed for the energy inputs to allow this to occur are from membrane energization and reversed electron flow in the respiratory chain (Haaker and Klugkist, 1987). Reduced pyridine nucleotides can be used to reduce membrane-located carriers when boosted by the membrane proton-motive force (Fig. 2a), and lead to reduction of flavodoxin or ferredoxins in Azotobacter (Haaker and Klugkist, 1987). In photosynthetic diazotrophs (Fig. 2b), ATP generated by photophosphorylation boosts the reducing potential of electrons flowing from ubiquinone to ferredoxin (Vignais et al., 1985).
These mechanisms emphasize the close relationship between coupled electron transport and the provision of reductants for nitrogenase (Fig. 2).

(iii) Tight coupling between ATP production and reductants. During steady state N_2 fixation, bacteroids from soybean nodules respond to increased O_2 fluxes with corresponding increases in rates of O_2 consumption and N_2 fixation (Bergersen

TABLE 1. Mid-point reduction potentials for components in the pathways of reduction of nitrogenase in Clostridium and Azotobacter spp.

Electron donor/acceptor	Organism	E'm (mV)	Reference
Nitrogenase Fe-protein	Azotobacter vinelandii	-475	Braaksma et al., 1982
	Clostridium pasteurianum	-400	Zumft et al., 1974
Ferredoxin II	A. vinelandii	-500	Klugkist et al., 1986b
Ferredoxin 1	C. pasteurianum	-420	Yoch and Arnon, 1972
Ferredoxin 2	C. pasteurianum	-460	
Pyruvate/ferredoxin	C. pasteurianum	-400 to -500	Haaker et al., 1988
NAD(P)H	-	-320	Mahler and Cordes, 1966
Ubiquinone	A. vinelandii	-160	Klugkist and Haaker, 1987
	Rhodopseudomonas capsulata	-100	Vignais et al., 1985

and Turner, 1990a). Analogous results were obtained with A. vinelandii where ATP/ADP and membrane potential ($\Delta\psi$) remained unaltered despite increasing O_2 consumption and nitrogenase activity (Klugkist et al., 1986a). The organisms in such experiments are clearly O_2-limited. Given an excess of oxidizable substrate and that the increased flux of electrons through the respiratory chain in response to increased O_2 consumption is tightly coupled to ATP production and to consumption of ATP and reductant by nitrogenase, increased nitrogenase activity would match increased ATP production and no change in ATP/ADP would be expected. Such tight coupling of related activities seems logical for a model of membrane-located activities such as that proposed for A. vinelandii (Fig. 2a). Similarly, the balance between oxidized and reduced forms of the electron carriers would remain constant and no change in membrane potential would occur.

(iv) Are there other electron donors? In A. vinelandii, a double mutant lacking both flavodoxin and ferredoxin I grew more slowly than the wild type on NH_4^+ as N-source (Martin et al., 1989). With N_2 as sole N-source, the mutant had almost the same growth rate as the wild type, suggesting that there must be another electron donor(s) for nitrogenase and that flavodoxin and ferredoxin I have other roles in growth on combined N.

2.4 Regulation by O_2

A range of strategies is found to reconcile the need for O_2 for ATP

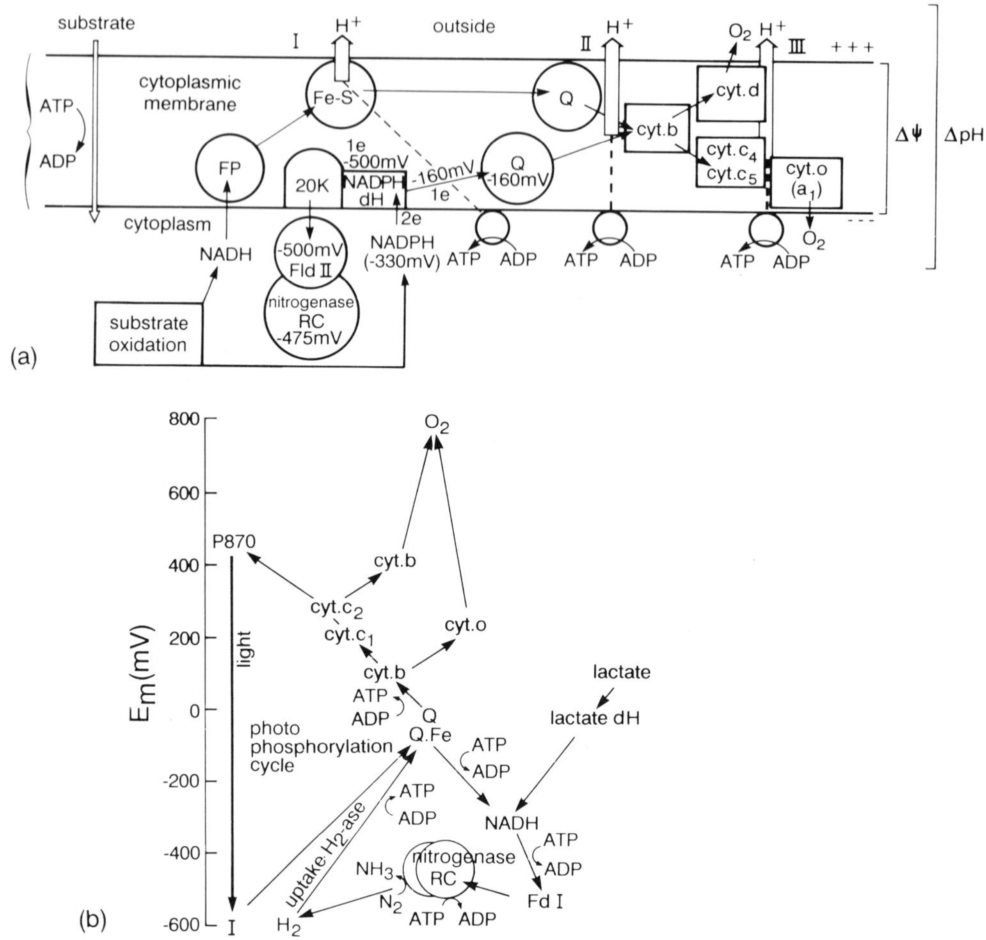

Fig 2. Pathways linking electron transport, ATP production and N_2 fixation.

(a) A model of the membrane-bound systems of electron transport linking substrate oxidation, respiration and N_2 fixation in Azotobacter vinelandii. (Adapted from Jones, 1977, and Haaker and Klugkist, 1987).

(b) Pathways linking photophosphorylation, nitrogenase, substrate utilization and respiration in Rhodobacter capsulatus. (Adapted from Vignais et al., 1985).

Thin arrows indicate electron flow: broad, unfilled arrows indicate proton ejection segments of the respiratory chains and ATP production via the transmembrane ATPase complex is also indicated at these segments. FdI = ferredoxin I; FldII = flavodoxin II; NADPHdH - a membrane-bound NADP-oxido-reductase; P_{870} is the reaction centre of bacteriochlorophyll; I is bacterio-pheophytin. Other abbreviations are as in Jones (1977). The component 20K is a peptide synthesized only in nitrogenase-derepressed cells.

generation and the sensitivity to O_2 of nitrogenase (see Hill, 1988). Anaerobes avoid the conflict completely. Aerobic and microaerobic diazotrophs adopt one or more of the following mechanisms.

(i) Protection: O_2 supply and demand. The O_2 status of a microbial environment is governed by the O_2 supply conditions (partial pressure of O_2 in the atmosphere, rate of transfer between gas and liquid phases, solubility of O_2 in the medium) and the O_2 demand, which depends on population density, O_2 consumption characteristics and O_2 concentration. When O_2 demand exceeds supply, O_2 concentration near to the organisms falls to a value close to the K_m for the terminal oxidase with the highest affinity for O_2. Indeed, many diazotrophs have terminal oxidases of very high O_2 affinity, which can sustain high rates of O_2 consumption, even though O_2 concentration in solution near the bacteria is vanishingly small (e.g. 10 nM for soybean bacteroids; Bergersen and Turner, 1980, 1990a). Consequently, despite a possibly high availability of gaseous O_2 (for example air above a stirred culture), inactivation of nitrogenase may be negligible. Many experiments describing the protective effects of various treatments against inactivation of nitrogenase by O_2 have given insufficient attention to these matters.

Respiratory demand as a protective mechanism is most highly developed in Azotobacter spp. These bacteria have extremely high rates of O_2 consumption, and adapt to a wide range of O_2 supply rates by mechanisms involving a branched electron transport pathway (Fig. 2a), tightly coupled to ATP production at lower O_2 concentration, but uncoupled at sites I and II at high O_2 with site III bypassed (see Yates, 1988).

The supply and demand effects are utilized in behavioural adaptations to O_2, such as when diazotrophs form associations with aerobic heterotrophs, or when they form clusters or pellicles, thus intensifying O_2 demand and minimising ingress of O_2 (Bergersen, 1983). Such aggregates may then face O_2 depletion, which may be alleviated by increasing the 'porosity' of an aggregate, e.g. by separation of the cells with polysaccharide capsules (Bergersen, 1983). In the case of the intracellular bacterial aggregates in cells of root nodules, the problem is overcome by developing specialized host structures and leghaemoglobin to transport O_2 (see Appleby, 1984).

In some bacteria, O_2 demand can vary in response to the concentration of ambient O_2. For example, suspensions of soybean nodule bacteroids experiencing a steady rate of supply of dissolved O_2 adjusted their demand for O_2 in response to a slow rise in O_2 concentration which occurred when O_2 supply slightly exceeded O_2 demand (Fig. 3). These changes were quite small, but the O_2 concentration stopped rising and returned to a range more favourable to N_2 fixation, an apparently homeostatic mechanism.

(ii) Conformational protection. This term is used to describe the O_2 tolerance of the nitrogenase proteins of Azotobacter spp. (in the oxidized state) when complexed in the presence of Mg^{2+} with the Fe_2S_2 (Shethna) protein (see Hill, 1988; Yates, 1988). This catalytically-inactive complex is formed in crude extracts or intact bacteria on exposure to O_2. In intact bacteria, conformational protection is indicated by a rapid "switch off" of nitrogenase

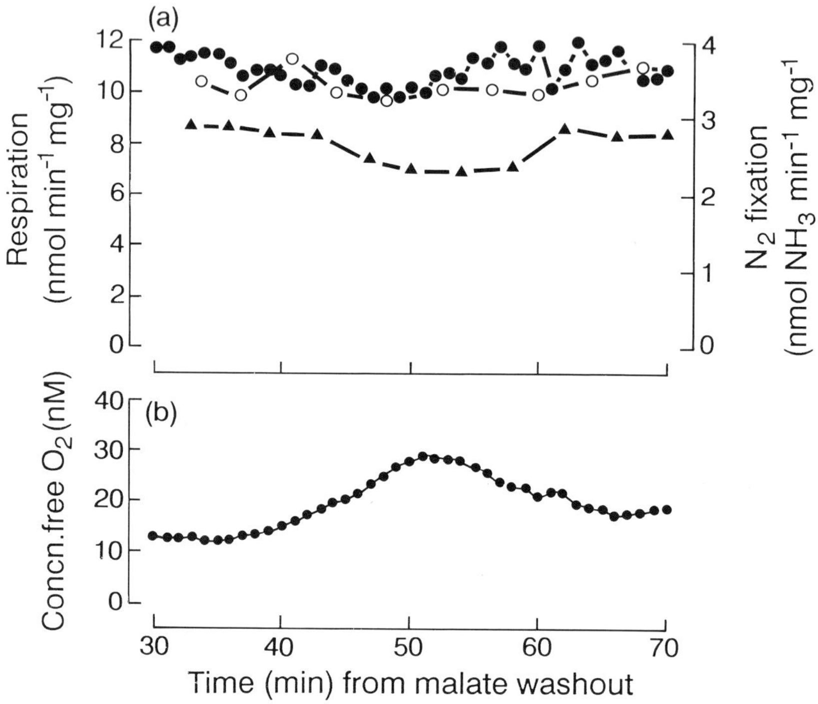

Fig. 3. Homeostasis in bacteroids from soybean root nodules, in a flow reaction chamber with constant O_2 supply.

At zero time, malate (supplied initially at 10 mM) had been completely washed out of the chamber by a flow of nutrient-free medium.

(a) The time course of oxygen consumption (●), CO_2 evolution (○) and N_2 fixation (▲).

(b) The time course of change in concentration of free, dissolved O_2.

The reaction took place in a stirred chamber (Bergersen and Turner, 1990a), containing 26.0 mg of bacteroid protein, supplied with a constant flow (1.00 ml min^{-1}) of medium containing dissolved air (226 µM free O_2) and 87 µM oxyleghaemoglobin. There was no gas phase in the chamber and the carbon source was endogenous poly-ß-hydroxbutyrate (Bergersen and Turner, 1990b).

activity upon exposure to excess O_2 and by a rapid return of activity upon removal of O_2 in the presence of inhibitors of protein synthesis. Reversible inhibition of nitrogenase by O_2 occurs in 13 species in 10 genera (Hill, 1988), though not all of these involve complex formation with an iron-sulphur protein (Hill, 1988).

(iii) Protection/regulation involving structural features. In some filamentous, aerobically-grown cyanobacteria, heterocysts are specialized cells developing when combined nitrogen is limiting and in which there is a suitable environment for nitrogenase. Heterocysts have thickened, multi-layered cell walls, believed to restrict access of O_2 (Haury and Wolk, 1978); they also have high respiratory activity (Tabita et al., 1988), thus maintaining low concentrations of intracellular O_2. Endogenous production of O_2 in heterocysts is avoided by the elimination of photosystem II (Tel-Or and Stewart, 1977). Nitrogenase is normally confined to the heterocysts when cultures are grown in air, but is also found in both proheterocysts and vegetative cells when they are grown microaerobically. Maximum levels of nitrogenase Fe protein coincided with maximum structural differentiation in heterocysts of the symbiotic cyanobacteria from Azolla spp. (Braun-Howland et al., 1988). Here also, nitrogenase was confined to the heterocysts.

Frankia spp., the nodule microsymbionts of non-leguminous angiosperms such as Alnus spp., bear terminal vesicles on some hyphae. These vesicles are analogous to heterocysts, in that they form in cultures deprived of NH_4^+, have thick walls restricting access of O_2 and contain the nitrogenase (Tjepkema et al., 1980; Torrey, 1985; Meesters, 1987). The thickness of the vesicle walls increases with raised O_2 concentrations (Parsons et al., 1987; Silvester et al., 1988a,b). Differences between the fatty acid composition of walls of vegetative cells and of vesicles (Tunlid et al., 1989) may alter the permeability to O_2. In nodules, cells of the central tissue frequently contain large clusters of vesicles (Torrey, 1985), but in other species (e.g. Casuarina spp.) the endophytes bear no vesicles (Torrey, 1985), and other structural features of the nodules restrict access of O_2 to the endophyte (Tjepkema et al., 1988).

The root nodules of legumes (see Dakora and Atkins, 1989), actinorhizal plants and Parasponia spp. show the most highly-developed morphological adaptations to regulate and protect nitrogenase from O_2. These outwardly-aerobic N_2-fixing systems are actually microaerobic because of the high O_2 demand generated by the closely-packed bacteria and the restricted entry of O_2 through the nodule tissues. In most legume nodules, (e.g. soybean; Tjepkema and Yocum, 1973; Bergersen and Goodchild, 1973) and in some actinorhizal nodules (e.g. Casuarina cunninghamiana; Tjepkema et al., 1988), there is a zone in the inner cortex in which gas-filled, intercellular spaces are few and narrow, or absent. This constitutes a narrow zone of high diffusion resistance, the properties of which may vary in response to stress (Witty et al., 1986). In addition, the cytoplasm of infected cells of the central nodule tissue contains O_2-carrying haemoglobins (see Appleby, 1984). The low diffusive flux of O_2 resulting from the very low concentrations of free, dissolved O_2 in these cells (10 nM in soybean nodule cells; Bergersen, 1982) is greatly augmented by a flux of oxyhaemoglobin, thus providing sufficient O_2 for the endophytes. However, the actinorhizal nodules of Datisca glomerata (Tjepkema et al., 1988) appear to be significantly different.

(iv) Superoxide dismutase and catalase. Interactions between biological systems and free O_2 frequently produce the superoxide radical (O_2^-) and H_2O_2, which are detoxified by superoxide dismutase (SOD) and catalase, respectively. Nitrogenase derepression and SOD activity are positively correlated in some diazotrophs (see Yates, 1988) and SOD may play an essential role in the protection of the nitrogenase system (Puppo and Rigaud, 1986). However, O_2^- is typically pro-duced by excessive exposure to free O_2; with N_2 fixation in natural systems, this is normally rare, because of the adaptations limiting O_2 access. Yates (1988) concluded that there was insufficient evidence to establish protective roles for SOD and catalase.

2.5 Regulation by combined nitrogen

Most diazotrophs and symbiotic systems containing them assimilate combined nitrogen from their environment in preference to fixing N_2. Although the primary mechanism for this is regulation of nif gene expression (Chapters 6 and 7), the activity of pre-existing nitrogenase is also regulated by combined nitrogen.

(i) Regulation by NH_3 (NH_4^+). Rapid inhibition of nitrogenase activity upon exposure to NH_4^+ has been reported in 18 species of diazotrophs, including free living heterotrophs (Azotobacter sp. and Azospirillum spp), methanotrophs (Methylosinus trichosporium and Methylococcus capsulatus), sulphate reducers (Desulphovibrio gigas), sulphide oxidisers (Beggiotoa alba) and in eight genera of phototrophs (see Fu and Burris, 1989).

(a) Reversible ADP ribosylation. Supply of an NH_4^+ salt to cultures of Rhodospirillum rubrum growing in the presence of glutamate (which does not repress nif expression) produces rapid inhibition of nitrogenase activity due to covalent modification of the nitrogenase Fe protein (Kanemoto and Ludden, 1984). The modification is reversible and is due to ADP-ribosylation of one of the arginine-100 residues of the dimeric Fe protein (Pope et al., 1985). The ADP ribosylation reaction (see Ludden et al., 1988) is catalyzed by dinitrogenase reductase ADP-ribosyl transferase (DRAT) and is dependent upon MgADP and divalent cations (usually Mn^{2+}). Reactivation of modified nitrogenase (see Ludden et al., 1988) involves dinitrogenase reductase activating glycohydrolase (DRAG), which hydrolyses ADP-ribose from the inactivated Fe protein in the presence of MgATP and divalent cations (Mn^{2+}). The MgATP binds to the Fe protein and is not hydrolyzed stoichiometrically in the DRAG reaction. DRAG has been purified as an extremely O_2-labile monomeric protein of 32 kD and DRAT as an O_2-stable monomer of 29 kD. DRAT from R. rubrum reacts with Fe proteins from other diazotrophs, provided an arg-100 is present and accessible. The draG and draT genes are contiguous in both R. rubrum and A. lipoferum and map near the nif genes in both organisms; draTG are not co-regulated with nif (Ludden et al., 1988). DraG and draT probes should prove powerful tools for identifying DRAT-DRAG modification systems for the rapid regulation of nitrogenase activity by NH_4^+ in other diazotrophs.

(b) Other mechanisms. Rapid inhibition of nitrogenase activity in the presence of NH_4^+ is found in many organisms lacking a DRAT-DRAG system (e.g. Herbaspirillum seropedicae (Fu and Burris, 1989). There may be several different mechanisms involved. Laane et al. (1980) showed that effects of NH_4^+ on nitrogenase activity in A. vinelandii correlated with a decrease in membrane potential. The NH_4^+ effect on nitrogenase was not seen in pea nodule bacteroids which did not show a change in membrane potential or accumulate NH_4^+. Neither covalent modification nor effects on membrane potential seem to be involved in rapid regulation of nitrogenase activity in Rhodopseudomonas sphaeroides (Haaker et al., 1982).

(ii) Regulation by nitrate. Nitrate is the dominant dissolved nitrogen-containing ion in agricultural soils and exerts a correspondingly dominant effect on N_2 fixation in such environments. The mechanisms of nitrate repression involve many different genes across a wide range of organisms. In filamentous cyanobacteria, the number and frequency of heterocysts are reduced in the presence of nitrate (see Gallon, 1980), while Azospirillum spp. do not fix N_2 microaerobically in the presence of 1 mM nitrate (Döbereiner and De Polli, 1980). Paradoxically, some aerobic or microaerobic diazotrophs express nitrogenase activity when growing anaerobically with nitrate as terminal electron acceptor (Rigaud et al., 1973; Döbereiner and De Polli, 1980), but accumulation of nitrite, a potent inhibitor of nitrogenase, often limits the reaction (e.g. Rigaud et al., 1973).

Nitrate also has profound effects on N_2 fixation by symbiotic systems (Peoples and Gibson, 1989). For example, in field grown soybeans, a soil nitrate-N concentration of 38 p.p.m. at sowing almost completely prevented nodulation (Brockwell et al., 1989). In addition, application of nitrate to the roots of nodulated plants rapidly inhibits nitrogenase activity. Streeter (1988) distinguished three classes of response to nitrate: (i) inhibition of infection of roots by nodule bacteria; (ii) inhibition of nitrogenase activity per unit mass of nodule and its reversal when nitrate is removed; (iii) decrease in nodule mass/plant. Avoidance of the inhibitory effects of nitrate has become an objective for breeding new cultivars of soybean (Carroll et al., 1988; Herridge and Bergersen, 1988). The underlying mechanisms for the nitrate inhibition of nodule nitrogenase activity remain uncertain (see Streeter, 1988).

2.6 Regulation by other metabolism

The high demand for ATP and reductant for functioning nitrogenase may be the basis for regulatory controls imposed by other metabolic pathways. For example, Esposito and Wilson (1958) described an interaction between Ca^{2+} and acetate in N_2 fixation by *A. vinelandii* O, an effect attributed to an O_2-sensing mechanism (Peterson, 1989). A rapid decline in N_2 fixation by soybean bacteroids has been noted when malate or succinate is supplied to bacteroids using only endogenous substrates with a constant O_2 supply and a very low concentration of free, dissolved O_2 (Bergersen and Turner, 1990a). N_2 fixation was restored or even enhanced following washout of the exogenous substrate. Supply of malate or succinate under these conditions caused the bacteroids to accumulate excess carbon and synthesize poly-ß-hydroxybutyrate rather than use the substrate for increased N_2 fixation (Bergersen and Turner, 1990b).

3 UPTAKE HYDROGENASE

Bacterial oxidation of H_2 by O_2 has been known for 150 years (see Lee and Umbreit, 1940), but only after the work of P.W. Wilson and others (Wilson, 1940) was this reaction clearly related to N_2 fixation. However, there was confusion between the roles of hydrogenase and the evolution of H_2 by nitrogenase, until the latter was established as a separate process. Phelps and Wilson (1941) showed that O_2 consumption by *A. vinelandii* was stimulated 1.6-fold when H_2 was supplied. They also found hydrogenase activity in nodule bacteroids of a strain of *Rhizobium leguminosarum* which had none in culture, while neither cultures nor bacteroids of soybean rhizobia had hydrogenase activity. There the matter lay, until Dixon (1967, 1968, 1972) demonstrated the oxidation of H_2 produced from nitrogenase action. He proposed that this reaction alleviated inhibition of nitrogenase by H_2, protected nitrogenase from damage by excess O_2, and produced ATP, thus regaining some of the energy lost during the evolution of H_2 from nitrogenase.

3.1 Occurrence of uptake hydrogenases and their properties

Uptake hydrogenase (Hup) is present in many bacteria and most aerobic diazotrophs (see Yates, 1988), though absent from many strains of Rhizobium and Bradyrhizobium spp. Three types of uptake hydrogenase are recognized, on the basis of their cellular location (periplasmic, membrane-bound or cytoplasmic), their physiological properties, the nature of their reactive centres and the ratio between their rate constants for H_2 evolution and H_2 uptake. Uptake hydrogenases are usually about 100 times more active in H_2 uptake than in H_2 evolution.

In the anaerobe Desulphovibrio vulgaris, the three hydrogenases contain [Fe], [Fe-Ni] and [Ni-Fe-Se] centres, respectively (Lissolo et al., 1986). In C. pasteurianum, two hydrogenases with distinctly different properties both contain unusual H-clusters of 6 Fe atoms (Adams et al., 1989).

Amongst aerobic microorganisms, two groups of membrane-bound hydrogenases can be recognized, the first in the H_2-oxidizing bacteria and the second in the aerobic diazotrophs; despite large differences in their $K_m(H_2)$ (ca. 30 µM and 1 µM, respectively), they share a number of characteristics (Arp et al., 1985).

Uptake hydrogenases have been reviewed recently by Yates (1988) for aerobic diazotrophs, by Vignais et al. (1985) for phototrophic diazotrophs and by Evans et al. (1988) for legume root nodules. They are membrane-bound enzymes of two subunits, of approximately 30 and 60 kD. They contain FeS groups and require Ni for activity. Selenium stimulates uptake hydrogenase activity in Se-deficient media and appears to be part of purified hydrogenase (Hanus et al., 1988), despite the apparent lack of a codon for selenocysteine in the nucleotide sequence of the hup genes (Sayavendra-Soto et al., 1988). Although often regarded as O_2-sensitive enzymes, they can now be prepared and purified safely in air (Stults et al., 1986).

3.2 Mechanisms of uptake hydrogenase activity

Arp and Burris (1981) proposed that the kinetics of uptake hydrogenase involved a two site, ping-pong mechanism, in which electrons were accepted at one site on the enzyme and H_2 oxidized at the other in a reversible reaction:-

$$H_2 \underset{k_{-1}}{\overset{k_1}{\rightleftharpoons}} 2H^+ + 2e^- \ (E_o = -420 \text{ mV}) \qquad \ldots\ldots(2)$$

In the oxidative direction, electron flow is coupled to generation of ATP via the aerobic electron transport chain (Eisbrenner and Evans, 1982).

3.3 Genetics and regulation of uptake hydrogenase

The genetics of hup have been studied in a number of diazotrophs, but specially in Bradyrhizobium japonicum (see Lambert et al., 1985; Evans et al., 1988). The structural hup genes have been cloned in a cosmid (pHU52), expressed in E. coli (Zuber et al., 1986) and sequenced (Sayavendra-Soto et al., 1988). The gene encoding the smaller subunit (34,538 D) was located 34 bp upstream of that for the larger subunit (65,930 D). The transcriptional start site is located 13 nucleotides upstream from the ATG start codon of the small subunit leader peptide. There are substantial homologies with other hydrogenase genes, especially in the positions of the cysteine residues, which are probably involved in binding the Fe and/or Ni cofactors (Sayavendra-Soto et al., 1988).

The hup genes of B. japonicum do not appear to be plasmid-located (Cantrell et al., 1982), as they are in many H_2-oxidizing bacteria. However, in R. leguminosarum (Cunningham et al., 1985) and other fast-growing rhizobia, hup is located on the Sym plasmids. In A. chroococcum, hup genes are probably chromosomal (Robson et al., 1984) and have significant genetic homology with rhizobial hup genes (Tibelius et al., 1987).

Expression of hup genes is regulated by factors which affect the redox state of the organisms (see Yates, 1988) and in particular by NAD/NADH ratio in sucrose-limited continuous cultures of A. chroococcum. Although hup may be repressed by fixed nitrogen, nif and hup are not co-regulated. Generally speaking, hup is repressed by carbon substrates (sugars, carboxylic acids, glycerol) and expression promoted by H_2 and microaerobic conditions; some of these effects may involve interactions between such factors as O_2 availability and pH (van Berkum and Maier, 1988). With symbiotic bacteria, hup expression often depends on host genotype (Dixon, 1968; Murillo et al., 1989).

In the Frankia strain CpI1, immuno-gold localization (Lindblad and Sellstedt, 1989) showed uptake hydrogenase in the vesicles and hyphae of N_2-grown cultures and also in the hyphae of cultures grown with combined-N (no vesicles present), possibly associated with intracellular membranes. In contrast, no uptake hydrogenase activity was found in aerobic, NH_4^+-grown, undifferentiated filaments of Frankia strain ArI3 (Murray and Lopez, 1989). H_2 uptake by aerobic N_2-grown cultures may increase O_2 demand and thus protect nitrogenase against the effects of high O_2 in carbon-limited conditions (Murray and Lopez, 1989).

Strains of heterocystous cyanobacteria which contain phycoerythrin and accumulate storage carbohydrate do not have uptake hydrogenase activity, but strains not accumulating excess carbohydrate do (Chen et al., 1989).

3.4 Chemolithotrophy

An important characteristic of the H_2-oxidising bacteria is their ability to grow with CO_2, H_2, O_2, NO_3^- or NH_4^+ and mineral salts. Interestingly, hup$^+$ strains of B. japonicum grow in a similar fashion, using ribulose-bisphosphate carboxylase as the principal route for CO_2 assimilation, and H_2 oxidation as sole energy source (see Evans et al., 1980). However, such cultures apparently fixed no N_2. It was postulated that the capacity for chemolithotrophy would be an advantage for survival/growth of these bacteria in soil in the absence of the host. Evidence for this has been reported by Lambert et al. (1985), using soil incubations enriched with 1% H_2. The crucial point is whether such a concentration of H_2 would occur in natural soils. Results such as those of Cunningham et al. (1986) suggest that soil atmospheres may be significantly enriched in H_2 near nodulated alfalfa roots.

3.5 Contribution of uptake hydrogenase to efficient N_2 fixation

Since evolution of H_2 during nitrogenase action represents an energy cost to the cell and since H_2 may also decrease the efficiency of the enzyme, possession of uptake hydrogenase should, in principle, benefit the cell. Much of the supporting evidence arises from laboratory experiments based on observation of energy-related parameters. Aguilar et al. (1985) showed that under sucrose limitation at high dilution rates in continuous cultures, H_2 recycling was advantageous to N_2-fixing A. chroococcum. There are also many examples of the advantages of hup$^+$ strains of B. japonicum over hup$^-$ strains in laboratory and field experiments (see Lambert et al., 1985; Evans et al., 1988; Yates, 1988), but important exceptions have been noted with other rhizobia (Sorensen and Windaele, 1986). Nevertheless, the improvement of rhizobial inoculants by the incorporation of an active Hup system has been proposed (Emerich and Evans, 1984).

Pate et al. (1981) considered the energetics of H_2 recycling in legume root nodules, taking into account all the ATP-consuming processes associated with N_2 fixation. If 30% of electron flow through nitrogenase were directed to H_2 evolution (the minimum is 25%; Reaction 1), then H_2 recycling would only lower the ATP used per N_2 fixed from 30 to 26. Such a small energetic advantage would be difficult to measure experimentally. Yates (1988) pointed out, that if nodules were carbon-limited, Hup activity would be beneficial, but that if they were O_2-limited the H_2-dependent and carbon-dependent respiratory systems would compete for O_2. If the H_2-respiration was also less well-coupled to ATP formation, then H_2 recycling would be disadvantageous. Thus, it seems doubtful that the present experimental evidence is sufficient to attribute the demonstrated benefits of the hup phenotype solely to energetic economies, as is frequently done in the introductions to papers about uptake hydrogenases.

3.6 Protection of nitrogenase

If O_2 levels around a diazotroph are incipiently damaging to nitrogenase or repressive of its synthesis, any process decreasing O_2 concentration should improve nitrogease activity. Uptake hydrogenase-catalyzed O_2 reduction with H_2 should then be an advantage to a diazotroph. Many laboratory experiments attest to the reality of the effect (e.g. Walker and Yates, 1978).

3.7 Relief of H_2 inhibition of N_2 fixation

Dixon (1972) first suggested that uptake hydrogenases might prevent the concentration of H_2 in root nodule tissue rising to levels which would inhibit N_2 fixation.

First of all, what concentration of H_2 would be inhibitory? Estimates of K_i for inhibition of nitrogenase by H_2 *in vitro* are rather high (e.g. 12 kPa ≈ 80 μM at 25°C; Hwang et al., 1973). However, Rasche and Arp (1988) estimated K_i values of about 2 kPa H_2 (≈ 14 μM at 25°C) for N_2 fixation by soybean bacteroids.

Several authors have considered H_2 in models of diffusive fluxes and concentrations of gases in root nodules. Dixon et al. (1981) calculated that dissolved H_2 in the central tissue of a nodule (of about 2 mm diam.) could range from 100 μM near the cortex to 200 μM at the centre, but this model considered the central tissue to be essentially water-filled. Sheehy et al. (1987) used a simulation which incorporated the distribution and dimensions of gas-filled intercellular spaces in the central tissue and in the cortex and included an idealized treatment of diffusion in the infected cells, which took account of the structure of the central tissue (Sheehy and Bergersen, 1986). They concluded that, in a soybean nodule functioning normally in air, the solution concentration of H_2 near the bacteroids would be 12-13 μM). Another model (Hunt et al., 1988) used a different nodule structure in the simulation of gas exchange in legume nodules, and concluded that dissolved H_2 in the infected cells of the central tissue of a nodule in air would be *ca* 300 μM. One factor contributing to these different findings was the very different values for the diffusion coefficients of dissolved H_2 at 20°C used by the authors (Sheehy et al., 1987: 63.4×10^{-10} m^2 s^{-1}; Hunt et al., 1988: 310 mm^2 d^{-1} = 35.9×10^{-10} m^2 s^{-1}; *cf.* Perry, 1950: 51.3×10^{-10} m^2 s^{-1}).

It is perhaps of interest to compare these complex simulations with a much simpler one. It is now generally agreed that the diffusion of gases into and out of soybean root nodules is most probably limited by a water-filled layer in the inner cortex (e.g. Sinclair and Goudrian, 1981; Sheehy et al., 1985). Parsons (1989) has described the morphology of a "boundary layer", one cell thick (radial thickness 15-20 µm beneath the surface of a nodule of 2 mm diameter) which is completely devoid of intercellular spaces. For field levels of N_2 fixation and nodulation (Bergersen 1987), a nodule of these dimensions would fix 1.166×10^{-10} mol N_2 sec^{-1} and would produce at least the same amount of H_2 (Reaction 1). The flux of H_2 through the water-filled "boundary layer" is described (Crank 1975) by:

$$Q = \frac{4\pi Dab\ dC}{b-a}$$

where Q is the flux of H_2, D is the diffusion co-efficient for dissolved H_2 (Sheehy et al., 1987), a and b the radii of the inner and outer boundaries (respectively) of the layer and dC the concentration drop across it. For the dimensions and quantities given above, $dC = 9.7 \times 10^{-3}$ mol m^{-3} (9.7 µM). That is, if no H_2 uptake system were present, the concentration of H_2 in central tissue cells must be greater than 10 µM, close to the K_i for inhibition of N_2 fixation by soybean bacteroids measured by Rasche and Arp (1988). Reported $K_m(H_2)$ values for uptake hydrogenases, although differing (e.g. 0.86 µM for A. vinelandii, Seefeldt and Arp, 1985; 0.25 µM for R. capsulata, Vignais et al., 1985; 1-3 µM for soybean bacteroids, Arp and Burris, 1981), all indicate a high affinity for H_2, favourable for in vivo activity of uptake hydrogenase.

In summary then, H_2 in root nodule tissue probably reaches concentrations sufficient to inhibit N_2 fixation, if the value for K_i (H_2) found by Raschke and Arp (1988) applies generally. At such concentrations of H_2, uptake hydrogenases could be expected to consume H_2 at a significant rate, thereby lowering the H_2 concentration to less inhibitory levels. Similar conditions may apply in aggregations of other diazotrophic micoorganisms in which outward diffusion of H_2 is restricted. It is less likely that such conditions would apply when diazotrophs are dispersed in soil or other environments.

4 REFERENCES

Adams, M.W.W., Eccleston, E. and Howard J.B., 1989. Iron-sulfur clusters of hydrogenase I and hydrogenase II of Clostridium pasteurianum. Proc. Natl. Acad. Sci. USA, 86: 4932-4936.

Aguilar, O.M., Yates, M.G. and Postgate, J.R., 1985. The beneficial effects of hydrogenase in Azotobacter chroococcum under nitrogen fixing, carbon limiting conditions in continuous and batch cultures. J. Gen. Microbiol., 131: 3141-3145.

Appleby, C.A., 1969. Electron transport systems of Rhizobium japonicum 1. Haemoprotein P-450, other CO-reactive pigments, cytochromes and oxidases in bacteroids from N_2-fixing root nodules. Biochim. Biophys. Acta, 172: 71-87.

Appleby, C.A., 1984. Leghemoglobin and Rhizobium respiration. Ann. Rev. Plant Physiol., 35: 443-478.
Appleby, C.A., Turner, G.L. and MacNicol, P.K., 1975. Involvement of oxyleghaemoglobin and cytochrome P-450 in an efficient oxidative phosphorylation pathway which supports nitrogen fixation in Rhizobium. Biochim. Biophys. Acta, 387: 461-474.
Arp, D.J. and Burris, R.H., 1981. Kinetic mechanism of the hydrogen-oxidising hydrogenase from soybean nodule bacteroids. Biochem., 20: 2234-2240.
Arp, D.J., McCallum, L.C. and Seefeldt, L.C., 1985. Molecular and immunological comparison of membrane-bound, H_2-oxidising hydrogenases of Bradyrhizobium japonicum, Alcaligenes eutrophus, Alcaligenes latus and Azotobacter vinelandii. J. Bacteriol., 163: 15-20.
Bergersen, F.J., 1982. Root Nodules of Legumes: Structure and Functions. Research Studies Press/J. Wiley & Sons. Chichester, New York, Brisbane, Toronto, Singapore, 164 pp.
Bergersen, F.J., 1983. Oxygen and the physiology of diazotrophic microorganisms. In C. Veeger and W.E. Newton (Editors), Advances in Nitrogen Fixation Research. Martinus Nijhoff/W. Junk, The Hague, Boston, Lancaster, pp. 171-180.
Bergersen, F.J., 1987. Concluding remarks. Phil. Trans. Roy. Soc. London, B, 317: 295-297.
Bergersen, F.J. and Goodchild, D.J., 1973. Aeration pathways in soybean root nodules. Aust. J. Biol. Sci., 26: 729-740.
Bergersen, F.J. and Turner, G.L., 1975. Leghaemoglobin and the supply of O_2 to nitrogen-fixing root nodule bacteroids: presence of two oxidase systems and ATP production at low free O_2 concentration. J. Gen. Microbiol., 91: 345-354.
Bergersen, F.J. and Turner, G.L., 1980. Properties of terminal oxidase systems of bacteroids from root nodules of soybean and cowpea and of N_2-fixing bacteria grown in continuous culture. J. Gen. Microbiol., 118: 235-252.
Bergersen, F.J. and Turner, G.L., 1990a. Bacteroids from soybean root nodules: Respiration and N_2 fixation in flow chamber reactions with oxyleghaemoglobin. Proc. Roy. Soc. Lond. B., 238: 295-320.
Bergersen, F.J. and Turner, G.L., 1990b. Bacteroids from soybean root nodules: poly-ß-hydroxybutyrate accumulates during supply of malate and succinate in flow-chamber reactions, and then supports efficient nitrogen fixation. Proc. Roy. Soc. Lond., B., 238: in press.
Bishop, P.E., Premakumar, R., Joerger, R.D., Jacobson, M.R., Dalton, D.A., Chisnell, J.R. and Wolfinger, E.D., 1988. Alternative nitrogen fixation systems in Azotobacter vinelandii. In: H. Bothe, F.J. de Bruijn and W.E. Newton (Editors), Nitrogen Fixation: Hundred Years After. Gustav Fischer, Stuttgart, New York. pp. 71-79.
Bothe, H., Nelles, H., Hager, K-P., Papen, H. and Neuer, G., 1984. Physiology and biochemistry of N_2 fixation by cyanobacteria. In: C. Veeger and W.E. Newton (Editors) Advances in Nitrogen Fixation Research. Martinus Nijhoff/W. Junk, The Hague, pp. 199-210.
Braaksma, A., Haaker, H., Grande, H.J. and Veeger, C., 1982. The effect of the redox potential on the activity of the nitrogenase and on the Fe-protein of Azotobacter vinelandii. Eur. J. Biochem., 121: 483-491.
Braun-Howland, E.B., Lindblad, P., Nierzwicki-Bauer, S.A. and Bergman, B., 1988. Dinitrogenase reductase (Fe protein) of nitrogenase in the cyanobacterial symbionts of three Azolla species: Localization and sequence of appearance during heterocyst differentiation. Planta, 176: 319-332.
Brockwell, J., Gault, R.R., Morthorpe, L., Peoples, M.B., Turner, G.L. and Bergersen, F.J., 1989. Effects of soil nitrogen status and rate of inoculation on the establishment of populations of Bradyrhizobium japonicum and on the nodulation of soybeans. Aust. J. Agric. Res., 40: 753-762.

Burns, R.C. and Hardy, R.W.F., 1975. Nitrogen Fixation in Bacteria and Higher Plants. Springer-Verlag, Berlin, Heidelberg, New York, pp. 66-67.

Cantrell, M.A., Hickock, R.E. and Evans, H.J., 1982. Identification and characterisation of plasmids in hydrogen-uptake positive and hydrogen-uptake negative strains of Rhizobium japonicum. Arch. Microbiol., 131: 102-106.

Carroll, B.J., Hartley, R., Ashley, D.A., Boerma, H.R., Lawn, R., Byth, D. and Gresshoff, P., 1988. Field evaluation of supernodulating soybean mutants. In: H. Bothe, F.J. de Bruijn and W.E. Newton (Editors), Nitrogen Fixation: Hundred Years After. Gustav Fischer, Stuttgart, p. 808.

Chen, P-C., Almon, H. and Boger, P., 1989. Physiological factors determining hydrogenase activity in nitrogen-fixing heterocystous cyanobacteria. Plant Physiol., 89: 1035-1038.

Ching, T-M., Bergersen, F.J. and Turner, G.L., 1981. Energy status, growth and nitrogenase activity in continuous cultures of Rhizobium sp. strain CB756 supplied with NH_4^+ and various rates of aeration. Biochim. Biophys Acta, 636: 82-90.

Crank, J., 1975. The Mathematics of Diffusion, 2nd edn. Clarendon Press, Oxford, p. 89.

Cunningham, S.D., Kapulnik, Y. and Phillips, D.A., 1986. Distribution of hydrogen metabolizing bacteria in alfalfa field soil. Appl. Env. Microbiol., 52: 1091-1095.

Cunningham, S.D., Kapulnik, Y., Brewin, N.J. and Phillips, D.A., 1985. Uptake hydrogenase activity determined by plasmid pRL6JI in Rhizobium leguminosarum does not increase symbiotic nitrogen fixation. Appl. Env. Microbiol., 50: 791-794.

Daesch, G. and Mortenson, L.E., 1967. Sucrose catabolism in Clostridium pasteurianum and its relation to N_2 fixation. J. Bacteriol., 96: 346-351.

Dakora, F. and Atkins, C.A., 1989. Diffusion of oxygen in relation to structure and function in legume root nodules. Aust. J. Plant Physiol., 16: 131-140.

Dixon, R.O.D., 1967. Hydrogen uptake and exchange by pea root nodules. Ann. Bot., 31: 179-188.

Dixon, R.O.D., 1968. Hydrogenase in pea root nodule bacteroids. Arch. Microbiol., 62: 272-283.

Dixon, R.O.D., 1972. Hydrogenase in pea root nodule bacteroids: occurrence and properties. Arch. Microbiol., 85: 193-201.

Dixon, R.O.D., Blunden, E.A.G. and Searl, J.W., 1981. Intercellular space and hydrogen diffusion in pea and lupin root nodules. Plant Sci. Lett., 27: 109-116.

Döbereiner, J. and De Polli, H., 1980. Diazotrophic rhizocoenoses. In: W.D.P. Stewart and J.R. Gallon (Editors), Nitrogen Fixation. Academic Press, London, New York, pp. 301-333.

Downs, A.J. and Jones, C.W., 1975. Respiration-linked proton translocation in Azotobacter vinelandii. FEBS Letts., 60: 42-46.

Drozd, J.W., Tubb, R.S. and Postgate, J.R., 1972. A chemostat study of the effect of fixed nitrogen sources on nitrogen fixation, membranes and free amino acids in Azotobacter chroococcum. J. Gen. Microbiol., 73: 221-232.

Eady, R.R., Eldridge, M., Kennedy, C., Mitchenall, L.A., Pau, R.N., Walmsley, J. and Miller, R.W., 1988a. Biological significance of non-molybdenum nitrogenases. Ann. Rep. AFRC Inst. Pl. Sci. Res. and John Innes Inst., 107-108.

Eady, R.R., Robson, R.L. and Smith, B.E., 1988b. Alternative and conventional nitrogenases. In: J.A. Cole and S.J. Ferguson (Editors), The Nitrogen and Sulphur Cycles (42nd Symposium of the Society for General Microbiology) Cambridge University Press, Cambridge, pp. 363-382.

Eisbrenner, G. and Evans, H.J., 1982. Spectral evidence for a component involved in hydrogen metabolism of soybean nodule bacteroids. Plant Physiol., 70: 1667-1672.
Emerich, D.W. and Evans, H.J., 1984. Enhanced biological nitrogen fixation in crop plants. In: R.D. Hauck (Editor), Nitrogen in Crop Production. American Society of Agronomy, Crop Science Society of America, Soil Science Society of America, Madison, pp. 133-143.
Esposito, R.G. and Wilson, P.W., 1958. Acetate as a calcium-sparing factor in nitrogen fixation by Azotobacter vinelandii. Proc. Natl. Acad. Sci., USA, 44: 472-476.
Evans, H.J., Emerich, D.W., Lepo, J.E., Maier, R.J., Carter, K.R., Hanus, F.J. and Russell, S.A., 1980. The role of hydrogenase in nodule bacteroids and free-living rhizobia. In: W.D.P. Stewart and J.R. Gallon (Editors), Nitrogen Fixation. Academic Press, London, New York, pp.55-81.
Evans, H.J., Russell, S.A., Hanus, F.J., Papen, H., Sayavedra-Soto, L., Zuber, M. and Boursier, P., 1988. Hydrogenase and nitrogenase relationships in Rhizobium: some recent developments. In: H. Bothe, F.J. de Bruijn and W.E. Newton (Editors), Nitrogen Fixation: Hundred Years After. Gustav Fischer, Stuttgart, pp. 577-582.
Fu, H. and Burris, R.H., 1989. Ammonium inhibition of nitrogenase activity in Herbaspirillum seropedicae. J. Bacteriol., 171: 3168-3175.
Gallon, J.R., 1980. Nitrogen fixation by photoautotrophs. In: W.D.P. Stewart and J.R. Gallon (Editors), Nitrogen Fixation. Academic Press, London, New York, pp.197-238.
Gallon, J.R. and Chaplin, A.E., 1988. Recent studies on N_2 fixation by non-heterocystous cyanobacteria. In: H. Bothe, F.J. de Bruijn and W.E. Newton (Editors), Nitrogen Fixation: Hundred Years After. Gustav Fischer, Stuttgart, pp. 183-188.
Haaker, H. and Klugkist, J., 1987. The bioenergetics of electron transport to nitrogenase. FEMS Microbiol. Rev., 46: 57-71.
Haaker, H., Laane, C., Hellingwerf, K., Houwer, B., Konings, W. and Veeger, C., 1982. Short-term regulation of the nitrogenase activity in Rhodopseudomonas sphaeroides. Eur. J. Biochem., 127: 639-645.
Haaker, H., Wassink, H., Mensink, R. and Veeger, C., 1988. Regulation of whole cell nitrogenase activity in Azotobacter vinelandii. In: H. Bothe, F.J. de Bruijn and W.E. Newton (Editors), Nitrogen Fixation: Hundred Years After. Gustav Fischer, Stuttgart, pp. 243-248.
Häger, K-P., Hundeshagen, B. and Bothe, H., 1988. Nitrogen fixation and reversed respiratory electron transport in Azotobacter and Bradyrhizobium. Loc. cit., pp. 249-255.
Hanus, F.J., Papen, H., Becker, M., Boursier, P., Russell, S.A. and Evans, H.J., 1988. Effect of selenium on hydrogenase induction and autotrophic growth of Bradyrhizobium japonicum. In: H. Bothe, F.J. de Bruijn and W.E. Newton (Editors), Nitrogen Fixation: Hundred Years After. Gustav Fischer, Stuttgart, p.585.
Haury, J.F. and Wolk, C.P., 1978. Classes of Anabaena variabilis mutants with oxygen-sensitive nitrogenase activity. J. Bacteriol., 136: 688-692.
Heckman, M-O. Drevon, J.-J., Saglio, P. and Salsac, L., 1989. Effect of oxygen and malate inhibition of nitrogenase in soybean nodules. Plant Physiol., 90: 224-229.
Herridge, D.F. and Bergersen, F.J., 1988. Symbiotic nitrogen fixation. In: J.R. Wilson (Editor), Advances in Nitrogen Cycling in Agricultural Ecosystems. C.A.B. International, Wallingford, pp.46-65.
Hill, S., 1976. The apparent ATP requirement for nitrogen fixation in growing Klebsiella pneumoniae. J. gen. Microbiol., 95: 297-312.
Hill, S., 1988. How is nitrogenase regulated by oxygen? FEMS Microbiol. Rev., 54: 111-130.

Hoch, G.E., Little, H.N. and Burris, R.H., 1957. Hydrogen evolution from soybean root nodules. Nature, 179: 430-431.
Hunt, S., Gaito, S.T. and Layzell, D.B., 1988. Model of gas exchange and diffusion in legume nodules. Planta, 173: 128-141.
Hwang, J.C., Chen, C.H. and Burris, R.H., 1973. Inhibition of nitrogenase-calalyzed reductions. Biochim. Biophys. Acta, 292: 256-270.
Jones, C.W., 1977. Aerobic respiratory systems in bacteria. In: B.A. Haddock and W.A. Hamilton (Editors), Microbial Energetics (27th Symposium of the Society for General Microbiology). Cambridge University Press, Cambridge University Press, Cambridge, pp.23-59.
Kanemoto, R.H. and Ludden, P.W., 1984. Effect of ammonia, darkness and phenazine methosulfate on whole-cell nitrogenase activity and Fe-protein modification in Rhodospirillum rubrum. J. Bacteriol., 158: 713-720.
Klugkist, J., Haaker, H. and Veeger, C., 1986a. Studies on the mechanism of electron transport to nitrogenase. Eur. J. Biochem., 155: 41-46.
Klugkist, J., Voorberg, J., Haaker, H. and Veeger, C., 1986b. Characterization of three different flavodoxins from Azotobacter vinelandii. Eur. J. Biochem., 155: 33-40.
Klugkist, J., Wassink, H., Haaker, H. and Veeger, C., 1984. On the specific activity of Av2 in vivo. In: C. Veeger and W.E. Newton (Editors), Advances in Nitrogen Fixation Research. Martinus Nijhoff/W. Junk, The Hague, p.154.
Laane, C., Haaker, H. and Veeger, C., 1978. Involvement of cytoplasmic membrane in nitrogen fixation by Rhizobium leguminosarum bacteroids. Eur. J. Biochem., 87: 147-153.
Laane, C., Haaker, H. and Veeger, H., 1979. On the efficiency of oxidative phosphorylation in membrane vesicles of Azotobacter vinelandii and of Rhizobium leguminosarum bacteroids. Eur. J. Biochem., 97: 369-377.
Laane, C., Krone, W., Konings, W., Haaker, H. and Veeger, C., 1980. Short-term effect of ammonium chloride on nitrogen fixation by Azotobacter vinelandii and of bacteroids of Rhizobium leguminosarum. Eur. J. Biochem., 103: 39-46.
Lambert, G.R., Harker, A.R., Zuber, M., Dalton, D.A., Hanus, F.J., Russell, S.A. and Evans, H.J., 1985. Characterization, significance and transfer of hydrogen uptake genes from Rhizobium japonicum. In: H.J. Evans, P.J. Bottomley and W.E. Newton (Editors), Nitrogen Fixation Research Progress. Martinus Nijhoff, Dordrecht, pp. 209-215.
Lee, S.B. and Umbreit, W.W., 1940. Biological oxidation of molecular hydrogen. Zentr. Bakt. Parisitentk. Inf., 101: 354-363.
Leschine, S.B., Holwell, K. and Canale-Parola, E., 1988. Nitrogen fixation by anaerobic cellulolytic bacteria. Science, 242: 1157-1159.
Lindblad, P. and Sellstedt, A., 1989. Immunogold localization of hydrogenase in free-living Frankia CpI1. FEMS Microbiol. Lett., 60: 311-316.
Lissolo, T.,Choi, E.S., Le Gall, J. and Peck, H.D., 1986. The presence of multiple intrinsic membrane nickel-containing hydrogenase in Desulphovibrio vulgaris. Biochem. Biophys. Res. Comm., 139: 701-708.
Lowe, D.J. and Thorneley, R.N.F., 1984. The mechanism of Klebsiella pneumoniae nitrogenase action. Pre-steady-state kinetics of H_2 formation. Biochem. J., 224: 877-886.
Ludden, P.W. Roberts, G.P., Lowery, R.G., Fitzmaurice, W.P., Saari, L.L., Lehman, L., Lies, L., Wochle, D., Wirt, H., Murrell, S.A., Pope, M.R. and Kanemoto, R.H., 1988. Regulation of nitrogenase activity by reversible ADP-ribosylation of dinitrogenase reductase. In: H. Bothe, F.J. de Bruijn and W.E. Newton (Editors), Nitrogen Fixation: Hundred Years After. Gustav Fischer, Stuttgart, pp. 157-162.
Mahler, H.R. and Cordes, E.H., 1966. Biological Chemistry. Harper and Row, New York, Evanston, London, p. 207.

Martin, A.E., Burgess, B.K., Iismaa, S.E., Smart, C.T., Jacobsen, M.R. and Dean, D.R., 1989. Construction and characterization of an Azotobacter vinelandii strain with mutations in the genes encoding flavodoxin and ferredoxin I. J. Bacteriol., 171: 3162-3167.

Maryan, P.S., Eady, R.R., Chaplin, A.E. and Gallon, J.R., 1986. Nitrogen fixation by Gloeothece sp. PCC 6909: respiration and not photosynthesis supports nitrogenase activity in the light. J. Gen. Microbiol., 132: 789-796.

Meesters, T.M., 1987. Localisation of nitrogenase in vesicles of Frankia spp. CCl.17 by immunogold labelling on ultra-thin cryosections. Arch. Microbiol., 146: 327-331.

Mortenson, L.E., 1963. Nitrogen fixation: role of ferredoxin in anaerobic metabolism. Ann. Rev. Microbiol., 17: 115-138.

Murillo, J., Villa, A., Chamber, M. and Ruiz-Argueso, T., 1989. Occurrence of H_2 -uptake hydrogenases in Bradyrhizobium spp. (Lupinus) and their expression in nodules of Lupinus spp., and Ornithopus compressus. Plant Physiol., 89: 78-85.

Murray, M.A. and Lopez, M., 1989. Interaction between hydrogenase, nitrogenase and respiratory activity in a Frankia isolate from Alnus rubra. Canad. J. Microbiol., 35: 636-641.

Parsons, R., 1989. The Physiological Regulation of Nitrogen Fixation in Soybean Root Nodules. Ph.D. Thesis, Australian National University, Canberra.

Parsons, R., Silvester, W.B., Harris, S., Gruitjers, W.T.M. and Bullivant, S., 1987. Frankia vesicles provide inducible and absolute oxygen protection for nitrogenase. Plant Physiol., 83: 728-731.

Pate, J.S., Atkins, C.A. and Rainbird, R.M., 1981. Theoretical and experimental costing of nitrogen fixation and related processes in nodules of legumes. In: A.H. Gibson and W.E. Newton (Editors), Current Perspectives in Nitrogen Fixation. Aust. Acad. Sci. Canberra, pp. 105-116.

Peoples, M.B. and Gibson, A.H., 1989. Nitrogen assimilation, metabolism and utilization in soybean. In: A.J. Pascale (Editor), World Soybean Conference IV, Buenos Aires, pp. 196-211.

Perry, J.H., 1950. Chemical Engineers Handbook, McGraw-Hill, New York, Toronto, London, pp. 539-540.

Peterson, J.B., 1989. Respiratory differences associated with culture aeration in Azotobacter vinelandii. Canad. J. Microbiol., 35: 918-924.

Phelps, A.S. and Wilson, P.W., 1941. Occurrence of hyrogenase in nitrogen-fixing organisms. Proc. Soc. Exp. Biol. Med., 47: 473-476.

Pope, M.R., Murrell, S.A. and Ludden, P.W., 1985. Covalent modification of the iron protein of nitrogenase from Rhodospirillum rubrum by adenosine diphosphorylation of a specific arginyl residue. Proc. Natl. Acad. Sci. USA, 82: 3173-3177.

Puppo, A. and Rigaud, J., 1986. Hypothesis superoxide dismutase: an essential role in the protection of the nitrogen fixation process? FEBS Lett., 201: 187-189.

Rasche, M.E. and Arp, D.J., 1988. Hydrogen inhibition of nitrogen reduction by nitrogenase in isolated soybean bacteroids. In: H. Bothe, F.J. de Bruijn and W.E. Newton (Editors), Nitrogen Fixation: Hundred Years After. Gustav Fischer, Stuttgart, p. 558.

Rigaud, J., Bergersen, F.J., Turner, G.L. and Daniel, R.M., 1973. Nitrate dependent anaerobic acetylene reduction and nitrogen fixation by soybean bacteroids. J. Gen. Microbiol., 77: 137-144.

Robson, R.L., Chesshyre, J.A., Wheeler, C., Jones, R., Woodley, P.R. and Postgate, J.R., 1984. Genome size and complexity in Azotobacter chroococcum. J. Gen. Microbiol., 130: 1603-1612.

Sayavendra-Soto, L.A., Powell, G.K., Evans, H.J. and Morris, R.O., 1988. Nucleotide sequence of the genetic loci encoding subunits of Bradyrhizobium japonicum uptake hydrogenase. In: H. Bothe, F.J. de Bruijn and W.E. Newton (Editors), Nitrogen Fixation: Hundred Years After. Gustav Fischer, Stuttgart, p. 589.
Seefeldt, L.C. and Arp, D.J., 1985. Purification to homogeneity of Azotobacter vinelandii hydrogenase: a nickel and iron-containing αβ dimer. In: H.J. Evans, P.J. Bottomley and W.E. Newton (Editors), Nitrogen Fixation Research Progress. Martinus Nijhoff, Dordrecht, p. 445.
Shah, V.K., Stacey, G. and Brill, W.J., 1983. Electron transport to nitrogenase. Purification and characterization of pyruvate:flavodoxin oxidoreductase, the nif J gene product. J. Biol. Chem., 258: 12064-12068.
Sheehy, J.E. and Bergersen, F.J., 1986. A simulation study of the functional requirements and distribution of leghaemoglobin in relation to biological nitrogen fixation in legume root nodules. Ann. Bot., 58: 121-136.
Sheehy, J.E. Bergersen, F.J., Minchin, F.R. and Witty, J., 1987. A simulation study of gaseous diffusion resistance, nodule pressure gradients and biological nitrogen fixation in soybean nodules. Ann. Bot., 60: 345-351.
Sheehy, J.E., Minchin, F.R. and Witty, J.F., 1985. Control of nitrogen fixation in a legume nodule: an analysis of the role of O_2 diffusion in relation to nodule structure. Ann. Bot., 55: 549-562.
Silvester, W.B., Silvester, J.K. and Torrey, J.G., 1988a. Adaptation of nitrogenase to varying oxygen tension and the role of the vesicle in roots nodules of Alnus incana ssp. rugosa. Canad. J. Bot., 66: 1772-1779.
Silvester, W.B., Whitbeck, J., Silvester, J.K. and Torrey, J.G., 1988b. Growth, nodule morphology and nitrogenase activity of Myrica gale with roots grown at various oxygen levels. Canad. J. Bot., 66: 1762-1771.
Sinclair, T.R. and Goudrian, J., 1981. Physical and morphological constraints on transport in nodules. Plant Physiol., 67: 143-145.
Sorensen, G.M. and Windaele, R., 1986. Effect of transfer of symbiotic plasmid and of hydrogenase (hup) on the symbiotic efficiency of Rhizobium leguminosarum strains. J. Gen. Microbiol., 132: 317-324.
Stouthamer, A.H., 1983. Energy generation and hydrogen metabolism in Rhizobium. In: C. Veeger and W.E. Newton (Editors), Advances in Nitrogen Fixation Research. Martinus Nijhoff/W. Junk, The Hague, pp. 189-197.
Streeter, J., 1988. Inhibition of legume nodule formation and N_2 fixation by nitrate. CRC Critical Reviews in Plant Sciences, 7: 1-23.
Stults, L.W., Moshiri, F. and Maier, R.J., 1986. Aerobic purification of hydrogenase from Rhizobium japonicum by affinity chromatography. J. Bacteriol., 166: 795-800.
Tabita, F.R., Van Baalen, C., Smith, R.L. and Kumar, D., 1988. In vivo protection and recovery of nitrogenase from oxygen in heterocystous cyanobacteria. In: H. Bothe, F.J. de Bruijn and W.E. Newton (Editors), Nitrogen Fixation: Hundred Years After. Gustav Fischer, Stuttgart, pp. 189-194.
Tel-Or, E. and Stewart, W.D.P., 1977. Photosynthetic components and activities of nitrogen-fixing isolated heterocysts of Anabaena cylindrica. Proc. Roy. Soc. London, B, 198: 61-86.
Thorneley, R.N.F. and Lowe, D.J., 1984a. The mechanism of Klebsiella pneumoniae nitrogenase action. Pre-steady-state kinetics of an enzyme-bound intermediate in N_2 reduction and of NH_3 formation. Biochem. J., 224: 887-894.
Thorneley, R.N.F. and Lowe, D.J., 1984b. The mechanism of Klebsiella pneumoniae nitrogenase action. Simulation of the dependences of H_2 evolution rate on component-protein concentration and ratio and sodium dithionite concentration. Biochem. J., 224: 903-909.
Thorneley, R.N.F. and Lowe, D.J., 1985. Kinetics and mechanism of the nitrogenase enzyme system. In: T.G. Spiro (Editor), Molybdenum Enzymes. J. Wiley & Sons, New York, pp. 221-284.

Thorneley, R.N.F., Lowe, D.J. and Deistung, J., 1985. Kinetics and mechanism of the nitrogenase enzyme system. In: H.J. Evans, P.J. Bottomley and W.E. Newton (Editors), Nitrogen Fixation Research Progress. Martinus Nijhoff, Dordrecht, p. 615.

Tibelius, K.H. Robson, R.L. and Yates, M.G., 1987. Cloning and characterization of hydrogenase genes from Azotobacter chroococcum. Mol. Gen. Genet., 206: 285-290.

Tjepkema, J.D. and Yocum, C.S., 1974. Measurement of O_2 partial pressure within soybean nodules by oxygen microelectrode. Planta, 119: 351-360.

Tjepkema, J.D., Pathirana, M.S. and Zeng, S., 1988. The gas diffusion pathway and haemoglobin content of actinorhizal nodules. In: H. Bothe, F.J. de Bruijn and W.E. Newton (Editors), Nitrogen Fixation: Hundred Years After. Gustav Fischer, Stuttgart, p. 701.

Tjepkema, J.D., Ormerod, W. and Torrey, J.G., 1980. Vesicle formation and acetylene reduction activity in Frankia sp. CPI1 cultures in defined nutrient media. Nature, 287: 633-635.

Torrey, J.G., 1985. The site of nitrogense in Frankia in free-living culture and in symbiosis. In: H.J. Evans, P.J. Bottomley and W.E. Newton (Editors), Nitrogen Fixation Research Progress. Martinus Nijhoff, Dordrecht, pp. 293-299.

Tunlid, A., Schultz, N.A., Benson, D.R., Steele, D.B. and White, D.C., 1989. Differences in fatty acid composition between vegetative cells and N_2-fixing vesicles of Frankia sp. strain CpI1. Proc. Natl. Acad. Sci. USA, 86: 3399-3403.

Van Berkum, P. and Maier, R.J., 1988. Lack of carbon substrate repression of uptake hydrogenase activity in Bradyrhizobium japonicum S.R. J. Bacteriol., 170: 1962-1964.

Vignais, P.M., Colbeau, A., Willison, J.C. and Jonaneau, Y., 1985. Hydrogenase, nitrogenase and hydrogen metabolism in the photosynthetic bacteria. Adv. Microbiol. Physiol., 26: 155-234.

Walker, C.C. and Yates, M.G., 1978. The hydrogen cycle in nitrogen fixing Azotobacter chroococum. Biochimie, 60: 225-231.

Wilson, P.W., 1940. The Biochemistry of Symbiotic Nitrogen Fixation. Univ. of Wisconsin Press, 302 pp.

Witty, J.F., Minchin, F.R., Skφt, L. and Sheehy, J.E., 1986. Nitrogen fixation and oxygen in legume root nodules. In: B.J. Miflin (Editor), Oxford Surveys of Plant Molecular and Cell Biology. Oxford University Press, Oxford, **3**: pp. 275-314.

Yates, M.G., 1977. Physiological aspects of nitrogen fixation. In: W.E. Newton, J.R. Postgate and C. Rodriguez-Barrueco (Editors), Recent Developments in Nitrogen Fixation. Academic Press, London, pp. 219-270.

Yates, M.G., 1988. The role of oxygen and hydrogen in nitrogen fixation. In: J.A. Cole and S.J. Ferguson (Editors), The Nitrogen and Sulphur Cycles. 42nd Symp. Soc. Gen. Micribiol. Cambridge University Press, Cambridge, pp. 383-416.

Yoch, D.C. and Arnon, D.I., 1972. Two biologically active ferredoxins from the aerobic nitrogen-fixing bacterium, Azotobacter vinelandii. J. Biol. Chem., 247: 4514-4520.

Zuber, M., Harker, A.R., Sultana, M.A. and Evans, H.J., 1986. Cloning and expression of Bradyrhizobium japonicum uptake hydrogenase structural genes in Escherichia coli. Proc. Natl. Acad. Sci. USA, 83: 7668-7672.

Zumft, W.G., Mortenson, L.E. and Palmer, G., 1974. Electron-paramagnetic-resonance studies on nitrogenase. Investigation of the oxidation-reduction behaviour of azoferredoxin and molybdoferredoxin with potentiometric and rapid-freeze techniques. Eur. J. Biochem., 46: 525-535.

CHAPTER 6

GENETICS AND REGULATION OF Mo-NITROGENASE

C. ELMERICH
Unité de Physiologie Cellulaire and CNRS URA 1300, Département des Biotechnologies, Institut Pasteur, 25 rue du Dr. Roux, 75724 Paris Cedex 15, France.

1 GENES AND PRODUCTS REQUIRED FOR Mo-NITROGENASE ACTIVITY

The enzyme complex referred to as Mo-nitrogenase is highly conserved in free-living and symbiotic diazotrophs and consists of two components. Component 1, also designated MoFe protein, is an $\alpha_2\beta_2$ tetramer of 220,000 D. Component 2, also called the Fe protein, is a dimer of 68,000 D formed from identical subunits (see Chapter 2). The Mo nitrogenase polypeptides are encoded by three structural genes, nifDK for component 1 and nifH for component 2. The formation of a functional enzyme requires processing of the gene products and a specific cofactor, termed FeMoco, which is bound to component 1.

1.1 Klebsiella pneumoniae as a model system.

(i) Early genetics. The availability of genetic tools in the Enterobacteriaceae facilitated genetic analysis of nitrogen fixation in K. pneumoniae strain M5a1 in the early seventies. It was rapidly established, by transduction with the bacteriophage P1 (Streicher et al., 1972) and by chromosome mobilization with R factors (Dixon and Postgate, 1972), that all the genetic information for nitrogen fixation (nif genes) was clustered on the chromosome near the histidine operon. The transfer of K. pneumoniae nif genes to the non-nitrogen fixer Escherichia coli, which thereby acquired the property to fix nitrogen (Dixon and Postgate, 1972), allowed the construction of F' episomes and R plasmids containing the whole nif cluster. Thus plasmid pRD1, containing the entire nif and his clusters of K. pneumoniae was constructed from RP4 (Dixon et al., 1976). This plasmid was transferable to most Gram negative bacteria, including K. pneumoniae, permitting analysis of the regulation of the nif cistrons by construction of partial diploids.

(ii) Transcriptional organization of the nif genes. The genetic analysis of the nif cluster involved the isolation of several hundred chromosomal or pRD1 Nif^- mutants after chemical or transposon mutagenesis. Genetic complementation analysis and physical mapping of Mu and Tn5 insertions led to the characterization of 14/15 *nif* cistrons organized in 8 transcriptional units (*nifJ*, *nifHDK*, *nifEN*, *nifUSV*, *nifM*, *nifF*, *nifL/A*, *nifBQ*) and to the establishment of a genetic and physical map of the *nif* region (Fig. 1) (McNeil et al., 1978; Riedel et al., 1979; Merrick et al., 1980; Pühler and Klipp, 1981; Sibold, 1982). Development of techniques in molecular biology led to the subcloning of *nif* restriction fragments in multicopy vectors and to the construction of a fine physical map (Cannon et al., 1977, 1979; Pühler and Klipp, 1981). The complete nucleotide sequence of the 24 kb DNA region containing the *nif* genes is now established (Arnold et al., 1988). Promoter mapping led to the characterization of a novel type of promoter, whose activation depends on a specific sigma factor (σ^{54}) encoded by *ntrA*, and which is present upstream from *nif* operons (see Part 2).

The relationship between genes, products and functions is reported in Table 1. The products of the *nif* genes were identification by comparison of polypeptides synthesized under nitrogen-fixing conditions by the wild type strain and deletion or polar mutants (Roberts et al., 1978), and by analysis of polypeptides produced by cloned *nif* DNA fragments in minicells (Pühler and Klipp, 1981). The latter technique enabled the characterization of new polypeptides encoded by *nifY* and *nifX*, that had not previously been identified as *nif* cistrons. Three additional genes, *nifT*, *nifW* and *nifZ*, were identified on the basis of the nucleotide sequence (Paul and Merrick, 1987; Arnold et al., 1988; Beynon et al., 1988a). The nucleotide sequence confirmed the transcriptional organization in operons and showed that the two monocistronic units, *nifF* and *nifJ*, were transcribed in an orientation opposite to that of the other *nif* operons (Drummond, 1985; Arnold et al., 1988). It also revealed an interesting homology between the *nifDKY* and *nifENX* products, respectively, possibly suggesting comparable functions (Beynon et al., 1988a, Arnold et al., 1988). This finding was in agreement with a previous report of homology between the *nifDK* and *nifEN* products of *Azotobacter vinelandii* deduced from the nucleotide sequence (Dean and Brigle, 1985; Brigle et al., 1987).

As discussed in Part 2, regulation of *nif* gene expression is controlled by a cascade regulation system, which involves the *nifLA* operon and five genes playing a role in the regulation of nitrogen source utilization in enteric bacteria, *ntrA*, *ntrB*, *ntrC*, *glnB* and *glnD*.

TABLE 1. Features of products coded by _nif_ genes or ORFs in _K. pneumoniae_ and _A. vinelandii_.

nif gene or ORF	M_w deduced from nucleotide sequence (D) _K. pneumoniae_(1)	_A. vinelandii_(2)	Proposed function
ORF12	-	?	unknown (2)
J	128,038	-	Electron transport, pyruvate flavodoxin odxidoreductase (3,4)
H	31,902	31,496	Component 2 (5,6,7)
D	54,139	55,267	Component 1 α-subunit (5,6,7)
K	58,406	59,438	Component 1 β-subunit (5,6,7)
T	8,321	8,038	unknown (1,2)
Y	24,691	26,683	unknown (1,2)
ORF1	-	9,552	unknown (2)
ORF2	-	27,963	unknown (2)
E	50,153	52,141	Synthesis of FeMoco (8)
N	50,567	49,187	Synthesis of FeMoco (8)
X	18,229	17,270	Regulation (9)
ORF3	-	18,087	unknown (2)
ORF4	-	7,855	unknown (2)
ORF11	-	7,985	unknown (2)
ORF10	-	39,627	unknown (2)
ORF5	-	27,491	unknown (2,10)
ORF6	-	11,029	unknown (2,10)
U	29,494	33,251	Maturation of Component 2 (10)
S	43,175	43,578	Maturation of Component 2 (5,10)
V	41,078	41,633	Synthesis of FeMoco, homocitrate synthase (11,12)
ORF7	-	28,327	unknown (2,10)
ORF8	-	19,899	unknown (2,10)
W	10,179	13,399	Maturation of Component 1 (10,13)
Z	16,659	17,774	Maturation of Component 1 (10,13)
M	30,613	32,782	Maturation of Component 2 (10,14,15)
ORF9	-	47,961	unknown (10)

nif gene or ORF	M_w deduced from nucleotide sequence (D)		Proposed function
	K. pneumoniae(1)	A. vinelandii(2)	
F	19,112	19,643	Electron transport, flavodoxin (16,17)
L	55,213	ORF13?	Regulation (18,19,20)
A	58,632	58,100	Regulation (20,21,22)
B	50,855	544,447	Synthesis of FeMoco (5,23)
fdx	-	9,603	Ferredoxin (23)
ORF14	-	16,405	unknown (23)
Q	19,665	22,389	Processing of Mo (23,24,25)
ORF15	-	?	unknown (23)

(1) Arnold et al, (1980); (2) Jacobson et al, (1989a); (3) Bogusz et al, (1981); (4) Shah et al, (1983); (5) Roberts and Brill (1981); (6) Robson et al, (1983); (7) Eady (1986); (8) Paustian et al, (1989); (9) Gosink et al, (1990); (10) Jacobson et al, (1989b); (11) Hawkes et al, (1984); (12) Hover et al, (1987); (13) Paul and Merrick (1989); (14) Howard et al, (1986); (15) Paul and Merrick (1987); (16) Nieva Gomez et al, (1985); (17) Deistung and Thornley (1986); (18) Hill et al, (1981); (19) Merrick et al, (1982); (20) Bennett et al, (1988); (21) Dixon et al, (1977); (22) Buchanan-Wollaston et al, (1981b); (23) Joerger and Bishop (1988); (24) Imperial et al, (1984); (25) Buikema et al (1987).

1.2 The Mo-nitrogenase structural genes (nifHDK) in free-living and symbiotic nitrogen fixers.

(i) Homology between nitrogenase structural genes. In 1980, Ruvkun and Ausubel, and Mazur et al., using the nifHDK cluster of K. pneumoniae as a probe in Southern experiments, detected hybridization bands with total DNA from 13 nitrogen-fixers, including free-living and symbiotic eubacteria. This technique was subsequently utilized to identify and to clone nif DNA from a large variety of microorganisms and enabled identification of nif genes in archaebacteria (Table 2). Nitrogen fixation in methanogens was discovered only recently (Belay et al., 1984; Murray and Zinder, 1984). Homology to nifH probes from K. pneumoniae and Anabaena was detected by Southern hybridization in 14 species of methanogens (Sibold et al., 1985; Possot et al., 1986).

(ii) Organization of the nifHDK genes in diazotrophs. In most organisms, the nifHDK genes are contiguous and organized in a single transcriptional unit. However, several species have different arrangements and in some cases there are multiple copies of nifHDK or nifH (Table 2). Multiple hybridization bands corresponding to nifH-like sequences were first detected in Azotobacter (Ruvkun and Ausubel, 1980; Robson et al., 1986), Anabaena (Rice et al., 1982) and

TABLE 2. Organisation of the *nifH*, *D*, and *K* genes in several free-living and symbiotic nitrogen fixers.

Species	Gene organisation	Reiteration
Eubacteria		
Anabaena sp. 7120 (1)	HD' -' DK	H_2
Azorhizobium caulinodans (2)	HDK	H_2
Azospirillum brasilense (3)	HDK	-
Azotobacter chroococcum (4)	HDK	vnfH-DGK
Azotobacter vinelandii (5)	HDK	vnfH-DGK anfHDGK
Bradyrhizobium japonicum (6)	H-DK	-
Clostridium pasteurianum (7)	HDK	H_2-H_3-H_4-H_5-H_6
Desulfovibrio gigas (8)	H	
Frankia (9)	HDK	-
Klebsiella pneumoniae (10)	HDK	-
Rhizobium leguminosarum (11)	HDK	-
Rhizobium meliloti (12)	HDK	-
Rhizobium sp. [*Parasponia*] (13)	H-DK	-
Rhizobium phaseoli (14)	HDK	$H_2D_2K_2$-H_3
Rhizobium trifolii (15)	HDK	-
Rhodobacter capsulatus (16)	HDK	$H_2D_2K_2$ $H_3D_3K_3$
Thiobacillus ferrooxidans (17)	HDK	-
Archaebacteria		
Methanobacterium ivanovii (18)	ORFH	
Methanococcus thermolithotrophicus (19)	H-DK	
Methanococcus voltae (20)	ORFH	H_2

Ligated letters correspond to contiguous genes. Non contiguous genes are separated by a dash. Genes whose nucleotide sequence has been established are underlined. *vnf* and *anf* correspond to genes coding for alternative nitrogenases. (1) Mevarech et al, (1980), Mazur and Chui (1982), Lammers and Haselkorn (1983), Haselkorn (1986); (2) Denèfle et al,(1987), Norel and Elmerich (1987); (3) de Zamaroczy et al, (1989), Fani et al,(1989); (4) Robson et al, (1986), Robson et al, (1989); (5) Brigle et al, (1985), Joerger et al, (1989); (6) Fuhrmann and Hennecke (1984), Fischer and Hennecke (1984), Kaluza and Hennecke (1984), Thöny et al, (1985); (7) Chen et al, (1986), Wang et al, (1988); (8) Kent et al, (1989); (9) Normand et al, (1988), Normand and Bousquet (1989); (10) Sundaresan and Ausubel (1981), Scott et al (1981), Arnold et al, (1988); (11) Krol et al, (1982); (12) Török and Kondorosi (1981), Ruvkun et al, (1982), Corbin et al, (1983); (13) Scott et al, (1983a); (14) Quinto et al, (1982), Quinto et al, (1985); (15) Scott et al, (1983b); (16) Avtges et al, (1983), Scolnik and Haselkorn (1984), Schumann et al, (1986), Jones and Haselkorn (1988); (17) Pretorius et al, (1987); (18) Souillard et al, (1988); (19) Souillard et al, (1988), Soulliard and Sibold (1989); (20) Soulliard and Sibold (1986).

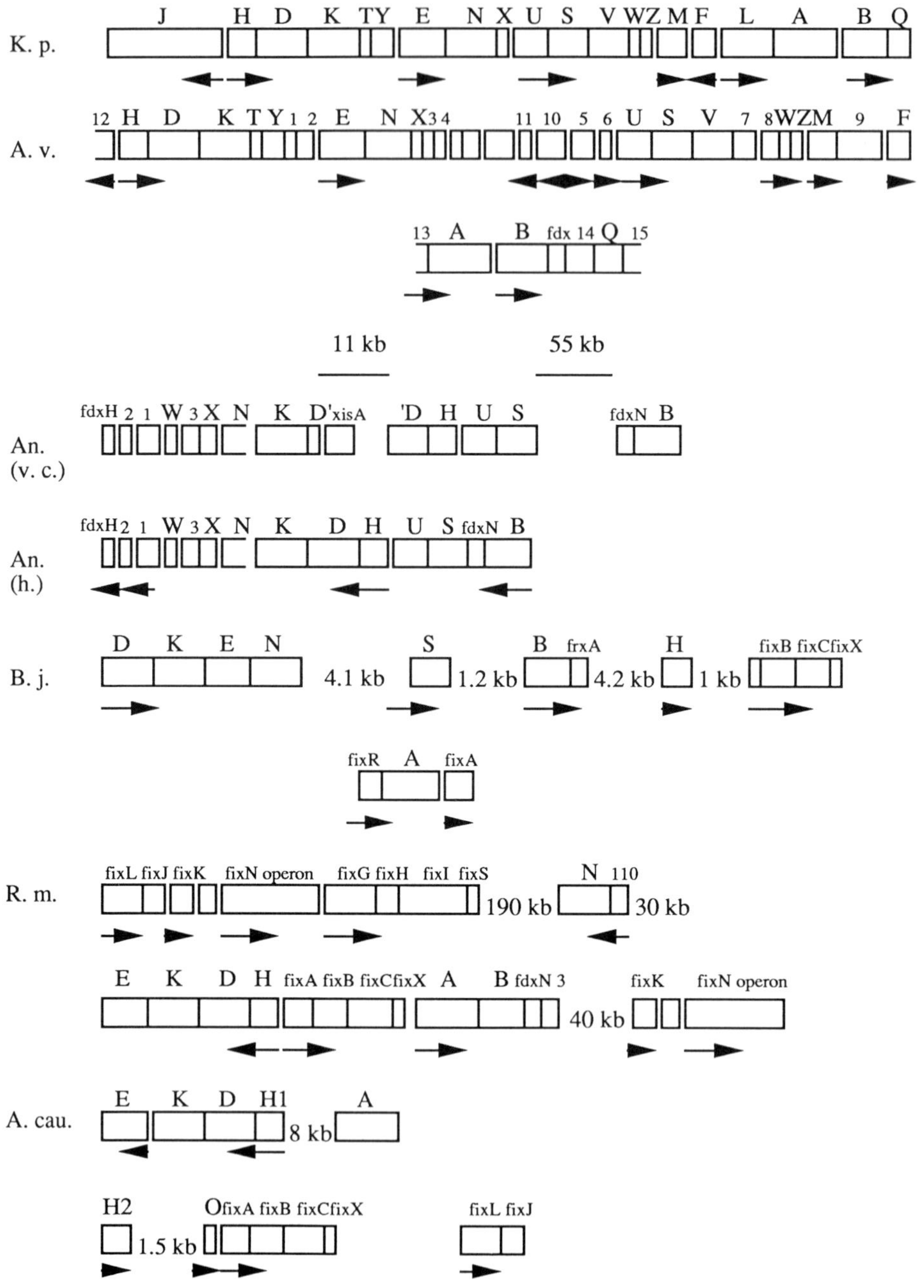
K. p.
J H D K TY E N X U S V WZ M F L A B Q
A. v.
12 H D K T Y 1 2 E N X 3 4 11 10 5 6 U S V 7 8 W Z M 9 F
13 A B fdx 14 Q 15
11 kb
55 kb
An. (v. c.)
fdxH 2 1 W 3 X N K D' xisA 'D H U S fdxN B
An. (h.)
fdxH 2 1 W 3 X N K D H U S fdxN B
B. j.
D K E N 4.1 kb S 1.2 kb B frxA 4.2 kb H 1 kb fixB fixC fixX
fixR A fixA
R. m.
fixL fixJ fixK fixN operon fixG fixH fixI fixS 190 kb N 110 30 kb
E K D H fixA fixB fixC fixX A B fdxN 3 40 kb fixK fixN operon
A. cau.
E K D H1 8 kb A
H2 1.5 kb O fixA fixB fixC fixX fixL fixJ

Fig. 1. Organization of nif and related genes in several free-living and symbiotic diazotrophs. Contiguous open boxes correspond to genes belonging to the same transcriptional unit. Capital letters correspond to nif genes. Identified ORFs are abbreviated by numbers (e.g. Anabaena ORF= 1). Bars above Anabaena (v.c.) genes indicate regions which are excised in heterocysts. Arrows indicate the position of the transcriptional start site and the direction of transcription.
K.p.: Klebsiella pneumoniae (Arnold et al., 1988); A.v.: Azotobacter vinelandii (Bennett et al., 1988; Joerger and Bishop, 1988; Jacobson et al., 1989a); An.(v.c.): Anabaena strain 7120 vegetative cells, (h): heterocysts (Rice et al., 1982; Golden et al., 1985; Böhme and Haselkorn, 1988; Mulligan et al., 1988; Borthakur et al., 1990; Haselkorn, unpublished); B.j.: Bradyrhizobium japonicum (Hennecke et al., 1988; Gubler et al., 1989); R.m.: Rhizobium meliloti Buikema et al., 1987; Batut et al., 1989; Klipp et al., 1989); A.cau.: Azorhizobium caulinodans (Denèfle et al., 1987; Pawlowski et al., 1987; Kaminski et al., 1988; C. Elmerich et al., unpublished).

Rhizobium leguminosarum bv. phaseoli (hereafter called R. phaseoli)(Quinto et al., 1982). The three bands in R. phaseoli corresponded to two nifHDK operons and one nifH gene. The three nifH copies were identical and at least two of themwere functional (Quinto et al., 1985). This was also the case for the two nifH copies of Azorhizobium caulinodans (Norel and Elmerich, 1987). In contrast, the six ORFnifH sequences of Clostridium pasteurianum are non-identical; though five of them are transcribed, only the copy adjacent to nifDK encodes the component 2 subunit (Wang et al., 1988). Only one of the nifHDK copies was functional in Rhodobacter capsulatus, but other copies could be activated after mutation (Scolnik and Haselkorn, 1984). Interestingly, the reiteration of nifH in Azotobacter corresponds to the alternative V nitrogenase Fe protein (Robson et al., 1986; see Chapter 7). The second nifH copy in Anabaena, ORFnifH*, had an amino acid sequence nearly identical to nifH and both were transcribed under conditions allowing nitrogen fixation (Haselkorn, 1986). Two ORFnifH sequences were cloned from Methanococcus thermolithotrophicus (Souillard et al., 1988; Souillard and Sibold, 1989); one of them (ORFnifH1) is separated from ORFs homologous to nifD and nifK by two ORFs of 105 and 128 bp, respectively, and nifH1 is not co-transcribed with nifDK (Souillard and Sibold, 1989). In addition, the second nifH copy (ORFnifH2), is not expressed under conditions suitable for nitrogen fixation (Souillard and Sibold, 1989).

(iii) DNA rearrangement in cyanobacteria. Nitrogen fixation in Anabaena takes place in differentiated cells termed heterocysts (Haselkorn, 1986). In strain 7120, nifH and nifD are contiguous and separated from nifK by 11 kb. During heterocyst differentiation, excision of the 11 kb fragment leads to the restoration of a nifHDK operon (Fig. 1) (Golden et al., 1985; Lammers et al., 1986). The rearrangement results from site-specific recombination between directly repeated 11 bp sequences located at each extremity of the excised DNA fragment (Lammers et al., 1986). The rearrangement was observed in E. coli carrying a plasmid containing the nifHD-K vegetative region. This led to the identification of a gene xisA, which is located in the excised region and is required for the rearrangement (Lammers et al., 1986). A mutant of Anabaena carrying an inactivated xisA could neither excise the 11 kb fragment during heterocyst differentiation nor fix nitrogen (Golden and Wiest, 1988). Rearrangements similar to that observed in Anabaena may occur in other filamentous cyanobacteria (see Chapter 16) but not in unicellular strains (Kallas et al., 1983; Haselkorn, 1986). A second rearrangement removing a 55 kb region upstream from nifS was also found in Anabaena (Golden et al., 1985; Golden et al., 1988) (Fig. 1). This rearrangement was not dependent on xisA (Golden and Wiest, 1988).

(iv) Conservation of nifH nucleotide sequences. The nucleotide sequence of nifH is known for more than 20 different species of nitrogen-fixing organisms (Table 2). Amino acid sequence comparisons reveal highly conserved regions, in particular the sites of cysteine residues and the sequences around cysteine residues, ATP binding sites and ADP-ribosylation sites (Fuhrmann and Hennecke, 1984; Robson, 1984; Eady, 1986; Souillard and Sibold, 1986; Normand et al., 1988; Souillard and Sibold, 1989). The phylogenic tree in eubacteria deduced from nifH homology is similar to that based on 16S rRNA homology, suggesting that nifH has evolved in a similar way (Hennecke et al., 1985). However, recent analysis of a large number of nifH sequences, including those from methanogens, Frankia, and Clostridium and the Azotobacter alternative Fe protein sequence, does not exclude the possibility of lateral transfer in Frankia and Anabaena about 100 Myr ago (Normand and Bousquet, 1989).

1.3 Common and additional genes involved in Mo nitrogenase-dependent nitrogen fixation in other free-living or symbiotic diazotrophs.

Current knowledge of nif gene organization and function is limited to a few prokaryotes. In most cases, genes equivalent to those found in K. pneumoniae were found, but additional genes were discovered both in free-living and symbiotic nitrogen-fixing bacteria. Because most of the letters of the alphabet had already been used for K. pneumoniae nif genes, there was a problem of nomenclature. In addition, in the case of symbiotic nitrogen fixers, it was difficult to decide whether a newly identified gene or open reading frame was encoding a product required for Mo nitrogenase activity or for nodule functioning. In general, those genes involved in symbiotic nitrogen fixation which do not correspond to a K. pneumoniae nif gene are designated fix. Genetic maps of nif clusters from Anabaena 7120, A. caulinodans, A. vinelandii,

Bradyrhizobium japonicum, K. pneumoniae and R. meliloti are presented in Fig. 1. Though several genes are common to all systems, the organization differs between genera.

(i) Azotobacter main nif cluster. The development of the genetics of Azotobacter was recently reviewed (Kennedy and Toukdarian, 1987). As well as the conventional Mo nitrogenase, a V nitrogenase was discovered in both A. vinelandii and Azotobacter chroococcum and a third nitrogenase in A. vinelandii (Bishop et al., 1988; Eady et al. 1988). I shall limit this review to Mo nitrogenase (but see Chapter 7). Azotobacter is an aerobic free-living diazotroph. Most strains harbour plasmids, but the nif genes are not plasmid-borne (Robson et al, 1984). The organization of nif genes in A. vinelandii (Fig. 1) is similar, if not identical, to that in A. chroococcum (Evans et al., 1988; Merrick, 1988a). The nucleotide sequence of a 29 kb DNA region containing the main nif cluster has been established (Brigle et al., 1985; Dean and Brigle, 1985; Beynon et al., 1987; Bennett et al., 1988; Jacobson et al., 1989a). The genes are in the same order as in K. pneumoniae and 12 new ORFs have been identified interspersed between known nif genes. In general, the amino acid sequences deduced from the nucleotide sequences of A. vinelandii nif genes share a high degree of homology with those of K. pneumoniae (see Jacobson et al., 1989a). However, in the case of nifM, identified on the basis of functional complementation of a K. pneumoniae nifM mutant (Kennedy et al., 1986), amino acid conservation was only 16 % (Jacobson et al., 1989a). Strains with deletions in nifE, nifN, nifU and nifM were Nif$^-$, suggesting that the four gene products are essential for nitrogen fixation, whereas the requirement for nifS and nifV was not as critical (Jacobson et al., 1989a, b).

(ii) Additional open reading frames. It is likely that the additional ORFs in the main cluster of A. vinelandii are not essential for nitrogen fixation. The locations of potential promoters matching the NtrA-dependent consensus promoter are arrowed in Fig. 1. Thus, ORFs 1, 2, 3, 4, 8 and 9 belong to operons containing nif genes. Deletion of nifT, nifY and ORFs 1 and 2 does not affect nitrogen fixation. ORFs 5, 6, 11 and 10 are preceded by NtrA-dependent promoters, suggesting that they may be transcribed under nitrogen-fixing conditions. Deletion of the DNA region between ORFs 4 and 5 had no effect on the Nif phenotype, suggesting that this region (including ORFs 10 and 11) is not involved in diazotrophic growth (Jacobson et al., 1989a). Similarly, it was shown that ORFs 5, 6, 7, 8 and 9 were not essential for diazotrophic growth (Jacobson et al., 1989a, b). However, it was found that polar mutations in ORFs 6 and 8 abolished nitrogen fixation, suggesting that potential NtrA-dependent promoters located upstream from these ORFs were used in transcription of nifUSV and nifWZM (Jacobson et al., 1989a, b). In addition, an ORF homologous to ORF3 was identified at the same location downstream from nifENX in R. capsulatus (Moreno-Vivian et al., 1989a), and ORF6 of Azotobacter corresponds to ORF2 (located near fdxH) in Anabaena (Borthakur et al., 1990). An ORF homologous to ORF6 was also found in B. japonicum (Jacobson et al., 1989b).

(iii) The second cluster of A. vinelandii. The second cluster contains nifA, nifB, nifQ and three ORFs, one of them encoding a ferredoxin (fdx) (Joerger and Bishop, 1988). Ferredoxin-encoding genes are also present downstream from nifB (Fig. 1) in Anabaena (fdxN) (Mulligan et al., 1988), B. japonicum (frxA) (Ebeling et al., 1988), R. meliloti (fdxN) (Buikema et al., 1987), Rhizobium leguminosarum bv. trifolii (hereafter R. trifolii (fdxN, not shown) (Iisma et al., 1989). The gene products are similar (Mulligan et al. 1988; Iisma et al., 1989). NifA is cotranscribed with an unidentified gene whose complete sequence has not yet been established (Bennett et al., 1988). However, the C-terminal part of the gene product shows homology with the primary sequences of K. pneumoniae NtrB and NifL, and of R. leguminosarum DctB (Bennett et al., 1988).

(iv) Nif genes in photosynthetic bacteria and in Azospirillum. Two main clusters are present in R. capsulatus and in A. brasilense (not shown). The first cluster of R. capsulatus contains nifHDK, ntrA, nifA, nifB and the second contains nifENXORF3, nifQ, nifS, nifA, and nifB (Klipp et al., 1988; Masepohl et al., 1988; Moreno-Vivian et al., 1989a, b). A 30 kb DNA region containing nifHDK, nifNE and nifUS from A. brasilense has been characterized by hybridization with heterologous probes and transposon mutagenesis (Galimand et al., 1989). Another region containing nifA and nifB has been identified (C. Elmerich et al., unpublished).

(v) Nif gene products required for symbiotic nitrogen fixation. Large plasmids have been detected in most Rhizobium species (see Prakash and Atherly, 1986). Hybridization experiments with K. pneumoniae and R. meliloti nifHDK genes localized nifHDK on plasmids of Rhizobium leguminosarum bv. viciae (hereafter R. leguminosarum), R. meliloti, R. phaseoli and R. trifolii (Nuti et al., 1979; Banfalvi et al. 1981; Prakash et al., 1981; Rosenberg et al., 1981). In addition, analysis of nodulation mutants of R. meliloti carrying plasmid deletions showed that the same plasmid (the symbiotic plasmid or pSym) carried both nitrogenase structural genes and nodulation genes (Banfalvi et al., 1981; Rosenberg et al. 1981). In contrast, nifHDK genes are not plasmid-borne in Bradyrhizobium (Prakash and Atherly, 1984) and no plasmid was found in the fast-growing A. caulinodans. Tn5 transposon site-directed mutagenesis, DNA hybridization experiments, and nucleotide sequencing led to the characterization of symbiotic nif genes corresponding to K. pneumoniae nifA, nifB, nifE, nifN or nifS (Fig. 1). A gene homologous to nifB was also detected in A. caulinodans (not shown) (Donald et al., 1986). Genes previously designated fixD, fixF and fixZ correspond to nifA, nifN and nifB respectively (Rossen et al., 1984; Aguilar et al., 1985; Weber et al., 1985).

(vi) Fix gene products required for Mo-nitrogenase activity. A fix locus which did not correspond to K. pneumoniae nif genes was identified in R. meliloti (Ruvkun et al., 1982; Corbin et al. 1983; Dusha et al., 1987). The fix region was preceded by an NtrA-dependent promoter and transcribed in the opposite orientation to nifHDK (Corbin et al., 1983). Analysis of the gene products in E. coli minicells led to the identification of three genes, termed fixABC (Pühler et al., 1984). The nucleotide sequence of fixABC (coding for polypeptides of 31,146, 37,786 and 47,228 D, respectively) was established, and a new gene fixX (encoding a polypeptide of 10,937 D) was identified downstream from fixC (Earl et al., 1987). The same organization was found in R. leguminosarum (Grönger et al., 1987) and R. trifolii (Iisma et al., 1989), where the order of nif and fix genes in the nifHDK region is similar to that in R. meliloti (Fig. 1). However, the nifHDKE cluster is separated from the fixABC-nifAB cluster in R. leguminosarum PRE (Hontelez et al., 1989). A novel gene (fixW, coding for a 24,9 kD polypeptide), precedes fixA. The fixABC operon can be transcribed from the fixA promoter or cotranscribed with fixW (Hontelez et al., 1989). As shown in Fig. 1, organization of the fixABCX genes differs also in B. japonicum and A. caulinodans. FixA is separated from fixBCX in B. japonicum and fixB is preceded by an ORF encoding a polypeptide of 35 amino acids (Fuhrmann et al., 1985; Gubler and Hennecke, 1988, Gubler et al., 1989). FixABCX are contiguous in A. caulinodans and a novel gene nifO (coding for a 12,136 D polypeptide) is located uspstream from fixA and transcribed from its own promoter (Kaminski et al., 1988; C. Elmerich et al., unpublished).

(vii) Other fix genes. Another fix region was characterized on the pSym of R. meliloti, separated by 200 kb or more from the nifHDK region (Fig. 1) (Batut et al., 1985; Renalier et al., 1987). Analysis of transcription during symbiosis revealed that the nifHDK region was transcribed under nifA control, whereas expression of the second fix cluster was independent of nifA (David et al., 1987). This suggested the existence of other regulatory genes which were identified as fixLJ (David et al., 1987). These two genes belong to the same family of sensor-transactivator proteins as ntrBC and nifLA (David et al., 1988). The fix region spans 12 kb and contains fixGHIS, fixK and the fixN operon (Kahn et al., 1989). These genes are required for the formation of an effective nodule but they may not be directly involved in Mo nitrogenase functioning. The nucleotide sequence of the fixGHIS operon suggests that the four products may be membrane bound, fixG encoding a redox protein, and fixI a cation ATPase (Kahn et al., 1989). Hybridization experiments show that this operon is conserved in most Rhizobium species, Bradyrhizobium, and Azorhizobium (Kahn et al., 1989). The second copies of fixK and the fixN operon are reiterations (Fig. 1). Deletion of the two copies is necessary to yield a Fix⁻ phenotype (Renalier et al., 1987; Batut et al., 1989). The fixK gene is similar to fnr and crp of E. coli (Batut et al., 1989). A DNA region containing genes highly homologous to R. meliloti fixLJ has been cloned from A. caulinodans (Fig. 1). The regulatory roles of fixLJ and fixK are discussed in Part 2.

1.4 Function of the Nif and Fix proteins.

(i) Nif proteins. The role of nif products was initially established in K. pneumoniae and Azotobacter, on the basis of biochemical complementation of crude extracts of Nif⁻ mutants with purified components 1 or 2, or FeMoco (see Brill, 1980; Roberts and Brill, 1981; Robson et al. 1983; Eady, 1986). The only nif products purified from K. pneumoniae are the nitrogenase components and the nifF and nifJ products; the latter are components of the electron transport chain and are presumably not conserved in other nitrogen fixers (Eady, 1986). Development of the genetics of A. vinelandii enabled further characterization of nif gene functions and comparison with K. pneumoniae (Table 1). Purification of NifN and NifE from A. vinelandii was recently achieved (Paustian et al., 1989). The native protein (NifNE) is a complex of M_r 200,000 D containing equimolar amounts of NifN and NifE subunits (Paustian et al., 1989). This protein is involved in synthesis of FeMoco, together with nifBQ, nifV and nifH products. K. pneumoniae nifV mutants cannot grow on N-free medium but are still able to reduce acetylene, due to the synthesis of a component 1 with a defective FeMoco (McLean and Dixon, 1981; Hawkes et al., 1984). It was subsequently found that homocitrate is required for FeMoco biosynthesis and that nifV encodes a homocitrate synthase (Hoover et al., 1987, 1988). Analysis of a nifV deletion mutant of A. vinelandii suggested that this gene plays a role similar to that in K. pneumoniae; addition of homocitrate to the growth medium does not restore the wild phenotype to nifV mutants (Jacobson et al., 1989b). Deletion of A. vinelandii nifUS led to mutants with very low component 2 activity, suggesting that NifU and NifS are required for activation and/or stability of the Fe protein, as proposed for NifS in K. pneumoniae (Roberts et al., 1978; Jacobson et al., 1989b). A role for K. pneumoniae NifM in Fe protein maturation has been proposed (Roberts et al., 1978) and confirmed by the development of an in vivo assay for component 2 maturation in E. coli (Howard et al., 1986; Paul and Merrick, 1987). A similar conclusion was reached in A. vinelandii (Jacobson et al., 1989b). NifZ and nifW are probably involved in the maturation of MoFe protein in both K. pneumoniae and A. vinelandii (Jacobson et al., 1989b; Paul and Merrick, 1989). It thus appears that the products of at least 10 genes are required for the functioning of Mo nitrogenase. They have identical functions in K. pneumoniae and A. vinelandii: nifB, Q, N, E, and V are involved in the synthesis of FeMoco; nifM, U, and S in maturation and/or stability of the Fe protein and nifZ, and W in maturation of the MoFe protein.

(ii) Role of FixABCX. At least four genes (fixABCX) are required for nitrogenase activity in Rhizobiaceae. With rhizobial fixABC genes as probes, homologous regions were identified in A. brasilense and Azotobacter (Fogher et al., 1985; Evans et al., 1988; Galimand et al., 1989). Site-directed transposon mutagenesis of the fixABC region in A. brasilense led to a Nif$^-$ phenotype (Galimand et al., 1989). Moreover, inactivation of fixABCX in B. japonicum and A. caulinodans led to a Nif$^-$ phenotype under free-living conditions (Gubler and Hennecke, 1986: Kaminski et al., 1988; C. Elmerich et al., unpublished). Thus, fixABCX genes are not limited to symbiotic nitrogen fixing bacteria and it is likely that the fixABCX genes encode proteins required for the functioning of Mo nitrogenase. Because fixX is a ferredoxin, it was postulated that the fixABCX operon might be involved in electron transport to nitrogenase (Gubler and Hennecke, 1986; Earl et al., 1987). However, nitrogenase activity in crude extracts of fixA, fixC or fixX mutants of A. caulinodans was very low in the presence of dithionite as electron donor (Kaminski et al., 1988; C. Elmerich et al., unpublished). Nitrogenase activity was restored in crude extracts of a fixC mutant by addition of pure Fe protein, suggesting that fixC might be involved in its maturation (Kaminski et al., 1988).

2 GENES AND PRODUCTS REQUIRED FOR Mo NITROGENASE SYNTHESIS

2.1 Modulation of glutamine synthetase activity and synthesis in K. pneumoniae.

(i) Regulation of utilization of nitrogen sources and its relation to nif gene regulation. Glutamine synthetase (GS) is a key enzyme in ammonia assimilation. In enteric and several other Gram negative bacteria, GS activity is regulated by a reversible covalent adenylylation of a Tyr residue and GS biosynthesis is controlled by the ntr (nitrogen regulation) system (see Ginsburg and Stadtman, 1973; Magasanik, 1982, 1988; Merrick, 1988b). In general, when the nitrogen source is in excess, GS is synthesized at a low level and is partially inactivated by adenylylation. In contrast, when the nitrogen source is limiting, as in the case of nitrogen fixation, GS is synthesized at a high level and is unadenylylated. The ntr system also controls the expression of genes involved in the utilization of nitrogen sources, particularly the nif genes. Regulation of

GS synthesis has been studied in E. coli, K. aerogenes, K. pneumoniae and Salmonella typhimurium. Of these, only K. pneumoniae fixes N_2, but the structural and regulatory genes in the four species are identical and play a similar role (Magasanik, 1982, 1988; Merrick, 1988b).

(ii) Covalent modification of GS by adenylylation. GS is a dodecamer of identical 50 kD subunits, encoded by glnA (Deleo and Magasanik, 1975). Each subunit can be adenylylated to an extent dependent on physiological conditions. The signals that turn adenylylation on or off are the intracellular concentrations of 2-ketoglutarate (2-KG) and glutamine (Gln), and the regulatory protein P_{II}, a tetramer composed of identical 11 kD subunits coded by glnB. As shown in Fig. 2, a uridylyltransferase (UTase, coded by glnD) uridylylates P_{II} when the ratio 2-KG/Gln is high (conditions of nitrogen deficiency) and deuridylylates P_{II}-UMP when this ratio is low (conditions of nitrogen excess). An adenylyltransferase (ATase, coded by glnE) adenylylates GS in the presence of P_{II}, and deadenylylates GS-AMP in the presence of P_{II}-UMP. Consequently, GS is fully active under conditions of nitrogen deficiency and partially inactivated under conditions of nitrogen excess (see Ginsburg and Stadtman, 1973, Magasanik, 1982, 1988).

(iii) Glutamine auxotrophs and ntr mutants. It has been known for some time that mutants impaired in GS synthesis or activity are also defective in nitrogen fixation and nitrogen source utilization (Ntr^- phenotype) (Streicher et al., 1974; Tubb, 1974). Mutations which affect GS synthesis have been mapped at

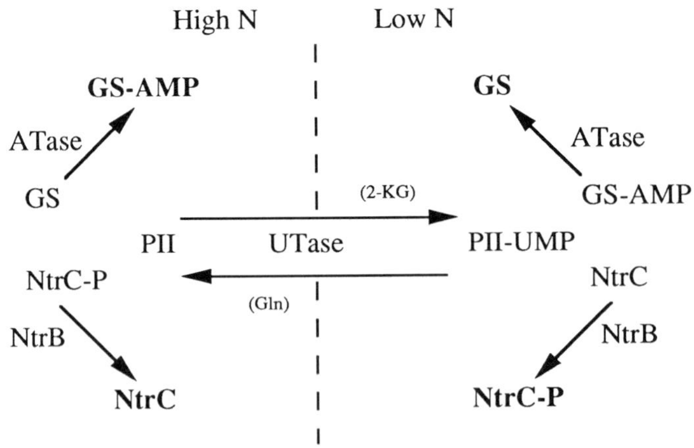

Fig. 2. System that regulates adenylylation of glutamine synthetase and phosphorylation of NtrC in enterobacteria in response to the availability of the nitrogen source. ATase adenylyltransferase (glnE); GS: glutamine synthetase (glnA); NtrB: ntrB (glnL) product; NtrC: ntrC (glnG) product; PII: protein PII (glnB); UTase: uridylyltransferase (glnD).

several unlinked loci. Two of them, glnB and glnD, code for enzymes of the adenylylation system. The other genes, termed glnL, glnG and glnF (then rpoN) in *E. coli* and *K. aerogenes* and ntrB, ntrC and ntrA in *K. pneumoniae* and *S. typhimurium*, respectively, code for regulatory proteins involved in the ntr system. This review concentrates on *K. pneumoniae* and follows the terminology used for this species.

NtrA mutants are pleiotropic, Gln^-, unable to use a range of nitrogen sources for growth (Garcia et al., 1977) and affected in nitrogen fixation (Leonardo and Golberg, 1980). Mutations which suppressed the glutamine auxotrophy of ntrA mutants did not suppress the Ntr^- (Nif^-) phenotype and were closely linked to glnA (Garcia et al., 1977; Pahel and Tyler, 1979; Leonardo and Golberg, 1980). This suggested that ntrA encoded a diffusible product required for activation of operons under ntr control; it has been identified as a novel type of RNA polymerase sigma factor (Hirschman et al., 1985). The ntrBC genes, adjacent to glnA, have been characterized by mutation and *in vivo* and *in vitro* cloning of the glnA region (Kustu et al., 1979; Pahel and Tyler, 1979; Bachman et al., 1981; de Bruijn and Ausubel, 1981; Espin et al., 1981, 1982; McFarland et al., 1981; Merrick, 1983). The isolation of mutants in ntrC capable of suppressing mutations in ntrA and restoring unregulated level of GS activity led to the conclusion that NtrC was a repressor of glnA transcription (Pahel and Tyler, 1979). However, since the suppressor mutants as well as most of the ntrC mutants had a Ntr^- (Nif^-) phenotype, it was proposed that NtrC was a bifunctional protein with activator and repressor activity and that the role of NtrB was to modulate NtrC activity (Kustu et al., 1979; Espin et al., 1982; McNeil et al., 1982; Merrick, 1983; Alvarez-Morales et al., 1984).

Mutants of glnB or glnD impaired in deadenylylation of GS-AMP were glutamine auxotrophs and were also defective in the derepression of GS synthesis under conditions of nitrogen limitation (Leonardo and Golberg, 1980; Bueno et al, 1985), as well as being Ntr^-. GlnB insertion mutants, isolated as suppressors of glnB or glnD mutations, recovered a constitutive level of GS activity and were prototrophs, but retained a wild Ntr phenotype (Leonardo and Golberg, 1980; Bueno et al., 1985; Holtel and Merrick, 1988, 1989). This suggested that P_{II} and UTase played a role in the regulation of GS synthesis.

(iv) Involvement of P_{II} in the ntr system. The nucleotide sequence of *K. pneumoniae* ntrB, ntrC and glnB genes has been established (Buikema et al., 1985; McFarlane and Merrick, 1985; Holtel and Merrick, 1988). The NtrB protein, a dimer composed of identical 38 kD subunits (Austin et al., 1987), belongs to the histidine protein kinase, or signal transducer, family. The NtrC protein, a dimer composed of identical 52 kD subunits (Hawkes et al., 1985), belongs to the response regulator protein family (Nixon et al., 1986; Stock et al., 1989). NtrB phosphorylates an aspartyl residue of NtrC to yield NtrC-P (Fig. 2), in the absence of P_{II} (or in the presence of P_{II}-UMP). The same protein activates the dephosphorylation of NtrC-P, in the presence of P_{II} (Ninfa and Magasanik, 1986; Keener and Kustu, 1988; Stock et al., 1989). Some ntrB mutants have been constructed (McFarlane and Merrick, 1987) with in-frame deletions, non-polar on ntrC and which constitutively phosphorylate NtrC and not respond to P_{II} (McFarlane and Merrick, 1987; Austin et al., 1987).

(v) The glnA ntrBC regulon. The glnA ntrBC regulon has three distinct promoters (Fig. 3). GlnAp1 (p1) and glnAp2 (p2) are located upstream from glnA, and ntrBCp (p3) is located in the intergenic region between glnA and ntrB, immediately upstream from ntrB (Alvarez-Morales et al., 1984; Dixon, 1984a). In addition, a rho-independent terminator sequence (T)is located between glnA and p3 (McFarlane and Merrick, 1985). The gene organization is very similar in other enteric bacteria. Under conditions of nitrogen excess, transcription proceeds at a low rate from p1 and p3, producing small amounts of GS and maintaining a low concentration of NtrB and NtrC. These promoters are dependent on the σ^{70} factor (Dixon, 1984a; Hunt and Magasanik, 1985). Under conditions of nitrogen deficiency, transcription from p2, one of the strongest promoters in the cell, is activated and produces increased amounts of GS, NtrB, and NtrC (Austin et al.,

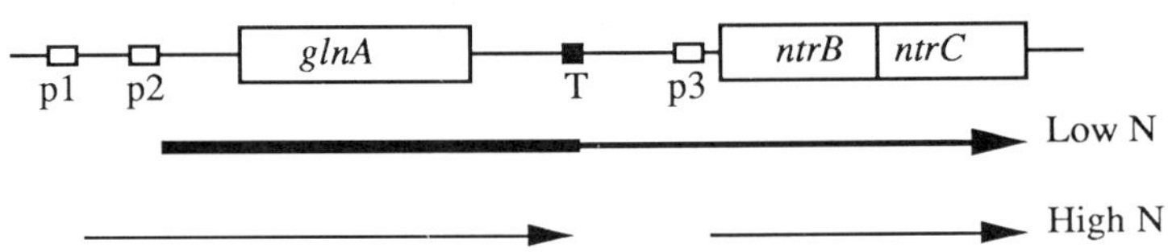

Fig. 3. Organization and transcription of the glnAntrBC operon. p1, p2, p3: glnAp1, ntrBp promoters, T: terminator. Arrows indicate the direction of transcription, and the size and relative amounts of the transcripts. (Adapted from Magasanik, 1988).

1987). This promoter is dependent on the σ^{54} factor and its consensus sequence is described below (2.2 (iv)).

In this system, NtrC acts both as a positive and a negative regulator. The p2 promoter is positively regulated by NtrC-P, which binds to a strong affinity site with the sequence TGCAC-N_5-TGGTGCA, located between positions -100 and -116 upstream from the transcriptional initiation site for p2 (Austin et al., 1987). Binding of NtrC-P at this site facilitates transcription by RNA polymerase-σ^{54} (Hirschman et al., 1985; Hunt and Magasanik, 1985). In contrast, p1 and p3 promoters are negatively regulated by NtrC. An NtrC binding site flanks the transcriptional initiation site from p3 and binding of NtrC impedes that of RNA polymerase-σ^{70} and thus affects ntrBC expression. This effect seems to be reinforced by NtrC-P (McFarlane and Merrick, 1987). Similarly, NtrC-binding to a site flanking the p1 promoter reduces transcription from this promoter by RNA polymerase-σ^{70}.

2.2 Regulation of nif gene expression in K. pneumoniae.

(i) Involvement of nifA in regulation of nif expression. Transcription of nif operons is prevented by sources of combined nitrogen (particularly ammonia), oxygen and temperatures above 34°C. Construction of lacZ (β-galactosidase) fusions to the nif operons showed that they are all repressed by ammonia and oxygen (Dixon et al., 1980; McNeil et al., 1981). However, the nifLA operon was derepressed in the presence of reduced concentrations of fixed nitrogen or oxygen, though these still repressed the other nif operons (Dixon et al., 1980, Merrick et al., 1982). NifA mutants were unable to synthesize any of the Nif polypeptides (pleiotropically-negative phenotype), suggesting that NifA is an activator of nif transcription (Dixon et al., 1977). This was confirmed by the isolation of ammonia-insensitive mutants with constitutive nifA expression (McNeil and Brill, 1980; Sibold et al., 1981; Sibold and Elmerich, 1982). Cloning of nifA under the control of an exogenous promoter in a multicopy vector also led to constitutive expression of nif operons in the presence of ammonia and oxygen (Buchanan-Wollaston et al., 1981b).

(ii) Involvement of nifL in regulation of nif expression. The role of nifL remained obscure for some time, since most nifL mutants were polar on nifA, so that the nifLA region appeared to be a single complementation unit (McNeil et al., 1978). However, Kennedy (1977) showed by transductional analysis that nifL was an independent gene. Analysis of non-polar nifL mutants led to the conclusion that NifL acted as a repressor of nif transcription, by modulating NifA activity in the presence of oxygen or ammonia (Hill et al., 1981; Merrick et al., 1982; Filser et al., 1983). This was also confirmed by the observation that overproduction of NifL by multicopy vectors prevented nif transcription (Buchanan-Wollaston et al., 1981a).

Temperature control is also mediated by nifLA. Non-polar nifL mutants are less sensitive to temperature than the wild type strain and the nifA product was found to be thermosensitive (Buchanan-Wollaston et al., 1981b; Merrick et al., 1982; Sibold et al., 1983; Brooks et al., 1984; Pühler et al., 1984).

(iii) Target of NtrC and NifA activity. It was shown that nif-lacZ fusions were not expressed in a ntrC (Ntr^-) mutant, suggesting that NtrC was required for nif expression (Dixon et al., 1980). Of the nif operons, NtrC activates only the nifL and nifH promoters, though activation of the latter was low (Buck et al., 1985). Indeed, NtrC was not necessary in a nifA-constitutive background, suggesting that the nifLA promoter was the only target of NtrC activation *in vivo* (McNeil and Brill, 1980; Buchanan-Wollaston et al., 1981b; Sibold et al., 1981). When overproduced by a multicopy vector, NifA can substitute for NtrC in activating promoters under NtrC positive control (Drummond et al., 1983; Merrick, 1983; Ow and Ausubel, 1983; Alvarez-Morales et al., 1984). In particular, a nifL-lacZ fusion was activated both by NtrC and NifA showing that NifA can autoregulate its own synthesis (Drummond et al., 1983). Thus, there is a striking similarity between ntrBC and nifLA functions. The amino acid sequences deduced from the nucleotide sequences of nifA and ntrC show considerable homology in their central domains but differ at their N-termini (Buikema et al. 1985; Drummond et al. 1986). At their C-termini, both proteins contain a helix-turn-helix motif characteristic of DNA-binding proteins (Stock et al., 1989).

(iv) Target of NtrA activity and identification of NtrA-dependent promoters. In an ntrA background, mutants carrying a constitutive nifA were both Nif^- and unable to express a nifH-lacZ fusion, suggesting that both NifA and NtrA are required for expression of nif operons (Sibold and Elmerich, 1982; de Bruijn and Ausubel, 1983; Merrick, 1983; Ow and Ausubel, 1983; Sundaresan et al., 1983a).

The genetic and physical mapping of the nif transcription units enabled an approximate localization of the nif promoters (Beynon et al., 1983). Promoters of the nif operons are devoid of the typical consensus sequence of *E. coli* promoters located at positions -35 and -10 upstream from the transcriptional start site. However, they showed a sequence conservation around positions -24 and -12 (Beynon et al., 1983; Drummond et al., 1983). Alignments led to the establishment of a consensus sequence for nif promoters, 5' TGGYRYR-N4-TTGCA 3', the GG and GC conserved residues being referred to here as positions -24 and -12, which on average is the spacing from the transcriptional start site (Ausubel, 1984; Dixon, 1984a, b; Buck, 1986; Gussin et al., 1986). The importance of conserved nucleotides in the consensus was demonstrated by site directed mutagenesis. Buck (1986) showed that the residue spacing (GC-N10-GG) between the -12 and -24 conserved regions was critical. Mutations of nifH and nifL promoters at positions -24, -13 and -12 led to a lack of activation by NifA and NtrC (Brown and Ausubel, 1984; Buck et al., 1985; Ow et al., 1985; Khan et al., 1986). However, a C to A mutation at -12 had no effect on nifH expression (Buck et al., 1985). It now appears that similar promoters are present upstream from many other genes not involved in nitrogen regulation, reflecting their dependence on a particular sigma factor, σ^{54}, which is encoded by ntrA (see Thöny and Hennecke,

1989). These promoters will be referred to here as NtrA-dependent promoters.

(v) Upstream activator. Buck et al. (1985) showed that a 112 bp deletion starting at position -72 from the nifH promoter resulted in a lack of activation by NifA, and in low activation by NtrC. This suggested that a DNA region located well upstream from the nifL promoter was also required for activation by NifA. Further analysis of sequences upstream from nif promoters led to the identification of a conserved sequence (TGT-N10-ACA), designated UAS (upstream activator sequence) (Buck et al., 1986). This region, located 100 bp or more upstream from the promoter region, appears to be required for specific activation by NifA (Buck et al., 1986). Upstream activator sequences are present in front of all nif operons except nifLA and nifM (Buck et al., 1986; Buck et al., 1987; Arnold et al., 1988); there are two upstream from nifJ and three upstream from nifU (Arnold et al., 1988; Cannon et al., 1990).

(vi) Model of nif gene activation. In the current model of regulation of nif gene expression, the nifLA operon remains repressed under conditions of nitrogen excess. Under conditions of nitrogen deficiency, NtrC-P activates nifLA transcription and NifA in turn activates the transcription of other nif operons, by RNA polymerase-σ^{54}. If oxygen or ammonia becomes available, the nifL product prevents further activation by NifA (Dixon, 1984a; Merrick, 1988a). The structures of the nifLA and nifH promoter regions are shown in Fig. 4. Two

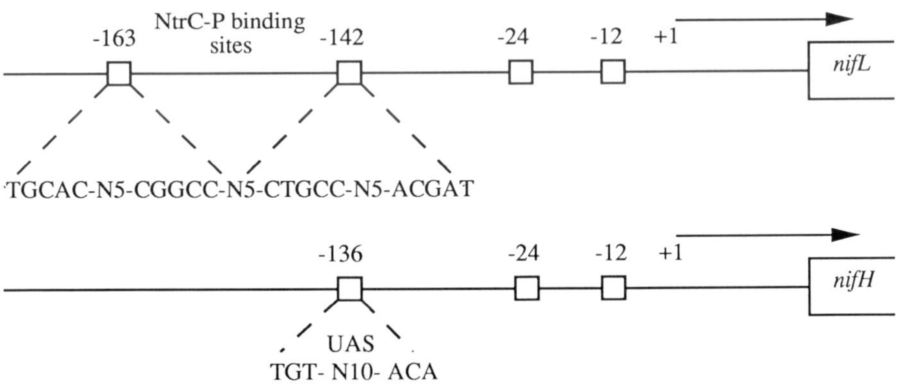

Fig. 4. Regulatory regions of nifLA and nifHDK operons from K. pneumoniae. +1 indicates the transcriptional start site. Boxes at -24 and -12 correspond to the consensus promoter elements of RNA polymerase-σ^{54}. UAS: NifA-dependent upstream activator sequence. (Adapted from Buck et al., 1987 and Minchin et al., 1988).

tandemly organized NtrC-binding sites have been characterized around positions -142 and -163 upstream from the nifLA transcriptional start site (Minchin et al., 1988). These sites show a limited homology with the strong affinity site located up-stream from glnAp2 and already described. Indeed, a higher concentration of NtrC-P is necessary for transcriptional activation of nifLA than of glnA (Austin et al., 1987; Wong et al., 1987). Further analysis of this region, based on the use of mutants and on *in vitro* binding experiments, showed that mutations at -148 and -169 within the two NtrC-binding sites reduce the activation of the nifLA operon *in vivo*, and reduce the affinity of NtrC for these sites *in vitro*. In addition, phosphorylation of NtrC is required for efficient binding (Minchin et al., 1988). Activation by NtrC-P involves the formation of a DNA loop. High concentrations of NtrC cause direct activation at the -24 to -12 region (Minchin et al., 1989).

A model explaining the activation of nifHDK transcription by NifA-induced loop formation in DNA has been proposed by Buck et al. (1987). In addition, it has been proposed that NifA activates the nifU promoter by NifA by catalyzing the formation of a DNA-RNA polymerase-σ^{54} open complex there (Cannon et al., 1990). The role of NifL in modulating the activity of NifA is not yet fully understood, but it has been proposed that NifL antagonizes NifA by formation of a protein complex (Henderson et al., 1989).

2.3 Regulation of nitrogen fixation in other free-living and symbiotic nitrogen fixers.

The finding that most diazotrophs contain potential NtrA-dependent promoters implied that regulation of nitrogen fixation in these organisms involves mechanisms similar to those in *K. pneumoniae*. However, although genes corresponding to ntrA, ntrB or ntrC have been isolated from several organisms, the regulation pathways differ from one species to another.

(i) Conservation of ntrA and NtrA-dependent promoters. Several nitrogen-fixers contain ntrA homologues. The nucleotide sequences of ntrA genes from *A. vinelandii* (Merrick et al., 1987), *R. meliloti*, (Ronson et al., 1987) and *R. capsulatus* (Jones and Haselkorn, 1989) have been established; an ntrA gene has also been identified in *Rhizobium* sp. NGR234 (Stanley et al., 1989). Further, analysis of the nucleotide sequence upstream from nif operon coding regions of most diazotrophs led to the identification of sequences matching the consensus established for NtrA-dependent promoters and nif UAS's in *K. pneumoniae* (Buck et al., 1986; Gussin et al., 1986; Thöny and Hennecke, 1989). There are, however, exceptions: *Anabaena* (Haselkorn et al., 1983; Borthakur et al., 1990), *C. pasteurianum* (Wang et al., 1986) and methanogens (Souillard and Sibold, 1986, 1989). The nifH promoter sequences of *R. phaseoli* CNF42, *R. trifolii* ANU843 and *R. leguminosarum* PRE differ from the NtrA-dependent consensus by the presence of an A residue at position -12 instead of a C (GG-N10-GA) (Quinto et al., 1985;

Watson and Schofield, 1985; Roelvink et al., 1990). This is consistent with the phenotypes of K. pneumoniae nifH promoter mutants at -12, where a C residue appears not to be essential for nifH expression (Buck et al., 1985). The transcriptional start site for nifH mRNA has been mapped about 10 bp downstream from the NtrA-dependent promoter sequence in several nitrogen fixers, including A. brasilense (de Zamaroczy et al., 1989), B. japonicum (Fuhrmann and Hennecke, 1984), R. leguminosarum (Roelvink et al., 1990), R. meliloti (Ow et al. 1983; Sundaresan et al., 1983a), R. trifolii (Watson and Schofield, 1985) and Rhodobacter capsulatus (Pollock et al., 1988).

(ii) Activation of nifH-lacZ fusions in E. coli. Activation of R. meliloti nifH- and B. japonicum nifH- and nifD-lacZ fusions by E. coli NtrC and NtrA, or by K. pneumoniae NifA and NtrC, was studied in E. coli (Sundaresan et al., 1983a, b; Alvarez-Morales and Hennecke, 1985). The gene from R. meliloti was activated both by NifA and NtrC, whereas genes from B. japonicum were activated only by NifA. In both cases, expression was dependent on NtrA. Mutagenesis of the NtrA-dependent consensus sequence showed that the -24 region and the TGT-N10-ACA conserved sequences upstream from nifH and nifD promoters in B. japonicum were essential for NifA activation (Kaluza et al., 1985; Alvarez-Morales et al., 1986).

(iii) Identification and cloning of nifA-like genes. NifA-like mutants have been isolated in A. caulinodans (Denèfle et al., 1987; Pawlowski et al. 1987), A. brasilense (Pedrosa and Yates, 1984; Singh et al., 1989), A. vinelandii (Kennedy and Robson, 1983), B. japonicum (Fischer et al., 1986), R. leguminosarum (Shetgens et al., 1985), R. meliloti (Szeto et al. 1984; Weber et al. 1985), and R. capsulatus (Ahombo et al., 1986; Kranz and Haselkorn, 1985). Identification of nifA mutants was more difficult in R. capsulatus, because it contains two copies of nifA (Klipp et al., 1988), but inactivation of the two copies of nifA led to a Nif^- phenotype (Masepohl et al., 1988). NifA mutants of Azorhizobium and Bradyrhizobium, which can derepress nitrogenase under free-living conditions, were impaired in nitrogen fixation both in planta and ex planta. The Azotobacter NifA-like mutant, identified by complementation with K. pneumoniae NifA, was later shown to carry a mutation in another regulatory gene, nfrX (Santero et al., 1988). Interestingly, in B. japonicum and A. vinelandii, nifA is preceded by an ORF (fixR in B. japonicum), which is co-transcribed with nifA and may play a role in regulation (Thöny et al., 1987, 1989; Bennett et al., 1988).

(iv) Functional analysis of NifA. To date, no homologue of nifL has been identified outside nitrogen-fixing enteric bacteria. The nucleotide sequence of nifA genes has been established in A. caulinodans (Nees et al., 1988; Ratet et al., 1989), A. vinelandii (Bennett et al., 1988), B. japonicum (Thöny et al.,

1987), R. leguminosarum (Grönger et al., 1987; Roelvink et al., 1989), R. meliloti (Buikema et al., 1985), R. trifolii (Iisma and Watson, 1989) and R. capsulatus (Masepohl et al., 1988). The fixRnifA operon of B. japonicum is preceded by an NtrA-dependent promoter (Thöny et al., 1987), but the nifA genes of R. leguminosarum (Grönger et al., 1987), R. meliloti (Buikema et al., 1985), R. trifolii (Iisma and Watson, 1989) and R. capsulatus (Masepohl et al., 1988) are not. The nifA gene of A. caulinodans is preceded by an NtrA-dependent promoter and by an fnr box, which characterizes promoters of genes induced in E. coli by anaerobiosis (Nees et al., 1988; Ratet et al., 1989). The polypeptide sequences coded by these nifA genes are highly homologous with that from K. pneumoniae NifA (Drummond et al., 1986). These authors have identified four domains in K. pneumoniae NifA and have suggested that NifL interacts with the N-terminal domain, while the C-terminal domain, which contains a helix-turn-helix motif, is involved in DNA-binding. The N-terminal domains of R. meliloti and B. japonicum NifA are not essential for NifA activity (Beynon et al., 1988b; Hennecke et al., 1988; Huala and Ausubel, 1989); the central domain is sufficient (Huala and Ausubel, 1989). The NifA polypeptide of R. trifolii has a M_r of only 38 kD compared to about 50 kD for NifA proteins from other organisms and lacks the N-terminal domain (Iisma and Watson, 1989). The central domain of NifA from Rhizobium spp. contains four cysteine residues which may be involved in the sensitivity of the protein to oxygen (Hennecke et al., 1988). These cysteine residues are not conserved in K. pneumoniae NifA, where oxygen sensitivity is mediated through NifL (Drummond and Wooten, 1987).

(v) The three GS of R. meliloti are not involved in symbiotic nitrogen fixation. In addition to the GS coded by glnA, the Rhizobiaceae contain another GS of eukaryote-type coded by glnII (Sommerville and Kahn, 1983; Carlson et al., 1985; Carlson and Chelm, 1986). Moreover, ntrBC genes are not linked to glnA in B. japonicum (Martin et al., 1988), R. leguminosarum (Colonna-Romano et al., 1987 and R. meliloti (Szeto et al., 1987). GlnA is constitutively expressed in B. japonicum, but glnII is under ntr control (Martin et al., 1988). A third GS has been described in R. meliloti (de Bruijn et al., 1989), although none of the three GS species is essential for symbiotic nitrogen fixation (de Bruijn et al., 1989).

(vi) GlnB is linked to glnA in several species. GlnA and ntrBC form an operon in enteric bacteria and Azotobacter (Toukdarian and Kennedy, 1986). Organization of glnA and ntr genes differs in other genera. The glnA gene is preceded by glnB in A. brasilense (de Zamaroczy et al., submitted), B. japonicum (Martin et al., 1989) and R. leguminosarum (Colonna-Romano et al., 1987). Transcription of glnB is regulated by tandem promoters in the first two species.

(vii) NtrC does not play a role in activation of nif expression in several diazotrophs. Genes corresponding to ntrBC have been identified in A. caulinodans (Pawlowski et al., 1987), Azotobacter (Toukdarian and Kennedy, 1986), Bradyrhizobium sp. (Parasponia) (Nixon et al., 1986), B. japonicum (Martin et al., 1988), R. meliloti (Szeto et al., 1987) and R. capsulatus (Jones and Haselkorn, 1989, see below). The phenotype of ntrC mutants in these organisms differs from that of enteric bacteria. Azotobacter ntrC is involved in nitrate utilization but not in nitrogen fixation (Toukdarian and Kennedy, 1986). This is also the case in B. japonicum (Martin et al., 1988) and R. meliloti (Szeto et al., 1987), where ntrC mutants have a Fix^+ phenotype in planta. NtrC mutants of A. caulinodans are Nif^- in the free-living state (Pawlowski et al., 1987).

(viii) NtrC plays a role in nitrogen fixation in R. capsulatus. Four regulatory genes organized into two unlinked clusters have been identified in R. capsulatus (Kranz and Haselkorn, 1985). Mutations in the four genes led to a Nif^- phenotype and to a lack of expression of a nifH-lac fusion, but utilization of other nitrogen sources was not impaired in these mutants. Cluster 1 contains three contiguous genes, designated nifR1-3, which are transcribed independently (Kranz and Haselkorn, 1985). The nucleotide sequences show that nifR1 corresponds to ntrC, and nifR2 to ntrB (Jones and Haselkorn, 1989). Cluster 2 contains nifR4 between nifHDK and nifA (Ahombo et al., 1986; Klipp et al. 1988). The nucleotide sequence of nifR4 shows that this gene corresponds to ntrA (Alias et al., 1989; Jones and Haselkorn, 1989).

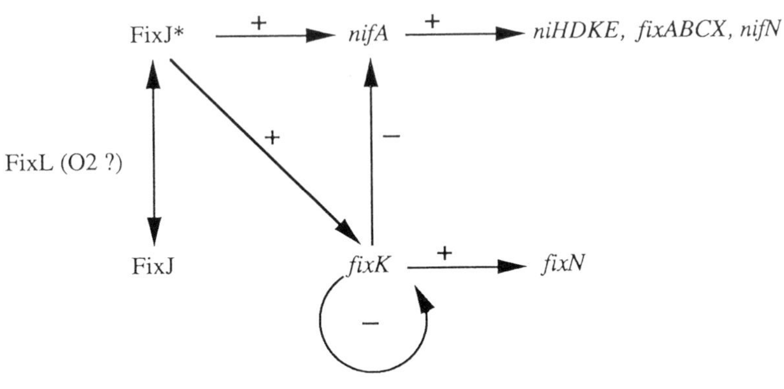

Fig. 5. Model of nif and fix gene regulation in R. meliloti. $FixJ^*$ is the active form of the fixJ product. Positive and negative controls are indicated by + and -, respectively. Adapted from Batut et al., 1989).

(ix) Model of regulation in *R. meliloti*. A reduction of oxygen tension to microaerobic levels results in *nifA* expression in *R. meliloti* in the free-living state (Ditta et al., 1987). Microaerobiosis is probably the primary physiological factor that controls activation of symbiotic nitrogen fixation (Ditta et al., 1987). As mentioned above, this control is independent of NtrC and depends on two newly identified regulatory genes, *fixL* and *fixJ* (David et al., 1988). *R. meliloti nif* and *fixABCX* genes are subject to a cascade regulation and a third regulatory gene, *fixK*, is also involved (Fig. 5). FixJ*, the active form of the *fixJ* product, positively regulates *nifA* expression. In turn, NifA activates other *nif* operons and *fixABCX* in the presence of NtrA. FixJ* also positively regulates *fixK*, which then activates the transcrition of the *fixN* operon. In addition, *fixK* negatively regulates both its own synthesis and *nifA* expression (Batut et al., 1989). FixL shows features of a transmembrane protein and belongs to the histidine kinase sensor protein family (David et al., 1988). It has been proposed that FixL acts as a sensor of an extracellular signal, possibly O_2 tension, and converts FixJ into its active form (David et al., 1988).

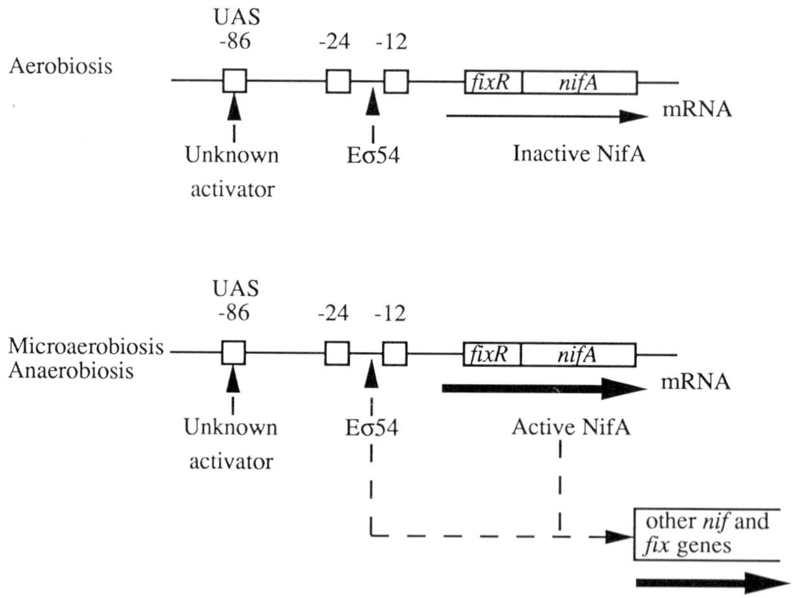

Fig. 6. Model of *nif* and *fix* gene regulation in *B. japonicum*. Boxes at -24 and -12 correspond to the consensus promoter elements of RNA polymerase-σ^{54}. UAS: upstream activator sequence. (Adapted from Thöny et al., 1989).

Inactivation of fixL or fixJ in A. caulinodans led to a Fix^- phenotype in planta and to a loss of 90% of the nitrogenase activity normally expressed under free-living conditions (C. Elmerich et al., unpublished). Thus, in addition to ntrC (Pawlowki et al., 1987), fixLJ appears to play a role both in symbiotic and non-symbiotic nitrogen fixation in this bacterium.

(x) Model of regulation in B. japonicum. As in R. meliloti, nif and fix structural operons of B. japonicum are derepressed under microaerobiosis. In contrast, the expression of the fixRnifA operon does not depend on oxygen tension, but at high oxygen tension the NifA protein is inactivated (Thöny et al., 1987). As shown in Fig. 6, transcription of fixRnifA depends on RNA polymerase-σ^{54} and on an unknown activator protein which binds to a site (UAS) located between positions -24 and -86 upstream from the transcriptional start site (Thöny et al., 1989). Once expressed, NifA activates the other nif and fix operons under microaerobiosis. NifA itself plays the role of oxygen sensor (Hennecke et al., 1988).

Acknowledgements

I wish to thank Drs. J-P. Aubert and M. de Zamaroczy for helpful discussions and Mrs M. Ferrand for help in typing the manuscript. Unpublished experiments from my laboratory were performed in collaboration with Dr. P. A. Kaminski and Messrs F. Arigoni and Y. Y. Liang.

DEDICATION

This review is dedicated to Lionel Sibold, my first PhD student, who contributed to the development of the genetics of nitrogen fixation in K. pneumoniae and in methanogens, and who died at the age of 35. His ashes were dispersed on a lawn of the Père Lachaise cemetery in Paris, on October 10, 1989.

"Frères humains qui après nous vivez,
N'ayez les coeurs contre nous endurcis....
De notre mal personne ne s'en rie;
Mais priez Dieu que tous nous veuille absoudre."
(François Villon, born in 1431)

3 REFERENCES

Aguilar, O.M., Kapp, D. and Pühler, A., 1985. Characterization of a Rhizobium meliloti fixation gene (fixF) located near the common nodulation region. J. Bacteriol., 164: 245-254.

Ahombo, G., Willison, J.C. and Vignais, P.M., 1986. The nifHDK genes are contiguous with a nifA-like regulatory gene in Rhodobacter capsulatus. Mol. Gen. Genet., 205: 442-445.

Alias, A., Cejudo, F.J., Chabert, J., Willison, J.C. and Vignais, P.M., 1989. Nucleotide sequence of wild-type and mutant nifR4 (ntrA) genes of Rhodobacter capsulatus: identification of an essential glycine residue. Nucl. Acids Res., 17: 5377.

Alvarez-Morales, A. and Hennecke, H., 1985. Expression of Rhizobium japonicum nifH and nifDK operons can be activated by the Klebsiella pneumoniae nifA protein but not by the product of ntrC. Mol. Gen. Genet., 199: 306-314.

Alvarez-Morales, A., Dixon, R. and Merrick, M., 1984. Positive and negative control of the glnA ntrBC regulon in Klebsiella pneumoniae. EMBO J., 3: 501-507.

Alvarez-Morales, A., Betancourt-Alvarez, M., Kaluza, K. and Hennecke, H., 1986. Activation of the Bradyrhizobium japonicum nifH and nifDK operons is dependent on promoter-upstream DNA sequences. Nucl. Acids Res., 14: 4207-4227.

Arnold, W., Rump, A., Klipp, W., Priefer, U.B. and Pühler, A., 1988. Nucleotide sequence of a 24,206-base-pair DNA fragment carrying the entire nitrogen fixation gene cluster of Klebsiella pneumoniae. J. Mol. Biol., 203: 715-738.

Austin, S., Henderson, N. and Dixon, R., 1987. Requirements for transcriptional activation in vitro of the nitrogen-regulated glnA and nifLA promoters from Klebsiella pneumoniae: dependence on activator concentration. Mol. Microbiol., 1: 92-100.

Ausubel, F.M., 1984. Regulation of nitrogen fixation genes. Cell, 37: 5-6.

Avtges, P., Scolnik, P.A. and Haselkorn, R., 1983. Genetic and physical map of the structural genes (nifH,D,K) coding for the nitrogenase complex of Rhodopseudomonas capsulata. J. Bacteriol., 156: 251-256.

Bachman, K., Chen, Y-M. and Magasanik, B., 1981. Physical and genetic characterization of the glnA-glnG region of the Escherichia coli chromosome. Proc. Natl. Acad. Sci. USA, 78: 3743-3747.

Banfalvi, Z., Sakanyan, V., Koncz, C., Kiss, A., Dusha, I. and Kondorosi, A., 1981. Location of nodulation and nitrogen fixation genes on a high molecular weight plasmid of R. meliloti. Mol. Gen. Genet., 184: 318-325.

Batut, J., Terzaghi, B., Ghérardi, M., Huguet, M., Terzaghi, E., Garnerone, A.M., Boistard, P. and Huguet, T., 1985. Localization of a symbiotic fix region on Rhizobium meliloti pSym megaplasmid more than 200 kilobases from the nod-nif region. Mol. Gen. Genet., 199: 232-239.

Batut, J., Daveran-Mingot, M-L., David, M., Jacobs, J., Garnerone, A.M. and Kahn, D., 1989. FixK, a gene homologous with fnr and crp from Escherichia coli, regulates nitrogen fixation genes both positively and negatively in Rhizobium meliloti. EMBO J., 8: 1279-1286.

Belay, N., Sparling, R. and Daniels, L., 1984. Dinitrogen fixation by a thermophilic methanogenic bacterium. Nature, 312: 286-288.

Bennett, L.T., Cannon, F. and Dean, D.R., 1988. Nucleotide sequence and mutagenesis of the nifA gene from Azotobacter vinelandii. Mol. Microbiol., 2: 315-321.

Beynon, J., Cannon, M., Buchanan-Wollaston, V. and Cannon, F., 1983. The nif promoters of Klebsiella pneumoniae have a characteristic primary structure. Cell, 34: 665-671.

Beynon, J., Ally, A., Cannon, M., Cannon, F., Jacobson, M., Cash, V. and Dean, D., 1987. Comparative organization of nitrogen fixation-specific genes from Azotobacter vinelandii and Klebsiella pneumoniae : DNA sequence of the nifUSV genes. J. Bacteriol., 169: 4024-4029.

Beynon, J., Cannon, M., Buchanan-Wollaston, V., Ally, A., Setterquist, R., Dean, D. and Cannon, F., 1988a. The nucleotide sequence of the nifT, nifY, nifX and nifW genes of K. pneumoniae. Nucl. Acids Res., 16: 9860.

Beynon, J.L., Williams, M.K. and Cannon, F.C., 1988b. Expression and functional analysis of the Rhizobium meliloti nifA gene. EMBO. J., 7: 7-17.

Bishop, P.E., Premakumar, R., Joerger, R.D., Jacobson, M.R., Dalton, D.A., Chisnell, J.R. and Wolfinger, E.D., 1988. Alternative nitrogen fixation systems in Azotobacter vinelandii. In: H. Bothe, F.J. de Bruijn and W.E. Newton (Editors), Nitrogen Fixation: Hundred Years After. Gustav Fischer, Stuttgart, New York, pp.71-79.

Bogusz, D., Houmard, J. and Aubert, J.-P., 1981. Electron transport to nitrogenase in Klebsiella pneumoniae. Purification and properties of the nifJ protein. Eur. J. Biochem., 120: 421-426.

Böhme, H. and Haselkorn, R., 1988. Molecular cloning and nucleotide sequence analysis of the gene coding for heterocyst ferredoxin from the cyanobacterium Anabaena sp. strain PCC 7120. Mol. Gen. Genet., 214: 278-285.

Borthakur, D., Basche, M., Buikema, W.J., Borthakur, P.B. and Haselkorn, R., 1990. Expression, nucleotide sequence and mutational analysis of two open reading frames in the nif gene region of Anabaena so. strain PCC7120. Mol. Gen. Genet., 221: 227-234.

Brigle, K.E., Newton, W.E. and Dean, D.R., 1985. Complete nucleotide sequence of the Azotobacter vinelandii nitrogenase structural gene cluster. Gene, 34:37-44.

Brigle, K.E., Weiss, M.C., Newton, W.E. and Dean, D.R., 1987. Products of the iron-molybdenum cofactor-specific biosynthetic genes, nifE and nifN, are structurally homologous to the products of the nitrogenase molybdenum-iron protein genes, nifD and nifK. J. Bacteriol., 169: 1547-1553.

Brill, W.J., 1980. Biochemical genetics of nitrogen fixation. Microbiol. Rev., 44: 449-467.

Brooks, S.J., Collins, J.J. and Brill, W.J., 1984. Repression of nitrogen fixation in Klebsiella pneumoniae at high temperature. J. Bacteriol., 157: 460-464.

Brown, S.E. and Ausubel F.M., 1984. Mutations affecting regulation of the Klebsiella pneumoniae nifH (nitrogenase reductase) promoter. J. Bacteriol., 157: 143-147.

Buchanan-Wollaston, V., Cannon, M.C. and Cannon, F.C., 1981a. The use of cloned nif (nitrogen fixation) DNA to investigate transcriptional regulation of nif expression in Klebsiella pneumoniae. Mol. Gen. Genet., 184: 102-106.

Buchanan-Wollaston, V., Cannon, M.C. Beynon, J.L. and Cannon, F.C., 1981b. Role of nifA gene product in the regulation of nif expression in Klebsiella pneumoniae. Nature, 294: 776-778.

Buck, M., 1986. Deletion analysis of the Klebsiella pneumoniae nitrogenase promoter: Importance of spacing between conserved sequences around position -12 and -24 for activation by the nifA and ntrC (glnG) products. J. Bacteriol., 166: 545-551.

Buck, M., Khan, H. and Dixon, R., 1985. Site-directed mutagenesis of the Klebsiella pneumoniae nifL and nifH promoters and in vivo analysis of promoter activity. Nucl. Acids Res., 13: 7621-7638.

Buck, M., Miller, S., Drummond, M. and Dixon, R., 1986. Upstream activator sequences are present in the promoters of nitrogen fixation genes. Nature, 320: 374-378.

Buck, M., Cannon, W. and Woodcock, J., 1987. Transcriptional activation of the Klebsiella pneumoniae nitrogenase promoter may involve DNA loop formation. Mol. Microbiol., 1: 243-249.

Buck, M., Woodcock, J., Cannon, W., Mitchenall, L. and Drummond, M., 1987. Positional requirements for the function of nif-specific upstream activator sequences. Mol. Gen. Genet., 210: 140-144.

Bueno, R., Pahel, G. and Magasanik, B., 1985. Role of glnB and glnD gene products in regulation of the glnALG operon in Escherichia coli. J. Bacteriol., 164: 816-822.
Buikema, W.J., Szeto, W.W., Lemley, P.V., Orme-Johnson, W.H. and Ausubel, F.M., 1985. Nitrogen fixation specific regulatory genes of Klebsiella pneumoniae and Rhizobium meliloti share homology with the general nitrogen regulatory gene ntrC of K. pneumoniae. Nucl. Acids Res., 13: 4539-4555.
Buikema, W.J., Klingensmith, J.A., Gibbons, S.L. and Ausubel, F.M., 1987. Conservation of structure and location of Rhizobium meliloti and Klebsiella pneumoniae nifB genes. J. Bacteriol., 169: 1120-1126.
Cannon, F.C., Riedel, G.E. and Ausubel, F.M., 1977. Recombinant plasmid that carries part of the nitrogen fixation (nif) gene cluster of Klebsiella pneumoniae. Proc. Natl. Acad. Sci. USA, 74: 2963-2967.
Cannon, F.C., Riedel, G.E. and Ausubel, F.M., 1979. Overlapping sequences of Klebsiella pneumoniae nif DNA cloned and characterized. Mol. Gen. Genet., 174: 59-66.
Cannon, W.V., Kreutzer, R., Kent, H.M., Morett, E. and Buck, M., 1990. Activation of the Klebsiella pneumoniae nifU promoter: identification of multiple and overlapping upstream NifA binding sites. Nucl. Acids Res., 18: 1693-1701.
Carlson, T.A. and Chelm, B.K., 1986. Apparent eukaryotic origin of glutamine synthetase II from the bacterium Bradyrhizobium japonicum. Nature, 322: 568-570.
Carlson, T.A., Guérinot, M.L. and Chelm, B.K., 1985. Characterization of the gene encoding glutamine synthetase I (glnA) from Bradyrhizobium japonicum. J. Bacteriol., 162: 698-703.
Chen, K.C-K., Chen, J-S. and Johnson, J.L., 1986. Structural features of multiple nifH-like sequences and very biased codon usage in nitrogenase genes of Clostridium pasteurianum. J. Bacteriol., 166: 162-172.
Colonna-Romano, S., Riccio, A., Defez, R., Lamberti, A., Iaccarino, M., Arnold, W., Priefer, U. and Pühler, A., 1987. Tight linkage of glnA and a putative regulatory gene in Rhizobium leguminosarum. Nucl. Acids Res., 5: 1951-1963.
Corbin, D., Barran, L. and Ditta, G., 1983. Organization and expression of Rhizobium meliloti nitrogen fixation genes. Proc. Natl. Acad. Sci. USA, 80: 3005-3009.
David, M., Dommergue, O., Pognonec, P. and Kahn, D., 1987. Transcription patterns of Rhizobium meliloti symbiotic plasmid pSym: identification of nifA-independent fix genes. J. Bacteriol., 169: 2239-2244.
David, M., Daveran, M-L., Batut, J., Dedieu, A., Domergue, O., Ghai, J., Hertig, C., Boistard, P. and Kahn, D., 1988. Cascade regulation of nif gene expression in Rhizobium meliloti. Cell, 54: 671-683.
de Bruijn, F.J. and Ausubel, F.M., 1981. The cloning and transposon Tn*5* mutagenesis of the glnA region of Klebsiella pneumoniae: identification of glnR, a gene involved in the regulation of the nif and hut operons. Mol. Gen. Genet., 183: 289-297.
de Bruijn, F.J. and Ausubel, F.M., 1983. The cloning and characterization of the glnF (ntrA) gene of Klebsiella pneumoniae: role of glnF (ntrA) in the regulation of nitrogen fixation (nif) and other nitrogen assimilation genes. Mol. Gen. Genet., 192: 342-353.
de Bruijn, F.S., Rossbach, S., Schneider, M., Ratet, P., Messmer, S., Szeto, W.W., Ausubel, F.M. and Schell, J., 1989. Rhizobium meliloti 1021 has three differentially regulated loci involved in glutamine biosynthesis, none of which is essential for symbiotic nitrogen fixation. J. Bacteriol., 171: 1673-1682.
de Zamaroczy, M., Delorme, F. and Elmerich, C., 1989. Regulation of transcription and promoter mapping of the structural genes for nitrogenase (nifHDK) of Azospirillum brasilense Sp7. Mol. Gen. Genet., 220: 88-94.

de Zamaroczy, M., Delorme, F. and Elmerich, C., 1990. Identification and characterization of three different nitrogen regulated promoters for the expression of glnB and glnA in Azospirillum brasilense. Submitted.
Dean, D.R. and Brigle, K.E., 1985. Azotobacter vinelandii nifD- and nifE-encoded polypeptides share structural homology. Proc. Natl. Acad. Sci. USA, 82: 5720-5723.
Deistung, J. and Thorneley, R.N.F., 1986. Electron transfer to nitrogenase. Biochem. J., 239: 69-75.
Deleo, A.B. and Magasanik, B., 1975. Identification of the structural gene for glutamine synthetase in Klebsiella aerogenes. J. Bacteriol., 121: 313-319.
Denèfle, P., Kush, A., Norel, F., Paquelin, A. and Elmerich, C., 1987. Biochemical and genetic analysis of the nifHDKE region of Rhizobium ORS571. Mol. Gen. Genet., 207: 280-287.
Ditta, G., Virts, E., Palomares, A. and Kim, C-H., 1987. The nifA gene of Rhizobium meliloti is oxygen regulated. J. Bacteriol., 169: 3217-3223.
Dixon, R., 1984a. Tandem promoters determine regulation of the Klebsiella pneumoniae glutamine synthetase (glnA) gene. Nucl. Acids Res., 12: 7811-7830.
Dixon, R.A. and Postgate, J.R., 1972. Genetic transfer of nitrogen fixation from Klebsiella pneumoniae to Escherichia coli. Nature, 237: 102-103.
Dixon, R.A., Cannon, F.C. and Kondorosi, A., 1976. Construction of a P plasmid carrying nitrogen fixation genes from Klebsiella pneumoniae. Nature, 260: 268-271.
Dixon, R.A., Kennedy, C., Kondorosi, A., Krishnapillai, V. and Merrick, M., 1977. Complementation analysis of Klebsiella pneumoniae mutants defective in nitrogen fixation. Mol. Gen. Genet., 157: 189-198.
Dixon, R.A., Eady, R., Espin, G., Hill, S., Iaccarino, M., Kahn, D. and Merrick, M., 1980. Analysis of regulation of the Klebsiella pneumoniae nitrogen fixation (nif) gene cluster with gene fusions. Nature, 286: 128-132.
Donald, R.G.K., Nees, D.W., Raymond, C.K., Loroch, A.I. and Ludwig, R.A., 1986. Characterization of three genomic loci encoding Rhizobium sp. strain ORS571 N_2 fixation genes. J. Bacteriol., 165: 72-81.
Drummond, M.H., 1985. The base sequence of the nifF gene of Klebsiella pneumoniae and homology of the predicted amino acid sequence of its protein product to other flavodoxins. Biochem. J., 232: 891-896.
Drummond, M.H. and Wooton, J.C., 1987. Sequence of nifL from Klebsiella pneumoniae: mode of action and relationship to two families of regulatory proteins. Mol. Microbiol., 1: 37-44.
Drummond, M., Clements, J., Merrick, M. and Dixon, R., 1983. Positive control and autogenous regulation of the nifLA promoter in Klebsiella pneumoniae. Nature, 301: 302-307.
Drummond M., Whitty, P. and Wootton, J., 1986. Sequence and domain relationships of ntrC and nifA from Klebsiella pneumoniae: homologies to other regulatory proteins. EMBO J., 5: 441-447.
Dusha, I., Kovalenko, S., Banfalvi, Z. and Kondorosi, A., 1987. Rhizobium meliloti insertion element ISRm2 and its use for identification of the fixX gene. J. Bacteriol., 169: 1403-1409.
Eady, R.R., 1986. Enzymology in free-living diazotrophs. In: W.J. Broughton and A. Pühler (Editors), Nitrogen Fixation (Molecular Biology). Clarendon Press, Oxford, pp.1-49.
Eady, R.R., Robson, R.L., Pau, R.N., Woodley, P., Lowe, D.J., Miller, R.W., Thorneley, R.N.F., Smith, B.E., Gormal, C., Fisher, K., Eldridge, M. and Bergström, J., 1988. The vanadium nitrogenase of Azotobacter chroococcum. In: H. Bothe, F.J. de Bruijn and W.E. Newton (Editors), Nitrogen Fixation: Hundred Years After. Gustav Fischer, Stuttgart, New York, pp.81-86.
Earl, C.D., Ronson, C.W. and Ausubel, F.M., 1987. Genetic and structural analysis of the Rhizobium meliloti fixA, fixB, fixC, and fixX genes. J. Bacteriol., 169: 1127-1136.

Ebeling, S., Noti, J.D. and Hennecke, H., 1988. Identification of a new Bradyrhizobium japonicum gene (frxA) encoding a ferredoxinlike protein. J. Bacteriol., 170: 1999-2001.
Espin, G., Alvarez-Morales, A. and Merrick, M., 1981. Complementation analysis of glnA-linked mutations which affect nitrogen fixation in Klebsiella pneumoniae. Mol. Gen. Genet., 184: 213-217.
Espin, G., Alvarez-Morales, A., Cannon, F., Dixon, R. and Merrick, M., 1982. Cloning of the glnA and ntrC genes of Klebsiella pneumoniae and studies of their role in regulation of the nitrogen fixation (nif) gene cluster. Mol. Gen. Gent., 186: 518-524.
Evans, D., Jones, R., Woodley, P. and Robson, R., 1988. Further analysis of nitrogen fixation (nif) genes in Azotobacter chroococcum: identification and expression in Klebsiella pneumoniae of nifS, nifV, nifM, and nifB genes and localization of nifE/N-, nifU-, nifA- and fixABC-like genes. J. Gen. Microbiol., 134: 931-942.
Fani, R., Allotta, G., Bazzicalupo, M., Ricci, F., Schipani, C. and Polsinelli, M., 1989. Nucleotide sequence of the gene encoding the nitrogenase iron protein (nifH) of Azospirillum brasilense and identification of a region controlling nifH transcription. Mol. Gen. Genet., 220: 81-87.
Filser, M., Merrick, M. and Cannon, F., 1983. Cloning and characterisation of nifLA regulatory mutations from Klebsiella pneumoniae. Mol. Gen. Genet., 191: 485-491.
Fischer, H-M and Hennecke, H., 1984. Linkage map of the Rhizobium japonicum nifH and nifDK operons encoding the polypeptides of the nitrogenase enzymes complex. Mol. Gen. Genet., 196: 537-540.
Fischer, H.M., Alvarez-Morales, A. and Hennecke, H., 1986. The pleiotropic nature of symbiotic regulatory mutants: Bradyrhizobium japonicum nifA gene is involved in control of nif gene expression and formation of determinate symbiosis. EMBO J., 5:1165-1173.
Fogher, C., Dusha, I., Barbot, P. and Elmerich, C., 1985. Heterologous hybridization of Azospirillum DNA to Rhizobium nod and fix genes. FEMS Microbiol. Lett., 30: 245-249.
Fuhrmann, M. and Hennecke, H., 1984. Rhizobium japonicum nitrogenase Fe protein gene (nifH). J. Bacteriol., 158: 1005-1011.
Fuhrmann, M., Fischer, H-M. and Hennecke, H., 1985. Mapping of Rhizobium japonicum nifB-, fixBC-, and fixA-like genes and identification of the fixA promoter. Mol. Gen. Genet., 199: 315-322.
Galimand, M., Perroud, B., Delorme, F., Paquelin, A., Vieille, C., Bozouklian, H. and Elmerich, C., 1989. Identification of DNA regions homologous to nitrogen fixation genes nifE, nifUS and fixABC in Azospirillum brasilense Sp7. J. Gen. Microbiol., 135: 1047-1059.
Garcia, E., Bancroft, S., Goo Rhee, S. and Kustu S., 1977. The product of a newly identified gene, glnF, is required for synthesis of glutamine synthetase in Salmonella. Proc. Natl. Acad. Sci. USA, 74: 1662-1666.
Ginsburg, A. and Stadtman, E.R., 1973. Regulation of glutamine synthetase in Escherichia coli. In: S. Prusiner and E.R. Stadtman (Editors), The Enzymes of Glutamine Metabolism. (164th National Meeting of the American Chemical Society, New York City, August 30-31, 1972) Academic Press, New York and London, p.9.
Golden, J.W. and Wiest, D.R., 1988. Genome rearrangement and nitrogen fixation in Anabaena blocked by inactivation of xisA gene. Science, 242: 1421-1423.
Golden, J.W., Robinson, S.J. and Haselkorn, R., 1985. Rearrangement of nitrogen fixation genes during heterocyst differentiation in the cyanobacterium Anabaena. Nature, 314: 419-423.
Golden, J.W., Carrasco, C.D., Mulligan, M.E., Schneider, G.J. and Haselkorn, R., 1988. Deletion of a 55-kilobase-pair DNA element from the chromosome during heterocyst differentiation of Anabaena sp. strain PCC 7120. J. Bacteriol., 170: 5034-5041.

Gosink, M.M., Franklin, N.M. and Roberts, G.P., 1990. The product of the Klebsiella pneumoniae nifX gene is a negative regulator of the nitrogen fixation (nif) regulon. J. Bacteriol., 172: 1441-1447.
Grönger, P., Mannian S.S., Reiländer, H., O'Connell, M., Priefer, U.B. and Pühler A., 1987. Organization and partial sequence of a DNA region of the Rhizobium leguminosarum symbiotic plasmid pRL6JI containing the genes fixABC, nifA, nifB and a novel open reading frame. Nucl. Acids Res., 15: 31-49.
Gubler, M. and Hennecke, H., 1986. FixA, B and C genes are essential for symbiotic and free-living, microaerobic nitrogen fixation. FEBS Lett., 200: 186-192.
Gubler, M. and Hennecke, H., 1988. Regulation of the fixA gene and fixBC operon in Bradyrhizobium japonicum. J. Bacteriol., 170: 1205-1214.
Gubler, M., Zürcher, T. and Hennecke, H., 1989. The Bradhyrhizobium japonicum fixBCX operon: identification of fixX and of a 5' mRNA region affecting the level of the fixBCX transcript. Mol. Microbiol., 3: 141-148.
Gussin, G.N., Ronson, C.W. and Ausubel, F.M., 1986. Regulation of nitrogen fixation genes. Ann. Rev. Genet., 20: 567-591.
Haselkorn, R., 1986. Organization of the genes for nitrogen fixation in photosynthetic bacteria and cyanobacteria. Ann. Rev. Microbiol., 40: 525-547.
Haselkorn, R., Rice, D., Curtis, S.E. and Robinson, S.J., 1983. Organization and transcription of genes important in Anabaena heterocyst differentiation. Ann. Microbiol. (Inst. Pasteur), 134 B: 181-193.
Hawkes, T.R., McLean, P.A. and Smith, B.E., 1984. Nitrogenase from nifV mutants of Klebsiella pneumoniae contains an altered form of the iron-molybdenum cofactor. Biochem. J., 217: 317-321.
Hawkes, T., Merrick, M. and Dixon, R., 1985. Interaction of purified NtrC protein with nitrogen regulated promoters from Klebsiella pneumoniae. Mol. Gen. Genet., 201: 492-498.
Henderson, N., Austin, S. and Dixon, R.A., 1989. Role of metal ions in negative regulation of nitrogen fixation by the nifL gene product from Klebsiella pneumoniae. Mol. Gen. Genet., 216: 484-491.
Hennecke, H., Kaluza, K., Thöny, B., Fuhrmann, M., Ludwig, W. and Stackebrandt E., 1985. Concurrent evolution of nitrogenase genes and 16S rRNA in Rhizobium species and other nitrogen fixing bacteria. Arch. Microbiol., 142: 342-348.
Hennecke, H., Fischer, H-M., Gubler, M., Thöny, B., Anthamatten, D., Kullik, I., Ebeling, S., Fritsche, S. and Zürcher, T., 1988. Regulation of nif and fix genes in Bradyrhizobium japonicum occurs by a cascade of two consecutive gene activation steps of which the second one is oxygen sensitive. In: H. Bothe, F.J. de Bruijn and W.E. Newton (Editors), Nitrogen Fixation: Hundred Years After. Gustav Fischer, Stuttgart, New York, pp.339-344.
Hill, S., Kennedy, C., Kavanagh, E., Goldberg, R.B. and Hanau, R., 1981. Nitrogen fixation gene (nifL) involved in oxygen regulation of nitrogenase synthesis in K. pneumoniae. Nature, 290: 424-426.
Hirschman, J., Wong, P-K., Sei, K., Keener, J. and Kustu, S., 1985. Products of nitrogen regulatory genes ntrA and ntrC of enteric bacteria activate glnA transcription in vitro: evidence that the ntrA product is a sigma factor. Proc. Natl. Acad. Sci. USA, 82: 7525-7529.
Holtel, A. and Merrick, M.J., 1988. Identification of the Klebsiella pneumoniae glnB gene: nucleotide sequence of wild-type and mutant alleles. Mol. Gen. Genet., 215: 134-138.
Holtel, A. and Merrick, M.J., 1989. The Klebsiella pneumoniae P_{II} protein (glnB gene product) is not absolutely required for nitrogen regulation and is not involved in NifL-mediated nif gene regulation. Mol. Gen. Genet., 217: 474-480.

Hontelez J.G.J., Lankhorst, R.K., Katinakis, P., van den Bos, R. and van Kammen, A., 1989. Characterization and nucleotide sequence of a novel gene fixW upstream of the fixABC operon in Rhizobium leguminosarum. Mol. Gen. Genet., 218: 536-544.
Hoover, T.R., Robertson, A.D., Cerny, R.L., Hayest, R.N., Imperial, J., Shah, V.K. and Ludden, P.W., 1987. Identification of the V factor needed for synthesis of the iron-molybdenum cofactor of nitrogenase as homocitrate. Nature, 329: 855-857.
Hoover, T.R., Imperial, J., Ludden, P.W. and Shah, V.K., 1988. Homocitrate cures the NifV$^-$ phenotype in Klebsiella pneumoniae. J. Bacteriol., 170: 1978-1979.
Howard, K.S., McLean, P.A., Hansen, F.B., Lemley, P.V., Kobla,. K.S. and Orme-Johnson, W.H., 1986. Klebsiella pneumoniae nifM gene product is required for stabilization and activation of nitrogenase iron protein in Escherichia coli. J. Biol. Chem., 261: 772-778.
Huala, E. and Ausubel, F.M., 1989. The central domain of Rhizobium meliloti NifA is sufficient to activate transcription from the R. meliloti nifH promoter. J. Bacteriol., 171: 3354-3365.
Hunt, T. and Magasanik, B., 1985. Transcription of glnA by purified Escherichia coli components: core RNA polymerase and the products of glnF, glnG, and glnL. Proc. Natl. Acad. Sci. USA, 82: 8453-8457.
Iisma, S.E. and Watson, J.M., 1989. The nifA gene product from Rhizobium leguminosarum biovar trifolii lacks the *N*-terminal domain found in other NifA proteins. Mol. Microbiol., 3: 943-955.
Iisma, S.E., Ealing, P.M., Scott, K.F. and Watson, J.M., 1989. Molecular linkage of the nif/fix and nod gene regions in Rhizobium leguminosarum biovar trifolii. Mol. Microbiol., 3: 1753-1764.
Imperial, J., Ugalde, R.A., Shah, V.D. and Brill, W.J., 1984. Role of the nifQ gene product in the incorporation of molybdenum into nitrogenase in Klebsiella pneumoniae. J. Bacteriol., 158: 187-194.
Jacobson M.R., Brigle, K.V., Bennett, L.T., Setterquist R.A., Wilson, M.S., Cash, V.L., Beynon, J., Newton, W.E. and Dean, D.R., 1989a. Physical and genetic map of the major nif gene cluster from Azotobacter vinelandii. J. Bacteriol., 171: 1017-1027.
Jacobson, M.R., Cash, V.L., Weiss, M.C., Laird, N.F., Newton, W.E. and Dean, D.R., 1989b. Biochemical and genetic analysis of the nifUSVWZM cluster from Azotobacter vinelandii. Mol. Gen. Genet., 219: 49-57.
Joerger, R.D. and Bishop, P.E., 1988. Nucleotide sequence and genetic analysis of the nifB-nifQ region from Azotobacter vinelandii. J. Bacteriol., 170: 1475-1487.
Joerger, R.D., Jacobson, M.R., Premakumar, R., Wolfinger, E.D. and Bishop, P.E., 1989. Nucleotide sequence and mutational analysis of the structural genes (anfHDGK) for the second alternative nitrogenase from Azotobacter vinelandii. J. Bacteriol., 171: 1075-1086.
Jones, R. and Haselkorn, R., 1988. The DNA sequence of the Rhodobacter capsulatus nifH gene. Nucl. Acids Res., 16: 8735.
Jones, R. and Haselkorn, R., 1989. The DNA sequence of the Rhodobacter capsulatus ntrA, ntrB and ntrC gene analogues required for nitrogen fixation. Mol. Gen. Genet., 215: 507-516.
Kahn, D., David, M., Domergue, O., Daveran, M-L., Ghai, J., Hirsch, P.R. and Batut, J., 1989. Rhizobium meliloti fixGHI sequence predicts involvement of a specific cation pump in symbiotic nitrogen fixation. J. Bacteriol., 171: 929-939.
Kallas, T., Rebière, M-C., Ripka, R. and Tandeau de Marsac, N., 1983. The structural nif genes of the cyanobacteria Gloeothece sp. and Calothrix sp. share homology with those of Anabaena sp. but the Gloeothece genes have a different arrangement. J. Bacteriol., 155: 5034-5041.

Kaluza, K. and Hennecke, H., 1984. Fine structure analysis of the nifDK operon encoding the alpha and beta subunits of dinitrogenase from Rhizobium japonicum. Mol. Gen. Genet., 196: 35-42.
Kaluza, K., Alvarez-Morales, A. and Hennecke, H., 1985. Oligonucleotide-directed mutagenesis of the Rhizobium japonicum nifH promoter. FEBS, 188: 37-42.
Kaminski, P.A., Norel, F., Desnoues, N., Kush, A., Salzano, G. and Elmerich, C., 1988. Characterization of the fixABC region of Azorhizobium caulinodans ORS571 and identification of a new nitrogen fixation gene. Mol. Gen. Genet., 214: 496-502.
Keener, J. and Kustu, S., 1988. Protein kinase and phosphoprotein phosphatase activities of nitrogen regulatory proteins NTRB and NTRC of enteric bacteria: roles of the conserved amino-terminal domain of NTRC. Proc. Natl. Acad. Sci. USA, 85: 4976-4980.
Kennedy, C., 1977. Linkage map of the nitrogen fixation (nif) genes in Klebsiella pneumoniae. Mol. Gen. Genet., 157: 199-204.
Kennedy, C. and Robson, R.L., 1983. Activation of nif gene expression in Azotobacter by the nifA gene product of Klebsiella pneumoniae. Nature, 301: 626-628.
Kennedy, C. and Toukdarian, A., 1987. Genetics of Azotobacters: applications to nitrogen fixation and related aspects of metabolism. Ann. Rev. Microbiol., 41: 227-258.
Kennedy, C., Gamal, R., Humphrey, R., Ramos, J., Brigle, K. and Dean, D., 1986. The nifH, nifM and nifN genes of Azotobacter vinelandii: characterization by Tn*5* mutagenesis and isolation from pLAFR1 gene banks. Mol. Gen. Genet., 205: 318-325.
Kent, H.M., Buck, M. and Evans, D.J., 1989. Cloning and sequencing of the nifH gene of Desulfovibrio gigas. FEMS Microbiol. Lett., 61: 73-78.
Khan, H., Buck, M. and Dixon, R., 1986. Deletion loop mutagenesis of the nifL promoter from Klebsiella pneumoniae: role of the -26 to -12 region in promoter function. Gene, 45: 281-288.
Klipp, W., Masepohl, B. and Pühler, A., 1988. Identification and mapping of nitrogen fixation genes of Rhodobacter capsulatus: duplication of a nifA-nifB region. J. Bacteriol., 170: 693-699.
Klipp, W., Reiländer, H., Schlüter, A., Krey, R. and Pühler, A., 1989. The Rhizobium meliloti fdxN gene encoding a ferredoxin-like protein is necessary for nitrogen fixation and is cotranscribed with nifA and nifB. Mol. Gen. Genet., 216: 293-302.
Kranz, R. and Haselkorn, R., 1985. Characterization of nif regulatory genes in Rhodopseudomonas capsulata using lac gene fusions. Gene, 40: 203-215.
Krol, A.J.M., Hontelez, J.G.J., Roozndaal, B. and van Kammen, A., 1982. On the operon structure of the nitrogenase genes of Rhizobium leguminosarum and Azotobacter vinelandii. Nucl. Acids Res., 10: 4147-4157.
Kustu, S., Burton, D., Garcia, E., McCarter, L. and McFarland, N., 1979. Nitrogen control in Salmonella: regulation by the glnR and glnF gene products. Proc. Natl. Acad. Sci. USA, 76: 4576-4580.
Lammers, P.J. and Haselkorn, R., 1983. Sequence of the nifD gene coding for the alpha subunit of dinitrogenase from the cyanobacterium Anabaena. Proc. Natl. Acad. Sci. USA, 80: 4723-4727.
Lammers, P.J., Golden, J.W. and Haselkorn, R., 1986. Identification and sequence of a gene required for a developmentally regulated DNA excision in Anabaena. Cell, 44: 905-911.
Leonardo, J.M. and Goldberg, R.B., 1980. Regulation of nitrogen metabolism in glutamine auxotrophs of Klebsiella pneumoniae. J. Bacteriol., 142: 99-110.
McFarland, N., McCarter, L., Artz, S. and Kustu, S., 1981. Nitrogen regulatory locus "glnR" of enteric bacteria is composed of cistrons ntrB and ntrC: identification of their protein products. Proc. Natl. Acad. Sci. USA, 78: 2135-2139.

McFarlane, S.A. and Merrick, M., 1985. The nucleotide sequence of the nitrogen regulation gene (ntrB) and the glnA-ntrBC intergenic region of Klebsiella pneumoniae. Nucl. Acids Res., 13: 7591-7606.
McFarlane, S.A. and Merrick, M.J., 1987. Analysis of the Klebsiella pneumoniae ntrB gene by site-directed in vitro mutagenesis. Mol. Microbiol., 1: 133-142.
McLean, P.A. and Dixon, R.A., 1981. Requirement of nifV gene for production of wild-type nitrogenase enzyme in Klebsiella pneumoniae. Nature, 292: 655-656.
McNeil, D. and Brill, W.J., 1980. Mutations in nif genes that cause Klebsiella pneumoniae to be derepressed for nitrogenase synthesis in the presence of ammonium. J. Bacteriol., 144: 744-751.
McNeil, T., McNeil, D., Roberts, G.P., Supiano, M.A. and Brill, W.J., 1978. Fine structure mapping and complementation analysis of nif (nitrogen fixation) genes in Klebsiella pneumoniae. J. Bacteriol., 136: 253-266.
McNeil, D., Zhu, J. and Brill, W.J., 1981. Regulation of nitrogen fixation in Klebsiella pneumoniae: Isolation and characterization of strains with nif-lac fusions. J. Bacteriol., 145: 348-357.
McNeil, T., McNeil, D. and Tyler B., 1982. Fine-structure deletion map and complementation analysis of the glnA-glnL-GlnG region in Escherichia coli. J. Bacteriol., 150: 1302-1313.
Magasanik, B., 1982. Genetic control of nitrogen assimilation in bacteria. Ann. Rev. Genet., 16: 135-168.
Magasanik, B., 1988. Reversible phosphorylation of an enhancer binding protein regulates the transcription of bacterial nitrogen utilization genes. TIBS, 13: 475-479.
Martin, G.B., Chapman, K.A. and Barry, K.C., 1988. Role of the Bradyrhizobium japonicum ntrC gene product in differential regulation of the glutamine synthetase II gene (glnII). J. Bacteriol., 170: 5452-5459.
Martin, G.B., Thomashow, M.F. and Chelm, B.K., 1989. Bradyrhizobium japonicum glnB, a putative nitrogen-regulatory gene, is regulated by NtrC at tandem promoters. J. Bacteriol., 171: 5638-5645.
Masepohl, B., Klipp, W. and Pühler, A., 1988. Genetic characterization and sequence analysis of the duplicated nifA/nifB gene region of Rhodobacter capsulatus. Mol. Gen. Genet., 212: 27-37.
Mazur, B.J. and Chui, C-F., 1982. Sequence of the gene coding for the beta-subunit of dinitrogenase from the blue-green alga Anabaena. Proc. Natl. Acad. Sci. USA, 79: 6782-6786.
Mazur, B.J., Rice, D. and Haselkorn, R., 1980. Identification of blue green algal nitrogen fixation genes by using heterologous DNA hybridization probes. Proc. Natl. Acad. Sci. USA, 77: 186-190.
Merrick, M.J., 1983. Nitrogen control of the nif regulon in Klebsiella pneumoniae: involvement of the ntrA gene and analogies between ntrC and nifA. EMBO J., 2: 39-44.
Merrick, M.J., 1988a. Organisation and regulation of nitrogen fixation genes in Klebsiella pneumoniae and Azotobacter. In: H. Bothe, F.J. de Bruijn and W.E. Newton (Editors), Nitrogen Fixation: Hundred Years After. Gustav Fischer, Stuttgart, New York, pp. 293-302.
Merrick, M.J., 1988b. Regulation of nitrogen assimilation by bacteria. In: J.A. Cole and S.J. Ferguson (Editors), The Nitrogen and Sulphur Cycles. Cambridge University Press, Cambridge, pp. 331-361.
Merrick, M.J., Gibbins, J.R. and Toukadavian, A., 1987. The nucleotide sequence of the sigma factor gene nitrA (rpoN) of Azotobacter vinelandii; Analysis of conserved sequences in nitrA proteins. Mol. Gen. Genet. 210: 323-330.
Merrick, M., Filser, M., Dixon, R., Elmerich, C., Sibold, L. and Houmard, J., 1980. Use of translocatable genetic elements to construct a fine structure map of the Klebsiella pneumoniae nitrogen fixation (nif) gene cluster. J. Gen. Microbiol., 117: 509-520.

Merrick, M., Hill, S., Henneke, H., Kahn, D., Dixon, R. and Kennedy, C., 1982. Repressor properties of the nifL gene product in Klebsiella pneumoniae. Mol. Gen. Genet., 185: 75-81.
Mevarech, M., Rice, D. and Haselkorn, R., 1980. Nucleotide sequence of a cyanobacterial nifH gene coding for nitrogenase reductase. Proc. Natl. Acad. Sci. USA, 77: 6476-6480.
Minchin, S.D., Austin, S. and Dixon, R.A., 1988. The role of activator binding sites in transcriptional control of the divergently transcribed nifF and nifLA promoters from Klebsiella pneumoniae. Mol. Microbiol., 2: 433-442.
Minchin, S.D., Austin, S. and Dixon, R.A., 1989. Transcriptional activation of the Klebsiella pneumoniae nifLA promoter by NTRC is face-of-the-helix dependent and the activator stabilizes the interaction of sigma 54-RNA polymerase with the promoter. EMBO J., 8: 3491-3499.
Moreno-Vivian, C., Hennecke, S., Pühler, A. and Klipp, W., 1989a. Open reading frame 5 (ORF5), encoding a ferredoxin-like protein, and nifQ are cotranscribed with nifE, nifN, nifX, and ORF4 in Rhodobacter capsulatus. J. Bacteriol., 171: 2591-2598.
Moreno-Vivian, C., Schmehl, M., Masepohl, B., Arnold, W. and Klipp, W., 1989b. DNA sequence and genetic analysis of the Rhodobacter capsulatus nifENX gene region: homology between NifX and NifB suggests involvement of NifX in processing of the iron-molybdenum cofactor. Mol. Gen. Genet., 216: 353-363.
Mulligan M.E., Buikema, W.J. and Haselkorn R., 1988. Bacterial-type ferredoxin genes in the nitrogen fixation regions of the cyanobacterium Anabaena sp. strain PCC 7120 and Rhizobium meliloti. J. Bacteriol., 170: 4406-4410.
Murray, P.A. and Zinder, S., 1984. Nitrogen fixation by a methanogenic archaebacterium. Nature, 312: 284-286.
Nees, D.W., Stein, A. and Ludwig, R., 1988. The Azorhizobium caulinodans nifA genes: identification of upstream-activating sequences including a new element, the "anaerobox". Nucl. Acids Res., 16: 9839-9851.
Nieva-Gomez, D., Roberts, G.P., Klevickis, S. and Brill, W.J., 1985. Electron transport to nitrogenase in Klebsiella pneumoniae. Proc. Natl. Acad. Sci. USA, 77: 2555-2558.
Ninfa, A.J. and Magasanik, B., 1986. Covalent modification of the glnG product, NR_I, by the glnL product, NR_{II}, regulates the transcription of the glnALG operon in Escherichia coli. Proc. Natl. Acad. Sci. USA, 83: 5909-5913.
Nixon, B.T., Ronson, C.W. and Ausubel, F.M., 1986. Two-component regulatory systems responsive to environmental stimuli share strongly conserved domains with the nitrogen assimilation regulatory genes ntrB and ntrC. Proc. Natl. Acad. Sci. USA, 83: 7850-7854.
Norel, F. and Elmerich, C., 1987. Nucleotide sequence and functional analysis of the two nifH copies of Rhizobium ORS571. J. Gen. Microbiol., 133: 1563-1576.
Normand, P. and Bousquet, J., 1989. Phylogeny of nitrogenase sequences in Frankia and other nitrogen-fixing microorganisms. J. Mol. Evol., 29: 436-447.
Normand, P., Simonet, P. and Bardin, R., 1988. Conservation of nif sequences in Frankia. Mol. Gen. Genet., 213: 238-246.
Nuti, M.P., Lepidi, A.A., Prakash, R.K., Schilperoort, R.A. and Cannon, F.C., 1979. Evidence for nitrogen fixation (nif) genes on indigenous Rhizobium plasmids. Nature, 282: 533-535.
Ow, D.W. and Ausubel, F.M., 1983. Regulation of nitrogen metabolism genes by nifA gene product in Klebsiella pneumoniae. Nature, 301: 307-313.
Ow D.W., Sundaresan, V., Rothstein, D.M., Brown, S.E. and Ausubel, F.M., 1983. Promoters regulated by the glnG (ntrC) and nifA gene products share a heptameric concensus sequence in the -15 region. Proc. Natl. Acad. Sci. USA, 80: 2524-2528.
Ow, D.W., Xiong, Y., Gu, Q. and Shen, S-C., 1985. Mutational analysis of the Klebsiella pneumoniae nitrogenase promoter: sequences essential for positive control by nifA and ntrC (glnG) products. J. Bacteriol., 161: 868-874.

Pahel, G. and Tyler, B., 1979. A new glnA-linked regulatory gene for glutamine synthetase in Escherichia coli. Proc. Natl. Acad. Sci. USA, 76: 4544-4548.
Paul, W. and Merrick, M., 1987. The nucleotide sequence of the nifM gene of Klebsiella pneumoniae and identification of a new nif gene: nifZ. Eur. J. Biochem., 170: 259-265.
Paul, W. and Merrick, M., 1989. The roles of the nifW, nifZ and nifM genes of Klebsiella pneumoniae nitrogenase biosynthesis. Eur. J. Biochem., 178: 675-682.
Paustian T.D., Shah, V.K. and Roberts, G.P., 1989. Purification and characterization of the nifN and nifE gene products from Azotobacter vinelandii mutant UW45. Proc. Natl. Acad. Sci. USA, 86: 6082-6086.
Pawlowski, K., Ratet, P., Schell, J. and de Bruijn, F.J., 1987. Cloning and characterization of nifA and ntrC genes of the stem nodulating bacterium ORS571, the nitrogen fixing symbiont of Sesbania rostrata: regulation of nitrogen fixation (nif) genes in the free living versus symbiotic state. Mol. Gen. Genet., 206: 207-219.
Pedrosa, F.O. and Yates, M.G., 1984. Regulation of nitrogen fixation (nif) genes of Azospirillum brasilense by nifA and ntr (gln) type gene products. FEMS Microbiol. Lett., 23: 95-101.
Pollock, D., Bauer, C.E. and Scolnik, P.A., 1988. Transcription of the Rhodobacter capsulatus nifHDK operon is modulated by the nitrogen source. Construction of plasmid expression vectors based on the nifHDK promoter. Gene, 65: 269-275.
Possot, O., Henry, M. and Sibold, L., 1986. Distribution of DNA sequences homologous to nifH among archaebacteria. FEMS Microbiol. Lett., 34: 173-177.
Prakash, R.K. and Atherly, A.G., 1984. Reiteration of genes involved in symbiotic nitrogen fixation by fast-growing Rhizobium japonicum. J. Bacteriol., 160: 785-787.
Prakash, R.K. and Atherly, A., 1986. Plasmids of Rhizobium and their role in symbiotic nitrogen fixation. Int. Rev. Cytol., 104: 1-24.
Prakash, R.K., Schilperoort, R.A. and Nuti, M.P., 1981. Large plasmids of fast-growing rhizobia: homology studies and location of structural nitrogen fixation (nif) genes. J. Bacteriol., 145: 1129-1136.
Pretorius, I.M., Rawlings, D.E., O'Neill, E.G., Jones, W.A., Kirby, R. and Woods, D.R., 1987. Nucleotide sequence of the gene encoding the nitrogenase iron protein of Thiobacillus ferrooxidans. J. Bacteriol., 169: 367-370.
Pühler, A. and Klipp, W., 1981. Fine structure analysis of the gene region for N_2-fixation (nif) of Klebsiella pneumoniae. In: H. Bothe and A. Trebst (Editors), Biology of Inorganic Nitrogen and Sulfur. Springer Verlag, Berlin, Heidelberg, New York, pp. 276-286.
Pühler, A., Aguilar, M.O., Hynes, M., Müller, P., Klipp, W., Priefer, U., Simon, R. and Weber, G., 1984. Advances in the genetics of free-living and symbiotic nitrogen fixing bacteria. In: C. Veeger and W.E. Newton (Editors), Advances in Nitrogen Fixation Research. Martinus Nijhoff, Dr. W. Junk Publishers, The Hague, p.609.
Quinto, C., de la Vega, H., Flores, M., Fernandez, L., Ballado, T., Soberon, G. and Palacios, R., 1982. Nitrogen fixation genes are reiterated in Rhizobium phaseoli. Nature, 299: 724-726.
Quinto, C., de la Vega, H., Flores, M., Leemans, J., Cevallos, M.A., Pardo, M.A., Azpiroz, R., Girard, M.D.L., Calva, E. and Palacios R., 1985. Nitrogenase reductase: a functional multigene family in Rhizobium phaseoli. Proc. Natl. Acad. Sci. USA, 82: 1170-1174.
Ratet, P., Pawlowski, K., Schell, J. and de Bruijn, F.J., 1989. The Azorhizobium caulinodans nitrogen-fixation regulatory gene, nifA, is controlled by the cellular nitrogen and oxygen status. Mol. Microbiol., 3: 825-838.

Renalier, M-H., Batut, J., Ghai, J., Terzaghi, B., Gherardi, M., David, M., Garnerone, A-M., Vasse, J., Truchet, G., Huguet, T. and Boistard, P., 1987. A new symbiotic cluster on the pSym megaplasmid of Rhizobium meliloti 2011 carries a functional fix gene repeat and a nif locus. J. Bacteriol., 169: 2231-2238.

Rice, D., Mazur, B.J. and Haselkorn, R., 1982. Isolation and physical mapping of nitrogen fixation genes from the cyanobacterium Anabaena 7120. J. Biol. Chem., 257: 13157-13163.

Riedel, G.E., Ausubel, F.M. and Cannon, F.C., 1979. The physical map of nitrogen fixation (nif) genes of Klebsiella pneumoniae. Proc. Natl. Acad. Sci. USA, 76: 2866-2870.

Roberts, G.P., and Brill, W.J., 1981. Genetics and regulation of nitrogen fixation. Ann. Rev. Microbiol., 35: 207-235.

Roberts, G.P., McNeil, T., McNeil, D. and Brill, W.J., 1978. Regulation and characterization of protein products coded by the nif (nitrogen fixation) genes of Klebsiella pneumoniae. J. Bacteriol., 136: 267-279.

Robson, R.L., 1984. Identification of possible adenine nucleotide-binding sites in nitrogenase Fe- and MoFe-proteins by amino acid sequence comparison. FEBS, 173: 394-398.

Robson, R., Kennedy, C. and Postgate, J.R., 1983. Progress in comparative genetics of nitrogen fixation. Can. J.. Microbiol., 29: 954-967.

Robson, R., Jones, R., Kennedy, C.K., Drummond, M., Ramos, J., Woodley, P.R., Wheeler, C., Chesshyre, J. and Postgate, J., 1984. Aspects of genetics of Azotobacter. In: C. Veeger and W.E. Newton (Editors), Advances in Nitrogen Fixation Research. The Hague: M. Nijhoo and Junk Publishers and Wageningen: Pudoc, pp.643-651.

Robson, R., Woodley, P. and Jones, R., 1986. Second gene (nifH*) coding for a nitrogenase iron protein in Azotobacter chroococcum is adjacent to a gene coding for a ferredoxin-like protein. EMBO J., 5: 1159-1163.

Robson, R.L., Woodley, P.R., Pau, R.N. and Eady, R.R., 1989. Structural genes for the vanadium nitrogenase from Azotobacter chroococcum. EMBO J., 8: 1217-1224.

Roelvink, P.W., Hontelez, J.G., van Kammen, A. and van den Bos, R.C., 1989. Nucleotide sequence of the regulatory nifA gene of Rhizobium leguminosarum PRE: transcriptional control sites and expression in Escherichia coli. Mol. Microbiol., 3: 1441-1447.

Roelvink, P.W., Harmsen, M., van Kammen, A. and van den Bos, R., 1990. The nifH promoter region of Rhizobium leguminosarum: nucleotide sequence and promoter elements controlling activation by NifA protein. Gene, 87: 31-36.

Ronson, C.W., Nixon, B.T., Albright, L.M. and Ausubel, F.M., 1987. Rhizobium meliloti ntrA (rpoN) gene is required for diverse metabolic functions. J. Bacteriol., 169: 2424-2431.

Rosenberg, C., Boistard, P., Dénarié, J. and Casse-Delbart, F., 1981. Genes controlling early and late functions in symbiosis are located on a megaplasmid in Rhizobium meliloti. Mol. Gen. Genet., 184: 326-333.

Rossen, L., Ma, Q.S., Mudd, E.A., Johnston, A.W.B. and Downie, J.A., 1984. Identification and DNA sequence of fixZ and nifB-like gene from Rhizobium leguminosarum. Nucl. Acids Res., 12: 7123-7134.

Ruvkun, G.B. and Ausubel, F.M., 1980. Interspecies homology of nitrogenase genes. Proc. Natl. Acad. Sci. USA, 77: 191-195.

Ruvkun, G.B., Sundaresan, V. and Ausubel, F.M., 1982. Directed transposon Tn5 mutagenesis and complementation analysis of Rhizobium meliloti symbiotic nitrogen fixation genes. Cell, 29: 551-559.

Santero, E., Toukdarian, A., Humphrey, R. and Kennedy, C., 1988. Identification and characterization of two nitrogen-fixation regulatory regions, nifA and nfrX, in Azotobacter vinelandii and Azotobacter chroococcum. Mol. Microbiol., 2: 303-314.

Schetgens R.M.P., Hontelez, J.G.J., van den Bos, R.C. and van Kammen, A., 1985. Identification and phenotypical characterization of a cluster of fix genes, including a nif regulatory gene, from Rhizobium leguminosarum PRE. Mol. Gen. Genet., 200: 368-374.
Schumann, J.P., Waitches, G.M. and Scolnik, P.A., 1986. A DNA fragment hybridizing to a nif probe in Rhodobacter capsulatus is homologous to a 16S rRNA gene. Gene, 48: 81-92.
Scolnik, P.A. and Haselkorn, R., 1984. Activation of extra copies of genes coding for nitrogenase in Rhodopseudomonas capsulata. Nature, 307: 289-292.
Scott, K.F., Rolfe, B.G. and Shine, J., 1981. Biological nitrogen fixation: Primary structure of the Klebsiella pneumoniae nifH and nifD genes. J. Mol. Appl. Genet., 1: 71-81.
Scott, K.F., Rolfe, B.G. and Shine, J., 1983a. Nitrogenase structural genes are unlinked in the nonlegume symbiont Parasponia Rhizobium. DNA, 2: 141-148.
Scott, K.F., Rolfe, B.G. and Shine, J., 1983b. Biological nitrogen fixation: Primary structure of the Rhizobium trifolii iron protein gene. DNA, 2: 149-155.
Shah, V.K., Stacey, G. and Brill, W.J., 1983. Electron transport to nitrogenase: Purification and characterization of pyruvate: flavodoxin oxidoreductase, the nifJ gene product. J. Biol. Chem., 258: 12064-12068.
Sibold, L., 1982. The polar effect on nifM of mutations in the nifU, -S, -V genes of Klebsiella pneumoniae depends on their plasmid or chromosomal location. Mol. Gen. Genet., 186: 569-571.
Sibold, L. and Elmerich, C., 1982. Constitutive expression of nitrogen fixation (nif) genes of Klebsiella pneumoniae due to a DNA duplication. EMBO J., 1: 1551-1558.
Sibold, L., Melck, D. and Elmerich, C., 1981. A nif mutant of Klebsiella pneumoniae fixing nitrogen in the presence of ammonia. FEMS Microbiol. Lett., 10: 37-41.
Sibold, L., Quiviger, B., Charpin, N., Paquelin, A. and Elmerich, C., 1983. Cloning and expression of a DNA fragment carrying a his nifA fusion and the nifBQ operon from a nif constitutive mutant of Klebsiella pneumoniae. Biochimie, 65: 53.
Sibold, L., Pariot, D., Bhatnagar, L., Henriquet, M. and Aubert, J.-P., 1985. Hybridization of DNA from methanogenic bacteria with nitrogenase structural genes (nifHDK). Mol. Gen. Genet., 200: 40-46.
Singh, M., Tripathi, A.K. and Klingmüller, W., 1989. Identification of a regulatory nifA type gene and physical mapping of cloned new nif regions of Azospirillum brasilense. Mol. Gen. Genet., 219: 235-240.
Somerville, J.E. and Kahn, M.L., 1983. Cloning of the glutamine synthetase I gene from Rhizobium meliloti. J. Bacteriol., 156: 168-176.
Souillard, N. and Sibold, L., 1986. Primary structure and expression of a gene homologous to nifH (nitrogenase Fe protein) from the archaebacterium Methanococcus voltae. Mol. Gen. Genet., 203: 21-28.
Souillard, N. and Sibold, L., 1989. Primary structure, functional organization and expression of nitrogenase structural genes of the thermophilic archaebacterium Methanococcus thermolithotrophicus. Mol. Microbiol., 3: 541-551.
Souillard, N., Magot, M., Possot, O. and Sibold, L., 1988. Nucleotide sequence of regions homologous to nifH (nitrogenase Fe protein) from the nitrogen-fixing Archaebacteria Methanococcus thermolithotrophicus and Methanobacterium ivanovii: evolutionary implications. J. Mol. Evol., 27: 65-76.
Stanley J., van Slooten J., Dowling, D.N., Finan, T. and Broughton, W.J., 1989. Molecular cloning of the ntrA gene of the broad host-range Rhizobium sp. NGR234, and phenotypes of a site-directed mutant. Mol. Gen. Gent., 217: 528-532.
Stock, J.B., Ninfa, A.J. and Stock, A.M., 1989. Protein phosphorylation and regulation of adaptive responses in bacteria. Microbiol. Rev., 53: 450-490.

Streicher, S.L., Gurney, E. and Valentine R.C., 1972. The nitrogen fixation genes in Klebsiella pneumoniae. Nature, 239: 495-499.

Streicher, S.L., Shanmugam, K.T., Ausubel, F., Morandi, C. and Goldberg, R.B., 1974. Regulation of nitrogen fixation in Klebsiella pneumoniae: evidence of a role of glutamine synthetase as a regulator of nitrogenase synthesis. J. Bacteriol., 120: 815-821.

Sundaresan, V. and Ausubel, F.M., 1981. Nucleotide sequence of the gene coding for the nitrogenase iron protein from Klebsiella pneumoniae. J. Biol. Chem., 256: 2808-2812.

Sundaresan, V., Jones, J.D.G., Ow, D.W. and Ausubel, F.M., 1983a. Klebsiella pneumoniae nifA product activates the Rhizobium meliloti nitrogenase promoter. Nature, 301: 728-731.

Sundaresan, V., Ow, D.W. and Ausubel, F.M., 1983b. Activation of Klebsiella pneumoniae and Rhizobium meliloti nitrogenase promoters by gln (ntr) regulatory proteins. Proc. Natl. Acad. Sci. USA, 80: 4030-4034.

Szeto W.W., Zimmerman, J.L., Sundaresan, V. and Ausubel, F.M., 1984. A Rhizobium meliloti symbiotic regulatory gene. Cell, 36: 1035-1043.

Szeto, W.W., Nixon, B.T., Ronson, C.W. and Ausubel, F.M., 1987. Identification and characterization of the Rhizobium meliloti ntrC gene: R. meliloti has separate regulatory pathways for activation of nitrogen fixation genes in free-living and symbiotic cells. J. Bacteriol., 169: 1423-1432.

Thöny, B. and Hennecke, H., 1989. The -24/-12 promoter comes of age. FEMS Microbiol. Rev., 63: 341-358.

Thöny, B., Kaluza, K. and Hennecke, H., 1985. Structural and functional homology between the alpha and beta subunits of the nitrogenase MoFe protein as revealed by sequencing the Rhizobium japonicum nifK gene. Mol. Gen. Genet., 198: 441-448.

Thöny, B., Fischer, H-M., Anthamatten, D., Bruderer, T. and Hennecke, H., 1987. The symbiotic nitrogen fixation regulatory operon (fixRnifA) of Bradyrhizobium japonicum is expressed aerobically and is subject to a novel, nifA-independent type of activation. Nucl. Acids Res., 15: 8479-8499.

Thöny, B., Anthamatten, D. and Hennecke, H., 1989. Dual control of the Bradyrhizobium japonicum symbiotic nitrogen fixation regulatory operon fixR nifA: analysis of cis- and trans-acting elements. J. Bacteriol., 171: 4162-4169.

Török, I. and Kondorosi, A., 1981. Nucleotide sequence of the R. meliloti nitrogenase reductase (nifH) gene. Nucl. Acids Res., 9: 5711-5723.

Toukdarian, A. and Kennedy, C., 1986. Regulation of nitrogen metabolism in Azotobacter vinelandii: Isolation of ntr and glnA genes and construction of ntr mutants. EMBO J., 5: 399-407.

Tubb, R.S., 1974. Glutamine synthetase and ammonium regulation of nitrogenase synthesis in Klebsiella. Nature, 251, 481-485.

Wang, S-Z., Chen J-S. and Johnson, J., 1988. The presence of five nifH-like sequences in Clostridium pasteurianum: sequence divergence and transcription properties. Nucl. Acids Res., 16: 439-545.

Watson, J.M. and Schofield, P.R., 1985. Species-specific, symbiotic plasmid-located repeated DNA sequences in Rhizobium trifolii. Mol. Gen. Genet., 199: 279-289.

Weber, G., Reiländer, H. and Pühler, A., 1985. Mapping and expression of a regulatory nitrogen fixation gene (fixD) of Rhizobium meliloti. EMBO J., 4: 2751-2756.

Wong, P-K., Popham, D., Keener, J. and Kustu, S., 1987. In vitro transcription of the nitrogen fixation regulatory operon nifLA of Klebsiella pneumoniae. J. Bacteriol., 169: 2876-2880.

CHAPTER 7

THE GENETICS AND REGULATION OF ALTERNATIVE NITROGENASES

ROBERT L. ROBSON
Department of Biochemistry, University of Georgia, Athens, Georgia 30602, U.S.A.

1 INTRODUCTION

The term "alternative nitrogen fixation system" was first used to describe a form of nitrogenase active in pseudorevertants of several mutants of Azotobacter vinelandii unable to fix N_2 using the "conventional" Mo- and Fe-requiring nitrogenase system (Bishop et al., 1980). It is now clear that this phenomenon is explained by the existence in Azotobacter of genetically independent nitrogenases which differ in their metal requirements. One system in A. vinelandii apparently requires only Fe and probably corresponds to the system first described in 1980 by Bishop and colleagues. Another system found in A. chroococcum and A. vinelandii also requires Fe together with V, but not Mo, and is known as the V-nitrogenase.

The object of this review is to describe, mainly from a genetic standpoint, the fascinating and at one time controversial history of the discovery of these new members of the nitrogenase family.

2 EARLY GENETIC STUDIES LEADING TO THE DISCOVERY OF THE ALTERNATIVE NITROGENASE SYSTEMS

Evidence for N_2-fixation in the absence of added Mo is found in early studies of N_2-fixation in Azotobacter, which also consistently reported that V stimulated N_2-fixation, though to a lesser degree than Mo (Burk, 1934; Bortels, 1936; Horner et al., 1942; Becking, 1962).

Studies of the "vanadium effect", as it became known, led to the partial purification and characterization of V-nitrogenases from both A. chroococcum ATCC 480 and A. vinelandii OP (also known as CA)(Burns et al., 1971; Benemann et al., 1972). With hindsight, it is difficult to understand why these reports did not stimulate further interest, though it was unclear at the time whether the explanation was anything more than replacement of V for Mo at, or close to, the active site of the same protein. The problem could well have remained unresolved without advances in the genetics of N_2 fixation in Azotobacter.

Bishop et al. (1980) provided the first genetic evidence for alternative N_2-fixation systems in chemically induced Nif⁻ mutants derived from A. vinelandii OP. Two mutants, UW6 and UW10, had earlier been characterized by Shah et al. (1973) as making inactive MoFe protein but active Fe protein. Bishop and colleagues described revertants of these mutants (CA2 and CA4, respectively), which could fix N_2 in media containing Mo, though they grew more slowly than the wild type on N_2 and had very low levels of acetylene reducing activity. Ammonia-repressible proteins, which often include those specifically associated with N_2 fixation, produced different patterns in extracts from CA2 and CA4 than in those from the wild-type.

N_2 fixation in these strains was also insensitive to inhibition by high levels of tungstate. Tungsten normally inhibits N_2 fixation in A. vinelandii (Keeler and Varner, 1957; Takahashi and Nason, 1957) and is incorporated into an inactive form of Mo-nitrogenase (Benemann et al., 1973). Strain CA6, a spontaneous tungstate-resistant (Tun^r) mutant derived from the wild-type strain CA (OP), behaved like CA2 and CA4, implying that N_2 fixation in this strain was not dependent on Mo. CA2, CA6 and several other Nif^- mutants defective in Fe protein (UW91), and MoFe protein (UW38) or both (UW1) were capable of fixing N_2 if Mo was not added to the medium. Nitrogenase in organisms grown with Mo was called "type A" and that present under Mo-deficient conditions was termed "type B" (Bishop et al., 1980).

A radioactive DNA probe made from the Klebsiella pneumoniae nitrogenase structural genes hybridized to genomic DNA from 19 diazotrophic bacteria (Ruvkun and Ausubel, 1980). The observation in A. vinelandii UW10 (a Nif^- mutant) of multiple bands hybridizing especially to a nifHD probe went unremarked at the time, but can now be cited as the first evidence for reiteration of one or more nitrogenase structural genes in this organism. These results also provided the basis for the later identification and cloning of homologous genes from Azotobacter.

Evidence for the existence of two Mo-independent nitrogenases in A. vinelandii CA was presented by Bishop et al. (1982), who studied the influence of Mo, V and Re on the appearance of ammonia-repressible polypeptides in cells fixing N_2 in a Mo-deficient medium. The wild-type strain grew at comparable rates irrespective of Mo concentration and neither V nor Re stimulated its growth under Mo-deficiency. However, ammonia-repressible polypeptides were different when V was added to Mo-deficient cultures, leading to the division of "nitrogenase B" into groups B1 and B2, the former being induced by V and therefore correlated with the partially purified V-nitrogenase studied earlier.

A Tun^r mutant (WD2) of A. vinelandii ATCC 12837 (also known as strain O) was resistant to high levels of tungstate (10 mM) when fixing N_2 in a Mo-sufficient medium (Riddle et al., 1982) and was therefore similar to strain CA6 of Bishop et al. (1980). The cells then contained very low concentrations of Mo, displayed low rates of acetylene reduction, lacked the EPR signal characteristic of FeMoco and showed an altered pattern of ammonia-repressible polypeptides.

Page and Collinson (1982) reexamined the phenotypic reversion of Nif^- mutants of A. vinelandii. In those capable of phenotypic reversion (UW3, UW6, UW10 and UW38), nitrogenase activity in a Mo-deficient medium was stimulated by addition of V, but partly inhibited (except for UW3) by Mo addition. They suggested that "type B" nitrogenase might result from the combination of the conventional Mo system with factors induced by Mo starvation.

These ideas found some support in the observation that although extracts of mutants UW1, UW3 and UW91 lacked Fe protein activity when cells were grown with adequate Mo, it was present in cells grown under Mo-deficient conditions (Premakumar et al., 1983). More importantly, these data implied the existence of two genetically distinct but biochemically similar Fe proteins, whose expression was regulated by Mo concentration.

Despite these data, the concept of alternative nitrogenases did not find ready acceptance, because of a number of unresolved questions. These included the low growth rates of the phenotypic revertants, the significance of low residual nitrogenase activities in most of the UW series Nif^- mutants even with adequate Mo, low rates of acetylene reduction, and the lack of the g = 3.65 EPR signal (see, for example, Shah et al., 1984). There were suggestions that the novel polypeptides induced under Mo deficiency could simply represent more efficient Mo-scavenging systems. The purification and characterisation of the enzyme or enzymes were clearly urgent.

3 CLONING OF THE Mo-NITROGENASE STRUCTURAL GENES IN AZOTOBACTER AND CONSTRUCTION OF SPECIFIC DELETION STRAINS

The nif genes from A. vinelandii were first cloned by Medhora et al. (1983), using a K. pneumoniae nitrogenase structural gene probe. They reported cloning two different genomic regions (on cosmids pMP1 and pMP2) and suggested that "there may be more than one copy of the nif gene sequence" in this organism, consistent with the earlier hybridization results of Ruvkun and Ausubel (1980).

Studies with A. chroococcum NCIMB 8003, a strain known to respond to V (Nicholas et al., 1960), began with the construction of a recombinant cosmid (pACB1) carrying about 40 kb of the genome homologous to nifHDK from K. pneumoniae (Jones et al., 1984). A sub-fragment of this region carrying all the Mo nitrogenase structural gene homologues complemented K. pneumoniae mutants with transposons inserted in nifH, nifD and nifK. This DNA fragment also enhanced acetylene reduction in A. vinelandii UW1 ($Av1^-Av2^-$), UW10($Av1^-$), UW38 ($Av1^-$), but not UW91 ($Av2^-$) grown under Mo-sufficient conditions.

Specific K. pneumoniae nif gene probes demonstrated two sequences in the genome of A. chroococcum with good homology to nifH (though only one was cloned in pACB1). Although this implied the presence of two Fe protein genes, no reiteration of nifD or nifK genes was detected (Jones et al., 1984).

Bishop et al. (1985) cloned small fragments of the A. vinelandii genome bearing homology to the K. pneumoniae nifHDK cluster. The recombinant plasmids restored Nif^+ phenotypes under Mo-sufficient conditions to three of the strains (UW6, UW10 and UW38) capable of phenotypic reversion, providing clear evidence that the mutations in these strains were in Mo nitrogenase structural genes. This conclusion was further strengthened when the nucleotide sequence for the Mo nitrogenase structural gene cluster (Brigle et al., 1985) was found to encompass the fragments cloned by Bishop et al. (1985).

Strains of both A. vinelandii and A. chroococcum deleted for nifHDK structural genes were constructed by gene replacement techniques (Bishop et al., 1986; Robson, 1986). These deletion strains were capable of fixing $^{15}N_2$, though only under Mo-deficient conditions. They had low growth rates and disproportionately low acetylene reduction activities. Whichever enzyme was responsible for N_2 fixation in these strains was entirely independent of the Mo nitrogenase structural genes.

In the Tun^r deletion strain MCD1155 of A. chroococcum, growth and nitrogenase activity in vivo were dependent on trace levels of V, allowing the characterization of a V nitrogenase essentially devoid of Mo (Robson et al., 1986a). In A. vinelandii deletion strains, growth appeared independent of added V. However, Hales et al. (1986) isolated a V-containing nitrogenase from their nifHDK deletion strain (LS15) grown in medium containing V. Chisnell et al. (1988) purified a different nitrogenase from their Tun^r deletion strain (CA11.6) grown without added Mo or V; it contained Fe but only low levels of Mo or V. This led to the conclusion that there were three nitrogenases, Mo nitrogenase, V nitrogenase and nitrogenase-3 (alternative nitrogenase), present in A. vinelandii, detailed comparisons of which are presented in Chapter 3.

4 CLONING AND CHARACTERIZATION OF ALTERNATIVE NITROGENASE STRUCTURAL GENES

The first alternative nitrogenase gene to be cloned was the second copy of nifH (called nifH*) from A. chroococcum (Robson et al., 1986b). This gene was transcribed only when cells were fixing N_2 in Mo-deficient conditions and encoded a protein with 89% sequence identity to the Fe protein from the Mo nitrogenase

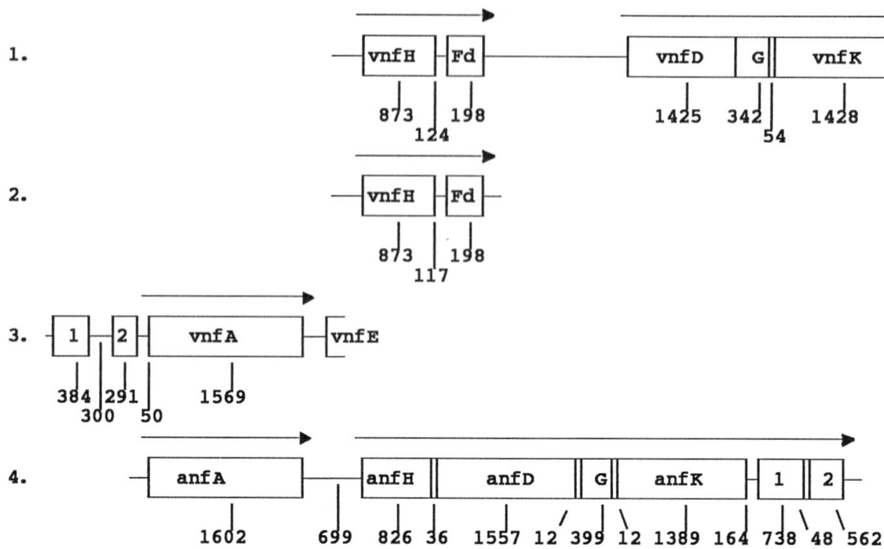

Fig. 1. Physical and genetic maps of cloned regions carrying genes required for alternative nitrogenases from Azotobacter.

1. Structural gene cluster for the vanadium nitrogenase from A. chroococcum (Robson et al., 1986, 1989). 2. Structural gene for the Fe protein of the vanadium nitrogenase from A. vinelandii (Raina et al., 1988). 3. Potential activator gene specific for the vanadium nitrogenase from A. vinelandii (Joerger et al., 1989a). 4. Structural genes and potential activator gene for nitrogenase-3 from A. vinelandii (Joerger et al., 1989a,b). Arrows represent direction of transcription and likely operon structures. Numbers below the maps give the sizes of genes and intergenic regions in base pairs. Open reading frames of unknown function are indicated by numbers within boxes.

system. It was co-transcribed with a small ORF similar to the small low-potential [8Fe-8S] ferredoxins found in obligate anaerobes.

With the construction of the nifHDK deletion strain (MCD1155), another sequence bearing very weak homology to the Azotobacter nifK gene was detected in its genome. The region bearing this nifK-like sequence was cloned, sequenced (Fig. 1) and used to construct double deletion strains. DNA sequence analysis revealed two open reading frames with homology to nifD and nifK, separated by a small ORF. The N-terminal sequences of the nifD and nifK homologues matched those determined directly for the α and β subunits of the VFe protein. The deduced sequence of the third ORF matched that of a small polypeptide co-purified in equimolar amounts with the other subunits of this component. Genetic and biochemical evidence suggested that this might be a subunit type peculiar to the V nitrogenase. The genetic nomenclature vnf was adopted to define genes unique to the V system; the three genes encoding the sub-units of the VFe protein in were named vnfD, vnfG and vnfK in order of transcription. Northern analysis suggested that the three genes were co-transcribed. Restriction maps for the cloned regions containing vnfDGK and nifH* showed an overlap locating nifH* 2.5 kb upstream of vnfD. NifH* was shown to encode the Fe protein of the V nitrogenase by again matching predicted and determined N-terminal amino acid sequences, and was re-designated vnfH. The role of the vnfDGK cluster was proven by constructing, in the nifHDK deletion strain MCD1155, a double deletion strain (MCD1301) lacking all of vnfG and parts of vnfD and vnfK. Strain MCD1301 was incapable of N_2 fixation under any conditions.

A comparable region from A. vinelandii present on one (pMP1) of the two original recombinant plasmids (Medhora et al., 1983) hybridizing to nifHDK from K. pneumoniae proved to carry Fe protein and ferredoxin-like genes when sequenced (Raina et al., 1988)(see Fig. 1). The involvement of this Fe protein gene (called nifH2) in the V nitrogenase in this organism was demonstrated by transferring a Tn5 mutation close to the N-terminal coding region of the gene into the chromosome of CA12, a nifHDK deletion strain. The resultant strain (RR29) was incapable of fixing N_2 in the presence of V, though significantly it still displayed some acetylene reducing activity. A mutant with Tn5 inserted approximately 1 kb 3' to the ferredoxin gene was unaffected in V-dependent N_2 fixation, indicating that no essential genes are located at this position.

Strains of A. vinelandii deleted in genes for the VFe protein were constructed by an unusual procedure in which a deletion constructed in vitro within the vnfDGK gene cluster of A. chroococcum was recombined in vivo with the homologous region of A. vinelandii (Pau et al., 1989). That this technique was effective argues for a high degree of homology at the DNA level between genes for the VFe protein of both organisms. Indeed, many restriction sites are conserved, though the distance between vnfH and vnfDGK may be about 1 kb less in A. vinelandii. The single vnfDGK deletion strain (RP114) was capable of diazotrophic growth in medium with or without added Mo, but growth in V-containing medium was poor. A double deletion strain (RP206) resembling RP114, but also lacking the Mo nitrogenase genes, only fixed N_2 well in medium devoid of Mo and V. However, very slow growth was observed in the presence of V. These results established that the residual nitrogenase activity in the double deletion strain was independent of structural genes for both Mo and V nitrogenases and was repressed by both Mo and V.

A third region bearing homology to nifH was cloned from A. vinelandii by Joerger et al. (1989b). The nucleotide sequence of a region of 6.1 kb was obtained and found to contain ORFs with homology to, in order, nifH, nifD, vnfG and nifK (Fig. 1). Strains CA70, CA71 and CA72 containing deletions and/or insertions in the chromosome within this region in the wild-type strain CA grew on medium to which either Mo or V had been added. Double deletions constructed in CA11 (the nifHDK deletion) grew slowly in V-supplemented medium. Joerger et al. (1989b) concluded that this region contained the structural genes for the third nitrogenase purified by Chisnell et al. (1988). They used the genotype anf (alternative nitrogen fixation) to denote these genes. The organization of the genes, coupled with Northern analysis, suggested that the genes were probably co-transcribed on a 6.6 kb message in the order anfHDGK, together with at least two other potential genes whose functions are not known.

It was thus unequivocally established that in A. vinelandii OP there are three distinct sets of structural genes encoding three nitrogenases which differ with respect to their metal requirements. In A. chroococcum NCIMB 8003, only two of these systems are found. A specific anfDGK probe finds only the vnf counterparts in the genome of this organism (E. Fallik and R.L. Robson, unpublished). Other strains of A. vinelandii or A. chroococcum may lack the V-nitrogenase, since Becking (1962) found that V did not stimulate N_2 fixation in all strains.

5 GENETIC INTER-RELATIONSHIPS BETWEEN NIF, VNF AND ANF GENES

Mutations within any one set of structural genes do not exert a marked pleiotropic effect on expression of the other system or systems, but there are pleiotropic mutations and these indicate a degree of genetic interdependence between the systems. These relationships are illustrated in Fig. 2.

Thus A. vinelandii UW71 makes active Fe protein but no MoFe protein and does not reduce acetylene whatever the metal status of the medium. It may be defective in Mo storage, but the lesion remains unidentified (Page and Collinson, 1982).

Kennedy et al. (1986) described a Tn5 mutant of A. vinelandii (MV21) which could not fix N_2 in the presence or absence of added Mo. The influence of V on the growth of this strain was not tested. Tn5 appears to be inserted in or near nifM within the major nif gene cluster. They suggested that nifM, essential for activity of the nifH gene product, may play an analogous role in maturation of the Fe protein required for nitrogenase-3 synthesis in Mo- and V-deficient medium. However, MV21 exhibits properties inconsistent with the nifM$^-$ phenotype of K. pneumoniae, since it shows high acetylene reducing activities (33% of wild type) in vivo, implying active Fe protein in this strain. Now that the major nif gene cluster has been sequenced and characterized in A. vinelandii (Jacobson et al., 1989), it should be possible to locate the mutation precisely.

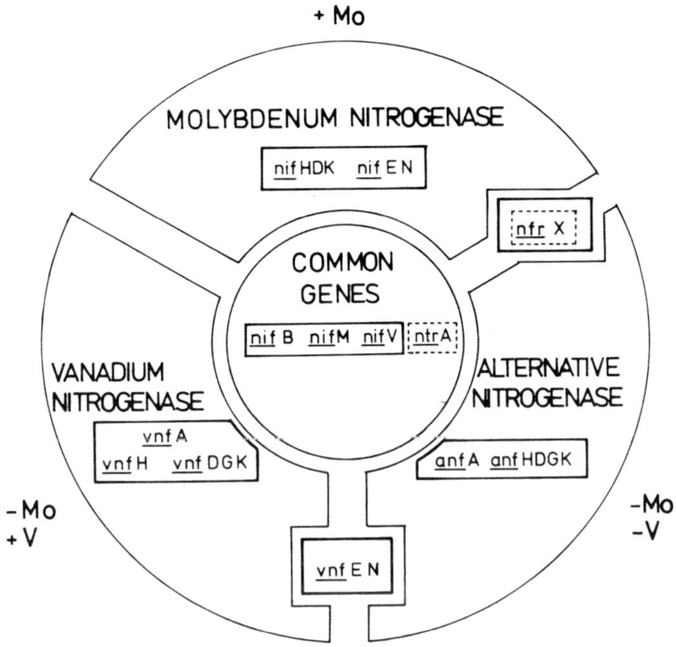

Fig. 2. Genetic interrelationships between nitrogenase systems in Azotobacter.

Genes that have been identified so far in A. vinelandii strain OP as specific to each system are shown within the central circle. Genes required by two systems are shown within the boxes placed between adjoining arcs. Expression of each system occurs only when the metal availability corresponds to that shown outside its arc. In A. chroococcum NCIMB 8003 only the Mo and V nitrogenases are found.

UW45 is an important member of the original set of Nif^- mutants of A. vinelandii OP (Shah et al., 1973) and is defective in FeMoco biosynthesis (Nagatani et al., 1974). It does not undergo phenotypic reversion under Mo-deficiency. The mutation in UW45 lies in nifB and this strain and a nifB Tn5 insertion mutant (CA30) isolated by Joerger et al. (1986) do not fix N_2 under any conditions. NifB is therefore probably required for processing or synthesis of analogous co-factors for all three systems.

Although dependence of the alternative nitrogenases on nif genes has not been exhaustively studied, it is clear that some genes which, like nifHDK, are essential for the Mo system are not required for the other systems. Mutations in nifE or nifN, which are also required for FeMoco synthesis or insertion, only affect the Mo-dependent system (Joerger et al., 1986; Kennedy et al., 1986).

However, there may be an analogous set of genes specific for the alternative nitrogenase systems, because nifE and nifN appear to be reiterated in both A. chroococcum (Evans et al., 1988) and A. vinelandii (Joerger et al., 1989b).

UW1, a nif regulatory mutant of A. vinelandii (Shah et al., 1973; Kennedy and Robson, 1983), is capable of N_2 fixation in Mo-deficient conditions (Bishop et al., 1980). The mutation in UW1 is now known to lie in nifA (Bennett et al., 1988), which codes for an activator of transcription of nif genes (Chapter 6). This implies that expression of alternative nitrogenase systems is not dependent upon nifA.

6 REGULATION OF THE ALTERNATIVE NITROGENASE SYSTEMS IN AZOTOBACTER

Alternative nitrogenases in Azotobacter are subject to regulation by N status, O_2 and metal availability. Mo is required for expression of nifHDK, but represses formation of the alternative system (Jacobson et al., 1986; Robson et al., 1986b). Surprisingly, V may not be important for expression of vnfH in either A. chroococcum (Robson et al., 1986b) or A. vinelandii (Jacobson et al., 1986). However, it is apparently required for induction of the vnfDGK operon in A. vinelandii (Bishop et al., 1982), but not in A. chroococcum (L. Kerber and R.L. Robson, unpublished). In turn, V represses formation of the third system (Joerger et al., 1986; Pau et al., 1989). At present, little is known about the mechanism of metal regulation except that it seems to be exerted at the transcriptional level.

More is known about the regulation of alternative systems in response to N availability. It is relevant to ask how regulation of these systems compares to that of the conventional Mo nitrogenase system, our knowledge of which derives mostly from studies with K. pneumoniae (Chapter 6). In almost all cases the model is essentially the same: nif genes require the products of ntrA and nifA for expression. NtrA is an alternative sigma factor, also required for transcription of other genes involved in N assimilation. It allows recognition of promoters with the consensus sequence GG-n10-GC situated between -24 and -12 to the transcription initiation site. NifA is a nif-specific transcriptional activator which binds to a site about 100 bp further upstream, with the consensus sequence TGT-n10-ACA. Other factors which may be important for nif promoter activity include Interference Host Factor (IHF)(Santero et al., 1989) and the degree of supercoiling of the DNA (Kranz and Haselkorn, 1986).

Several lines of evidence can be cited which indicate that transcriptional activation of at least the structural gene operons for both the V nitrogenase and nitrogenase-3 follows essentially the same pattern, in probably requiring NtrA and an analogue of NifA specific to each system.

In A. vinelandii, mutations in ntrA lead to loss of N_2-fixing capacity in both Mo-sufficient and -deficient conditions (Toukdarian and Kennedy, 1986). By itself, this is not conclusive evidence that NtrA is directly required for transcription of vnf or anf genes, since some nif genes (e.g. nifB or nifM), which presumably are dependent on NtrA, are essential to all systems. However, the promoter regions for vnfH and vnfD in A. chroococcum (Robson et al., 1986b; Robson et al., 1989) and vnfH (nifH2)(Raina et al., 1988) and anfH (Joerger et al., 1989a) from A. vinelandii all contain sequences which conform well to NtrA binding sites, though in no case has the transcription initiation site been determined.

Azotobacter nif genes are dependent on nifA for expression. Kennedy and Robson (1983) showed that in K. pneumoniae, constitutively-expressed NifA corrected Nif^- regulatory mutants of A. vinelandii and A. chroococcum under Mo-sufficient conditions. Conversely, Azotobacter nif genes express from their own promoters and complement Nif^- mutants of K. pneumoniae (Jones et al., 1984; Kennedy et al., 1986; Evans et al., 1988). NifA homologues have been cloned from both A. vinelandii (Bennett et al., 1988; Joerger and Bishop, 1988) and A. chroococcum (Evans et al., 1988; Santero et al., 1988). The nifA gene from A. chroococcum, when expressed from a constitutive promoter, activated nifH::lac fusions in E. coli (Santero et al., 1988). Finally, Azotobacter nif structural gene promoters contain the nifA consensus sequence (Buck et al., 1986).

By comparison, NifA binding sites are apparently missing from vnf and anf structural gene promoters, leading to the suggestion that nifA is not required for their activation (Robson et al., 1986b; Raina et al., 1988; Joerger et al., 1989b; Robson et al., 1989). Indeed, the vnfH promoter, unlike the nifH promoter from A. chroococcum, was not activated by NifA from K. pneumoniae (Robson et al., 1986b). The prediction that expression of alternative nitrogenase systems would require transcriptional activators other than NifA was confirmed by Joerger et al. (1989a), who cloned two nifA homologues from A. vinelandii (Fig. 1). One was characterized in the Tn5 mutant CA46, which grew normally in Mo-sufficient medium but more slowly in Mo-deficient conditions with or without V. Furthermore, V did not stimulate acetylene reduction in this strain. N_2 fixation was absent and neither alternative nitrogenase protein pattern was expressed under Mo-deficient conditions. The region containing the mutation was cloned and sequenced and Tn5 was found to be inserted in an ORF with a high degree of identity to nifA. The inability of this strain to express polypeptides associated with the V nitrogenase led to the designation of this gene as vnfA.

However, the effect of insertions into vnfA are complex in two respects. Firstly, the mutation in CA46 also affects expression of the third system. This may mean that vnfA is required for expression of one or more anf specific genes, though not anfHDGK. Another possibility is that it is required for genes common to vnf and anf, but not nif. Alternatively, the Tn5 insertion may exert a polar effect on downstream genes: the N-terminal sequence of the deduced gene product of the adjacent gene resembles nifE and may be a gene for cofactor biosynthesis for alternative nitrogenases. Secondly, the mutation in CA46 also seems to allow expression of the Mo nitrogenase even under Mo deficiency. This suggests that vnfA (or some other gene activated by it) is involved in controlling lowered expression of the nifHDK genes under low Mo status.

Joerger et al. (1989b) also found an ORF with a strong resemblance to nifA upstream of anfH. A mutant (CA66) in this presumptive gene grew well when supplied with V and Mo but not in their absence. Accordingly, this gene was called anfA implying its necessity for transcription of genes specifically involved in the third system.

VnfA and AnfA are 61% identical, but only 37% identical with NifA from A. vinelandii. All three proteins appear to have the four domain structure predicted for the nifA product by Drummond et al. (1986), in which the C-terminal domain is thought to contain a helix-turn-helix DNA-binding motif. Differences in amino acid sequences between these regions in NifA, VnfA and AnfA may determine activator specificity, but how their activities are controlled is not known.

All three activators may be expressed under N-starvation conditions, but their activity could be modulated directly by metal status. Alternatively, expression of each activator may be regulated at the transcriptional level so that only one activator at a time is expressed. In K. pneumoniae (Chapter 6), regulation of nif expression occurs at both levels. NifA activity can be blocked by NifL, which mediates repression by O_2 and abundant N. At the same time, expression of the nifAL operon is in turn dependent upon the ntrC gene product, which acts as a positive activator of transcription analogous to NifA itself. NtrC is itself subject to regulation by N-status at two levels. It is only transcribed at high levels under conditions of N starvation. Moreover, it is co-transcribed with nifB, whose gene product can block NtrC activity if N starvation is relieved. This cascade probably ensures that relatively poor N sources like amino acids will be used in preference to N_2.

Whereas there is a direct analogue of ntrC in A. vinelandii capable of complementing ntrC mutants of E. coli, it is not required for N_2-fixation under Mo-sufficient conditions (Toukdarian and Kennedy, 1986). However, Santero et al. (1988) characterised another locus called nfr (nitrogen fixation regulation)X, which seems to be part of a cascade system in A. vinelandii and may be analogous to ntrC. A mutation in this locus affects diazotrophy under Mo-sufficient and Mo-deficient conditions, but interestingly does not affect N_2 fixation with V. That this locus is involved in regulating expression of nif genes was demonstrated in two ways. The nifH promoter is not active in an NfrX$^-$ background and, secondly, these mutations are corrected in vivo by provision of nifA from K. pneumoniae, though strangely not from A. chroococcum.

Although regulation of genes specific for each system may be controlled by specific activator genes, it is not clear how genes common to both or all systems are regulated. Whereas nifB in A. vinelandii has consensus sequences consistent with its activation by nifA, nifA mutants such as UW1 are still capable of fixing N_2 without Mo or with V. This suggests that "common" genes may have complex promoters subject to activation by more than one activator, or by another as yet unidentified activator. Much work remains to be done to sort out the regulation of these systems, particularly the factors mediating regulation by metal status.

7 EVIDENCE FOR ALTERNATIVE NITROGENASES IN BACTERIA OTHER THAN AZOTOBACTER

Definitive proof for the existence of alternative nitrogenases ultimately rests on the purification and characterization of the enzyme, as was the case in Azotobacter. At present there is no formal proof that such enzymes exist outside Azotobacter, although it would be surprising if they did not. Evidence for the existence of such enzymes in other genera rests on one or more diagnostic features, each of which may have alternative explanations.

(a) The ability of an organisms to fix N_2 in the absence of added Mo may be misleading because many organisms are effective scavengers of Mo.

(b) The ability of V (or other metals) to stimulate N_2 fixation in Mo-deficient conditions may be an indicator, but would only be apparent if that metal serves no other vital role in the cell. For example, a role for nitrogenase-3 in A. vinelandii could not be established by studying responses to Fe.

(c) Changes in the catalytic properties of the enzyme in cells starved for Mo, such as a disproportionately low ratio between acetylene and N_2 reduction, or ethane formation from acetylene (Dilworth et al., 1987), may be valuable indicators.

(d) Evidence for reiteration of nif structural gene sequences may also be helpful, but need not imply the existence of alternative nitrogenases. Some examples of nif structural gene reiteration are most likely associated with the Mo-based system. In future, specific vnf or anf probes may prove useful.

(e) Phenotypic reversion of Nif$^-$ mutants under conditions of Mo-deficiency may also be a good indicator, but requires that such mutants are fairly well characterised and within the nif structural genes.

(f) The ability of organisms or mutants to fix N_2 in the presence of high concentrations of W may also be indicative, but may simply reflect the presence of a Mo uptake system not antagonized by W.

Organisms like K. pneumoniae M5A1 probably do not contain alternative nitrogenases, judging by the above criteria. Yet in the examples discussed below one or more of the above attributes has been observed. The existence of alternative nitrogenases seems highly likely in clostridia, cyanobacteria and possibly purple photosynthetic bacteria. They seem unlikely to be common, if present at all, in rhizobia, while the situation in archaebacteria is unclear.

In Clostridium species, several criteria have been met. Jensen and Spencer (1947) showed that V at low levels (<40 nM) stimulated N_2 fixation in a Mo-deficient medium in 4 or 5 out of 8 strains of Clostridium butyricum and C. pasteurianum strain W but not in C. acetobutylicum strain B. Cultures of C. pasteurianum W5 produce ethane and ethylene from acetylene only when grown in Mo-deficient media, and the proportion of ethane is increased when V is added (Dilworth et al., 1987). Furthermore, nifH genes are present in six copies in this strain (Wang et al., 1988), four or five of which appear to be transcribed during diazotrophic growth in the presence of Mo. Only one (nifH3) does not appear to be transcribed under these conditions.

In cyanobacteria, evidence for alternative nitrogenases is also increasing. Bortels (1940) first demonstrated that V could stimulate N_2 fixation in Anabaena and Nostoc species, but later Fay and de Vasconcelos (1974) could not find evidence for a "V-effect" in Anabaena cylindrica. However, more recently, Kentemich et al. (1988) reported stimulation of N_2-dependent growth of Anabaena variabilis by V in a Mo-deficient medium and low levels of ethane formation from acetylene which were suppressed by Mo. A report of V stimulation of N_2 fixation in Anabaena circinalis may be genuine, but was apparently conducted with a culture reportedly not free of contaminating bacteria (Sahay and Sankaran, 1968). Singh et al. (1978) reported interesting mutants of Nostoc muscorum selected as resistant to W or Cr when fixing N_2. These mutants no longer required Mo for growth, but surprisingly required either W or Cr, respectively. These mutants may be similar to the Tun^r mutants of Azotobacter and suggest that Cr may be a useful analogue when looking for alternative systems. Additionally, there is evidence for nif structural gene reiteration in several cyanobacteria, particularly Anabaena strain 7120, where two closely related nifH genes have been sequenced (Mevarech et al., 1980; Haselkorn et al., 1984), and in Calothrix (Kallas et al., 1983). The function of these reiterations is not known and their expression has not been tested in media lacking Mo.

In the photosynthetic bacterium Rhodobacter capsulata, insertion and deletion mutants in the nifHDK cluster were obtained by Scolnik and Haselkorn (1984). These mutants were Nif^-, but pseudorevertants were obtained which grew slowly on N-free medium and also had only 10% of wild-type acetylene reduction activity. The authors suggested that the phenotypic reversion occurred through the activation of normally silent copies of nitrogenase structural genes elsewhere in the genome. No experiments were carried out to test whether the Nif^- strains fixed N_2 in the absence of Mo. Hybridization studies revealed reiteration of nif structural gene sequences in the genome. One of these may have been the nifH-like sequence surprisingly located in a photosynthetic gene cluster sequenced by Youvan et al. (1984). In Rhodospirillum rubrum, Paschinger (1974) reported that nitrogenase exhibited different properties in Mo-deficient medium or if W was supplied.

Whereas there is evidence for nif structural gene reiteration in Rhizobium leguminosarum bv. phaseoli (Quinto et al., 1982, 1985), Azorhizobium caulinodans (Rhizobium ORS571)(Norel and Elmerich, 1987), Rhizobium meliloti (Barran and Bromfield, 1988) and cowpea Rhizobium ANU240 (Badenoch-Jones et al., 1989), this does not seem to be associated with alternative nitrogenases. In each case examined in greater detail, the reiterated genes show a high degree of identity, are expressed under Mo-sufficient conditions and appear to be functionally interchangeable. In a sense, these examples represent alternative nitrogenases, but of a different class to those considered here.

In archaebacteria, there is also evidence for nifH gene reiteration in Methanosarcina ivanovii and Methanosarcina barkeri (Sibold and Souillard, 1988). The thermophile Methanococcus thermolithotrophicus contains two copies of nifH which have both been sequenced (Souillard and Siebold, 1989). It is unclear whether these reiterations adumbrate alternative nitrogenases. In M. thermolithotrophicus, only nifH is expressed during N_2 fixation with Mo. Rates of acetylene reduction are very low in this organism even with Mo, but comparative N_2 reduction data are needed before the significance of this can be judged. Perhaps the most compelling reason for thinking that this organism may not have an alternative nitrogenase is that Mo appears obligatory for N_2 fixation.

8 EVOLUTION OF ALTERNATIVE NITROGENASES

The evolutionary origins of the alternative nitrogenases should be clearer if and when other examples are characterized. At present, conflicting conclusions can be drawn from the genetic data. If alternative nitrogenases are not confined to Azotobacter and exist in diverse eubacteria, then their origin is probably ancient. If this is the case, the three nitrogenases may have arisen from a common set of ancestral genes, and alternative systems arose as one early evolutionary development which later diverged to give the two enzyme forms seen in Azotobacter. This picture arises from sequence comparisons of the respective D and K gene products (Joerger et al., 1989a; Robson et al., 1989). AnfD and VnfD, AnfK and VnfK have 54% and 56% identity, respectively, while in each case the identity with Azotobacter NifD and NifK is only about 30%. VnfG and anfG gene products seem unique to the alternative nitrogenases and show 39% identity. Furthermore, both G genes are located in comparable positions in the respective structural gene operons.

An altogether different scenario follows from comparisons of nifH, vnfH and anfH gene products (Normand and Bousquet, 1989; Souillard and Sibold, 1989). In Azotobacter, AnfH is strikingly different from NifH and VnfH and most similar to NifH3 from C. pasteurianum, and less so to NifH1 in the archaebacterium M. thermolithotrophicus. This raises the possibility that nifH3 in C. pasteurianum may be associated with an alternative nitrogenase. Though nifH3 is apparently not expressed during diazotrophic growth (Wang et al., 1988), it may be in Mo-deficient media or if V is added. Souillard and Sibold (1989) also suggest that the third system may be most closely related to the archaebacterial nitrogenase system which, though seemingly requiring Mo, exhibits surprisingly low acetylene reducing activities. However, the reverse is true for VnfH, which is most closely related to NifH from Azotobacter, and only distantly related to the clostridial or archaebacterial examples. What is clear is that the evolution of vnfH in Azotobacter has followed a different path to that of the vnfDGK gene cluster and may have arisen more recently from a gene duplication following speciation of Azotobacter. Detailed comparative analyses of respective D and K genes may provide more clues.

In Azotobacter, alternative nitrogenases most likely provide a contingency against first Mo and then V starvation. The regulatory hierarchy ensures expression of only the most catalytically active system given the prevailing metal status. If this is true, it must be assumed that metal availability in the environment has been a selective pressure favouring nitrogenase evolution. Postgate and Eady (1988) summarised considerations about the evolution of the different metal requirements and came to the reasonable conclusion that the ancestral "nitrogenase" may have had optional metal requirements. Gene reiteration may then have allowed specialised and better nitrogenases to appear. Postgate and Eady (1988) also pointed out that "it is a characteristic of discussions of evolution that they rarely permit firm conclusions" as is certainly the case with this system.

9 CONCLUDING REMARKS AND FUTURE DIRECTIONS

Genetic studies of N_2 fixation in Azotobacter played an important part in uncovering the existence of alternative nitrogenases. Genetically engineered strains lacking one or more sets of structural genes will continue to be essential for routine purification of the different enzymes. Rapid progress has been made in understanding the genetics of these systems, but much work still remains. We need to know how many other genes are specific for each system. Studies of genetic regulation of the different systems should reveal how metals exert their effects. An understanding about the interrelationships between the systems may give new insights into the general processes of metal assimilation, cofactor biosynthesis and maturation of the enzymes. The isolation of complete sets of vnf or anf genes may allow us to transfer the ability to fix N_2 without Mo to other diazotrophs.

It is important that other examples of alternative nitrogenases be sought and studied. At present all our knowledge arises from Azotobacter. The perception that alternative nitrogenases are inferior to Mo nitrogenase comes from two observations. Expression of the alternative systems is prevented by Mo and alternative nitrogenases have relatively poor activities for N_2 reduction. However, these relationships may not hold in other organisms. Hence, the global importance to the biological nitrogen cycle of systems not dependent on Mo can only be assessed when their occurrence and properties in other organisms are known. Clues to their existence in other species may be provided by use of specific vnf and anf probes, much as nifHDK probes have proved valuable in studies of Mo nitrogenases. More importantly, additional members of the nitrogenase enzyme family may emerge from such studies. The fascinating evolutionary relationships within the nitrogenase family should be clearer when more examples of alternative enzymes are uncovered and characterised.

10 REFERENCES

Badenoch-Jones, J., Holton, J.A., Morrison, C.M., Scott, K.F. and Shine, J., 1989. Structural and functional analysis of nitrogenase genes from the broad-host range Rhizobium strain ANU 240. Gene, 77:141-153.

Barran, L.R. and Bromfield, E.S.P., 1988. Symbiotic gene probes hybridize to cryptic plasmids of indigenous Rhizobium meliloti. Canad. J. Microbiol., 34: 703-707.

Becking, J.H., 1962. Species differences in molybdenum and vanadium requirements and combined nitrogen utilization by Azotobacteriaceae. Plant and Soil, 2: 171-201.

Benemann, J.R., McKenna, C.E., Lie, R.F., Traylor, T.G. and Kamen, M.N., 1972. The vanadium effect in nitrogen fixation by Azotobacter. Biochim. Biophys. Acta, 264: 25-38.

Benemann, J.R., Smith, G.M., Kostel, P.J. and McKenna, C.E., 1973. Tungsten incorporation into Azotobacter vinelandii nitrogenase. FEBS Lett., 29: 219-221.

Bennett, L.T., Cannon, F. and Dean, D.R., 1988. Nucleotide sequence and mutagenesis of the nifA gene from Azotobacter vinelandii. Mol. Microbiol., 2: 315-321.
Bishop, P.E., Jarlenski, D.M.L. and Hetherington, D.R., 1980. Evidence for an alternative nitrogen fixation system in Azotobacter vinelandii. Proc. Natl. Acad. Sci. USA, 77: 7342-7346.
Bishop, P.E., Jarlenski, D.M.L. and Hetherington, D.R., 1982. Expression of an alternative nitrogen fixation system in Azotobacter vinelandii. J. Bacteriol., 150: 1244-1251.
Bishop, P.E., Premakumar, R., Dean, D.R., Jacobson, M.R., Chisnell, J.R., Rizzo, T.M. and Kopczynski, J., 1986. Nitrogen fixation by Azotobacter vinelandii strains having deletions in structural genes for nitrogenase. Science, 232: 92-94.
Bishop, P.E., Rizzo, T.M. and Bott, T.M., 1985. Molecular cloning of nif DNA from Azotobacter vinelandii nitrogenase structural gene cluster. J. Bacteriol., 162: 21-28.
Bortels, H., 1936. Weitere Untersuchungen über die Bedeutung von Molybdän, Wolfram und anderen Erdchenstoffen für stickstoffbindene und andere Mikroorganismen. Zentr. Bakteriol. Parasitenkde 95: 193-218.
Bortels, H., 1940. Über die Bedeutung des Molybdäns für stickstoffbindene Nostocaceen. Arch. Mikrobiol., 11: 155-186.
Brigle, K.E., Newton, W.E. and Dean, D.R., 1985. Complete nucleotide sequence of the Azotobacter vinelandii nitrogenase structural gene cluster. Gene, 37: 37-44.
Buck, M., Miller, S., Drummond, M. and Dixon, R., 1986. Upstream activator sequences are present in promoters of nitrogen fixation genes. Nature, 320: 374-378.
Burk, D., 1934. Azotase and nitrogenase in Azotobacter. Ergeb. Enzymforsch., 3: 25-36.
Burns, R.C., Fuchsman, W.H. and Hardy, R.W.F., 1971. Nitrogenase from vanadium-grown Azotobacter: isolation, characteristics and mechanistic implications. Biochem. Biophys. Res. Commun., 41: 353-358.
Chisnell, J.R., Premakumar, R. and Bishop, P.E., 1988. Purification of a second alternative nitrogenase from a nifHDK deletion strain of Azotobacter vinelandii. J. Bacteriol., 170: 27-33.
Dilworth, M.J., Eady, R.R., Robson, R.L. and Miller, R.W., 1987. Ethane formation from acetylene as a potential test for vanadium nitrogenase in vivo. Nature, 327: 167-168.
Drummond, M., Whitty, P. and Wootton, J., 1986. Sequence and domain relationships of ntrC and nifA from Klebsiella pneumoniae: homologies to other regulatory proteins. EMBO J., 5: 441-447.
Evans, D., Jones, R., Woodley, P and Robson, R., 1988. Further analysis of nitrogen fixation (nif) genes in Azotobacter chroococcum: identification and expression in Klebsiella pneumoniae of nifS, nifV, nifM and nifB genes and localization of nifE/N-, nifU-, nifA- and fixABC-like genes. J. Gen. Microbiol., 134: 931-942.
Fay, P. and de Vasconcelos, L., 1974. Nitrogen metabolism and ultrastructure in Anabaena cylindrica. Arch. Microbiol., 99: 221-230.
Hales, B.J., Case, E.E., Morningstar, J.E., Dzeda, M.F. and Mauterer, L.A., 1986. Isolation of a new vanadium-containing nitrogenase from Azotobacter vinelandii. Biochemistry, 25: 7251-7255.
Haselkorn, R., Lammers, P.J., Rice, D. and Robinson, S.J., 1984. Organization and transcription of nitrogenase genes in the cyanobacterium Anabaena. In C. Veeger and W.E. Newton (Editors), Advances in Nitrogen Fixation Research. Martinus Nijhoff/Dr W. Junk, Pudoc, Wageningen, pp. 653-659.
Horner, C.K., Burk, D., Allison, F.E. and Sherman, M.S., 1942. Nitrogen fixation by Azotobacter as influenced by molybdenum and vanadium. J. Agric. Res., 65: 175-193.

Jacobson, M.R., Brigle, K.E., Bennett, L.T., Setterquist, R.A., Wilson, M.S., Cash, V.L., Beynon, J., Newton, W.E. and Dean, D.R., 1989. Physical and genetic map of the major nif gene cluster from Azotobacter vinelandii. J. Bacteriol., 171: 1017-1027.
Jacobson, M.R., Premakumar, R. and Bishop, P.E., 1986. Transcriptional regulation of nitrogen fixation by molybdenum in Azotobacter vinelandii. J. Bacteriol., 167: 480-486.
Jensen, H.L. and Spencer, D., 1947. The influence of molybdenum and vanadium on nitrogen fixation by Clostridium butyricum and related organisms. Proc. Linn. Soc. N.S.W., 72: 73-86.
Joerger, R.D. and Bishop, P.E., 1988. Nucleotide sequence and genetic analysis of the nifB-nifQ region from Azotobacter vinelandii. J. Bacteriol., 170: 1475-1487.
Joerger, R.D., Jacobson, M.R. and Bishop, P.E., 1989a. Two nifA-like genes required for expression of alternative nitrogenases by Azotobacter vinelandii. J. Bacteriol., 171: 3258-3267.
Joerger, R.D., Jacobson, M.R., Premakumar, R., Wolfinger, E.D. and Bishop, P.E., 1989b. Nucleotide sequence and mutational analysis of the structural genes (anfHDGK) for the second alternative nitrogenase from Azotobacter vinelandii. J. Bacteriol., 171: 1075-1086.
Joerger, R.D., Premakumar, R. and Bishop, P.E., 1986. Tn5-induced mutants of Azotobacter vinelandii affected in nitrogen fixation under Mo-deficient and Mo-sufficient conditions. J. Bacteriol., 168: 673-682.
Jones, R., Woodley, P. and Robson, R.L., 1984. Cloning and organization of some genes for nitrogen fixation from Azotobacter chroococcum and their expression in Klebsiella pneumoniae. Mol. Gen. Genet., 197: 318-327.
Kallas, T., Rebiere, M.G., Rippka, R. and Tandeau de Marsac, N., 1983. The structural nif genes of cyanobacteria Gloeothece sp. and Calothrix sp. share homology with those of Anabaena sp., but the Gloeothece genes have a different arrangement. J. Bacteriol., 155: 427-431.
Keeler, R.F. and Varner, J.E., 1957. Tungstate as an antagonist of molybdate in Azotobacter vinelandii. Arch. Biochem. Biophys., 70: 585-590.
Kennedy, C., Gamal, R., Humphrey, R., Ramos, J., Brigle, K. and Dean, D.R., 1986. The nifH, nifM and nifN genes of Azotobacter vinelandii: characterization by Tn5 mutagenesis and isolation from pLAFR1 gene banks. Mol. Gen. Genet., 205: 318-325.
Kennedy, C.K. and Robson, R.L., 1983. Activation of nif gene expression in Azotobacter by the nifA gene product of Klebsiella pneumoniae. Nature, 301: 626-628.
Kentemich, T., Danneburg, G., Hundeshagen, B. and Bothe, H., 1988. Evidence for the occurrence of the alternative vanadium-containing nitrogenase in the cyanobacterium Anabaena variabilis. FEMS Microbiol. Lett., 51: 19-24.
Kranz, R.G. and Haselkorn, R., 1986. Anaerobic regulation of nitrogen fixation genes in Rhodopseudomonas capsulata. Proc. Natl. Acad. Sci. USA, 83: 6805-6809.
Medhora, M., Phadnis, S.H. and Das, H.K., 1983. Construction of a gene library from the nitrogen-fixing aerobe Azotobacter vinelandii. Gene, 25 355-360.
Mevarech, M., Rice, D. and Haselkorn, R., 1980. Nucleotide sequence of a cyanobacterial nifH gene coding for nitrogenase reductase. Proc. Natl. Acad. Sci. USA, 77: 6476-6480.
Nagatani, H.H., Shah, V.K. and Brill, W.J., 1974. Activation of inactive nitrogenase by acid-treated component 1. J. Bacteriol., 120: 697-701.
Nicholas, D.J.D., Fisher, D.J., Redmond, W.J. and Wright, M.A., 1960. Some aspects of hydrogenase activity and nitrogen fixation in Azotobacter spp. and in Clostridium pasteurianum. J. Gen. Microbiol., 22: 191-205.
Norel, F. and Elmerich, C., 1987. Nucleotide sequence and functional analysis of the two nifH copies of Rhizobium ORS 571. J. Gen. Microbiol., 133: 1563-1576.

Normand, P. and Bousquet, J., 1989. Phylogeny of nitrogenase sequences in Frankia and other nitrogen-fixing organisms. J. Mol. Evol., 29: 436-447.
Page, W.J. and Collinson, S.K., 1983. Molybdenum enhancement of nitrogen fixation in a molybdenum-starved Azotobacter vinelandii Nif^{-1} mutant. Canad. J. Microbiol., 28: 1173-1180.
Premakumar, R., Lemos, E.M. and Bishop, P.E., 1984. Evidence for two dinitrogenase reductases under regulatory control by molybdenum in Azotobacter vinelandii. Biochim. Biophys. Acta, 797: 64-70.
Paschinger, H., 1974. A changed nitrogenase activity in Rhodospirillum rubrum after substitution of tungsten for molybdenum. Arch. Microbiol., 101: 379-389.
Pau, R.N., Mitchenall, L.A. and Robson, R.L., 1989. Genetic evidence for an Azotobacter vinelandii nitrogenase lacking molybdenum and vanadium. J. Bacteriol., 171: 124-129.
Postgate, J.R. and Eady, R.R., 1988. The evolution of biological nitrogen fixation. In: H. Bothe, F.J. de Bruijn and W.E. Newton (Editors), Nitrogen Fixation: Hundred Years After. Gustav Fischer, Stuttgart, pp. 31-40.
Quinto, C., de la Vega, H., Flores, M., Fernandez, L., Ballado, T., Soberon, G. and Palacios, R., 1982. Reiteration of nitrogen fixation gene sequences in Rhizobium phaseoli. Nature, 299: 724-726.
Quinto, C., de la Vega, H., Flores, M., Leemans, J., Cevallos, M.A., Pardo, M.A.,Aspiroz, R., de Lourdes Girard, M., Calva, E. and Palacios, R., 1985. Nitrogenase reductase: a functional multigene family in Rhizobium phaseoli. Proc. Natl. Acad. Sci. USA, 82: 1170-1174.
Raina, R., Reddy, M.A., Ghosal, D. and Das, H.K., 1988. Characterization of a gene for the Fe protein of the vanadium dependent alternative nitrogenase of Azotobacter vinelandii and construction of a Tn5 mutant. Mol. Gen. Genet., 214: 121-127.D
Riddle, G.D., Simonson, J.G., Hales, B.J. and Braymer, H.D., 1982. Nitrogen fixation system of tungsten-resistant mutants of Azotobacter vinelandii. J.Bacteriol., 152: 72-80.
Robson, R.L., 1986. Nitrogen fixation in strains of Azotobacter chroococcum bearing deletions of a cluster of genes coding for nitrogenase. Arch. Microbiol., 146:74-79.
Robson, R.L., Eady, R.R., Richardson, T.H., Miller, R.W., Hawkins, M. and Postgate, J., 1986a. The alternative nitrogenase of Azotobacter chroococcum is a vanadium enzyme. Nature, 322: 388-390.
Robson, R.L., Woodley, P. and Jones, R., 1986b. Second gene (nifH*) coding for a nitrogenase iron protein is adjacent to a gene coding for a ferredoxin-like protein. EMBO J., 5: 1159-1163.
Robson, R.L., Woodley, P., Pau, R.N. and Eady, R.R., 1989. Structural genes for the vanadium nitrogenase from Azotobacter chroococcum. EMBO J., 8:1217-1224.
Ruvkun, G.B. and Ausubel, F., 1980. Interspecies homology of nitrogenase genes. Proc. Natl. Acad. Sci. USA, 77: 191-195.
Santero, E.,. Hoover, T., Keener, J. and Kustu, S., 1989. In vitro activity of the nitrogen fixation regulatory protein NifA. Proc. Natl. Acad. Sci. USA, 86: 7346-7350.
Santero, E. Toukdarian, A. and Kennedy, C.K., 1988. Identification and characterization of two nitrogen fixation regulatory regions, nifA and nfrX, in Azotobacter vinelandii and Azotobacter chroococcum. Mol. Microbiol., 2 303-314.
Sahay, M.N. and Sankaram, A., 1968. Effect of trace elements and growth promoting substances on Anabaena circinalis RABH. Ind. J. Microbiol., 88: 47-48.
Scolnik, P.A. and Haselkorn, R., 1984. Activation of extra copies of genes coding for nitrogenase in Rhodopseudomonas capsulata. Nature, 307: 289-292.

Shah, V.K., Davis, L.C., Gordon, J.K., Orme-Johnson, W.H. and Brill, W.J., 1973. Nitrogenase III. Nitrogenaseless mutants of Azotobacter vinelandii: activities, cross reactions and EPR spectra. Biochim. Biophys. Acta, 292: 246-265.

Shah, V.K., Ugalde, R.A., Imperial, J. and Brill, W.J., 1984. Molybdenum in nitrogenase. Ann. Rev. Biochem., 53: 231-257.

Sibold, L. and Souillard, N., 1988. Genetic analysis of nitrogen fixation in methanogenic archaebacteria. In: H. Bothe, F.J. de Bruijn and W.E. Newton (Editors), Nitrogen Fixation: Hundred Years After. Gustav Fischer, Stuttgart, pp. 705-710.

Singh, H.N., Vaishanpaya, A. and Singh, R.K., 1978. Evidence for the involvement of a genetic determinant controlling functional specificity of group VIB elements in the metabolism of N_2 and NO_3^- in the blue-green alga Nostoc muscorum. Biochem. Biophys. Res. Commun., 81: 61-74.

Souillard, M. and Sibold, L., 1989. Primary structure, functional organization and expression of nitrogenase structural genes of the thermophilic archaebacterium Methanococcus thermolithotrophicus. Mol. Microbiol., 2: 541-551.

Takahashi, H. and Nason, A., 1957. Tungstate as a competitive inhibitor of molybdate in nitrate assimilation and in N_2 fixation by Azotobacter. Biochim. Biophys. Acta, 23: 433-435.

Toukdarian, A. and Kennedy, C., 1986. Regulation of ntr and glnA genes and construction of ntr mutants. EMBO J., 5: 399-407.

Wang, S.Z., Chen, J.S. and Johnson, J.L., 1988. The presence of five nifH-like sequences in Clostridium pasteurianum: sequence divergence and transcription properties. Nucl. Acids Res., 16: 439-454.

Youvan, D.C., Bylina, E.J., Alberti, M., Begusch, H. and Hearst, J.E., 1984. Nucleotide and deduced polypeptide sequences of the photosynthetic reaction center, B870 antenna, and flanking polypeptides from R. capsulata. Cell, 37: 949-957.

CHAPTER 8

FREE-LIVING DIAZOTROPHS

WARWICK B. SILVESTER and DAVID R. MUSGRAVE
Department of Biological Sciences, University of Waikato, Hamilton, New Zealand

1 INTRODUCTION

The free-living or asymbiotic diazotrophs are the loose assemblage of organisms remaining after the root nodules, rhizocoenoses and cyanobacteria are dealt with. The term asymbiotic is technically inappropriate since most "other" diazotrophs live in relatively close nutritional association with other organisms. Despite the drawbacks of the terms "free-living" and "asymbiotic" they will be used here to describe organisms not dealt with elsewhere, particularly the heterotrophs and non-photosynthetic autotrophs.

After the first isolations of free-living diazotrophs; Clostridium pastorianum (now C. pasteurianum) by Winogradsky in 1893 and Azotobacter chroococcum (Beijerinck, 1901), there followed 40 years in which these two genera were virtually the only known types. New techniques and the recognition of the physiological and ecological importance of oxygen in diazotrophy has seen a rapid increase in both the number of organisms identified and in the variety of their habitats. This chapter focuses on that range of organisms, their relatedness and origins, their habitat diversity and their N_2-fixing potential.

2 ORGANISMS

2.1 General

In 1980 Postgate published a list of authenticated nitrogen-fixing species. Comparison with one published in 1964 showed the very rapid growth in the numbers of known diazotrophs. Table 1 updates that list; the last decade has added one new kingdom, 15 new genera and at least 32 new species. Some of the more significant additions are briefly discussed.

2.2 Archaebacteria

The Archaebacteria are the most significant addition to the list. Murray and Zinder (1984) provided unequivocal proof by using $^{15}N_2$ uptake to confirm nitrogen fixation in Methanosarcina barkeri 227 in a pure culture, while in the same year nitrogen fixation was reported in a thermophile, Methanococcus

TABLE 1

List of acceptable heterotrophic diazotrophs in 1989. Those new to the list since 1979 (Postgate, 1980) are identified by an asterisk.

Kingdom Habitat	Family	Genus	Species
Archaebacteria*	Methanobacteraceae*	*Methanobacterium**	*bryantii** *ivanovii**
Anaerobic	Methanococcaceae*	*Methanococcus**	*thermolithotrophicus** *maricaludis**
	Methanosarcinaceae*	*Methanosarcina**	*barkeri**
		*Methanolobus**	*tindarius**
Eubacteria			
Aerobic	Acetobacteraceae*	*Acetobacter**	*nitrocaptans**
	Azotobacteraceae	*Azomonas**	*beijerinckii* *insignis* *macrocytogenes*
		Azotobacter	*beijerinckii* *chroococcum* *halophilus* *miscellus* *paspali* *vinelandii*
		Azotococcus	*agilis*
		Beijerinckia	*derxii* *fluminensis* *indica* *nobilis*
		Derxia	*gummosa*
		Xanthobacter	*autotrophicus* *flavus*
	Frankiaceae*	*Frankia**	taxonomy uncertain
Microaerobic	Methylmonadaceae	*Methylococcus**	*capsulatus**
		*Methylocystis**	*parvus**
		Methylosinus	*sporium** *trichosporium*
	Spirillaceae	*Azospirillum*	*amazonense** *brasilense* *lipoferum* *halopraeferens**
		Aquaspirillum	*fasciculus* *perigrenum*
		Campylobacter	*nitrofigilis**
		*Herbaspirillum**	*seropedicae**
	Corynebacteriaceae	*Arthrobacter**	*fluorescens**
	Thiobacteriaceae	*Thiobacillus*	*ferro-oxidans*
	Rhizobiaceae	*Rhizobium*	many
	Pseumonadaceae*	*Pseudomonas**	*diazotrophicus***pseudoflava** *saccharophylla** *stutzeri**
	Beggiatoaceae*	*Beggiatoa**	*alba**
		*Thiothrix**	*nivea**

Facultative anaerobic	Enterobacteriaceae	Citrobacter	freundii intermedius
		Enterobacter	aerogenes agglomerans cloacae
		Erwinia	herbicola
		Escherichia	intermedia
		Klebsiella	aerogenes oxytoca pneumoniae
	Bacillaceae	Bacillus	azotofixans* macerans polymyxa
	Vibrionaceae	Vibrio*	diazotrophicus* natriegens* cincinnatiensis* pelagius*
Anaerobic	Bacillaceae	Clostridium	acetobutylicum beijerinckii butylicum butyricum felsinium kluyveri lactoacetophilum madisonii pasteurianum pectinovorum saccharo-butyricum tetanomorphum tyrobutyricum
		Desulfatomaculum	ruminis orientis
	Uncertain	Desulfovibrio	africanus* baculatus* desulfuricans gigas salexigens* vulgaris

thermolithotrophicus (Belay et al., 1984). The presence of nitrogenase in the Archaebacteria and in thermophiles has far-reaching implications for the study of both the biochemistry and evolution of nitrogenase.

2.3 Frankia

Although Frankia, the endophyte of non-legume root nodules like those of Comptonia (Callaham et al., 1978), is a common soil inhabitant (Chapter 15), it is not known if it fixes nitrogen asymbiotically in the soil.

2.4 Methanotrophs

Considerable confusion has existed over the status of the methanotrophs (obligate methane-oxidising bacteria). These diazotrophic chemoautotrophs, termed the "Cinderellas of the nitrogen-fixing world" (Dalton, 1980), can be readily isolated from soils by methane enrichment. Prior to 1980, Methanosinus (now Methylosinus) was the only reported diazotroph in this group, probably because in other methanotrophs added acetylene also inhibited methane oxidation so that the organisms rapidly became starved of energy and failed to reliably reduce acetylene. Subsequently, methanol has been used as a substrate (Dalton, 1980). Up to five genera have been claimed to be diazotrophs (see Benson, 1985). However, the apparently definitive work of Murrell and Dalton (1983) concluded, on the basis of acetylene reduction tests and growth on N-free media, that Type I organisms including Methylobacter and Methylomonas were not diazotrophs, despite many strains showing homology to a nifH probe (Oakley and Murrell, 1988). However, all Type II strains, including Methylosinus and Methylocystis, and the Type X strain Methylococcus capsulatus, grow well on nitrogen-free media and reduce acetylene.

2.5 Pseudomonads

Although early reports of N_2 fixation in pseudomonads were doubted because of the uncertain taxonomic status of the isolates (see Palleroni, 1984), four separate reliable reports (see Jenni et al., 1989) established diazotrophy in Pseudomonas though only in a few strains.

2.6 Root zone organisms

The root zone of higher plants is a rich source of new diazotrophs including species of Pseudomonas (see above) isolated from wetland rice and Campylobacter nitrofigilis (McLung et al., 1983) from the root-zone of the sea grass Spartina (McLung and Patriquin, 1980). The latter genus is normally pathogenic or commensal in animals, with one species (C. pylori) the causal agent of enteric ulcers in humans. C. nitrofigilis is associated with plant roots and may be part of a rhizocoenosis, it was also present in adjacent sediments.

Studies on cereals and sugar cane have unearthed new associated species; a diazotroph resembling Acetobacter and Gluconobacter has consistently been isolated from the roots and stems of sugar cane. It was originally designated Saccharobacter nitrocaptans (Cavalcante and Döbereiner, 1988), but is now regarded as a new species of Acetobacter (A. nitrocaptans) (Döbereiner et al., 1988). It apparently lives in high sugar concentrations in the stems and roots of sugar cane, is very acid tolerant (pH 3.8-5.5) and expresses nitrogenase in the presence of high (10mM) nitrate (Döbereiner et al., 1988). Other newly isolated root zone diazotrophs (Table 1) include new species of Azospirillum (A. amazonense, A. halopraeferens), a new genus (Herbaspirillum (Herbazospirillum) seropedicae) and a new species of Bacillus, B. azotofixans (Döbereiner et al., 1988).

2.7 Gliding bacteria

Beggiatoa, a filamentous, microaerophilic sulphide-oxidiser fixes nitrogen (Nelson et al., 1982) and is found in waters, in flooded soils and in the root zones of rice and Spartina (Polman and Larkin, 1988). The related genera Thiothrix and Vitreoscilla also have nitrogen-fixing strains (Polman and Larkin, 1988).

2.8 Other genera

In 1981, Guerinot and Patriquin reported nitrogenase in a facultatively anaerobic *Vibrio* isolated from the intestinal tract of sea urchins. It was identified as *V. diazotrophicus*; a survey of 27 species of *Vibrio* (Urdaci et al., 1988) showed that strains from four species showed nitrogenase activity. The majority of N_2-fixing strains were isolated from sea-water or inland saline rivers and lakes (Urdaci et al., 1988).

TABLE 2

Relatedness, nitrogenase activity and isolation method for six strains of *Pseudomonas pseudoflava* (after Jenni et al., 1989).

Strain	Phenotypic[2] similarity		DNA:DNA homology	Nitrogenase[3]	*nif*HDK[4] hybridisation	Isolation[5] method
	92%	82%				
2111[1]				0	-	-N
2008				0	-	-N
2252				0	+	-N
2226				+	+	+N
2220				+	+	+N
2222				+	+	+N

[1]: Strain 2111 is the type strain
[2]: Based on growth on 150 different C sources
[3]: Acetylene reduction
[4]: *Klebsiella pneumoniae nif*HDK
[5]: -N isolation is for H2-chemolithoautotrophs in a medium without N source
+N isolation is for H2-chemolithoheterotrophs in a medium with NH_4^+

Postgate and Kent (1985) summarised all the data on diazotrophy in *Desulfovibrio*, showing that 18 out of 22 pure strains (covering six species) were diazotrophic.

2.9 The search for new diazotrophs

The new discoveries over the past decade raise challenges to the classical isolation methods used for diazotrophs and the use of *nif* probes. Traditional techniques used for isolation of diazotrophs rely on nitrogen-free media containing a narrow range of carbon substrates and the culture collections of the world reflect those techniques rather than the real world. This is elegantly illustrated by the work of Jenni et al. (1989) on Pseudomonas, which quite unexpectedly showed that diazotrophic strains of P. alba were isolated only in media containing combined nitrogen, while closely related strains isolated on media without combined nitrogen all lacked nitrogenase activity (Table 2). We may therefore need to widen our view of isolation conditions required for diazotrophs. An example is the recent isolation of cellulolytic diazotrophs (Leschine et al., 1988) using cellulose as the fermentable substrate.

The use of nif probes to determine the distribution of diazotrophy is an increasingly important tool. However, some isolates of Pseudomonas (Jenni et al., 1989) (Table 2), methanotrophs (Oakley and Murrell, 1988) and Desulfovibrio (Postgate et al., 1986) with clear homology to nifH probes do not express nif phenotypes. In some cases, this is attributable to incorrect conditions for the expression of nitrogenase; in Desulfovibrio these are N-limitation prior to exposure to derepressing media containing pyruvate and sulphate. In the case of methanogens, however, the nifH homology and lack of nitrogenase expression across the majority of type I species is thought to be due to a loss of nitrogen-fixing ability in an ancestral strain (Oakley and Murrell, 1988).

3 RELATEDNESS AND EVOLUTION

The widespread and haphazard distribution of diazotrophy can best be seen when diazotrophic organisms are placed on a phylogenetic tree based on 16S rRNA nucleotide sequences (Woese, 1987). The debate still continues as to whether this distribution represents loss of nif genes from populations of organisms diverging in environments in which positive selection no longer occurrs, or whether continual lateral transfer of the nif genes has allowed organisms to exploit changing conditions. These two possible scenarios are not mutually exclusive, however, and both may be necessary to explain the particularly haphazard distribution of the nif genes both in the genome and across the bacteria.

Before the discovery of plasmids, the predominant view on the evolution of diazotrophy was that its haphazard distribution resulted from the loss of nif genes during evolutionary divergence. Higashi (1967) led the speculation that plasmids may have had a role in the spread of nif genes, based on the fact that a plasmid-encoded gene can be easily lost. Nodulation instability in Rhizobium species was explained when plasmids encoding nod, nif, fix and tra were identified (Nuti et al., 1977; Eckhardt, 1978). Postgate (1974, 1982) also challenged the divergent loss hypothesis on the basis that geochemical conditions for selection of diazotrophy had not occurred for the first 2-3 aeons of the biosphere's existence and on the ease with which nif genes could be transferred in the laboratory. The distribution of nif genes among bacterial species was therefore seen to be analogous to the distribution of antibiotic resistance genes in bacteria containing conjugative plasmids.

In general, natural environments have been seen to resist the spread of foreign genes. However, unsuccessful attempts to introduce laboratory strains into natural populations may tell us more about our poor knowledge of the effects of long term laboratory selection on the bacterial genome than on any natural process. The advent of pulsed field gel electrophoresis (Schwartz and Cantor, 1984) with its ability to analyse bacterial genomes as large restriction fragments, is providing evidence that large scale genetic rearrangements allowed in laboratory culture may preclude an organism's survival in the environment (Smith et al., 1987). The survival of large conjugative plasmids requires selection for their conjugative ability or some other linked gene and/or their replication and distribution to their progeny. The presence of conjugative plasmids has often been used to determine the likelihood that genes would be transferred laterally among species. The discovery of conjugative transposons (reviewed by Clewell and Gawron-Burke, 1986; Hecht and Malamy, 1989) has shed new light on DNA transfer in the absence of plasmids. Conjugative transposons are large DNA elements ranging from 16 kb to 60 kb in size and have been implicated in the dissemination of antibiotic resistance among pathogenic organisms that lack detectable plasmids (Horaud et al., 1987). In Gram-positive organisms, conjugative transposons are especially significant because the host range of transfer is such that the donor and recipient need not be of the same genus. Ultimately, it will be sequence analysis of nif genes which will reveal if some or all of these DNA transfer mechanisms have played a part in their evolution. Indeed, as Woese (1987) has so lucidly pointed out, significant interspecies transfer during evolution would mean that an organism would not have a history of its own but would be a collection of genes or operons each with its own history. If this were so, a study of the sequence relatedness of different characters would give different values of relatedness, as if the genes had evolved separately.

A molecular analysis of the putative nifH genes cloned by Souillard et al. (1988) from two archibacteria, *Methanobacterium ivanovii* and *M. thermolithotrophicus*, is sure to fuel the origin debate. They suggested that the phylogenetically distant Archaebacteria encode putative nifH polypeptides in which 30% of the amino acids are common to all of the twelve eubacterial nifH proteins for which the amino acid sequence has been deduced. It is significant that when pairs of nifH polypeptide sequences were compared, the data suggested a similar distant phylogenetic position for the Archaebacteria as that determined from analysis of 16S rRNA sequences. Nucleotide sequence matches between the three archaebacterial nifH genes are low (53%, 58% and 59%) when compared to the ten eubacterial sequence matches (61% - 88%). Postgate and Eady (1988) have down-played the evolutionary importance of these findings in the archaebacteria by saying that "....they do not advance the present discussion because they are consistent with either divergent loss or lateral acquisition of nif during evolution."

While the low nucleotide sequence homology of the archaebacterial nifH genes suggest a wide divergence and therefore an early evolutionary origin, other supporting evidence can be deduced from the sequence data. The lack of a promoter sequence adjacent to the archaebacterial nifH ORF's suggests that these genes are part of a polycistronic mRNA transcript. However, the ORF's upstream of nifH, which would form part of this transcript, show no sequence homology to the eubacterial nifD or nifK genes (Souillard et al., 1988). This suggests that either the nifH, nifD and nifK genes in the Archaebacteria do not form an operon, or that the archaebacterial nifD and nifK genes have changed faster than nifH, to the point where there is no longer any homology to their eubacterial counterparts. If the latter hypothesis is supported by an analysis of the encoded proteins, it would provide further evidence for the ancient evolutionary origin of the archaebacterial nif genes. Two features of the *M. thermolithotrophicus* nifH gene are suggestive. The first is the conservation of the ribosome binding site sequence of the nifH and hisA genes (Weil et al., 1987), which suggests that these two genes have co-evolved. The second is the discovery of a 36 bp putative or vestigial intron, corresponding to twelve amino acids, not found in any other known nifH gene. This sequence has 5' and 3' boundaries which are similar to eukaryotic consensus splice sites and possibly represents the first intron in an archaebacterial coding sequence.

An important consideration is whether the molybdenum and alternative nitrogenases have evolved from a common or separate precursor. The latter route is supported by genetic complementation studies (Smith et al., 1988); DNA sequence homology between nifD and nifE, and between nifK and nifN, suggests that gene duplication has played a significant role in their evolution (Brigle et al., 1987). It is not known if the clustering of nif genes is a more or less primitive arrangement. A decison between these alternatives has been overly influenced by the dispersed nif gene arrangement in *Azotobacter* and *Rhizobium*, where the presence of sophisticated symbiotic and O_2 exclusion systems suggests that dispersed genes are a more advanced feature. In the cyanobacteria the development of heterocysts is accompanied by a genetic rearrangement in which a contiguous nifHDK gene cluster is re-formed (Golden et al., 1985; Chapter 16).

It is also important to consider here, as Postgate has pointed out, that the predominance of Mo nitrogenases is due to selection pressure instigated by microbiologists rather than evolutionary processes and that the lack of excessive functional constraint on the evolution of the nitrogenase genes would allow a diverse repertoire of enzymes to evolve.

In summary, elucidation of the mechanisms by which evolutionary diversity of the nif genes has occurred will only be achieved by the use of novel approaches for the isolation of new diazotrophs and the comparison of the nucleotide sequence of their nif genes with those already established.

4 HOW MUCH NITROGEN CAN BE FIXED? SOME FIRST PRINCIPLES

Rates of N_2 fixation by legumes are of the same order (30 to 300 kg N ha^{-1} yr^{-1}) as the requirements for plant growth (La Rue and Patterson, 1981). It is useful to compare the actual and potential inputs from the "other" diazotrophs with these rates. In situ measurements of nitrogen fixation that can realistically be converted to a kg N ha^{-1} yr^{-1} basis are notoriously difficult to make, especially in the spatially and temporally discontinuous soil environment. Nevertheless a vast number of measurements of nitrogen increment have been made and extrapolations to the hectare/year basis have proven irresistible. The resulting spread of figures from 0.1 - 1000 kg N ha^{-1} yr^{-1} (Rice et al., 1967; Knowles, 1977; Waughman et al., 1981; Boring et al., 1988) covers a range from the insignificant to the improbable. While it is not appropriate here to explore the methodology that has given rise to these figures, the principles that control diazotroph numbers and nitrogenase activity need to be considered as a basis for realistic projections of nitrogen fixation potential in various habitats.

The major factor limiting asymbiotic nitrogenase is carbon/energy supply. Most soils respond dramatically to the addition of sugar and much work on diazotroph physiology has been conducted on sugar-amended soils (Knowles and O'Toole, 1975). Other nutrients, oxygen tension and pH also play important parts in controlling nitrogenase activity, but it is the availability of soluble carbon compounds that sets the upper limit for nitrogen fixation. N_2 fixation is highly energy-demanding, with 28 moles of ATP required for every mole of N_2 fixed (Dixon and Wheeler, 1986); in the absence of H_2 recycling this requires the complete oxidation of 0.75 moles of glucose (or very much more anaerobically). This is usually expressed as the efficiency term mg N fixed g^{-1} sugar, which for the above calculation is 211 mg N g^{-1} (glucose) or 528 mg N g^{-1} C*.

* The term mg N g^{-1} (carbohydrate) is a terribly awkward one and is used here for comparative and historical reasons only. Enormous confusion exists in the literature, with authors using mg N g^{-1} C and mg N g^{-1} (carbohydrate) as if they were equivalent. The term mg N g^{-1} (C-source) is even more confusing.

Efficiencies as high as 270 mg N g^{-1} C have been demonstrated for the pea symbiosis (Minchin and Pate, 1973) and even higher for clover (Gibson, 1966), but the measured efficiencies for asymbiotic bacteria are generally one-tenth those of legume nodules (Table 3). This apparently low efficiency in asymbiotic diazotrophs is a product of two important constraints not confronted by symbiotic bacteria. Firstly, asymbiotic bacteria only fix N_2 while actively growing, in contrast to symbiotic bacteria which are non-dividing, and thus a great deal of carbon is required for growth. Secondly, the aerobes invest a significant amount of carbon in mechanisms to reduce the impact of oxygen. In Azotobacter, this takes the form of excessive uncoupled respiration to "burn off" the oxygen, while in others it may be extra structural material to provide an oxygen diffusion resistance. The anaerobes avoid the oxygen problem and their equally low efficiencies result from the low energy yield of anaerobic metabolism. Highest efficiencies in culture are apparently obtained at optimal oxygen concentrations and when the diazotrophs are carbon-limited (Table 3). The common asymbiotic diazotrophs are unable to utilise complex carbohydrates, and this further complicates the efficiency formula, as they rely on associated micro-organisms to liberate sugars from cellulose (Jensen, 1940). Despite this complication, the apparent efficiencies in co-cultures of diazotrophic and cellulolytic organisms are as good as those in pure culture (Table 3), possibly reflecting the lowered O_2 conditions and carbon limitation of such cultures.

A reasonable average for the efficiencies is 20 mg N g^{-1} C which, on the basis of 40% carbon in plant dry matter, converts to 8 mg N g^{-1} (carbohydrate) or 8 kg N t^{-1}. The carbon to support this comes from plant residues and their input into soil varies enormously; for wheat stubble in Australia, this averages 4 to 7t ha^{-1} (Halsall et al., 1985). Such an input at the optimum pH could yield a predicted 32-56 kg N ha^{-1} and N_2 fixation rates of this order have been measured in the field.

Other methods of predicting N gains from asymbiotic organisms rely on cell counts from soils and a prediction of the likely turnover time. It has been calculated (Jensen, 1940) that the production of 10^9 Azotobacter cells requires the fixation of 0.1 mg of N_2. Thus for a soil maintaining 10^6 g^{-1} the fixation rate would be 0.1 mg N kg^{-1} (soil) or 0.2 kg N ha^{-1} (ca. 2 x 10^6 kg soil ha^{-1}) per bacterial turnover. Chang and Knowles (1965) showed that the turnover time of diazotrophs in soil was considerably less than 1 month and probably closer to a week. Assuming 50 turnovers and a steady 10^6 cells g^{-1}, asymbiotic fixation could account for 10 kg N ha^{-1} yr^{-1}. A similar approach was made for Beijerinckia in paddy soils in the tropics (Becking, 1978). This calculation proposed that the total cell nitrogen of Beijerinckia was derived from N_2 fixation and predicted 50 generations per year would give 45 kg N ha^{-1} for a population of 10^6 cells. However, the figures used are in error, as the cell nitrogen should be 0.8% of fresh weight rather than 8.0%; thus a more reasonable figure of 4.5 kg N ha^{-1} might be expected from a cell population of 10^6 g^{-1} soil.

TABLE 3

Efficiency of nitrogen fixation in asymbiotic bacteria in axenic culture on sugars or in mixed cultures on straw or cellulose or in soil

Species	Conditions	Efficiency $mgN.g^{-1}C$	Reference
Axenic culture			
Azotobacter chroococcum	air	7.4	Parker, 1954
	4.1 kPa O_2	21.6	Parker, 1954
Azotobacter chroococcum	different soils	16.6-33.2	Mahmoud et al., 1978
Beijerinckia indica	2% glucose	39	Becking, 1978
	0.2% glucose	50	Becking, 1978
Clostridium pasteurianum		8.8-21.7	Mishustin and Yemtsev, 1973
Clostridium butyricum		20	Line and Loutit, 1971
Klebsiella pneumoniae		9.5	Line and Loutit, 1971
Azospirillum lipoferum	0.2% glucose	50-120	Volpon et al., 1981
Mixed culture			
Azospirillum brasilense Cellulomonas	straw	160*(38)	Halsall and Gibson, 1986
Azotobacter sp. Corynebacterium	cellulose	15-35	Jensen and Swaby, 1941
Clostridium butyricum Trichoderma	straw	30	Lynch and Harper, 1985
Soils			
Azotobacter	clay loam	22-87	
	sand	27-75	Abd-el-Malek, 1971

* per g C utilised (per g C added)

It is apparent then from both empirical studies and from applying some reasonable first principles that the amount of N_2 fixed in asymbiotic systems will be small in most cases. However, in some instances where large amounts of carbon are returned to soil, where the pH is close to neutrality and where oxygen and moisture are optimal, bacterial numbers as high as 10^8 g^{-1} soil can be generated, and ecologically, perhaps agronomically, significant amounts of N_2 are fixed.

5 HABITAT

5.1 Introduction

The physiological diversity of the diazotrophs is matched by the diversity of the environments which these properties allow them to exploit. In the habitats of the non-photosynthetic diazotrophs, two variables appear to be in control. These are the need for large amounts of respirable carbon compounds and an ecological niche where the oxygen tension is lowered by a physical diffusion barrier coupled with respiration.

5.2 Soil

Natural soil is not an ideal environment for diazotrophs. Soil organic matter consists largely of relatively stable humic complexes and the pH, especially of forest soils, is often suboptimal. Shortage of readily available carbon compounds is the main limitation, as demonstrated by the vast increase in cell population after the addition of 2% glucose to soil (O'Toole and Knowles, 1973). Similarly, Mishustin and Yemtsev (1973) showed incorporation of raw organic matter to water-logged soil increased clostridia from 10^3 g^{-1} to 10^9 g^{-1} and addition of crop residues to irrigated Egyptian and Iraqi soils increased Azotobacter numbers to 10^7 g^{-1} or more (Hamdi et al., 1978; Mahmoud et al., 1978). Under natural conditions, however, diazotroph populations average 10^2-10^3 g^{-1} and seldom exceed 10^4 g^{-1} (Knowles, 1977).

Clostridia are almost ubiquitous, especially in wet organic soils, in numbers ranging from 10^2 to 10^5 g^{-1} (DiMenna, 1966; Mishustin and Yemtsev, 1973), while Azotobacter is restricted to lighter sand or silty soils of neutral pH (Ross, 1960; DiMenna, 1966; Mahmoud et al., 1978). Beijerinckia is a common tropical soil organism tolerant of low pH and high aluminium (Becking, 1961), while forest and high country soils yield a variety of microaerophiles and facultative anaerobes including Bacillus, Enterobacter and Klebsiella (Line and Loutit, 1969, 1971; Jones and Bangs, 1985).

Soil diazotrophs seldom add more than 1 kg N ha^{-1} yr^{-1} (Mishustin and Shil'nikova, 1971; Knowles, 1977; Waughman et al., 1981). The majority of tests rely on disturbed samples treated with acetylene and problems with spurious ethylene production (Hendrickson, 1989), variable ratios between acetylene reduction and N_2 fixed (Roskoski, 1981) and spatial and temporal variations in activity (Silvester, 1983) make extrapolations to annual rates extremely difficult if not impossible.

A number of assumptions on the role of soil diazotrophs have been drawn from the growth of these organisms on chemically defined media in the laboratory, when in the field they grow on a physical and chemical complex that bears little relation to laboratory flasks. The morphology of Azotobacter grown on dialyzed soil media is totally different from the large pleiomorphic rods found in laboratory cultures (Gonzalez-Lopez et al., 1984). Although Azotobacter in soil may utilise a whole range of phenolic acids as carbon sources for growth (Wu et al., 1987), only 4-hydroxybenzoate supports nitrogenase activity (Moreno et al., 1988). These results suggest that high numbers of diazotrophs in soil may not be a good indicator of high rates of nitrogen fixation.

Azotobacter is a good competitor for soil nitrate (Green and Wilson, 1953) and it is likely that it will make an ecologically significant nitrogen input from N_2 only when large quantities of plant residues, with high C/N ratios, are returned to soil.

5.3 Plant residues

(i) General. Herbaceous and woody residues on the soil surface (LFH) or incorporated into the soil provide an ideal chemical and physical habitat for diazotrophs and much work on so-called soil nitrogen fixation probably represents activity in plant residues on and within soils. This is particularly true of forests where a majority of reports show that a very high proportion of observed nitrogenase activity is located in the A_0 horizons (Silvester, 1978; Nohrstedt, 1985).

Three main types of organic residues have been studied; forest leaf litter, decaying woody debris and crop residues.

(ii) Leaf litter. The first reports of asymbiotic nitrogen fixation were from decaying leaf litter (Henry, 1897); those rather equivocal results were confirmed and extended by careful experiments (Olsen, 1932), showing that during a year of decay diazotrophs in oak and beech leaves may add up to 1.7 mg N g^{-1} leaf (4.2 mg N g^{-1} C). The phenomenon, widely studied in forest species, is especially important in mor-forming litter, where leaves remain intact for long periods. A wide range of conifers, hardwoods and mangroves support relatively high levels of nitrogenase (Gotto and Taylor, 1976; Silvester, 1978; Nohrstedt, 1985, 1988; Heath et al., 1988; Silvester, 1989). Many of the species studied produce an acid litter and variability in nitrogenase activity is explained to a large degree by pH (Nohrstedt, 1985). In kauri (Agathis australis), the pH of decaying litter rises to 6.0 during the early stages of decay (Silvester, 1978), before falling to pH 3.8; nitrogenase activity is strongly correlated with the high pH phase. Moisture is also important, insofar as it maintains an oxygen diffusion barrier into the leaf interior, and moisture levels in excess of 200% are optimal (Heath et al., 1988).

It is significant that addition of sugars to decaying leaves results in only small increases in nitrogenase, while phosphate stimulates nitrogenase in kauri litter (Silvester, 1978) and Mo is very deficient in Douglas-fir leaves (Silvester, 1989).

(iii) Woody debris. Coarse woody debris is a feature of the floor of many forest types and accounts for over 50% of litter fall in Douglas fir forest (Sollins, 1982). Decomposition of this wood can be extraordinarily slow due to its low content of N and P. Wood-invading fungi are able to concentrate N from the decaying wood and transport it into the wood from the soil and external litter layers. Nitrogen fixation occurs in decaying wood from many species (Cornaby and Waid, 1973; Sharp and Millbank, 1973; Todd et al., 1975; Larsen et al., 1978; Roskoski, 1980; Silvester et al., 1982) and is always associated with wood-rotting fungi which are assumed to liberate sugars from the wood polysaccharides (Larsen et al., 1978; Jurgensen et al., 1989).

Wood decay organisms can be separated into two main types; white-rotting fungi which remove cellulose, hemicellulose and lignin at the same rate, and the brown-rotting fungi which remove cellulose and leave lignin virtually unchanged. Nitrogenase activity in wood containing white-rot fungi is invariably higher than in that containing brown-rot fungi (Jurgensen et al., 1989). These higher activities may be due to the higher residual soluble carbon output from white rotting fungi (Jurgensen et al., 1989), the possible N input from associated diazotrophs then accelerating wood breakdown by the white rotting fungi.

Nitrogenase activities in decaying hardwoods (Cornaby and Waid, 1973; Roskoski, 1980) are consistently higher than in softwoods, possibly because of a higher incidence of white rot in decaying hardwoods (Jurgensen et al., 1989). The major diazotrophs extracted from decaying wood are facultative and strict anaerobes, including *Enterobacter*, *Klebsiella* and *Clostridium* (Aho et al., 1976; Spano et al., 1982), with numbers of diazotrophs up to 10^4 ml^{-1} of expressed water in rotting Douglas fir wood (Spano et al., 1982) and in the range 10^4-10^6 ml^{-1} in white fir (Aho et al., 1976). All attempts to isolate microaerophiles such as *Azospirillum* have been unsuccessful (Jurgensen et al., 1989) and this leaves a dilemma. Since the early isolations had yielded anaerobes, all workers had carried out nitrogenase assays anaerobically. However, anaerobic incubations completely inhibit nitrogenase activity after 10 hours, while aerobic incubations were active for 60 hours or more (Silvester et al., 1982). Even small pieces (< 5mm thick) of wood showed high rates of aerobic acetylene reduction over many hours. The optimum pO_2 for nitrogen fixation was shown to be 5 kPa oxygen with a significant inhibition of activity at 20 kPa O_2 in wood at less than field-capacity. The responsible diazotrophs may therefore be micro- aerophiles despite the isolation of anaerobes from some decaying wood. The obligately aerobic nature of the wood-rotting fungi and their likely close association with the diazotrophs makes it unlikely that the agents are strict aerobes. This problem obviously needs more attention.

Several reports exist of large bacterial populations in the wet-wood of living trees. Nitrogenase activity associated with bacterial populations up to 10^6 ml^{-1} have been found in living trees of elm and poplar (Schink et al., 1981; van der Kamp, 1986). The nitrogenase activities in this material are every bit as high as those in decaying wood and Clostridium and Klebsiella have been isolated from it. It is interesting that nitrogenase activity increases from the base of the tree (where it is near zero) to the top of the tree, and that the wetwood has a foetid odour throughout, indicating anaerobiosis (van der Kamp, 1986). The carbon supply to nitrogenase is postulated to come from pectin breakdown or from the sap stream.

In conclusion, wood decay often supports significant populations of diazotrophs which appear to be symbiotrophic with a wide variety of decay fungi. Some diazotrophs are also found in living trees. Contributions to ecosystem nitrogen have been reliably estimated to be between 0.1 and 2 kg N ha^{-1} yr^{-1}.

(iv) Crop residues. Crop residues such as cereal straw and sugar cane slash have normally been burnt off. Such materials are low in nitrogen due to its withdrawal into the grain and the traditional wisdom has been that these materials lock up soil nitrogen when decomposing in soil (Lynch and Harper, 1985). Interest in straw as a natural substrate for N_2 fixation (Jensen and Swaby, 1941) waned when it became obvious that no nitrogen-fixing bacteria could utilise cellulose directly and that the inherent inefficiency of carbon use in mixed cultures would preclude significant nitrogen inputs. As recently as 1965, Jensen concluded that N_2 fixation in crop residues was of little agronomic significance. However, large N inputs occur during crop residue breakdown (Patriquin, 1982; Halsall and Gibson, 1986; Hill and Patriquin, 1988) and the process may be improved by inoculation and selection of organisms (Halsall and Gibson, 1989).

As in leaf litter the activity of diazotrophs is dependent on associated cellulolytic organisms which in laboratory co-cultures give very favourable fixation efficiencies (Table 3). In the field, the incorporation of straw into neutral pH soils of Egypt and Iraq (Abd-el-Malek, 1971) results in Azotobacter counts of over 10^7 g^{-1} and apparent nitrogen yields of over 100 kg N ha^{-1} yr^{-1}.

The enormous amounts of wheat, maize and rice straw that are burnt each year represent a large loss of potential energy for asymbiotic nitrogen fixation. Considerable potential exists for the artificial manipulation of this system by inoculation of both partners in the straw breakdown (Halsall and Gibson, 1989).

(v) Phyllosphere. Reports of significant N_2 fixation in the phyllosphere have yet to be fully explained despite the isolation of Beijerinckia from leaf surfaces (Ruinin, 1956). The foliage of Douglas fir is reported to support diazotrophs contributing up to 14 kg N ha^{-1} yr^{-1} (Jones, 1970; Jones et al., 1974; Balloni et al., 1988), while the foliage of various crop plants, including rice and cotton, show acetylene reduction rates between 30 and 230 nmole C_2H_4 g^{-1} h^{-1} (Murty, 1983, 1984a, 1984b). On the assumption of an efficiency of 10 mg N g^{-1} (carbohydrate), the reported nitrogen fixation in Douglas-fir would require the loss of 1.4 tonnes of soluble carbohydrate per hectare from the leaves in forests that are poor in nitrogen and low in productivity. In the case of crop plants, the organism is reported to be Beijerinckia in populations of less than 10^3 g^{-1} of leaf, and at the stated efficiency of the isolate of 1.5 mg N g^{-1} (carbohydrate) would require the exudation of a mg of soluble carbohydrate every hour to support an acetylene reduction rate of 150 nmoles g^{-1} h^{-1}.

It has been stated (Sadykov and Umarov, 1980) that 0.1 - 10% of total plant nitrogen may derive from leaf surface diazotrophs. In the absence of a proven mechanism for delivery of the vast amount of carbohydrate needed and in the absence of a mechanism to provide a suitably moist, low oxygen environment on the leaf surface, claims for such high rates must be more widely confirmed before they can be credible.

(vi) Animal associations. Nitrogen-fixing bacteria can be isolated from the gut contents of many animals ranging from worms and insects through to humans (Waughman et al., 1981). While the bacteria remain viable in many cases, there are few examples in which the nitrogen input is significant or where the presence of the bacteria has been proven to be more than ingestion in the feed. The rumen appears to be an ideal environment for diazotrophs; it is anaerobic, releases a whole range of soluble carbon products and may have a high C/N ratio depending on the feedstuff. Sheep rumen contents show very significant nitrogenase activity when fed on natural grass mixtures, but this disappears when they are transferred to a sterile diet (Jones and Thomas, 1974), suggesting the rumen diazotrophs are continually replenished from the diet rather than a constitutive feature of the rumen. A diet supplemented with molasses and inoculated with Bacillus macerans resulted in increased nitrogen fixation in the sheep rumen, but measurements thus far account for less than 2% of ruminant nitrogen coming from N_2, with most measurements equating to 0.1% of the N intake.

It is in the lower animals, however, that gut diazotrophs such as Citrobacter and Enterobacter may be both constitutive and significant. Many of the xyllophagous insects, particularly the termites, are able to continue their life cycles entirely on a wood diet containing vanishingly low quantities of N. However, nitrogenase activities in these insects are generally low (Benemann, 1973), and when converted to time to double nitrogen content of the organism or colony (TDN), give average figures of 40 plus years, indicating very little ecological significance (Breznak, 1982). Some termites, however, show very high nitrogenase activities (Benemann, 1973) which are inversely correlated with their nitrogen intake (Prestwich et al., 1980). When whole Nasutitermes nests (up to 500,000 individuals) are tested for nitrogenase, the activities are very much higher than for disturbed individuals (Prestwich and Bentley, 1981). Whole colony nitrogenase gives a TDN of 200 - 600 days, which could account for 25 - 100% of colony N coming from the atmosphere. $^{15}N_2$ techniques show that fixed nitrogen is rapidly transferred to the body of the termite, and also between castes by trophallaxis (Bentley, 1984).

Perhaps the most interesting of the animal associates is the cellulose-degrading diazotroph isolated from the gland of Deshayes in wood-boring shipworms (Waterbury et al., 1983). This organism is implicated both in cellulose hydrolysis and nitrogen fixation in shipworms and although not identified, is an obligately marine, Gram-negative, aerobe which fixes nitrogen under microaerobic conditions. There appears to have been no follow-up to this 1983 report. The whole field of animal-associated diazotrophs is virtually untouched and while a majority of gut diazotrophs may well be ingested organisms there is still good reason to believe that some associations may be quite significant.

(vii) Aquatic. The productivity of oceans and lakes, like that of most soils, is nitrogen-limited and nitrogen fixation by asymbiotic heterotrophs in aquatic systems displays many parallels with soil systems. Cyanobacteria are the dominant diazotrophs in aquatic systems (Paerl et al., 1981), but recent evidence shows that heterotrophic bacteria are ubiquitous, even in the oceans (Guerinot and Colwell, 1985; Paerl et al., 1988). While their numbers in oceans are only of the order of 10^3 ml^{-1}, these Vibrio- and Oceanospirillum-like bacteria have an active nitrogenase (Guerinot and Colwell, 1985). As in soils, the supply of carbon is critical and association with phytoplankton (Martinez et al., 1983; Guerinot and Colwell, 1985) or organic matter (Paerl and Prufert, 1987; Paerl and Carlton, 1988) is essential to nitrogenase activity.

The maintenance of O_2-depleted microzones around and within organic matrices is important to aquatic diazotrophs. Added soluble carbohydrate is most effective in stimulating nitrogenase only when the organic matrix is supplied

(Paerl and Carlton, 1988). Under these conditions soluble carbon compounds leached from phytoplankton or derived from organic matter are apparently utilised by the diazotrophs on or in the organic flocs or particles.

Associations of diazotrophic bacteria with planktonic or benthic photosynthetic organisms are apparently more common than previously recognised (Martinez et al., 1983; Bebout et al., 1987). Nitrogenase activity is inhibited during the day by photosynthetic oxygen production and then rises rapidly upon darkening. Net nitrogen inputs of 2-8 μg N m^{-3} h^{-1} have been measured for the diatom/bacterial association in the eastern Pacific. While low, this activity may represent the only significant nitrogen input into oceanic water, but could support a primary productivity of only ca. 10 kg ha^{-1} yr^{-1} for the top metre of the ocean.

(viii) Bark. The warty lenticellate bark of a variety of tropical and temperate trees, including mangroves, have been shown to have acetylene reduction activity (Yatazawa et al., 1983a; Uchino et al., 1984). This activity is responsive to pO_2 with an optimum at 3 - 5 kPa (Yatazawa et al., 1983b). Some isolates from mangrove bark are diazotrophic strains of *Enterobacter* and *Klebsiella* (Uchino et al., 1984). Some estimates of *in situ* N_2 fixation put the contribution at "several 10 kg N ha^{-1} yr^{-1} in temperate and tropical wet forests" (Yatazawa et al., 1983a). The mutualistic relationship between bacteria and host needs careful investigation as does the possibility that this is a widespread and significant phenomenon.

6 NUTRITION

Asymbiotic diazotrophs utilise a wide range of low molecular weight carbon compounds including sugars, organic acids, simple aromatics and hydrocarbons. The traditional dogma that, apart from starch utilization by *A. chroococcum* (Jensen, 1965), diazotrophs do not utilise polysaccharides, has now been overturned by the reports of cellulose-utilising aerobes from the Deshayes gland of shipworms (Waterbury et al., 1983) and anaerobic cellulose fermenters from mud and soil (Leschine et al., 1988). Agar is now reported to support N_2-fixing bacteria (Shieh et al., 1988); they are facultative anaerobes, resembling *Vibrio* spp., and grow on a wide variety of carbon sources including starch, inulin and agar. As yet there are no reports of lignin-degrading diazotrophs, but this may only be a matter of time. The ability to degrade complex polysaccharides and other polymers may be more widespread among diazotrophs than previously thought.

The potential for using straw as a basis for extensive nitrogen fixation (Jensen, 1940) by way of mixtures of cellulolytic and diazotrophic organisms (Jensen and Swaby, 1941; Lynch and Harper, 1983) has stimulated a wide interest in the carbon nutrition of the asymbiotic diazotrophs. A survey of their ability to utilise straw, carboxymethyl cellulose and their breakdown products cellobiose and xylan (Halsall and Gibson, 1989) suggests that some strains of Azospirillum, Bradyrhizobium and Pseudomonas are cellulolytic and that a majority of asymbiotic diazotrophs utilise xylan. However, in all cases, nitrogenase activity on straw is greatly enhanced by the presence of the cellulolytic bacterium Cellulomonas.

7 CONCLUSION

The wealth of new organisms and new information discovered in the last decade emphasise the fact that much of our knowledge of free-living diazotrophs has been artifactual. In the past, the range of organisms used has been determined by the specific techniques available and limited to a group of laboratory adapted strains. Recent discoveries have served to demolish a significant amount of dogma pertaining to the distribution, ecology and evolution of these organisms and it is certain that there are many and varied new organisms to be isolated. The use of novel isolation techniques (Jenni et al., 1989), including the use of refractory substrates (Leschine et al., 1988) and close observation of new habitats (Guerinot and Colwell, 1985) have provided dramatic advances. The reported ecological and agronomic importance of free-living diazotrophs has shown extreme shifts of support. Early reports of the efficacy of inoculating soil with bacteria have largely been discounted (Mishustin and Shil'nikova, 1971) and recent work has played down any agronomic importance. However, a resurgence of interest in the management of crop residues such as straw, including the recent Australian work, shows considerable promise for manipulating this system for significant nitrogen gain.

Whatever the contribution of these organisms to ecology and agriculture might be, their contribution to the debate on the origins of nitrogenase and the mobility of the nif genes is disproportionately great. With further genetic analysis, especially of the archaebacterial systems, some resolution of that debate can now be expected.

8 REFERENCES

Abd-el-Malek, Y., 1971. Free-living nitrogen fixing bacteria in Egyptian soils and their possible contribution to soil fertility. Plant and Soil, Special Volume, 423-442.

Aho, P.E., Seidler, R.J., Evans, H.J. and Nelson, A.D., 1976. Association of nitrogen-fixing bacteria with decay in white fir. In: W.E. Newton and C.J. Nyman (Editors), Proceedings of the 1st International Symposium on Nitrogen Fixation. Washington State University Press, Pullman, pp. 629-640.

Balloni, W., Favilli, F. and Giannini, R., 1988. Phyllospheric nitrogen fixation in douglas fir in Italy. In: H. Bothe, F.J. de Bruijn and W.E. Newton (Editors), Nitrogen Fixation: Hundred Years After. Gustav Fischer, Stuttgart, p. 806.
Becking, J.H., 1961. Studies on nitrogen-fixing bateria of the genus Beijerinckia. II Mineral nutrition and resistance to high levels of certain elements in relation to soil type. Plant Soil, 14: 297-322.
Becking, J.H., 1978. Beijerinckia in irrigated rice soil. In: U. Granhall (Editor), Environmental Role of Nitrogen Fixing Blue-green Algae and Asymbiotic Bacteria. Ecol. Bull., Stockholm, 26: 116-129.
Beijerinck, M.W., 1901. Über oligonitrophilen Mikroben. Zentralbl. Bakteriol. 7: 561-582.
Belay, N., Sparling, R. and Daniels, L., 1984. Dinitrogen fixation by a thermophilic methanogenic bacterium. Nature, 312: 286-288.
Benemann, J.R., 1973. Nitrogen fixation in termites. Science, 181: 164-165.
Benson, D.R. 1985. Consumption of atmospheric nitrogen. In: E.R. Leadbetter and J.S. Poindexter (Editors), Bacteria in Nature, 1. Plenum, New York, pp. 155-198.
Bentley, B.L., 1984. Nitrogen fixation in termites: fate of newly fixed nitrogen. J. Insect Physiol., 30: 653-655.
Bebout, B.M., Paerl, H.W., Crocker, K.M. and Prufert, L., 1987. Diel interactions of oxygenic photosynthesis and N_2 fixation (acetylene reduction) in a marine microbial mat community. Appl. Environ. Microbiol., 53: 2353-2362.
Boring, L.R., Swank, W.T., Waide, J.B. and Henderson, G.S., 1988. Sources, fates and impacts of nitrogen inputs to terrestrial ecosystems: review and synthesis. Biogeochem. 6: 119-159.
Breznac, J.A., 1982. Intestinal microbiota of termites and other xyllophagous insects. Ann. Rev. Microbiol., 36: 323-343.
Brigle, K.E., Weiss, M.C., Newton, W.E. and Dean, D.R., 1987. Products of the iron-molybdenum co-factor specific biosynthetic genes nifE and nifN are structurally homologous to the products of the molybdenum iron protein genes nifD and nifK. J. Bacteriol., 169: 1547-1553.
Callaham, D., de Tredici, P. and Torrey J.G., 1978. Isolation and cultivation in vitro of the actinomycete causing root nodulation in Comptonia. Science, 199: 899-902.
Cavalcante, V.A. and Döbereiner, J., 1988. A new acid-tolerant nitrogen fixing bacterium associated with sugar cane. Plant Soil, 108: 23-31.
Chang, P.C. and Knowles, R., 1965. Non-symbiotic nitrogen fixation in some Quebec soils. Can. J. Microbiol., 11: 29-38.
Clewell, D.B. and Gawron-Burke, C., 1986. Conjugative transposons and the dissemination of antibiotic resistance in streptococci. Ann. Rev. Microbiol., 40: 635-639.
Cornaby, B.W. and Waide, J.B., 1973. Nitrogen fixation in decaying chestnut logs. Plant Soil, 39: 445-448.
Dalton, H., 1980. Chemoautotrophic nitrogen fixation. In: W.D.P. Stewart and J.R. Crallen (Editors), Nitrogen Fixation. Academic Press, London, pp. 177-196.
Dixon, R.O.D. and Wheeler, C.T., 1986. Nitrogen fixation in plants. Blackie, Glasgow.
Di Menna, M.E., 1966. The incidence of Azotobacter and of nitrogen-fixing clostridia in some New Zealand Soils. N.Z. J. Ag. Res., 9: 218-226.
Döbereiner, J., Reis, V.M. and Lorzini, A.C., 1988. New nitrate-fixing bacteria in association with cereals and sugar cane. In: H. Bothe, F.J. de Bruijn and W.E. Newton (Editors), Nitrogen Fixation: Hundred Years After. Gustav Fischer, Stuttgart, pp. 717-722.
Eckhardt, T., 1978. Rapid method for identification of plasmid deoxyribonucleic acid in bacteria. Plasmid, 1: 584-588.

Gibson, A.H., 1966. The carbohydrate requirements for symbiotic nitrogen fixation. A 'whole-plant' growth analysis approach. Aust. J. Biol. Sci., 19: 499-515.
Golden, J.W., Robinson, S.J. and Haselkorn, R., 1985. Rearrangement of the nitrogen fixation genes during heterocyst differentiation in the cyanobacterium Anabaena. Nature, 314: 419-423.
Gonzalez-Lopez, J., Moreno, J., Ballasteros, F. and Ramos-Cormenzana, A., 1984. Nitrogen fixation (acetylene reduction) by Azotobacter vinelandii ATCC12837 in dialysed and normal soil media. Microbios Lett., 27: 121-126.
Gotto, J.W. and Taylor, B.F., 1976. N_2 fixation associated with decaying leaves of the red mangrove (Rhizophora mangle). Appl. Environ. Microbiol., 31: 781-783.
Green, M. and Wilson, P.W., 1953. The utilisation of nitrate nitrogen by the azotobacter. J. Gen. Microbiol., 9: 89-96.
Guerinot, M.L. and Colwell, R.A., 1985. Enumeration isolation and characterisation of N_2-fixing bacteria from seawater. Appl. Environ. Microbiol., 50: 350-355.
Guerinot, M.L. and Patriquin, D.G., 1981. N_2-fixing vibrios isolated from the digestive tract of sea urchins. Can. J. Microbiol., 27: 311-317.
Halsall, D.M., Turner, G.L. and Gibson, A.H., 1985. Straw and xylan utilisation by pure cultures of Azospirillum spp. App. Environ. Microbiol., 49: 423-428.
Halsall, D.M. and Gibson, A.H. 1986. Comparison of two Cellulomonas strains and their interactions with Azospirillum brasilense in degradation of wheat straw and associated nitrogen fixation. Appl. Environ. Microbiol., 51: 855-861.
Halsall, D.M, and Gibson, A. H. 1989. Nitrogenase activity of a range of diazotrophic bacteria on straw, straw breakdown products and related compounds. Soil Biol. Biochem., 21: 291-298.
Hamdi, Y.A., Yousef, A.N., Al Azawi, S., Al-Tai, A. and Al-Baquari, M.S., 1978. Distribution of certain non-symbiotic nitrogen fixing organisms in Iraqi soils. In: U. Granhall (Editor), Environmental Role of Nitrogen Fixing Blue-green Algae and Asymbiotic Bacteria. Ecol. Bull., Stockholm, 26: 110-115.
Heath, B., Sollins, P., Perry, D.A. and Cromack, Jr. K., 1988. Asymbiotic nitrogen fixation in litter from Pacific Northwest forests. Can. J. For. Res., 18: 68-74.
Hecht D. and Malamy M.H., 1989. Tn4399, a conjugal mobilizing transposon of Bacteroides fragilis. J. Bacteriol., 171: 3603-3608.
Hendrickson, O.Q., 1989. Implications of natural ethylene cycling for forest soil acetylene reduction assays. Can. J. Microbiol., 35: 713-718.
Henry, E., 1897. L'azote et la vegetation forestière. Revue d'eaux et forêts, 36: 641.
Higashi, S., 1967. Transfer of clover infectivity of Rhizobium trifolii to Rhizobium phaseoli as mediated by an episomic factor. J. Gen. Appl. Microbiol., 13: 391-403.
Hill, N.M. and Patriquin, D.G., 1988. Induction of aerobic nitrogen fixation and enhancement of wheat straw decomposition by inoculation with a crude culture from sugar cane litter. Soil Biol. Biochem., 20: 613-618.
Horaud, T., Le Bouguenec, C. and De Cespedes, G., 1987. Genetic and molecular analysis of streptococcal and enterococcal chromosome-borne antibiotic resistance markers. In: J.J. Ferretti and R. Curtiss (Editors), Streptococcal Genetics. American Soc. Microbiol. Washington, D.C., pp. 74-78.
Jenni, B., Isch, C. and Aragno, M., 1989. Nitrogen fixation by new strains of Pseudomonas pseudoflava and related bacteria. J. Gen. Microbiol., 135: 461-467.
Jensen, H.L., 1940. Nitrogen fixation and cellulose decomposition by soil microorganisms. I. Aerobic cellulose decomposers in association with Azotobacter. Proc. Linn. Soc. NSW, 65: 543-556.

Jensen, H.L., 1965. Nonsymbiotic nitrogen fixation. In: W.V. Bartholomew and F.E. Clark (Editors), Soil Nitrogen. American Soc. of Agronomy, Madison, pp. 440-485.
Jensen, H.L. and Swaby, R.J., 1941. Nitrogen fixation and cellulose decomposition by soil microorganisms. II The association between Azotobacter and facultative cellulose decomposers. Proc. Linn. Soc. NSW, 66: 89-106.
Jones, K., 1970. Nitrogen fixation in the phyllosphere of Douglas fir, Pseudotsuga douglasii. Ann. Bot., 34: 239-244.
Jones, K. and Bangs, D., 1985. Nitrogen fixation by free-living heterotrophic bacteria in an oak forest: the effect of liming. Soil Biol. Biochem., 17: 705-709.
Jones, K., King, E. and Eastlick, M. 1974. Nitrogen fixation by free-living bacteria in the soil and in the canopy of Douglas fir. Ann. Bot., 38: 765-722.
Jones, K. and Thomas, J.G., 1974. Nitrogen fixation by the rumen contents of sheep. J. Gen. Microbiol., 85: 97-101.
Jurgensen, M.F., Larsen, M.J., Wolosiewicz, M. and Harvey, A.E., 1989. A comparison of dinitrogen fixation rates in wood litter decayed by white-rot and brown-rot fungi. Plant Soil, 115: 117-122.
Knowles, R., 1977. The significance of asymbiotic dinitrogen fixation by bacteria. In: R.W.F. Hardy and A.H. Gibson (Editors), A Treatise on Dinitrogen Fixation. Section IV: Agronomy and Ecology. John Wiley, New York, pp. 33-84.
Knowles, R. and O'Toole, P., 1975. Acetylene-reduction assay at ambient pO_2 of field and forest soils:laboratory and field core studies. In: W.D.P. Stewart (Editor), Nitrogen Fixation by Free-Living Micro-organisms, IBP Vol. 6, John Wiley, New York, pp. 285-294.
Larsen, M.J., Jurgensen, M.T. and Harvey, A.L., 1978. N_2 fixation associated with wood decayed by some common fungi in Western Montana. Can. J. For. Res., 8: 341-345.
La Rue, T.A. and Patterson, T.G., 1981. How much nitrogen do legumes fix? Adv. Agron., 34: 15-38.
Leschine, S.B., Holwell, K. and Canali-Perves, E., 1988. Nitrogen fixation by anaerobic cellulolytic bacteria. Science, 242: 1157-1159.
Line, M.A. and Loutit, M.W., 1969. Occurrence of Azotobacter in some soils of the South Island, New Zealand. N.Z. J. Ag. Res., 12: 630-638.
Line, M.A. and Loutit, M.W., 1971. Non-symbiotic nitrogen-fixing organisms for some New Zealand Tussock-Grassland Soils. J. Gen. Microbiol., 66: 309-318.
Lynch, J.M. and Harper, S.H.T., 1983. Straw as a substrate for cooperative nitrogen fixation. J. Gen. Microbiol., 129: 251-253.
Lynch, J.M. and Harper, S.H.T., 1985. The microbial upgrading of straw for agricultural use. Phil. Trans. Roy. Soc. Lond., B.310: 221-226.
McLung, C.R. and Patriquin, D.G., 1980. Isolation and properties of nitrogen-fixing Campylobacter isolated from the roots of Spartina alterniflora Loisl. Can. J.Microbiol., 26: 881-886.
McLung, C.R., Patriquin, D.G. and Davis, R.E., 1983. Campylobacter nitrofigilis sp. nov., a nitrogen-fixing bacterium associated with the roots of Spartina alterniflora Loisl. Int. J. Syst. Bacteriol., 33: 605-612.
Mahmoud, S.A.Z., El-Sawy, M., Ishac, Y.A. and El-Safty, M.M., 1978. The effects of salinity and alkalinity on the distribution and capacity of N_2-fixation by Azotobacter in Egyptian soils. In: U. Granhall (Editor), Environmental Role of Nitrogen Fixing Blue-green Algae and Asymbiotic Bacteria. Ecol. Bull., Stockholm, 26: 99-109.
Martinez, L., Silver, M.W. King, J.M. and Alldredge, A.L., 1983. Nitrogen fixation by floating diatom mats. A source of new nitrogen in oligotrophic ocean waters. Science, 221: 152-154.
Minchin, F.R. and Pate, J.S., 1973. The carbon balance of a legume and the functional economy of its nodules. J. Exp. Bot., 24: 259-271.

Mishustin, E.N. and Shil'nikova, V.K., 1971. Biological fixation of atmospheric nitrogen. MacMillan, London.
Mishustin, E.N. and Yemtsev, V.T., 1973. Anaerobic nitrogen-fixing bacteria in U.S.S.R. soils. Soil Biol. Biochem., 5: 97-107.
Moreno J., Gonzalez-Lopez, J., Martinez-Toledo, M.V., de la Pubia, T., Ramos-Cormenzana, A. and Vela, G.R. 1988. Growth and nitrogenase activity of Azotobacter vinelandii in the presence of phenolic acids. Arch. Microbiol., 50: 113-116.
Murray, P.A. and Zinder, S.H., 1984. Nitrogen fixation by a methanogenic archaebacterium. Nature, 312: 284-286.
Murrell, J.C and Dalton, H., 1983. Nitrogen fixation in obligate methanotrophs. J. Gen. Microbiol., 129: 3481-3486.
Murty, M.G., 1983. Nitrogen fixation (acetylene reduction) in the phyllosphere of some economically important plants. Plant Soil, 73: 151-153.
Murty, M.G., 1984a. Nitrogenase activity (acetylene reduction) associated with leaves of different cultivars of paddy (Oryza sativa L.). Plant Soil, 77: 253-261.
Murty, M.G., 1984b. Phyllosphere of cotton as a habitat for diazotrophic microorganisms. Appl. Environ. Microbiol., 48: 713-718.
Nelson, D.C., Waterbury, J.B. and Jannasch, H.W., 1982. Nitrogen fixation and nitrate utilisation by marine and freshwater Beggiatoa. Arch. Microbiol., 133: 172-177.
Nohrstedt, H.O., 1985. Nonsymbiotic nitrogen fixation in the topsoil of some forest stands in central Sweden. Can. J. For. Res., 15: 715-722.
Nohrstedt, H.O., 1988. Effect of liming and N-fertilisation on denitrification and N_2 fixation in an acid coniferous forest floor. Forest Ecol. Management, 24: 1-13.
Nuti, M.P., Ledehoer, A.M., Lepidi, A.A. and Schilperoort, R.A., 1977. Large plasmids in different Rhizobium species. J. Gen. Mirobiol., 100: 241-248.
O'Toole, P. and Knowles, R., 1973. Oxygen inhibition of acetylene reduction (nitrogen fixation) in soil:effect of glucose and oxygen concentrations. Soil Biol. Biochem., 5: 783-788.
Oakley, C.J. and Murrell, J.C., 1988. nifH genes in the obligate methane oxidising bacteria. FEMS Microbiology Letters, 49: 53-57.
Olsen, C., 1932. Studies of nitrogen fixation. I. Nitrogen fixation in the dead leaves of forest beds. Compt. rend. du laboratoire Carlsberg, 19: 1-36.
Paerl, H.W., Webb, K.L., Baker, J. and Wiebe, W.J., 1981. Nitrogen fixation in waters. In: W.J. Broughton (Editor), Nitrogen Fixation, Vol. 1, Ecology. Oxford University Press, Oxford, pp. 193-240.
Paerl, H.W. and Prufert. L., 1987. Oxygen-poor microsites as potential sites of microbial N_2-fixation in nitrogen-depleted aerobic marine waters. Appl. Environ. Microbiol., 53: 1078-1087.
Paerl, H.W. and Carlton, R.G., 1988. Control of nitrogen fixation by oxygen depletion in surface-associated microzones. Nature, 332: 260-262.
Paerl, H.W., Bebout, B.M. and Prufert, L.E., 1988. Physiological ecology and environmental control of N_2-fixation in marine and freshwater microbial associations, In: H. Bothe, F.J. de Bruijn and W.E. Newton (Editors), Nitrogen Fixation: Hundred Years After. Gustav Fischer, Stuttgart, pp. 711-716.
Palleroni, N.J., 1984. Gram-negative aerobic rods and cocci. In: N.R. Krieg (Editor), Bergey's Manual of Systematic Bacteriology, Vol. 1. Williams and Wilkins, Baltimore, pp. 141-199.
Parker, C.A., 1954. Effect of oxygen on the fixation of nitrogen by Azotobacter. Nature, 173: 780-781.
Patriquin, D.G., 1982. Nitrogen fixation in sugar cane litter. Biol. Agr. Hort., 1: 39-64.
Polman, J.K. and Larkin, J.M., 1988. Properties of in vivo nitrogenase activity in Beggiatoa alba. Arch. Microbiol., 150: 126-130.

Postgate, J.R., 1974. Evolution within nitrogen-fixing systems. In: M. Carlisle and J. Sketel (Editors), Evolution in the Microbial World. Symp. Soc. Gen. Microbiol. Cambridge University Press, Cambridge, pp 263-292.
Postgate, J., 1980. Foreword. In: N.S. Subba Rao (Editor), Recent Advances in Biological Nitrogen Fixation. Edward Arnold, London, pp. i-xiv.
Postgate, J.R., 1982. The Fundamentals of Nitrogen Fixation. Cambridge University Press.
Postgate, J. and Kent, H.M., 1985. Diazotrophy within Desulfovibrio. J. Gen. Microbiol., 131: 2119-2122.
Postgate, J., Kent, H.M. and Robson, R.L., 1986. DNA from diazotrophic Desulfovibrio strains is homologous to Klebsiella pneumoniae structural nif DNA and can be chromosomal or plasmid-borne. FEMS Microbiol. Lett., 33: 159-163.
Postgate, J.R. and Eady, R.R., 1988. The evolution of biological nitrogen fixation. In: H. Bothe, F.J. de Bruijn and W.E. Newton (Editors), Nitrogen Fixation: Hundred Years After. Gustav Fischer, Stuttgart, pp. 31-40.
Prestwich, G.D., Bentley, B.L. and Carpenter, E.J., 1980. Nitrogen sources of neotropical nasute termites: fixation and selective foraging. Oecologia, 46: 397-401.
Prestwich, G.D. and Bentley, B.L., 1981. Nitrogen fixation by intact colonies of the termite Nasutitermes corniger. Oecologia, 49: 249-251.
Rice, W.A., Paul, E.A. and Wetter, L.R., 1967. The role of anaerobiosis in asymbiotic nitrogen fixation. Can. J. Microbiol., 13: 829-836.
Roskoski, J.P., 1980. Nitrogen fixation in hardwood forests of north eastern United States. Plant Soil, 54: 33-44.
Roskoski, J.P., 1981. Comparative C_2H_2 reduction and $^{15}N_2$ fixation in deciduous wood litter. Soil Biol. Biochem., 13: 83-85.
Ross, D.J., 1960. Biological studies of some tussock-grassland soils. XVI. Non-symbiotic nitrogen-fixing bacteria of two cultivated soils. N.Z. J. Ag. Res., 3: 224-229.
Ruinin, J., 1956. Occurrence of Beijerinckia species in the phyllosphere. Nature, 177: 220-221.
Sadykov, B.F. and Umarov, M.M., 1980. Detection of nitrogen-fixing activity in the plant phyllosphere. Microbiology, 49: 125-128.
Sharp, R.F. and Millbank, J.W., 1973. Nitrogen fixation in deteriorating wood. Experientia, 29: 895-896.
Schink, B., Ward, J.C. and Zeikus, J.G., 1981. Microbiology of wetwood: role of anaerobic bacterial populations in living trees. J. Gen. Microbiol., 123: 313-322.
Schwartz, D.C. and Cantor, B.C.R., 1984. Separation of yeast chromosome-sized DNA's by pulse field gradient gel electrophoresis. Cell, 37: 67-75.
Shieh, W.Y., Simidu, U. and Maruyama, Y., 1988. Nitrogen fixation by marine agar-degrading bacteria. J. Gen. Microbiol., 134: 1821-1825.
Silvester, W.B., 1978. Nitrogen fixation and mineralisation in kauri (Agathis australis) forest in New Zealand. In: M.W. Loutit and J.A.R. Miles (Editors). Microbial Ecology (Proceedings Life Sciences). Springer-Verlag, Berlin, pp. 138-143.
Silvester, W.B., Sollins, P., Verhoeven, T. and Cline, S.P., 1982. Nitrogen fixation and acetylene reduction in decaying conifer boles: effects of incubation time, aeration, and moisture content. Can. J. For. Res., 12: 646-652.
Silvester, W.B., 1983. Analyses of nitrogen fixation. In: J.C. Gordon and C.T. Wheeler (Editors), Biological Nitrogen Fixation in Forest Ecosystems: Foundations and Applications. Martinus Nijhoff. The Hague, pp 173-212.
Silvester, W.B., 1989. Molybdenum limitation of asymbiotic nitrogen fixation in forests of Pacific Northwest America. Soil Biol. Biochem., 6: 1-7.
Smith, B.E., Eady, R.R., Lowe, D.J. and Gormal, C., 1988. The vanadium iron protein of vanadium nitrogenase from Azotobacter chroococcum contains an iron vanadium co-factor. Biochem. J., 250: 299-302.

Smith, C.L., Econome, J.G., Schutt, A., Klco, S. and Cantor, C.R., 1987. A physical map of the E. coli K12 genome. Science, 236: 1448-1453.
Sollins, P., 1982. Input and decay of coarse woody debris in coniferous forests of Western Oregon and Washington. Can J. For. Res., 12: 18-28.
Souillard, N., Magot, M., Possot, O. and Sibold, L., 1988. Nucleotide sequence of regions homologous to nifH (nitrogenase Fe protein) from the nitrogen-fixing Archaebacteria Methanococcus thermolithotrophicus and Methanobacterium ivanovii: evolutionary implications. J. Mol. Evol., 27: 65-76.
Spano, S.D., Jurgensen, M.F., Larsen, M.J. and Harvey, A.E., 1982. Nitrogen-fixing bacteria in Douglas-fir residue decayed by Fomitopsis pinicola. Plant Soil, 68: 117-123.
Todd, R.L., Waide, J.B. and Cornaby, B.W., 1975. Significance of biological nitrogen fixation and denitrification in a deciduous forest ecosystem. In: F.G. Howell, J.B. Gentry and M.H. Smith (Editors), Mineral Cycling in Southeastern Ecosystems. ERDA Symposium Series, pp. 729-735.
Uchino, F., Hambali, G.G. and Yatazawa, M., 1984. Nitrogen-fixing bacteria from warty lenticellate bark of a mangrove tree, Bruguiera gymnorrhiza (L.) Lamk. Appl. Environ. Microbiol., 47: 44-48.
Urdaci, M.C., Stal, L.J. and Marchand, M., 1988. Occurrence of nitrogen fixation among Vibrio spp. Arch. Microbiol., 150: 224-229.
Van der Kamp, B.J., 1986. Nitrogen fixation in cottonwood wetwood. Can J. For. Res., 16: 1118-1120.
Volpon, G.T., De Polli, H. and Döbereiner, J., 1981. Physiology of nitrogen fixation in Azospirillum lipoferum Br17 (ATCC29709). Arch. Microbiol., 128: 371-375.
Waterbury, J.B., Calloway, C.B. and Turner, R.D., 1983. A cellulolytic nitrogen-fixing bacterium cultured from the Gland of Deshayes in shipworms (Bivalvia: Teredinidae). Science, 221: 1401-1403.
Waughman, G.J., French, J.R.S. and Jones K., 1981. Nitrogen fixation in some terrestrial environments. In W.J. Broughton (Editor), Nitrogen Fixation, Volume I: Ecology. Clarendon Press, Oxford, pp. 135-192.
Weil, C.F., Beckler, G.S. and Reeve, J.N., 1987. Structure and organisation of the hisP gene of the thermophilic archaebacterium Methanococcus thermolithotrophicus. J. Bacteriol., 169: 485-
Winogradsky, S., 1893. Sur l'assimilation de l'azote de l'atomosphere par les microbes. Comp. Rend. Acad. Sci. (Paris), 116: 1385-1388.
Woese, C.R., 1987. Bacterial Evolution. Microbiol. Rev., 51: 221-271.
Wu F.J., Moreno, J. and Vela, G.R., 1987. Growth of Azotobacter vinelandii on soil nutrients. Appl. Environ. Microbiol., 53: 489-494.
Yatazawa, M., Hamball, G.G. and Uchino, F., 1983a. Nitrogen fixing activity in warty lenticellate tree barks. Soil Sci. Plant Nutr., 29: 285-294.
Yatazawa, M., Hamball, G.G., and Uchino, F., 1983b. Oxygen partial pressure a affecting nitrogen fixing activity in warty tree barks. Soil Sci. Plant Nutr., 29: 483-487.

CHAPTER 9

ASSOCIATIVE NITROGEN FIXATION IN PLANTS

ROBERT V. KLUCAS
Biochemistry Department, University of Nebraska-Lincoln, Lincoln, NE 68583-0918, USA

1 INTRODUCTION

Nitrogen fixation by associations of grasses and diazotrophs is not a new concept. Interest in this process developed early this century from research on wheat (Lipman and Taylor, 1922) and was renewed by the development of a simple and sensitive assay for nitrogen fixation (Koch and Evans, 1966) and by research in Brazil that demonstrated the existence of a specific association between Azotobacter paspali and Paspalum notatum (Döbereiner, 1968). It is generally accepted that associations between diazotrophs and plant roots and herbage do exist, but a major question concerns their relative importance to the nitrogen economy of plants.

Associative nitrogen fixation is defined as nitrogen fixation by free-living diazotrophs under the direct influence of a host. The distinction between symbiotic, associative and free-living diazotrophic nitrogen fixation is not clear. Symbiosis is the living together in intimate association of two dissimilar organisms, the collaboration being mutually beneficial (Paul and Clark, 1989); for example, nitrogen fixation by rhizobia in legume root nodules. Associative nitrogen fixation is a protocooperative relationship in that both partners benefit, but the partnership is more casual than symbiosis. Nitrogen fixation by free-living diazotrophs (Chapter 8) requires an energy source, but not the involvement of a living host. Measurements of biological nitrogen fixation do not usually usually differentiate between associative systems versus free-living diazotrophs because the distinction is trivial. However, from a mechanistic viewpoint, the distinction is important. Research designed to understand and enhance nitrogen fixation by the two systems must be quite different.

In associative nitrogen-fixing systems, plants supply the diazotrophs with substrates and micro-environments in which to live. In return, the microorganism fixes atmospheric N_2 that is directly or indirectly transferred to the plant. For these associations to succeed, plants must supply substantial amounts of material to the diazotrophs and the diazotrophs must exist without parasitizing the plant.

Living plants are most likely to support growth of diazotrophs at sites where organic substrates are accessible. These include the rhizosphere, rhizoplane, histosphere (colonized tissues), and phyllosphere (surface of herbage). The rhizosphere is the soil region under the direct influence of plant roots, in which there is proliferation of microorganisms. This region, extending out about 3 mm from the root surface, is composed of soil particles, micro-organisms and a complex mixture of polymeric secretions, water-soluble exudates released by roots, lysates and gases. The outer edge of the rhizosphere is determined by diffusion limits of water-soluble compounds (Foster, 1986). Soluble exudates, root cap and mucigel available to microorganisms in the rhizosphere are in the range of 3 to 15 g per 100 g dry weight of root (Newman, 1985). The amount of substrates available to micro-organisms in the rhizosphere varies considerably. Factors affecting plant root exudation include plant species, plant age, temperature, light, plant nutrition, microorganisms, soil moisture and root damage (Rovira, 1969). In practice, the material which adheres to roots after their removal from the soil is considered as the rhizosphere (Newman, 1985) and this includes the rhizoplane. This region, defined as a zone of 1 to 10 mm around the root surface, is composed of a complex mixture of soil colloids and root- and microbially-derived gels. Because of substrate concentrations, a 10-fold greater bacterial population exists in the rhizoplane compared to the rhizosphere. Microbial populations of 1.2 to 1.6 x 10^{11} bacteria cm^{-3} have been observed at the rhizoplane, using transmission electron microscopy, suggesting that the root can be covered with microorganisms (Foster, 1986). Normally, however, only 5 - 10% of the root surface is reported to be covered with bacteria (Paul and Clark, 1989).

2 REQUIREMENTS FOR ASSOCIATIVE NITROGEN FIXATION

Requirements for associative nitrogen-fixing systems are similar to those for the legume nodule symbioses. The major difference is the degree of coupling between the host and the diazotroph. In the legume symbiosis, the differentiation events leading from initial recognition between roots and rhizobia to a final N_2-fixing nodule involve an integration of host and bacterial physiology and structure. In plants, associative systems range from casually coupled associations in the phyllosphere and outer rhizosphere to coupled associations in the inner rhizosphere, rhizoplane and histosphere. The requirements for associative systems are (i) a host, (ii) a diazotroph, (iii) an adequate supply of substrates from the host to the diazotroph, (iv) a micro-environment for the diazotroph that is conducive to nitrogen fixation and

(v) a positive nett transfer of fixed N from the diazotroph to the host. Because biological nitrogen fixation requires a substantial input of energy, the plant must divert resources to drive nitrogen fixation. Calculations by Albrecht and Gaskins (1982), in which the bioener- getics of associative nitrogen fixation were addressed, showed potential decreases of 10 to 20% in crop yields if sufficient photosynthate were diverted to the diazotroph to fix 40 kg of N ha^{-1}. Only under N-limiting conditions, such as low maintenance pastures, would diversion of this quantity of photosynthate lead to increased plant productivity.

A diversity of diazotrophs including strict anaerobes, facultative anaerobes and aerobes are readily isolated from a variety of environments (see Chapter 8). The presence and activity of particular diazotrophs will depend on the microsites in which they live. Micro-environmental conditions that are compatible with nitrogen fixation are low availability of combined nitrogen, appropriate O_2 levels and the presence of organic substrates. Anaerobic diazotrophs, such as _Clostridium pasteurianum_, are active under anaerobic conditions. Facultative anaerobes, such as _Klebsiella_, grow both aerobically and anaerobically, but only fix N_2 under anaerobic or microaerobic conditions. Aerobes, such as _Azotobacter_ and _Azospirillum_, require microaerobic conditions for maximal rates of nitrogen fixation.

It is those diazotrophs which can reach sufficient population density and which have high efficiencies (N_2 fixed to carbon substrate consumed) which would be expected in associative nitrogen-fixing systems. Since aerobic metabolism is energetically more efficient than anaerobic metabolism, aerobic diazotrophs are potentially more efficient than anaerobes. However, maximal efficiencies are complicated by the extreme O_2 sensitivity of nitrogenase and thus aerobes appear to be more efficient under microaerobic than under fully aerobic conditions. Measurements on free-living cultures of diazotrophs appear to support this conclusion. Microaerobes, such as _Azospirillum brasilense_, have efficiencies as high as 92 mg N_2 fixed g^{-1} carbon source (Stephan et al., 1981), although values from 12 to 50 are more commonly reported (Albrecht and Gaskins, 1982; Giller and Day, 1985; Döbereiner and Pedrosa, 1987). Aerobes, such as _Azotobacter chroococcum_, have efficiencies around 40 mg N_2 fixed g^{-1} carbon source (Dalton and Postgate, 1969). Obligate and facultatively anaerobic diazotrophs such as _C. pasteurianum_ and _Klebsiella pneumoniae_ have efficiencies around 10 (Daesch and Mortenson, 1968) and 8 mg N_2 g^{-1} carbon source (Hill, 1976), respectively. Efficiencies of diazotrophs under field conditions are unknown. In legume nodules, efficient nitrogen fixation is accomplished by rhizobia residing inside plant cells where O_2 levels and fluxes are regulated, rhizobial population density is high, and competing microbes are excluded. Associative diazotrophs are apparently present within plant tissues, but their population densities are low. The critical population of diazotrophs required for efficient associative nitrogen fixation is yet to be determined. Large numbers of active diazotrophs might exist in the histosphere of some plants, but with higher populations, bacterial pathogenicity might be a problem.

3 EVIDENCE FOR ASSOCIATIVE NITROGEN FIXATION

3.1 Measurement of nitrogen fixation

Support for the occurrence of associative nitrogen fixation comes from a variety of sources ranging from general observations on continued plant growth under N-limiting conditions to evidence involving specific associations. Certain ecosystems in which legumes are absent or limited, such as pastures of *Paspalum*, sugar cane fields in Brazil, and kallar grass in Pakistan, allow perennial growth and harvesting with no obvious addition of N. Without a careful N balance study, neither the amount nor the source of the N is certain, but biological nitrogen fixation is the most likely answer. Sixteen N balance studies, including grasslands, agricultural soils and woodland (cited by Giller and Day, 1985), showed considerable gains in N, ranging from 2 kg of N ha^{-1} yr^{-1} in a savanna grassland in Ghana to 594 kg ha^{-1} yr^{-1} in a regenerating rain forest in Nigeria. Longer term studies in the United Kingdom generally showed nett N gains of around 50 kg ha^{-1} yr^{-1}. Again, biological nitrogen fixation is responsible for at least some of the net gain in nitrogen and associative nitrogen fixation is likely to be a contributor. Long term N balance experiments with wetland rice have also shown considerable nett N gain from biological nitrogen fixation (App et al., 1984).

Studies using ^{15}N (enrichment or dilution) suggest that associative nitrogen fixation does indeed occur (Boddey, 1987). With ^{15}N enrichment studies, the presence of excess ^{15}N in the plant after the system has been exposed to $^{15}N_2$ is presumptive evidence for associative N_2 fixation and N transfer. This method is very useful because it provides an absolute measurement of associative nitrogen fixation. However, its limitation is that the system must be enclosed and field experiments are therefore difficult to perform. Using soil cores containing *P. notatum* or *Digitaria decumbens* under greenhouse conditions, De Polli et al. (1977) demonstrated ^{15}N enrichment in roots, stems and leaves upon exposure to $^{15}N_2$. The enrichments were very low, especially in stems and leaves, but sufficient to substantiate that nitrogen fixation occurred and that ^{15}N-containing molecules were transferred to the plant. Other plants including rice (Eskew et al., 1981), sugarcane (Ruschel et al., 1976), sorghum (Giller et al., 1984) and grasses (Morris et al., 1985; Boddey, 1987) have also shown ^{15}N enrichment in roots, stems, or leaves of plants grown under conditions that favour associative nitrogen fixation.

The ^{15}N dilution method is useful for measuring associative nitrogen fixation because it integrates fixation over time, it is an absolute measurement of nitrogen fixation, and after the initial enrichment of the growth matrix, it does not require extensive manipulations of the environment during the course of an experiment (see Rennie and Rennie, 1983; Witty, 1983; Boddey, 1987; Shearer, 1989). In the ^{15}N dilution method, non-nitrogen-fixing control plants and putative nitrogen-fixing test plants are grown under normal atmospheric conditions on a growth support that is enriched in ^{15}N. Plant tissues are analyzed for ^{15}N and the difference in ^{15}N incorporation into test and control plants is determined. The test plant will have a lower level of ^{15}N than the control plant if nitrogen fixation has occurred and if the fixed N has been transferred to the plant. The amount of nitrogen fixation is calculated using this difference in ^{15}N. The sensitivity of the assay depends on the extent of fixation, the length of the growth period, the level of ^{15}N enrichment, and the extent of incorporation of ^{15}N into the plant. Selection of the appropriate control plant and uniform incorporation of the ^{15}N into the growth support are critical. Success in using this method for measuring associative nitrogen fixation depends on whether the fixed N goes into the plant or into the growth support.

The isotope dilution method has been used to demonstrate associative nitrogen fixation in a large number of ecosystems (Boddey, 1987). Rennie (1980), one of the first to use the method on associative systems, showed low levels of nitrogen fixation in N-deficient corn seedlings inoculated with A. brasilense and grown in test tubes. The ^{15}N dilution method was used extensively in Brazil to screen various genotypes and ecotypes of pasture grasses that were grown under simulated pasture conditions. Boddey et al. (1983) and Boddey and Victoria (1986) reported that 11% of N in P. notatum cv. batatais and 30 to 40% of the N in Brachiaria fumidicola and Brachiaria decumbens were derived from associative nitrogen fixation. Miranda and Boddey (1987) examined 11 ecotypes of Panicum maximum for associated nitrogen fixation over a 390-day period. Throughout most of the growing period, they found no evidence for any nitrogen fixation. However, during the last 3 summer months, when the temperatures and moisture levels were high, they detected significant isotope dilutions relative to their control plant (Brachiaria radicans) and a significant difference in the isotope dilutions between two of the P. maximum ecotypes. Over this 3-month period, they calculated that 24 to 38% of the total N incorporated was from nitrogen fixation (5 to 10 kg ha^{-1} per 30 days). Other examples in which the isotope dilution method has been used to demonstrate associative nitrogen fixation include bluegrasses (Haahtela and Kari, 1986) and kallar grass (Malik et al., 1987). Taken *in toto*, the isotope dilution data indicate that under certain N-limiting conditions some nitrogen fixation does occur, and that some of the fixed N is assimilated into the plant.

The reduction of acetylene to ethylene is a very simple and sensitive assay for measuring nitrogenase over short periods of time. Because of the nature of the assay, it is extremely valuable for detecting nitrogenase in initial surveys and model systems. The assay is of limited use for providing absolute evidence for associative nitrogen fixation because artificial conditions must be imposed on the system of interest and only nitrogenase activity is measured. Information is not gained on whether or not fixed N is incorporated into the plant.

3.2 Diazotrophs associated with plants

The list of heterotrophic diazotrophs commonly associated with plants is lengthy and continues to grow with discoveries of new genera and species. Because diazotrophs are ubiquitous in soils under vegetation, problems exist in differentiating between casual diazotrophs and true associative diazotrophs. This is normally done by determining which diazotrophs are dominant in a sample or microsite where fixation occurs. Genera of diazotrophs most commonly reported for associative systems are *Azospirillum*, *Azotobacter*, *Klebsiella*, *Enterobacter* and *Pseudomonas*. *Azospirillum* spp. have been isolated from over 50% of the samples tested from the tropics (Döbereiner and Pedrosa, 1987). Azospirilla are also reported in samples from many temperate areas but less frequently. Although its role in associative nitrogen-fixing systems is still being evaluated, *Azospirillum* appears to be an important rhizosphere organism in the tropics and of less importance in temperate regions, where it does not persist. Other recently identified diazotrophs that may be important in certain ecological niches are *Campylobacter nitrofigilis* in *Spartina* (McClung et al., 1983), *Herbaspirillum seropedicae* in Brazilian soils (Baldini et al., 1986), *Acetobacter nitrocaptans* in sugar cane (Stephan et al., 1988) and *Azospirillum halopraeferens* plus an unidentified Gram-negative rod in kallar grass (Reinhold et al., 1987). A high population density of diazotrophs on or in the host is required if significant nitrogen fixation is to occur. Azospirilla at numbers between 10^5 to 10^7 g^{-1} of soil or root sample are commonly reported in samples from the tropics (Döbereiner and Pedrosa, 1987). In temperate zones, such as Belgium and England, azospirilla are frequently isolated from soil and root samples but in much lower numbers of 10^3 g^{-1} soil or root (de Coninck et al., 1989; Harris et al., 1989). In the rhizoplane of kallar grass, populations of diazotrophs around 2×10^7 g^{-1} dry root are reported (Reinhold et al., 1986). The total population and most abundant diazotrophs in the rhizospheres of plants can be estimated using a "spermosphere" method first described for nitrogen-fixing systems by Thomas-Bauzon et al. (1982). This method uses non-specific exudates from germinating surface-sterilized seeds to support diazotrophic growth. In this way Bally et al. (1983) and Omar et al. (1989) found diazotrophs exceeding 10^8 g^{-1} dry weight of rice rhizosphere soil and the dominant diazotrophs were members of the Enterobacteriaceae.

3.3 Specific association between diazotrophs and plants

Although not a prerequisite for associative nitrogen-fixing systems, a degree of specificity between the plant and diazotroph might be expected, resulting in an enrichment for particular diazotrophs on, or in, the host. Many mechanisms could lead to specificity ranging from merely maintaining a nitrogen deficient micro-environment that favours diazotrophs to recognition sites and an infection process similar to the legume system. Other possibilities include the involvement of specific growth substances, microbial toxins and the involvement of other microbes. The sites for these specific interactions are the rhizosphere, rhizoplane, histosphere or phyllosphere.

Considerable evidence exists in the literature that diazotrophic populations are higher around and in plant roots. One of the first reported and more distinctive associations occurs between a pasture grass, *P. notatum* and *A. paspali* (Döbereiner, 1968). *A. paspali* is not detected around the roots of other plants nor on most other species of *Paspalum*. When plants carrying *A. paspali* on their roots are transplanted the population of *A. paspali* drops to nearly zero and then re-establishes itself around the roots after 6 months. Whether the *P. notatum*-*A. paspali* association results in nitrogen fixation is not clear. The evidence for nitrogen fixation is that *A. paspali* is a diazotroph, that acetylene reduction is detected in undisturbed cores, and that these long-established pastures have not been fertilized with N. This association is very interesting because it represents what appears to be a specific interaction between a plant and a diazotroph.

Other specific plant-diazotroph associations are thought to exist. The most widely studied, especially in Brazil, are associations involving the rhizosphere bacterium *Azospirillum*. Pot and field experiments in which corn and wheat were inoculated with a mixture of *Azospirillum* spp. and strains showed that *Azospirillum lipoferum* was the dominant isolate from surface-sterilized corn and *A. brasilense* was the dominant isolate from surface-sterilized wheat roots. This, and other evidence cited by Döbereiner and Pedrosa (1987), indicates that roots of plants with C-3 and C-4 photosynthetic pathways are preferentially infected with A. brasilense and A. lipoferum, respectively, and that infection of root tissues is not intracellular. Another extensively studied associative system is kallar grass from Pakistan. Root surfaces of this grass carry around 10^7 g^{-1} root dry weight of A. lipoferum and A. halopraeferens, a NaCl-tolerant bacterium (Reinhold et al., 1987). In addition, kallar roots contain an unidentified diazotrophic, motile rod at 10^8 g^{-1} root dry weight and in extra-cellular packets of 10^4 in the aerenchyma (Reinhold and Hurek, 1989). Another association of interest is that between sugar cane and A. nitrocaptans, which is reported at concentrations of 10^5 to 10^7 g^{-1} dry weight in roots and stems (Stephan et al., 1988).

Klebsiella or Enterobacter apparently attach non-specifically to the roots of temperate grasses like Poa pratensis (Haahtela et al., 1981) via fimbriae (Haahtela et al., 1985). Such interactions could be important for colonization of these grasses by these enterobacterial diazotrophs (Haahetela and Korhonen, 1985).

3.4 Inoculation studies

Inoculation with certain diazotrophs may produce beneficial effects on plants (Okon, 1985; Döbereiner and Pedrosa, 1987). In experiments to test the effects of inoculation, plants have been grown in the presence of live or heat-killed inoculants and various parameters of plant growth (dry weight, total N, protein and minerals) and nitrogen fixation measured. Growth conditions have ranged from environmentally controlled chambers and sterilized growth media, such as agar, to field conditions. Selection of the diazotroph for the inoculant is critical because it must colonize the root and sustain this colonization over a period of time. These conditions are achievable in plant growth chambers and greenhouses but are much more difficult to achieve under field conditions where competition with other microbes and environmental stresses may intervene. Contamination by indigenous diazotrophs can be a problem, because soil and even surface-sterilized seeds routinely carry a multitude of diazotrophs (Sundaram et al., 1988).

Under carefully controlled conditions, it is relatively easy to demonstrate both increases in growth and nitrogen fixation in plants inoculated with diazotrophs (Rennie, 1980). However, under these conditions, plant growth is seldom normal because of the restrictions imposed on the plants, such as extreme N-deficiency. From the large number of reports, very little doubt exists that under certain conditions in the field, plant growth and total N in plants are increased, but the major question of whether nitrogen fixation is involved remains unanswered. Under field conditions in Florida, using pearl millet and guinea grass, Smith et al. (1976) reported increases in crop yields resulting from inoculation with A. brasilense. The observed increase was about 40 kg N ha^{-1}. In Israel, where the indigenous population of azospirilla is low, fields of sorghum inoculated with Azospirillum showed increases of up to 20% in grain yield (Okon et al., 1988). Increases of up to 33% in grain yields of pearl millet after inoculation with Azospirillum were reported in India, but the response was not routinely observed (Wani et al., 1989). Substantial increases in nitrogen were also reported by Boddey et al. (1986) by inoculation of field grown wheat with Azospirillum, but the increases in nitrogen were probably due to increases in nitrate assimilation and not nitrogen fixation.

Clearly, diazotrophic inoculants do enhance plant growth under certain field conditions but a number of factors may be involved besides associative nitrogen fixation. The selection of the inoculant is important as well as low levels of indigenous diazotrophs. To establish that associative nitrogen fixation occurs, nitrogen fixation must be measured directly. Other factors including phytohormones, enhanced mineral uptake, and increased N utilization can influence plant growth. Many diazotrophs, including Azotobacter (Brown and Burlington, 1968) and Azospirillum (Tien et al., 1979) are capable of producing phytohormones that stimulate root development. Any plant inoculant that increases plant yields reproducibly is potentially important to agriculture, but a considerable difference exists between short term growth stimulation and longer-term biological nitrogen fixation.

4 CONCLUSIONS

Considerable controversy exists as to the relevance of associative nitrogen fixation. While it is generally accepted that associations between diazotrophs and plant roots and herbage do exist, questions remain concerning the extent of nitrogen fixation. Under controlled conditions, low levels of nitrogen fixation and assimilation into plants can be demonstrated. Under field conditions, it is more difficult to demonstrate associative nitrogen fixation, especially in temperate regions. Accumulated evidence supports a role for associative nitrogen fixation in tropical and subtropical zones but not in cool temperate regions. However, even in the tropics, associative nitrogen fixation is probably limited to short periods of time and to N-limiting growth conditions. Beneficial effects on plant growth can be observed when diazotrophs are used as inoculants, but while enhancements in plant growth may result from increased rates of nitrogen fixation, they may be caused by other factors. To establish that associative nitrogen fixation occurred, nitrogen fixation and assimilation must be measured directly. A degree of specificity is detected between diazotrophs and plant roots. Associative nitrogen fixation is a contributor of N to the biosphere but additional research is required to assess its quantitative significance.

5 REFERENCES

Albrecht, S.L. and Gaskins, M.H., 1982. The bioenergetics of asymbiotic nitrogen fixation. In: F. Vogt (Editor) Proceeding of the 2nd International Seminar on Energy Conservation and Use in the Bio-Industry. Pergamon Limited, Oxford, pp. 23-38.

App, A., Santiago, T., Menguito, C., Ventura, W., Tirol, A., Po, J., Watanabe, I. and Datto, S.K., 1984. Estimation of the nitrogen balance for irrigated rice and the contributiion of phototrophic nitrogen fixation. Field Crops Res., 9: 17-27.

Baldini, J.I., Baldini, V.L.D., Seldin, L. and Döbereiner, J., 1986. Characterization of Herbaspirillum seropedicae gen. nov., sp. nov., a root- associated nitrogen-fixing bacterium. Int. J. Syst. Bacteriol., 12: 86-93.

Bally, R., Thomas-Bauzon, D., Heulin, T. and Balandreau, J., 1983. Determination of the most frequent N_2-fixing bacteria in a rice rhizosphere. Can. J. Microbiol., 29: 881-887.

Boddey, R.M., 1987. Methods for quantification of nitrogen fixation with Gramineae. CRC Critical Rev. Plant Sci., 6: 209-266.

Boddey, R.M., Baldini, V.L.D., Baldini, J.I. and Döbereiner, J., 1986. Effect of inoculation of Azospirillum spp. on the nitrogen assimilation of field grown wheat. Plant Soil, 95: 109-121.

Boddey, R.M., Chalk, P.M., Victoria, R.L. and Matsui, E., 1983. The ^{15}N-isotope dilution technique applied to the estimation of biological nitrogen associated with Paspalum notatum cv. batatais in the field. Soil Biol. Biochem., 15: 25-32.

Boddey, R.M. and Victoria, R.L., 1986. Estimation of biological nitrogen fixation associated with Brachiaria and Paspalum grasses using ^{15}N labelled organic matter and fertilizer. Plant Soil, 90: 265-292.

Brown, M.E. and Burlington, S.K., 1968. Production of plant growth substances by Azotobacter chroococcum. J. Gen. Microbiol., 53: 135-144.

Daesch, G. and Mortenson, L.E., 1968. Sucrose catabolism in Clostridium pasteurianum and its relation to N_2 fixation. J. Bacteriol., 96: 346-351.

Dalton, H. and Postgate, J.R., 1969. Growth and physiology of Azotobacter chroococcum in continous culture. J. Gen. Microbiol., 56: 307-317.

de Conninck, K., Horemans, S., Randombage, S. and Vlassak, K., 1989 Occurence and survival of Azospirillum spp. in temperate regions. In: F.A. Skinner, R.M. Boddey and I. Fendrik (Editors), Nitrogen Fixation in Non-Legumes. Kluwer Academic, Dordrecht, pp. 109-114.

De-Polli, H., Matsui, E., Döbereiner, J. and Salati, E.,1977. Confirmation of nitrogen fixation in two tropical grasses by ^{15}N incorporation. Soil Biol. Biochem., 9: 119-123.

Döbereiner, J., 1968. Non-symbiotic nitrogen fixation in tropical soils. Pesqui. Agropecu. Bras., 3: 1-6.

Döbereiner, J. and Pedrosa, F.O., 1987 Nitrogen-fixing bacteria in nonleguminous crop plants. Science Tech, Madison, 155 pp.

Eskew, D.L., Eaglesham, A.R.J. and App, A.A., 1981. Heterotrophic $^{15}N_2$ fixation and distribution of newly fixed nitrogen in a rice-flooded soil system. Plant Physiol., 68: 48-52.

Foster, R.C., 1986. The ultrastructure of the rhizoplane and rhizosphere. Ann. Rev. Phytopathol., 24:

Giller, K.E. and Day, J.M., 1985 Nitrogen fixation in the rhizosphere: significance in nature and agricultural systems. In: A.H. Fitter, D. Atkinson, D.J. Read and M.B. Usher (Editors), Ecological Interactions in Soil. Blackwell Scientific, Oxford, pp. 127-147.

Giller, K.E., Day, J.M., Dart, P.J. and Wani, S.P., 1984. A method for measuring the transfer of fixed nitrogen from free-living bacteria to higher plant using $^{15}N_2$. J. Microbiol. Meth., 2: 307-316.
Haahtela, K. and Kari, K., 1986. The role of root-associated Klebsiella pneumoniae in the nitrogen nutrition of Poa pratensis and Triticum aestivum as estimated by the method of ^{15}N isotope dilution. Plant Soil, 90: 245-254.
Haahtela, K. and Korhonen, T.K., 1985. In vitro adhesion of N_2-fixing enteric bacteria to roots of grasses and cereals. Appl. Environ. Microbiol., 49: 1186-1190.
Haahtela, K., Tarkka, E. and Korhonen, T.K., 1985. Type 1 fimbria-mediated adhesion of enteric bacteria to grass roots. Appl. Environ. Microbiol., 49: 1182-1185.
Haahtela, K., Wartiovaara, T., Sundman, V. and Skujins, J., 1981. Root-associated N_2 fixation (acetylene reduction) by Enterobacteriaceae and Azospirillum strains in cold-climate sodosols. Appl. Environ. Microbiol., 41: 203-206.
Harris, J.M., Lucas, J.A., Davey, M.R. and Lethbridge, G., 1989. Establishment of Azospirillum inoculant in the rhizosphere of winter wheat. Soil Biol. Biochem., 21: 59-64.
Hill, S., 1976. The apparent ATP requirement for nitrogen fixation in growing Klebsiella pneumoniae. J. Gen. Microbiol., 95: 297-313.
Koch, B. and Evans, H.J., 1966. Reduction of acetylene to ethylene by soybean root nodules. Plant Physiol., 41: 1748-1749.
Lipman, C.B. and Taylor, J.K.,1922. Proof of the power of wheat plant to fix atmospheric nitrogen. Science, 56: 605-606.
Malik, K.A., Zafar, Y., Bilal, R. and Azam, F., 1987. Use of ^{15}N isotope dilution for quantification of N_2-fixation associated with roots of Kallar grass [Leptochloa fusca (L.)]. Biol. Fert. Soil, 4: 103-108.
McClung, C.R., Patriquin, D.G. and Davis, R.E., 1983. Campylobacter nitrofigilis sp. nov., a nitrogen-fixing bacterium associated with roots of Spartina alterniflora Loisel. Int. J. Syst. Bacteriol., 33: 605-612.
Miranda, C.H.B. and Boddey, R.M., 1987. Estimation of biological nitrogen fixation associated with 11 ecotypes of Panicum maximum grown in nitrogen-15-labeled soil. Agron. J., 79: 558-563.
Morris, D.R., Zuberer, D.A. and Weaver, R.W.,1985. Nitrogen fixation by intact grass-soil cores using $^{15}N_2$ and acetylene. Soil Biol. Biochem., 17: 87-91.
Newman, E.I., 1985. The rhizosphere: carbon sources and microbial populations. In: A.H. Fitter, D. Atkinson, D.J. Read and M.B. Usher (Editors), Ecological Interactions in Soil. Blackwell Scientific, Oxford, pp. 107-119.
Okon, Y., 1985. Azospirillum as a potential inoculant for agriculture. Trends Biotechnol., 3: 223-228.
Okon, Y., Fallik, S., Yahalom, E. and Tal, S., 1988. Plant growth promoting effects of Azospirillum. In: H. Bothe, F J. de Bruijn and W.E. Newton (Editors), Nitrogen Fixation: Hundred Years After. Gustav Fischer, New York, pp. 741-746.
Omar, A.M.N., Richard, C., Weinhard, P. and Balandreau, J., 1989. Using the spermosphere model technique to describe the dominant nitrogen-fixing microflora associated with wetland rice in two Egyptian soils. Biol. Fertil. Soil, 7: 158-163.
Paul, E.A. and Clark, F.E., 1989. Soil Microbiology and Biochemistry. Academic Press, San Diego, 273 pp.
Reinhold, B. and Hurek, T., 1989. Location of diazotrophs in the root interior with special attention to kallar grass association. In: F.A. Skinner, R.M. Boddey and I. Fendrik (Editors), Nitrogen Fixation in Non-Legumes. Kluwer Academic, Dordrecht, pp. 209-218.

Reinhold, B., Hurek, T., Fendrik, I., Pot, B., Gillis, M., Kersters, K., Thielemans, S. and De Ley, J., 1987. Azospirillum halopraeferens sp nov., a nitrogen fixing organism associated with roots of kallar grass (Leptochloa fusca (L.) Kunth). Int. J. Syst. Bacteriol., 37: 43-51.
Reinhold, B., Hurek, T., Niemann, E. and Fendrik, I., 1986. Close association of Azospirillum and diazotrophic rods with different root zones of Kallar grass. Appl. Environ. Microbiol., 52: 520-526.
Rennie, R.J., 1980. ^{15}N-isotope dilution as a measure of dinitrogen fixation by Azospirillum brasilense associated with maize. Can. J. Bot., 58: 21-24.
Rennie, R.J. and Rennie, D.A., 1983. Techniques for quantifying N_2-fixation in association with non legumes under field conditions. Can. J. Microbiol., 29: 1022-1035.
Rovira, A.D.,1969. Plant root exudates. Bot. Rev., 35: 35-57.
Ruschel, A.P., Henis, Y. and Salati, E., 1975. Nitrogen-15 tracing of N-fixation with soil-grown sugarcane seedings. Soil Biol. Biochem, 7: 181-182.
Shearer, G. and Kohl, D.H., 1989. Natural ^{15}N abundance as a method of estimating the contribution of biologically fixed nitrogen to N_2-fixing systems; potential for non-legumes. In: F.A. Skinner, R.M. Boddey and I. Fendrik (Editors), Nitrogen Fixation in Non-Legumes. Kluwer Academic, Dordrecht, pp. 289-299.
Smith, R.L., Bouton, J.H., Schank, S.C., Quesenberry, K.H., Tyler, M.E., Milam, J.R., Gaskins, M.H. and Littell, R.C., 1976. Nitrogen fixation in grasses inoculated with Spirillum lipoferum. Science, 193: 1003-1005.
Stephan, M.P., Pedrosa, F.O. and Döbereiner, J., 1981. Physiological studies with Azospirillum spp. In: P.B. Vose and A.P. Ruschel (Editors), Associative N_2-Fixation, Vol 1. CRC Press, Boca Raton, Fl, pp. 7-13.
Stephan, M.P., Teixeira, K.R.S. and Döbereiner, J., 1988. Nitrogen fixation physiology of Acetobacter nitrocaptans: effect of oxygen, pH and carbon source on respiration and nitrogen fixation activity. In: H. Bothe, F.J. de Bruijn and W.E. Newton (Editors), Nitrogen Fixation: Hundred Years After. Gustav Fischer, New York, p. 287.
Sundaram, S., Arunakamari, A. and Klucas, R.V., 1988. Characterization of azospirilla isolated from seeds and roots of turf grass. Can. J. Microbiol., 34: 212-217.
Thomas-Bauzon, D., Weinhard, P., Villecourt, P. and Balandreau, J., 1982. The spermosphere model. I. Its use in growing, counting, and isolating N_2-fixing bacteria from the rhizosphere of rice. Can. J. Microbiol., 28: 922-928.
Tien, T.M., Gaskins, M.H. and Hubbell, D.H., 1979. Plant growth substances produced by Azospirillum brasilense and their effect on the growth of pearl millet (Pennisetum americanum L.). Appl. Environ. Microbiol., 37: 1016-1024.
Wani, S.P., Chandrapalaih, S., Zambre, M.A. and Lee, K.K., 1989. Association between N_2-fixing bacteria and pearl millet plants: responses, mechanisms and persistence. In: F.A. Skinner, R.M. Boddey and I. Fendrik (Editors), Nitrogen Fixation in Non-Legume Plants. Kluwer Academic, Dordrecht, pp. 249-262.
Witty, J.F., 1983. Estimating N_2 fixation in the field using ^{15}N labelled fertilizer: Some problems and solutions. Soil Bioln. Biochem., 15: 631-639.

CHAPTER 10

ECOLOGY OF LEGUME ROOT-NODULE BACTERIA

Y.M. BARNET
School of Microbiology, University of New South Wales, P.O. Box 1, Kensington, N.S.W., Australia.

1 INTRODUCTION

Ecological research into the root-nodule bacteria has lagged behind that in genetics and physiology. Much of our existing information has been gained from studies aimed at increasing the productivity of crops and pastures by improving the success rate of legume inoculation programmes. This practical orientation has proved both a strength and a weakness. On the positive side, it has provided a large amount of data on the behaviour of the root-nodule bacteria and has emphasized the complexity of their response to their environment. However, a negative aspect has been the fact that practical success has caused the cessation of many investigations at the point of defining questions, and our knowledge of the ecological processes controlling the survival and growth of root-nodule bacteria in soil is still fragmentary.

The aim of this review is to present an overview of current information on the ecology of the free living root-nodule bacteria, and to raise problems of methodology and conceptual approach. Emphasis has been placed on more recent research and on data which come from field studies rather than *in vitro* experiments. Readers are referred to other reviews for additional discussion (Parker et al., 1977; Dommergues et al., 1978; Schmidt, 1978; 1979; Lowendorf, 1980; Alexander, 1985; Schmidt and Robert, 1985; Dowling and Broughton, 1986; Barnet et al., 1988).

2 PROBLEMS IN STUDYING THE ECOLOGY OF ROOT-NODULE BACTERIA

2.1 Methods for identification of strains

Ecological studies require rapid and reliable methods for strain identification; while the past decade has seen the development of an array of such techniques, problems still exist.

Serological techniques have improved in sensitivity and usefulness with the introduction of enzyme-linked immunosorbent assay (ELISA) (e.g. Kishinevsky and Gurfel, 1980; Morley and Jones, 1980; Rice and Olsen, 1984) and immunofluorescence (Bohlool and Schmidt, 1980). Their precision should improve further with the development of monoclonal antibodies.

A limitation of serological methods, however, is that little information can be obtained about strains which do not react with the available antisera. In addition, serogroups may not provide a sufficiently fine discrimination for ecological studies (Brockman and Bezdicek, 1989). A number of alternative approaches have been developed to try to overcome these problems. Intrinsic antibiotic resistance patterns (Beynon and Josey, 1980; Kingsley and Bohlool, 1983), phage typing (Lesley, 1982; Bromfield et al., 1986), electrophoretic techniques for separation of whole cell (Noel and Brill, 1980; Dughri and Bottomley, 1983) or periplasmic (Glenn et al., 1986) proteins and DNA restriction patterns (Schmidt et al., 1986) all provide means of "finger-printing" strains. These techniques, alone or in combination, have proved very useful in epidemiological studies to follow the fate of particular strains in soil. They have also been used to measure diversity in natural habitats, although interpretation of this type of data is more difficult. The degree of heteogeneity detected depends on the method used: organisms which group together on the basis of one test can appear heterogeneous when other methods are applied (Noel and Brill, 1980; Dughri and Bottomley, 1983; Kamicker and Brill, 1986), while different techniques can sometimes divide an assemblage of isolates into totally different groupings (Broughton et al., 1987a).

The term "strain" in this context has no unambiguous meaning. Provided the results from a study are internally consistent, this may be unimportant. However, it does mean that different studies are not strictly comparable. In addition, any attempt to describe and explain the variation in field populations must eventually deal with the genetic structure of the community. A number of workers (see Young et al., 1987) have pointed out that results from techniques such as serology or phage typing do not lend themselves to a straightforward genetic interpretation. Many research groups are now using enzyme polymorphism to identify lineages within populations of root-nodule bacteria (e.g. Young, 1985; Pinero et al., 1988; Harrison et al., 1989a,b; Eardly et al., 1990). Such work, combined with the older approaches, may eventually lead to concensus on what constitutes genetically and ecologically distinct strains and on the methods most useful for identifying them.

Among other genetic techniques, DNA hybridisation using total genomic DNA has been employed for genus and species identification (Hodgson and Roberts, 1983; Cooper et al., 1987; Wedlock and Jarvis, 1986; Jarvis et al., 1989), although the method does not distinguish between closely related strains. Recently, Bjourson and Cooper (1988) successfully used subtraction hybridization to obtain highly-specific probes for *R. loti* strains. This appears to be one of the most sensitive identification techniques currently available.

2.2 Methods for examination of root-nodule bacteria in soil

All the above techniques require prior isolation of strains, a major problem in the study of their free-living ecology. Most research has relied on isolation from nodules to delineate population composition and on most probable number plant counts to determine population size (Vincent, 1970), but these are not always satisfactory procedures. It has been clearly demonstrated that, although nodule occupancy may reflect the constitution of the soil population (Harrison et al., 1989b), it can be affected by host selection and environmental conditions and may not give a representative sample of strains actually present (e.g. Moawad et al., 1984; Bromfield et al., 1986; Almendras and Bottomley, 1987; Demezas and Bottomley, 1987; Turco and Bezdicek, 1987; Thurman and Bromfield, 1988; Valdivia et al., 1988). Harrison et al. (1989b) suggest that "baiting" of tube-grown plants with soil suspensions may give a better representation of the population than growing plants in soil, since the selective environmental stresses are likely to be less. This idea is worth further evaluation.

Another problem with using the host legume to select root-nodule bacteria is that non-nodulating strains will be totally missed. Young (1985) first suggested that some soil "rhizobia" may lack symbiotic plasmids. The existence of such strains has now been demonstrated by Jarvis et al. (1989) and Sõberon-Chávez and Nājera (1989). Although this has obvious implications for the ecology of the root-nodule bacteria, such non-nodulating organisms remain difficult to isolate and study.

Two approaches have been developed for enumerating strains directly. The use of immunofluorescence (Bohlool and Schmidt, 1980) has allowed comparison of the population dynamics in rhizospheres and bulk soil and it has been important in elucidating competitive relationships (e.g. Robert and Schmidt, 1983; Moawad et al., 1984; Demezas and Bottomley, 1986a,b). Its major restriction is the requirement that the strains of interest must first be obtained in pure culture for antiserum production, which limits the use of this technique in studies of diverse communities or minority populations. In addition, a proportion of the cells recorded by immunofluorescence is non-culturable (Bottomley and Dughri, 1989), implying a need for caution in interpreting counts.

The fate of strains added to soil can also be studied by using antibiotic-resistant mutants. This method has proved reliable and sensitive (Bushby, 1981), but is limited to the study of introduced strains. The technique can, however, be extended to investigate indigenous populations if the organisms to be studied possess intrinsic antibiotic resistance. Kuykendall et al. (1988) showed that B. japonicum of DNA homology group II possess uniformly high resistance to a number of antibiotics, while Sōberon-Chávez and Nājera (1989) found that all R. leguminosarum biovar phaseoli strains tested, including examples from geographically distant areas, were resistant to high levels of nalidixic acid. Such natural resistances of specific taxa could be used to assist in direct isolation from soil. Although Sōberon-Chávez and Nājera (1989) report that their method was not suited for quantitative purposes, the approach warrants further investigation.

Holben et al. (1988) successfully used specific DNA probes to examine total DNA extracted from soil populations for evidence of the presence of B. japonicum. Although this technique is not quantitative at present, it offers great potential.

2.3 Measurements of activity in soil

The assumption is usually made that if numbers of bacteria in soil decrease or remain static, no growth is occurring. This is not necessarily correct. Population size is determined by a dynamic balance between growth and death. The soil is a heterogeneous habitat where both processes can be occurring simultaneously at different microsites. Measures of growth-related activity, as distinct from measures of total population size, are needed to distinguish a dynamic balance from genuine stasis. No quantitative technique is available for specifically measuring the activity of root-nodule bacteria in the presence of other microorganisms. However, indirect phage analysis (Germida and Casida, 1983), a technique which uses changes in bacteriophage numbers as a specific, though qualitative, indicator of increases in bacterial multiplication rate, has been used in studies of Rhizobium (Zeph and Casida, 1986; Germida, 1988), and could be more widely applied. A variation of the method using a most probable number approach has been developed for counting soil bacteria (Makkar and Casida, 1987). Bacteriophages of root-nodule bacteria are known to be species-specific in their host-range (Staniewski, 1970; Patel, 1976; Lesley, 1982; Lindström et al., 1983; Lindström and Lehtomäki, 1988) and to multiply only when their host is actively growing (Y.M. Barnet, unpublished), so the technique offers a highly specific monitoring system.

2.4 Diversity of the root-nodule bacteria

Ecological knowledge is restricted by the fact that only a small fraction of the large number of types of root-nodule bacteria has been isolated and studied. Only about 57% of genera within the Leguminosae have been surveyed for nodulation

(Faria et al., 1989) and in only a small proportion of these cases have the root-nodule bacteria been examined. Thus, we have little knowledge of the true diversity within the Rhizobiaceae. Some attempts have been made to redress this situation with isolation of strains from woody legumes and non-agricultural ecosystems (Trinick, 1980; Lawrie, 1983; Barnet et al., 1985; Catt, 1985; Shoushtari and Pepper, 1985; Jenkins et al., 1987; Barnet, 1988; Jenkins et al., 1988; Catt and Barnet, 1990).

2.5 Taxonomy of the root-nodule bacteria

Ideally, a taxonomic system should provide a conceptual and practical framework for ecological studies. In the absence of an adequate taxonomy, data from genotypically different strains may be grouped together, so that it is difficult to perceive patterns and develop general theories. There is almost certainly a greater variation within the root-nodule bacteria than is indicated by the current taxonomy, a fact seldom recognized by ecologists.

Many of the currently recognised taxa within the Rhizobiaceae embrace a range of organisms which, in phylogenetic terms, are only distantly related. *Bradyrhizobium* shows a degree of divergence from *Rhizobium* which suggests that it should be removed from the Rhizobiaceae (Jarvis et al., 1986). Species definition within the slow-growing group is also far from satisfactory. The type species, *Bradyrhizobium japonicum*, comprises two distinct phenons (Hollis et al., 1981; Stanley et al., 1985; Kuykendall et al., 1988) which could well be separated into at least two different species. Strains of bradyrhizobia from hosts other than soybean have not been classified at the species level. Problems also exist with the fast-growing rhizobia. The recent proposal for a new genus, *Azorhizobium* (Dreyfus et al., 1988), means that the simple concept of division into two genera on the basis of fast or slow growth (*Rhizobium* or *Bradyrhizobium*, respectively; Jordan, 1984) can no longer be maintained. Other currently accepted taxa seem also to be inadequately defined. *R. leguminosarum* biovar *phaseoli*, for instance, is a heterogenous assemblage with a degree of genotypic divergence compatible with several distinct taxa (Pinero et al., 1988). *R. meliloti* is divisible into at least two primary divisions, which may represent two distinct evolutionary species (Eardly et al., 1990). *R. loti* is so distantly related to other rhizobia that it could be a separate genus (Jarvis et al., 1986). Some rearrangement within the *Rhizobium* group has already been proposed; Chen et al. (1988) suggest, on the basis of numerical taxonomic studies, that *R. fredii* be transferred to a new genus *Sinorhizobium*. This assignment needs verification since other workers have shown a close relationship between *R. fredii* and *R. meliloti* (Jarvis et al., 1986).

This degree of divergence within the family means that although two organisms may be currently included within the Rhizobiaceae, or be placed in the same genus or species, there is no guarantee that they will possess a large number of common properties. The root-nodule bacteria share a common habitat, the soil, and they must all overcome similar problems in order to survive and grow. However, their phylogenetic divergence is such that there is no reason to assume that strategies evolved for overcoming these problems will be the same for all members of the group.

3 PHYSIOLOGICAL ECOLOGY OF THE ROOT-NODULE BACTERIA

While understanding of the physiology of the root-nodule bacteria continues to expand, this knowledge has usually not been placed in an ecological context. However, since physiological properties represent the phenotypic diversity on which environmental selection can operate, knowledge of them is important for an understanding of the ecology of the root-nodule bacteria. This section is intended to give a brief overview of some salient features with only selected references.

3.1 Metabolic versatility

Knowledge of heterotrophic carbon metabolism in the root-nodule bacteria is now quite extensive (see Stowers, 1985; Chapter 12). There are significant differences between taxa. *Rhizobium* appears to use a broader range of hexoses, pentoses, disaccharides and organic acids than *Bradyrhizobium*. The newly described genus *Azorhizobium* (Dreyfus et al., 1988) oxidizes glucose but no other carbohydrate, although it grows well on a range of organic acids. Unclassified organisms, such as the stem-nodulating strains from *Aeschynomene* (Stowers and Eaglesham, 1983) and the isolates from the desert tree *Prosopis* (Waldon et al., 1989), have patterns of carbon catabolism intermediate between those of *Rhizobium* and *Bradyrhizobium*. Authors disagree about the use of substrates by particular taxa (cf. Stowers, 1985, with Dreyfus et al., 1988), perhaps indicating that there is variation between strains within each genus. This considerable diversity within the group could well have ecological implications.

Although sugars and organic acids are important energy substrates in the rhizosphere (Whipps and Lynch, 1986), aromatic carbon compounds and complex biopolymers are likely to predominate in soil (Heal and Ineson, 1984) and an ability to oxidize them may be essential for survival of root-nodule bacteria away from the root. An array of aromatic compounds can be utilized for growth (Chen et al.,1984; Parke and Ornston, 1984; Gajendiran and Mahadevan, 1988; Chen et al., 1989; Gajendiran and Mahadevan, 1990), with slow-growers utilizing a wider range than the fast-growers. The two DNA homology groups of _B. japonicum_ (Stanley et al., 1985) differ, with one group possessing the protocatechuate pathway, while the other can only degrade catechol.

Some root-nodule bacteria can also grow autotrophically. For example, some strains of _B. japonicum_ and cowpea bradyrhizobia are capable of autotrophic growth in mineral salts media when supplied with hydrogen and carbon dioxide (Lepo et al., 1980; Stowers and Elkan, 1984; Lambert et al., 1985).

A number of strains of both _Bradyrhizobium_ and _Rhizobium_ grow anaerobically in the presence of nitrate (Daniel et al., 1980, 1982; El Hassen et al., 1985). Autotrophic growth can also be supported under these conditions (Neal et al., 1983). Nitrate-dependent growth is strain-dependent and does not occur uniformly through any one taxon. Since the soil has many anaerobic microsites and the demands of the respiring root may make the rhizosphere itself a severely oxygen-limited habitat, an ability for anaerobic growth may be an advantage. Even from this limited survey, it is apparent that root-nodule bacteria use a variety of growth strategies, enabling them to occupy a wide range of ecological niches.

3.2 Tolerance of physical stress

It is still not possible to define which aspects of the physical environment are important in controlling the distribution and growth of root-nodule bacteria. The most important reason for this is the difficulty inherent in using measurements made on bulked soil to evaluate the environmental conditions encountered by any particular bacterium. The soil is a heterogeneous environment, and conditions may vary over micrometre distances; for instance, acidity in the rhizosphere may be different from that in bulk soil (Nye, 1981; Jarvis and Robson, 1983). Averaged physical parameters do not indicate their true range in any habitat.

In addition, strains under energy-limited conditions in the soil may be more sensitive to environmental stresses than they would be in rich laboratory media. Combating adverse conditions requires substantial energy expenditure. Maintenance of internal pH requires the pumping of protons across the cell membrane (Padan, 1984), while accommodation to high external solute concentrations necessitates the accumulation of osmoprotectant molecules (Brown et al., 1986; Imhoff, 1986), both energetically expensive operations. The difficulty of replicating soil conditions in the laboratory means that there may be poor correlation between the environment required for survival of root-nodule bacteria in the soil, and the conditions under which they will grow in laboratory media.

(i) Effect of soil acidity. Most work done on the tolerance of root-nodule bacteria to low pH has been concerned with developing methods of screening for acid tolerance which will be predictive of the behaviour of strains in acid soils (e.g. Wood and Cooper, 1984; Howieson et al., 1988; Wood and Cooper 1988a,b). There appears, however, to be little correlation between ability to survive in soils of low pH and growth on acid laboratory media (Wood and Shepherd, 1987; Karanja and Wood, 1988; Chapter 14).

Acid soils provide a complex array of environmental stresses, since they are usually low in available phosphorus and calcium and have high levels of soluble aluminium. Tolerance to high aluminium levels appears sometimes to be correlated with ability to grow at low pH (Karanja and Wood, 1988a), but in other strains to be an independent property (Wood et al., 1988). The mechanism of aluminium toxicity is unknown (Wood and Cooper, 1988b). Taxa differ in acid sensitivity; *R. loti* appears to be more resistant (Wood et al., 1988) than *R. trifolii* (Wood and Shepherd, 1987) while *R. meliloti* is generally the most sensitive (see e.g. Robson and Loneragan, 1970), although within each species there are considerable differences between individual strains (e.g. Howieson et al., 1988)

Soil pH levels affect the distribution of root-nodule bacteria but to what extent is unknown. There seems to be a pH level (around pH 4.0) below which root-nodule bacteria do not occur (e.g. Wood and Shepherd, 1987), but above this value the role of pH as a selective factor needs further definition.

(ii) Effect of reduced water potential. Water stress may be imposed through reduction in either matric or osmotic potential. In non-saline soil, reduced matric potential is probably the more important stress on bacteria, although as soil dries there will also be an increased contribution of solutes to the total water potential. Tolerance of reduced matric potential has been little studied though it is known that strains differ in their tolerance to drying (e.g Mahler and Wollum, 1981a; Al-Rashidi et al., 1982; Boonkerd and Weaver, 1982; Fuhrmann et al., 1986). However, without an understanding of the physiological basis of resistance or susceptibility to drying, it is impossible to explain these differences.

The mechanisms of tolerance to salt stress are better understood. While most strains of root-nodule bacteria seem comparatively salt-sensitive (e.g. Botsford, 1984), a few workers have obtained tolerant isolates, chiefly from desert or beachfront soils (Hua et al., 1982; Singleton et al., 1982; Yelton et al., 1983). The mechanisms involved in osmotolerance seem similar to those described for other genera. When subjected to high external concentrations of sodium chloride, root-nodule bacteria accumulate potassium ions concomitantly with glutamate (Hua et al., 1982; Yap and Lim, 1983; Yelton et al., 1983), while levels of organic molecules such as glycine betaine (Smith et al., 1988) or proline betaine (Gloux and Le Rudelier, 1989) may also increase. These substances serve as osmopro- tectants, balancing internal and external osmotic pressures and stabilizing cell proteins. Presumably strains which are able to compensate for high external solutes in this way are better fitted for surviving in saline soils, although there has been little attempt to directly correlate the two properties. Singleton et al. (1982) found some correlation between salinity at the site of isolation and salt-tolerance in laboratory media, but also isolated salt-sensitive strains from saline environments. The osmotic response and survival of strains may depend on the particular solute responsible for the stress (Mary et al., 1986; Elsheikh and Wood, 1989) and any extrapolation to the natural environment needs to take this into account.

The mechanisms of protection against desiccation could be similar to those involved in osmotolerance. Limited data suggest that strains of bacteria resistant to drying can also withstand saline conditions (Al-Rashidi et al., 1982; Singleton et al., 1982) and that prior exposure to high salt increases resistance to desiccation (Mary et al., 1986). However, other evidence indicates that the mechanism of osmoregulation may depend on the nature of the water stress (Busse and Bottomley, 1989). Further work is needed to determine whether the two properties are correlated.

(iii) Temperature. Whether high soil temperatures control the growth and distribution of the root-nodule bacteria is unclear. There are marked differences between strains in both the maximum temperature at which they can survive and the range over which they will grow (Munevar and Wollum, 1981; Hartel and Alexander, 1984; Jenjareontham and Barnet, 1986). However, there are few data directly connecting these differences with ecology. Isolates from tropical and exposed soils are generally more heat-tolerant than those from less stressed conditions (Hartel and Alexander, 1984; Jenjareontham and Barnet, 1986; Karanja and Wood, 1988b), providing indirect evidence that high soil temperatures may exert a selective pressure. However, this needs to be demonstrated more directly.

4 BIOTIC INTERACTIONS

4.1 Competition from other bacteria

The difficulty of establishing large numbers of root-nodule bacteria in the presence of other soil microorganisms has led to suggestions that root-nodule bacteria are poor competitors for soil nutrients (Alexander, 1985). A number of studies show that rhizobia do not attain the same levels in non-sterile as in sterile soil (Pena-Cabriales and Alexander, 1983; Li and Alexander, 1986).

However, this is also true of other bacteria (Acea et al., 1988) and presumably reflects the fact that nutrient limitation in soil imposes a maximimum "carrying capacity" so that, when the available substrates are shared between a number of species, each will be reduced in numbers. Indigenous R. leguminosarum, R. meliloti (Germida, 1988) and B. japonicum (Viteri and Schmidt, 1987) compete successfully with other soil bacteria for a wide range of added organic compounds. B. japonicum also grows chemoautotrophically in competition with other hydrogen-oxidizing soil bacteria (Viteri and Schmidt, 1989). It seems that any concept of the root-nodule bacteria as poor competitors, dependent on their unique haven in the nodule for survival, is unlikely to be correct.

4.2 Antagonistic or stimulatory interactions with other soil bacteria

Many soil bacteria have antagonistic (Pugashetti et al., 1982; Trinick et al., 1983; Furhmann and Wollum, 1989; Hicks and Loynachan, 1989) or stimulatory (Trinick et al., 1983; Hicks and Loynachan, 1989) effects on root-nodule bacteria when tested in vitro. Whether such interactions are significant in natural habitats is uncertain. Trinick et al. (1983) found that soils in which it was difficult to establish root-nodule bacteria had high levels of antagonistic micro-organisms, although no causal relationship was established. Hodgson et al. (1985) demonstrated that bacteriocin-producing strains of root-nodule bacteria were effective under natural conditions. However, Hicks and Loynachan (1989) showed that stimulatory effects seen in sterile conditions were not reproducible in non-sterile soil. The result of this type of interaction is unlikely to be generally predictable, but must be assessed for each situation.

4.3 Predation and parasitism

Both bacterial (Zeph and Casida, 1986) and protozoan (Heynen et al., 1988) predators in soil are known to multiply at the expense of root-nodule bacteria. The effect of predaceous bacteria on abundance of their prey is unknown. However, protozoans do have an important controlling role, since bacterial populations are significantly smaller in their presence (Habte and Alexander, 1977; Heynen et al., 1988; Sinclair and Alexander, 1989). It is unlikely, however, that protozoans would ever eliminate any one species in a soil. Sinclair and Alexander (1989) suggest that under intense protozoan grazing, slow-growing species might be eliminated, but this could only occur in a homogeneous habitat. In soil, bacteria can be protected from predation by the effects of clay (Heynen et al., 1988) or by seclusion inside soil aggregates (Vargas and Hattori, 1986). The practical importance of protozoa in limiting populations of root-nodule bacteria is still uncertain.

The effects of parasitic interaction are also obscure. There is no evidence that bacteriophages or bdellovibrios cause significant reductions in numbers of bacteria in natural habitats. Bacteriophage numbers do not rise during periods of their host's decline and in fact there appears to be a positive correlation between phage numbers and those of their bacterial hosts in field soils (Dhar et al., 1980; Lawson et al., 1987). The most significant effect of bacteriophages is likely to be the production of genetic change in host populations (see later discussion). In addition, phage-resistant mutants are often changed in symbiotic characteristics (Barnet, 1979) and have a strong competitive advantage when the selective bacteriophage is present (Lawson, 1987). This is unlikely to cause major changes in the composition of populations, but may be significant in introducing some heterogeneity.

5 DISTRIBUTION OF THE ROOT-NODULE BACTERIA

A major goal of any ecological research is the identification of the factors which determine species distribution and abundance. Knowledge of the distribution of a species helps to define which of a number of possible environmental constraints actually control its survival and growth. Distribution needs to be considered at a number of different levels, since different factors obviously determine a species' broad geographic range, its abundance at a given site and the detailed pattern of its occurrence within habitats. Climatic factors are likely to be important in controlling biogeography, while the conditions in microhabitats in a heterogenous environment will be important locally. Unfortunately, we know little about the distribution of the root-nodule bacteria at either the global or microsite level, but the importance of such data is being recognized.

5.1 Geographic distribution of natural populations

Only a few workers have attempted to delineate the natural global distribution of the root-nodule bacteria and to correlate this with the operation of selective environmental factors. Not all such attempts have been successful. Norris (1965) proposed that the alkali-producing, slow-growing bradyrhizobia are the primitive, ancestral type which have continued to occupy the tropical habitats in which legumes first evolved. Their ability to produce alkali was thought to provide a selective advantage in the weathered acidic soils of these zones. Acid-producing, fast-growing strains were postulated to be of later origin, and their shift to acid sensitivity a result of selection as they colonised the more neutral soils of the temperate areas. This theory is frequently quoted in the literature, although there is little evidence to support it. A number of workers have pointed out that growth rate and the pH change produced in media are unrelated and that acid or alkali production may be dependent on the particular medium and strain used (Date and Halliday, 1979; Cooper, 1982; Hernandez and Focht, 1984; Stowers and Elkan, 1984). Additionally, there appears no absolute correlation between acid production on artificial media and acid intolerance (Date and Halliday, 1979; Cooper, 1982; Halliday, 1985). Some fast-growers are more acid tolerant than slow-growing strains (e.g. Cooper, 1982). Recent reports of the isolation of fast-growing strains from tropical legumes (Trinick, 1980; Dreyfus and Dommergues, 1981) also show that the concept of exclusive bradyrhizobial occupancy of tropical soils is invalid. In the form originally proposed, the Norris theory appears untenable and demonstrates the need for thorough study before assignment of ecological roles to physiological characteristics.

Studies of the distribution of symbionts associated with woody legumes may help understand some of the environmental factors controlling distribution of fast- and slow-growing root-nodule bacteria. Surveys of strains from Australian acacias showed that fast-growing isolates were only obtained from two of six study areas with slow-growers obtained at all other sites (Barnet et al., 1985; Catt and Barnet, 1990). The two sites both had low soil organic matter, high soil surface temperatures over summer and usually low soil water levels (R. Jenjareontham and Y. M.Barnet, unpublished). This combination of factors produces a severely stressful environment in which bacterial growth would be expected only during short and infrequent periods. It may be that such conditions select for fast-growers. Support for this suggestion comes from a study of the root-nodule bacteria associated with woody legumes in arid areas of the United States, where fast-growing strains were dominant in dry surface soils (Jenkins et al., 1987, 1988). In none of these studies has the taxonomic position of the isolates been determined. It is therefore impossible to decide whether the predominance of fast-growers is due to an inherent advantage bestowed by a potentially short generation time, or whether the isolates represent taxonomic groups specifically adapted to withstand desert conditions. Jenkins et al. (1987) base their case for the first of these explanations on the expectation that variable and physically stressed environments select for physiological strategies which maximize growth during favourable periods (Heal and Ineson, 1984). However, more data are required before this point can be decided.

Other workers have also reported geographic localisation of species of root-nodule bacteria. In a study of soybean-nodulating strains from two sites in China, Dowdle and Bohlool (1985) reported that *R. fredii* predominated in one area, a soybean field, while only *B. japonicum* was found at the other site, a rice paddy. The rice paddy would presumably provide the more constant environment, although details of conditions at the two sites were not given. In a study of the distribution of root-nodule bacteria in Maui, Woomer et al. (1988) found that Bradyrhizobium sp. occurred in soils high in organic carbon and had numbers which showed a positive correlation with rainfall, supporting the idea that slow-growers do best under more stable conditions.

The distribution of host legumes is another factor controlling the distribution of root-nodule bacteria. Legumes newly introduced into an area where they do not occur naturally usually require inoculation (e.g. Slattery and Coventry, 1989). In Maui, Woomer et al. (1988) found that the presence of an appropriate host correlated strongly with that of particular species of symbiont. Research by Lie and co-workers (Lie, 1981; Lie and Goktan, 1984; Lie et al., 1987) showed that rhizobia capable of nodulating wild peas can be found only in the Middle East and Central Asia, regions where their hosts occur naturally. Isolates from European soils nodulate only the cultivated pea, not their wild progenitors. Lie et al. (1987) suggested that modern pea lines, which have a narrow genetic base, have selected a similarly narrow range of Rhizobium genotypes from amongst those casually transported during the migration of leguminous plants from the Middle East to Europe. There seems little doubt that this explanation is correct but the mechanism of the selective process is obscure. There is abundant evidence that the availability of the specific host is not necessary for a strain's continued presence in the short term (e.g. Nutman, 1969; Nutman and Ross, 1970; Bromfield et al., 1986; Kucey and Hynes, 1987; Rupela et al., 1987; Brunel et al., 1988; Germida, 1988). However, the presence of compatible hosts do boost the numbers of a strain present in soil probably through release from decaying nodules (Mahler and Wollum, 1981b,1982a; Kuykendall et al., 1982; Bushby, 1984; Moawad et al., 1984; Kucey and Hynes, 1987; Richardson and Simpson, 1988). In the absence of the host, numbers are usually low. It is conceivable that, without the symbiotic association, root-nodule bacteria cannot maintain themselves in the longer term in the face of competition from the soil population. An alternative explanation, however, is that it is the genes encoding host specificity which are lost, rather than the rhizobial strain. In the absence of the host, there would be no selective pressure to retain the ability to establish a symbiosis, although the strain could continue in the soil, unrecognized by our current identification methods.

5.2 Distribution of the root-nodule bacteria within habitats

In addition to the effects of macroclimate and geography, it is also necessary to consider distribution of root-nodule bacteria on a finer scale if we are to fully understand the functioning of populations. Habitats are heterogeneous. Each site will possess a number of different microclimates determined by meteorological conditions, topography, vegetation and other local features. At a finer level of resolution, the soil habitat is itself heterogeneous with discontinuous distribution of roots, organic matter, water and anaerobic zones, all of which may influence the occurrence and activity of the root-nodule bacteria. The presence of plants is particularly important in producing local discontinuities. Roots provide substantial amounts of both particulate and soluble organic carbon in their immediate vicinity (Whipps and Lynch, 1986) as well as changing physical conditions such as pH (Jarvis and Robson, 1983). Distinct habitats may also be distributed vertically through the soil profile. Organic matter will usually be higher at the surface, while temperature, pH and available nutrients may all change substantially with depth, producing a range of microhabitats for colonisation by specifically adapted organisms.

There is substantial variation within local populations of root-nodule bacteria (Beynon and Josey, 1980; Young, 1985; Bromfield et al., 1986; Kamicker and Brill, 1986; Broughton et al., 1987a; Young et al., 1987; Valdivia et al., 1988; Thurman and Bromfield, 1988; Kucey and Hynes, 1989) supporting the concept that each site provides a range of niches which can maintain a diverse population. The specific bacterial adaptations involved in producing this diversity have been little studied. Bromfield et al. (1986) and Thurman and Bromfield (1988) showed significant differences in the composition of the population at different sites, but did not identify the environmental factors responsible. Further work (Chan et al., 1989) showed that the spectrum of strains at two sites differed in their ability to denitrify, but no attempt was made to correlate this difference with environmental variables. Strain variation in R. leguminosarum was lowest in acidic habitats in Wales (Harrison et al., 1989b), but again this was not examined in detail.

A few studies have attempted to examine the relationship between microhabitat and strain distribution. The two serogroups of R. leguminosarum bv. viciae which occur in eastern Washington State have distributions which can be related to temperature and moisture (Mahler and Bezdicek, 1980). In an extension of this work, Brockman and Bezdicek (1989) showed that topography had a marked effect with strains which occurred in the warm dry conditions of south slopes differing from those found in the moist, cool environment of north slopes and hill bottoms. Glasshouse trials (Turco and Bezdicek, 1987) had shown that preconditioning soils to extremes of water potential before planting trap hosts altered the predominant serogroup forming nodules and it seems likely that soil water levels were also important in determining competitiveness and/or survival of particular groups of strains in the field.

Position in the soil profile can also determine the composition of the population in some circumstances. In a taxonomic study of root-nodule bacteria from desert trees, isolates from surface, intermediate and deep phreatic environments grouped into physiologically distinct phena (Waldon et al., 1989). Although it is known that each of these habitats provides a very different environment (Jenkins et al., 1987), the selective forces have not been identified. Rupela et al. (1987) showed that chick pea rhizobia are most numerous in the top 30 cm of the soil profile. Work with other soil bacteria suggests that this a common pattern, probably related to the concentration of organic matter in this region. A similar response to available substrate is seen with the increase in numbers of some species of root-nodule bacteria in the rhizosphere (Robert and Schmidt, 1983; Moawad et al., 1984), an effect only demonstrable when the bacteria in bulk soil, away from the root, are energy-limited. Where unplanted soils can support high populations, there is little rhizosphere stimulation (Demezas and Bottomley, 1986a). The concentration of root-nodule bacteria in the organically enriched top few centimetres of the soil may sometimes be due to lower acidity and exchangeable aluminium levels than those occurring lower down the profile. Mixing of acid soil by cultivation reduced rhizobial numbers, presumably because of the disruption of this favourable microhabitat (Richardson and Simpson, 1988).

Distribution of root-nodule bacteria at the microsite level has been very little studied. Data on other soil microorganisms, however, indicate that bacteria are not uniformly distributed, but occur in microcolonies surrounded by uncolonized space; even on the root surface only five percent of the area is occupied (Rovira and Campbell, 1974; Foster et al., 1983). There seems no reason to believe that root-nodule bacteria will be differently arranged. Such a discontinuous distribution means that interactions between soil organisms will be more limited than they would be in a more homogeneous habitat, a factor not always taken into account in evaluating ecological data. Physical protection may also play a role in determining interactions. Foster and Rovira (1978) demonstrated that bacteria may be tightly enclosed within clay plates. Organisms in such a position might be limited by space or nutrients, but could also be protected from adverse external factors.

Bacteria are normally most abundant in the rhizosphere with numbers decreasing rapidly within micrometres of the root surface (Foster and Rovira, 1978). Theoretical models of plant-root interactions (Newman and Watson, 1977) also show that the influence of the plant root may not extend very far into the soil. Although some rhizosphere effect may be evident 1-2 mm from the root, it seems that most associated bacteria are found within 50 μm of its surface. Cells in the greater proportion of the soil volume must be able to survive without direct input from the plant and this also seems likely to be true for the root-nodule bacteria.

5.3 Dispersal of root-nodule bacteria

Comparatively little is known of the dispersal mechanisms of the root-nodule bacteria. It is assumed that mechanical transfer, on seeds or in soil, is important for spread over long distances, but there have been no studies of this.

A little more work has been done on movement within the soil profile. Root-nodule bacteria appear to be comparatively immobile on a macroscale. The rhizoplane populations and proportion of nodules formed by introduced strains has been shown to decline with distance from the site of inoculation (Kamicker and Brill, 1987; McDermott and Graham, 1989) possibly indicating that their movement is minimal. Vertical carriage in infiltrating water appeared the major mechanism involved where movement did occur (Madsen and Alexander, 1982). An attractant role of roots seems important for motility of other soil organisms such as Azospirillum (Bashan and Lavanoy, 1987). It is unclear whether this also holds true for the root-nodule bacteria, although they are motile and chemotactic towards nutrients (Bowra and Dilworth, 1981; Bergman et al., 1988) and root exudates (Gitte et al., 1978; Caetano-Anolles et al., 1988). Some laboratory studies have used non-motile or non-chemotactic mutants to investigate the role of motility in promoting nodulation and have found that, in soil, motile strains have an advantage (Hunter and Fahring, 1980; Ames and Bergman, 1981; Mellor et al., 1987; Liu et al., 1989). Both Mellor et al. (1987) and Liu et al. (1989), however, found no difference in the rhizosphere multiplication of motile and non-motile strains, suggesting that rhizosphere colonization is unaffected by motility. However, the presence of other soil organisms may affect dispersion even of motile bacteria (Catlow et al., 1990). The authors of this study suggest that competition for nutrients in the low organic matter soil they used might restrict growth and with it mobility. It is possible that flagellar motility is not important in the dispersion of root-nodule bacteria, at least under severely nutrient-limited conditions.

Away from the root, movement on a microscale level might be important in allowing cells to reach new sources of energy substrate. Y.M.Barnet and K.A.Lawson (unpublished) showed that non-motile mutants of R. leguminosarum bv. trifolii grew less well than the motile parent when inoculated into sterile soil, despite the fact that generation times in shaken soil slurry were identical. This suggests that ability to colonize new microsites may be a survival trait in natural habitats. However, movement through soil, even over micrometre distances, is dependent on factors other than motility. Around 90 percent of cells of indigenous B. japonicum in an agricultural field soil were attached to soil particles (Ozawa and Yamaguchi, 1986), although the remaining ten percent of free cells may well be sufficient for continued migration. The effects of lowered soil water are also likely to be important.

As soil water is reduced over a range which has little effect on water activity, marked decreases occur in both bacterial growth rates and total population size (Mahler and Wollum, 1981a; Lawson and Barnet, 1984). The most likely explanation is that, as soil water levels decrease, thinning of water films, emptying of pores and reduction in the extent of continuous water-filled pathways will reduce bacterial movement and nutrient accessibility. Correlations exist between water potential and bacterial movement in soil (Griffin and Quail, 1968; Hamdi, 1971; Soby and Bergman, 1983; Lawson, 1987). A direct correlation between limitation of movement by lowered water potentials and reduced growth of R. leguminosarum bv. trifolii was shown by Lawson (1987). Although these data are not direct proof, they suggest that dispersion at the microsite level may be important for growth and survival of root-nodule bacteria.

6 ABUNDANCE AND POPULATION DYNAMICS OF THE ROOT-NODULE BACTERIA

The abundance of a strain at a site is not necessarily related to its ability to survive there. Delineation of factors producing variations in numbers over time and space would increase our understanding of the forces operating on root-nodule bacteria. There is considerable evidence that the proportional representation of strains, as well as the total size of the community of root-nodule bacteria, may vary with site and over time (e.g. Noel and Brill, 1980; Mahler and Wollum, 1981b, 1982a,b; Robert and Schmidt, 1983; Moawad et al., 1984; Kvien and Ham, 1985; Bromfield et al., 1986; Thurman and Bromfield, 1988). In most cases the environmental factors responsible for this have not been determined and only a few studies have directly addressed the problem.

General trends in the abundance of peanut-nodulating strains in Iraq were related to soil conditions such as organic carbon levels (Youssef et al., 1987). However, the only statistically significant correlation was with electrical conductivity and cation exchange capacity. The meaning of this is unclear, since the authors did not attempt to include other potentially important variables in their analysis. Meteorological data were listed but not analysed in detail and the proportion of variation in population size accounted for by the significant variable was not determined. The importance of mathematics in evaluating data of this kind cannot be overemphasized: the multivariate nature of environmental influences means that correlations will not readily be apparent without careful statistical analysis.

Two other studies have attempted to produce regression models to express numbers of root-nodule bacteria as a function of soil and climatic factors. Woomer et al. (1988) determined the average size of populations at fourteen sites on the island of Maui, while Lawson et al. (1987) followed the fluctuations in numbers of R. leguminosarum bv. trifolii in two Australian fields over an eighteen month period. In neither study did soil texture appear an important determinant of population size. Both studies, however, indicate that plants are important in controlling the abundance of root-nodule bacteria. On Maui, release from nodules may have been important in this respect. However, total legume cover was also significant, suggesting that rhizosphere stimulation may play a part, a conclusion supported by the Australian study. Solar radiation and plant size both showed positive correlation with rhizobial numbers. These factors would determine photosynthetic activity and influence the amount of root exudate and hence the size of the rhizosphere population. Energy limitation may be one of the chief factors controlling abundance of root-nodule bacteria.

Soil water levels also seemed to be an important determinant. A positive relationship between rainfall and numbers of root-nodule bacteria was apparent in the Maui data (Woomer et al.,1988). In the Australian soils, although no direct correlation of matric potential with bacterial population size was demonstrable, a rise in phage numbers with increasing soil water levels suggested that host activity was also greater. Soil water content could affect bacterial growth and survival either by means of desiccation-induced death or, at higher water levels, by controlling the availablity of energy sources.

Mean soil temperature correlated positively with numbers of root-nodule bacteria in the Hawaiian study, although the Australian data showed no such relationship. The two studies were, however, making different types of comparisons. In Maui, the annual average populations at different sites were compared, while the Australian study was of changes within the same area over time. Temperature differences between sites would presumably have a greater impact on the average soil population than would seasonal fluctuations, since they would operate consistently throughout the year. It remains to be seen whether this is universally true. It is clear, however, that environmental data of this kind must be very carefully evaluated before generalizations are made.

7 STABILITY OF POPULATIONS OF ROOT-NODULE BACTERIA

An understanding of genetic variation is an important aspect of ecology. The stability of the organisms, the possibilities for their genetic change, and the time frame over which such changes operate, are all important elements in defining the structure of populations and the temporal variations in them.

Root-nodule bacteria appear to be comparatively stable. Strains can survive in soil for periods of up to thirteen years with little change in their phenotypic properties (Van Rensburg and Stridjom, 1985; Brunel et al., 1988; Lochner et al., 1989). Population genetic studies provide more definitive evidence about the stability of populations of root-nodule bacteria. Where individual loci are considered, a strong genetic disequilibrium is evident; certain combinations of alleles tend to occur together while other theoretically possible combinations are rare or non-existent (Young, 1985; Young et al., 1987; Pinero et al., 1988; Harrison et al., 1989; Eardly et al., 1990). This suggests either that there is strong selection against the rare lineages, or that chromosomal recombination between particular genotypes is comparatively infrequent. In either case, the population structure will not be fluid, but should, in the absence of catastrophic selective forces and within the limits of normal seasonal variation, remain relatively unchanging. This is not to say that no genetic exchange occurs. Plasmid transfer would not necessarily disturb chromosomal lineages. Even transfer of small numbers of chromosomal genes would not break the basic clonal structure of populations if it occurred with only low frequency. In fact, population geneticists have pointed out that some gene transfer is necessary within a species if it is to remain coherent and genetically recognizable (Young, 1989). All the mechanisms of bacterial gene transfer - conjugation (e.g. Hooykas et al., 1981), transduction (Finan et al., 1984) and transformation (Kiss and Kalman, 1982) occur in the root-nodule bacteria *in vitro*. Although most work has dealt with the fast-growing group, limited information suggests that the slow-growers may also undergo genetic exchange (e.g. Pilacinski and Schmidt, 1981). There is some evidence that gene transfer is not simply a laboratory phenomenon, but does occur in natural habitats. *Sym* plasmid transfer in soil has been demonstrated between *R. leguminosarum* and *R. meliloti* (Broughton et al., 1987b), and between *Escherichia coli* and *R. fredii* (Richaume et al., 1989). In addition, the occurrence of indistinguishable plasmids in strains with different chromosmal backgrounds (Young and Wexler, 1988) and the presence of unrelated *Sym* plasmids in identical host strains (Schofield et al., 1987) suggests that some natural plasmid transfer occurs within field populations. This has important implications in the light of evidence that soil contains populations of organisms which lack *Sym* plasmids, but which have the capacity to acquire them (Jarvis et al., 1989; Sõberon-Chávez and Nājera, 1989). There has been no work demonstrating transformation or transduction of root-nodule bacteria in natural habitats. However, these mechanisms of gene exchange have been demonstrated in soil with Pseudomonas and Escherichia respectively (Carlson et al., 1983; Germida and Khachatourian, 1988) and there seems no reason to believe that they do not occur with the root-nodule bacteria.

8 CONCLUSION

It is obvious that our knowledge of the ecology of the root-nodule bacteria is still insufficient for it to be of practical use in the establishment of inoculant strains. Ecologists can offer no unequivocal answers to questions about the characteristics needed for success in the field. Current research only emphasizes the complexity of the problem. However, progress has been made and, with the new techniques now becoming available and careful formulation of the questions which need to be addressed, progress in the next decade should be appreciable.

9 REFERENCES

Acea, M.J., Moore, C.R. and Alexander, M., 1988. Survival and growth of bacteria introduced into soil. Soil Biol. Biochem., 20: 509-515.

Alexander, M., 1985. Ecological constraints on nitrogen fixation in agricultural ecosystems. In: K.C. Marshall (Editor), Advances in Microbial Ecology. Plenum Press, New York, London, pp. 163-303.

Almendras, A.S. and Bottomley, P.J., 1987. Influence of lime and phosphate on nodulation of soil-grown Trifolium subterraneum L. by indigenous Rhizobium trifolii. Appl. Environ. Microbiol., 53: 2090-2097.

Al-Rashidi, R.K., Loynachn, T.E. and Frederick, L.R., 1982. Desiccation tolerance of four strains of Rhizobium japonicum. Soil Biol. Biochem., 14: 489-493.

Ames, P. and Bergman, K., 1981. Competitive advantage provided by bacterial motility in the formation of nodules by Rhizobium meliloti. J. Bacteriol., 148: 728-729.

Barnet, Y.M., 1979. Properties of Rhizobium trifolii isolates surviving exposure to specific bacteriophages. Can. J. Microbiol., 25: 979-986.

Barnet, Y.M., 1988. Nitrogen-fixing symbioses with Australian native legumes. In: W.G. Murrell and I.R. Kennedy (Editors), Microbiology in Action. John Wiley and Sons, Chichester, New York, pp. 81-92.

Barnet, Y.M., Catt, P.C. and Hearne, D.A., 1985. Biological nitrogen fixation and root-nodule bacteria (Rhizobium sp. and Bradyrhizobium sp.) in two rehabilitating sand dune areas planted with Acacia spp.. Aust. J. Bot., 33: 595-610.

Barnet, Y. M., Trinick, M.J., Date, R.A. and Roughley, R.J., 1988. Ecology of the root-nodule bacteria. In: W.G. Murrell and I.R. Kennedy (Editors), Microbiology in Action. John Wiley and Sons, Chichester, New York, pp. 1-22.

Bashan, T. and Levanony, H., 1987. Horizontal and vertical movement of Azospirillum brasilense Cd in the soil and along the rhizosphere of wheat and weeds in controlled and field environments. J. Gen. Microbiol., 133: 3473-3480.

Bergman, K., Gulash-Hoffee, M., Hovestadt, R.E., Larosiliere, R.C., Ronco, P.G. and Su, L., 1988. Physiology of behavioral mutants of Rhizobium meliloti: evidence for a dual chemotaxis pathway. J. Bacteriol., 170: 3249-3254.

Beynon, J.L. and Josey, D.P., 1980. Demonstration of heterogeneity in a natural population of Rhizobium phaseoli using variation in intrinsic antibiotic resistance. J. Gen. Microbiol., 118: 437-442.
Bjourson, A.J. and Cooper, J.E., 1988. Isolation of Rhizobium loti strain-specific DNA sequences by subtraction hybridization. Appl. Environ. Microbiol., 54: 2852-2855.
Bohlool, B.B. and Schmidt, E.L., 1980. The immunofluorescence approach in microbial ecology. Advances in Microbial Ecology, 4: 203-241.
Boonkerd, N. and Weaver, R.W., 1982. Survival of cowpea rhizobia in soil as affected by soil temperature and moisture. Appl. Environ. Microbiol., 43: 585-589.
Botsford, J.L., 1984. Osmoregulation in Rhizobium meliloti: inhibition of growth by salts. Arch. Microbiol., 137: 124-127.
Bottomley, P.J. and Dughri, M.H., 1989. Population size and distribution of Rhizobium leguminosarum bv. trifolii in relation to total soil bacteria and soil depth. Appl. Environ. Microbiol., 55: 959-964.
Bowra, B.J. and Dilworth, M.J., 1981. Motility and chemotaxis towards sugars in Rhizobium leguminosarum. J. Gen. Microbiol., 126: 231-235.
Brockman, F.J. and Bezdicek, D.F., 1989. Diversity within serogroups of Rhizobium leguminosarum biovar viciae in the Palouse region of eastern Washington as indicated by plasmid profiles, intrinsic antibiotic resistance and topography. Appl. Environ. Microbiol., 55: 109-115.
Bromfield, E.S.P., Sinha, I.B. and Wolnetz, M.S., 1986. Influence of location, host cultivar, and inoculation on the composition of naturalized populations of Rhizobium meliloti in Medicago sativa nodules. Appl. Environ. Microbiol., 51: 1077-1084.
Broughton, W.J., Heycke, N., Priefer, U., Schneider, G.-M. and Stanley, J., 1987a. Ecological genetics of Rhizobium meliloti: diversity and competitive dominance. FEMS Microbiol. Lett., 40: 245-249.
Broughton, W.J., Samrey, U. and Stanley, J., 1987b. Ecological genetics of Rhizobium meliloti: symbiotic plasmid transfer in the Medicago sativa rhizosphere. FEMS Microbiol. Lett., 40: 251-255.
Brown, A.D., Mackenzie, K.F. and Singh, K.K., 1986. Selected aspects of microbial osmoregulation. FEMS Microbiol. Rev., 39: 31-36.
Brunel, B., Cleyet-Marel, J., Normand, P. and Bardin, R., 1988. Stability of Bradyrhizobium japonicum inoculants after introduction into soil. Appl. Environ. Microbiol., 54: 2636-2642.
Bushby, H.V.A., 1981. Quantitative estimation of rhizobia in non-sterile soil using antibiotics and fungicides. Soil Biol. Biochem., 13: 237-239.
Bushby, H.V.A., 1984. Colonization of rhizospheres and nodulation of two Vigna species by rhizobia inoculated onto seed: influence of soil. Soil Biol. Biochem., 16: 635-641.
Busse, M.D. and Bottomley, P.J., 1989. Growth and nodulation responses of Rhizobium meliloti to water stress induced by permeating and nonpermeating solutes. Appl. Environ. Microbiol., 55: 2431-2436.
Caetano-Anolles, G., Crist-Estes, D.K. and Bauer, W.D., 1988. Chemotaxis of Rhizobium meliloti to the plant flavone luteolin requires functional nodulation genes. J. Bacteriol., 170: 3164-3169.
Carlson, C.A., Pierson, L.S., Rosen, J.J.and Ingraham, J.L., 1983. Pseudomonas stutzeri and related species undergo natural transformation. J. Bacteriol., 153: 93-99.
Catlow, H.Y., Glenn, A.R. and Dilworth, M.J., 1990. The use of transposon-induced non-motile mutants in assessing the significance of motility of Rhizobium leguminosarum biovar trifolii for movement in soils. Soil Biol. Biochem., 22: 331-336.
Catt, P.C., 1985. Studies of root-nodule bacteria from acacias in New South Wales. Ph.D. thesis, University of New South Wales.

Catt, P.C. and Barnet, Y.M., 1990. Distribution and characteristics of root-nodule bacteria isolated from Australian Acacia spp. Plant Soil, in press.
Chan, Y.-K., Barran, L.R. and Bromfield, E.S.P., 1989. Denitrification activity of phage types representative of two populations of indigenous Rhizobium meliloti. Can. J. Microbiol., 35: 737-740.
Chen, Y.P., Glenn, A.R. and Dilworth, M.J., 1984. Uptake and oxidation of aromatic substrates by Rhizobium leguminosarum MNF 3841 and Rhizobium trifolii TA1. FEMS Microbiol. Lett., 21: 201-205.
Chen, Y.P., Dilworth, M.J. and Glenn, A.R., 1989. Degradation of mandelate and 4-hydroxymandelate by Rhizobium leguminosarum biovar trifolii TA1. Arch. Microbiol., 151: 520-523.
Chen., W.X., Yan, G.H. and Li, J.L. 1988. Numerical taxonomy study of fast-growing soybean rhizobia and a proposal that Rhizobium fredii be assigned to Sinorhizobium gen. nov. Int. J. Syst. Bacteriol., 38: 392-397.
Cooper, J.E., 1982. Acid production, acid tolerance and growth rate of Lotus rhizobia in the laboratory. Soil Biol. Biochem., 14: 127-131.
Cooper, J. E., Bjourson, A.J. and Thompson, J.K., 1987. Identification of Lotus rhizobia by direct DNA hybridization of crushed root nodules. Appl. Environ. Microbiol., 53: 1705-1707.
Daniel, R.M., Smith, I.M., Phillip, J.A., Ratcliffe, H.D., Drozd, J.W. and Bull, A.T.,1980. Anaerobic growth and denitrification by Rhizobium japonicum and other rhizobia. J. Gen. Microbiol., 120: 517-521.
Daniel, R.M., Limmer, A.W., Steele, K.W. and Smith, I.M., 1982. Anaerobic growth, nitrate reduction and denitrification in 46 Rhizobium strains. J. Gen. Microbiol., 128: 1811-1815.
Date, R.A. and Halliday, J., 1979. Selecting Rhizobium for acid, infertile soils of the tropics. Nature, 277:62.
Demezas, D.H. and Bottomley, P.J., 1986a. Autoecology in rhizospheres and nodulating behaviour of indigenous Rhizobium trifolii. Appl. Environ. Microbiol., 52:1014-1019.
Demezas, D.H. and Bottomley, P.J., 1986b. Interstrain competition between representatives of indigenous serotypes of Rhizobium trifolii. Appl. Environ. Microbiol., 52: 1020-1025.
Demezas, D.H. and Bottomley, P.J., 1987. Influence of soil and nonsoil environments on nodulation by Rhizobium trifolii. Appl. Environ. Microbiol., 53: 596-597.
Dhar, B., Singh, B.D., Singh, R.B., Srivastava, J.S. and Singh, R.M., 1980. Seasonal incidence of rhizobiophages in soils around Varanasi. Ind. J. Exp. Biol., 18: 1168-1170.
Dommergues, Y.R., Belser, L.W. and Schmidt, E.L., 1978. Limiting factors for microbial growth and activity in soil. Advances in Microbial Ecology, X: 49-104.
Dowdle, S.F. and Bohlool, B.B., 1985. Predominance of fast-growing Rhizobium japonicum in a soybean field in the People's Republic of China. Appl. Environ. Microbiol., 50: 1171-1176.
Dowling D.N. and Broughton, W.J., 1986. Competition for nodulation of legumes. Ann. Rev. Microbiol., 40: 131-137.
Dreyfus, B.L. and Dommergues, Y.R., 1981. Nodulation of Acacia species by fast and slow-growing tropical strains of Rhizobium. Appl. Environ. Microbiol., 41: 97-99.
Dreyfus, B., Garcia, J.L. and Gillis, M., 1988. Characterization of Azorhizobium caulinodans gen. nov., sp. nov., a stem-nodulating nitrogen-fixing bacterium isolated from Sesbania rostrata. Int. J. Syst. Bacteriol., 38: 89-98.
Dughri, M.H. and Bottomley, P.J., 1983. Complementary methodologies to delineate the composition of Rhizobium trifolii populations in root nodules. Soil Sci. Soc. Am. J., 47: 939-945.

Eardly, B.D., Materon, L.A., Smith, N.H., Johnson, D.A., Rumbaugh, M.D. and Selander, R.K., 1990. Genetic structure of natural populations of the nitrogen-fixing bacterium Rhizobium meliloti. Appl. Environ. Microbiol., 56: 187-194.
El Hassan, G.A., Zablowtowicz, R.M. and Focht, D.D., 1985. Kinetics of denitrifying growth by fast-growing cowpea rhizobia. Appl. Environ. Microbiol., 49: 517-521.
Elsheikh, E.A.E. and Wood, M., 1989. Response of chickpea and soybean rhizobia to salt: osmotic and specific ion effects of salts. Soil Biol. Biochem., 21: 889-895.
Faria, S.M. de, Lewis, G.P. Sprent, J.I. and Sutherland, J.M., 1989. Occurence of nodulation in the Leguminosae. New Phytol., 111: 607-619.
Finan, T.M., Hartwieg, E., LeMieux, K., Bergman, K., Walker, G.C. and Signer, E.R., 1984. General transduction in Rhizobium meliloti. J. Bacteriol., 159: 120-124.
Foster, R.C. and Rovira, A.D., 1978. The ultrastructure of the rhizosphere of Trifolium subterraneum L. In: M.W. Loutit and J.A.R. Miles (Editors), Microbial Ecology. Springer-Verlag, Berlin, Heidelberg, New York, pp. 278-290.
Foster, R.C., Rovira, A.D. and Cock, T.W., 1983. Ultrastructure of the Root-Soil Interface. American Phytopathological Society.
Fuhrmann, J. and Wollum, A.G., 1989. In vitro growth responses of Bradyrhizobium japonicum to soybean rhizosphere bacteria. Soil Biol. Biochem., 21: 131-135.
Fuhrmann, J., Davey, C.B. and Wollum, A.G., 1986. Desiccation tolerance of clover rhizobia in sterile soils. Soil Sci. Soc. Am. J., 50: 639-644.
Gajendiran, N. and Mahadevan, A., 1988. Utilization of catechin by Rhizobium sp. Plant Soil, 108: 263-266.
Gajendiran, N. and Mahadevan, A., 1990. Plasmid-borne catechol dissimilation in Rhizobium sp. FEMS Microbiol. Ecol., 73: 125-130.
Germida, J.J., 1988. Growth of indigenous Rhizobium leguminosarum and Rhizobium meliloti in soils amended with organic nutrients. Appl. Environ. Microbiol., 54: 257-263.
Germida, J.J. and Casida, L.E., 1983. Ensifer adhaerens predatory activity against other bacteria in soil, as monitored by indirect phage analysis. Appl. Environ. Microbiol., 45: 1380-1388.
Germida, J.J. and Khachatourian, G.G., 1988. Transduction of Escherichia coli in soil. Can. J. Microbiol., 34: 190-193.
Gitte, R.R., Rai, P.V. and Patil, R.B., 1978. Chemotaxis of Rhizobium spp. towards root exudate of Cicer arietinum L.. Plant Soil, 50: 553-566.
Glenn, A.R., Knuckey, R. and Dilworth, M.J., 1986. Periplasmic proteins of Rhizobium: variation with growth conditions and use in strain identification. FEMS Microbiol. Lett., 35: 65-69.
Gloux, K. and Le Rudulier, D., 1989. Transport and catabolism of proline betaine in salt-stressed Rhizobium meliloti. Arch. Microbiol., 151: 143-148.
Griffin, D.M. and Quail, G., 1968. Movement of bacteria in moist, particulate systems. Aust. J. Biol. Sci., 21: 583-585.
Habte, M. and Alexander, M., 1977. Further evidence for the regulation of bacterial populations in soil by protozoa. Arch. Microbiol., 113: 181-183.
Halliday, J., 1985. Biological nitrogen fixation in tropical agriculture. In: H.J. Evans, P.J. Bottomley and W.E. Newton (Editors), Nitrogen Fixation Research Progress. Martinus Nijhoff Publishers, Dordrecht, Boston, Lancaster, pp. 675-682.
Hamdi, Y.A., 1971., Soil-water tension and the movement of rhizobia. Soil Biol. Biochem., 3: 121-126.
Harrison, S.P., Jones, D.G. and Young, J.P.W., 1989a. Rhizobium population genetics: genetic variation within and between populations from diverse locations. J. Gen. Microbiol., 135: 1061-1069.

Harrison, S.P., Young, J.P.W. and Jones, D.G., 1989b. Rhizobium population genetics: host preference and strain competition effects on the range of Rhizobium leguminosarum biovar trifolii genotypes isolated from natural populations. Soil Biol. Biochem., 21: 981-986.
Hartel, P.G. and Alexander, M., 1984. Temperature and desiccation tolerance of cowpea rhizobia. Can. J. Microbiol., 30: 820-823.
Heal, O.W. and Ineson, P., 1984. Carbon and energy flow in terrestrial ecosystems: relevance to microflora. In: M.J. Klug and C.A. Reddy (Editors), Current Perspectives in Microbial Ecology. American Society for Microbiology, Washington D.C., pp. 394-404.
Hernandez, B.S. and Focht, D.D., 1984. Invalidity of the concept of slow growth and alkali production in cowpea rhizobia. Appl. Environ. Microbiol., 48: 206-210.
Heynen, C.E., Van Elsas, J.D., Kuikman, P.J. and Van Veen, J.A., 1988. Dynamics of Rhizobium leguminosarum biovar trifolii introduced into soil; the effect of bentonite clay on predation by protozoa. Soil Biol. Biochem., 20: 483-488.
Hicks, P.M. and Loynachan, T.E., 1989. Bacteria of the soybean rhizosphere and their effect on growth of Bradyrhizobium japonicum. Soil Biol. Biochem., 21: 561-566.
Hodgson, A.L.M. and Roberts, W.P., 1983. DNA colony hybridization to identify Rhizobium strains. J. Gen. Microbiol., 129: 207-212.
Hodgson, A.L.M., Roberts, W.P. and Waid, J.S., 1985. Regulated nodulation of Trifolium subterraneum inoculated with bacteriocin-producing strains of Rhizobium trifolii. Soil Biol. Biochem., 17: 475-478.
Holben, W.E., Jansson, J.K., Chelm, B.K. and Tiedje, J.M., 1988. DNA probes for the detection of specific microorganisms in the soil bacterial community. Appl. Environ. Microbiol., 54: 703-711.
Hollis, A.B., Kloos, W.E. and Elkan, G.H., 1981. DNA:DNA hybridization studies of Rhizobium japonicum and related Rhizobiaceae. J. Gen. Microbiol., 123: 215-222.
Hooykaas, P.J.J., van Brussel, A.A.N., den Dulk-Ras, H., van Slogteren, G.M.S. and Schilperoort, R.A., 1981. Sym plasmid of Rhizobium trifolii expressed in different rhizobial species and Agrobacterium tumefaciens. Nature, 291: 351-353.
Howieson, J.G., Ewing, M.A. and D'Antuono, M.F., 1988. Selection for acid tolerance in Rhizobium meliloti. Plant Soil, 105: 179-188.
Hua, S.T., Tsai, V.Y., Lichens, G.M. and Noma, A.T., 1982. Accumulation of amino acids in Rhizobium sp. strain WR1001 in response to sodium chloride salinity. Appl. Environ. Microbiol., 44: 135-140.
Hunter, W.J. and Fahring, C.J., 1980. Movement by Rhizobium and nodulation of legumes. Soil Biol. Biochem., 12: 537-542.
Imhoff, J.F., 1986. Osmoregulation and compatible solutes in eubacteria. FEMS Microbiol. Rev., 39: 57-66.
Jarvis, S.C. and Robson, A.D., 1983. A comparison of the cation/anion balance of ten cultivars of Trifolium subterraneum L., and their effects on soil acidity. Plant Soil, 75: 235-243.
Jarvis, B.D.W., Gillis, M. and De Ley, J., 1986. Intra- and intergeneric similarities between the ribosomal ribonucleic acid cistrons of Rhizobium and Bradyrhizobium species and some related bacteria. Int. J. Syst. Bacteriol., 36: 129-138.
Jarvis, B.D.W., Ward, L.J.H. and Slade, E.A., 1989. Expression by soil bacteria of nodulation genes from Rhizobium leguminosarum biovar trifolii. Appl. Environ. Microbiol., 55: 1426-1434.

Jenjareotham, R. and Barnet, Y.M., 1986. Ecotypic adaptation of root-nodule bacteria isolated from acacias in New South Wales. In: Proceedings of the Eighth Australian Nitrogen Fixation Conference. (A.I.A.S. Occasional Publication No. 25), Australian Institute of Agricultural Science, Melbourne, pp. 103-104.

Jenkins, M.B., Virginia, R.A. and Jarrell, W.M., 1987. Rhizobial ecology of the woody legume mesquite (Prosopis glandulosa) in the Sonoran desert. Appl. Environ. Microbiol., 53: 36-40.

Jenkins, M.B., Virginia, R.A. and Jarrell, W. M., 1988. Rhizobial ecology of the woody legume Psorothamnus spinosus in a Sonoran desert arroyo. Plant Soil, 105: 113-120.

Jordan, D.C., 1984. Rhizobiaceae Conn 1938 321. In: N.R. Krieg (Editor), Bergey's Manual of Systematic Bacteriology., 1. Williams and Wilkins. Baltimore and London, pp. 234-244.

Kamicker , B.J. and Brill, W.J., 1986. Identification of Bradyrhizobium japonicum nodule isolates from Wisconsin soy-bean farms. Appl. Environ. Microbiol., 51: 487-492.

Kamicker, B.J. and Brill, W.J., 1987. Methods to alter the recovery and nodule location of Bradyrhizobium japonicum inoculant strains on field-grown soybeans. Appl. Environ. Microbiol., 53: 1737-1742.

Karanja, N.K. and Wood, M., 1988a. Selecting Rhizobium phaseoli strains for use with beans (Phaseolus vulgaris L.) in Kenya: infectiveness and tolerance of acidity and aluminium. Plant Soil, 112: 7-13.

Karanja, N.K. and Wood, M., 1988b. Selecting Rhizobium phaseoli strains for use with beans (Phaseolus vulgaris L.) in Kenya: tolerance of high temperature and antibiotic resistance. Plant Soil, 112: 15-22.

Kingsley, M.T. and Bohlool, B.B., 1983. Characterization of Rhizobium sp. (Cicer arietinum L.) by immunofluorescence, immunodiffusion and intrinsic antibiotic resistance. Can. J. Microbiol., 29: 518-526.

Kishinevsky, B. and Gurfel, D., 1980. Evaluation of enzyme-linked immunosorbent assay (ELISA) for serological identification of different Rhizobium strains. J. Appl. Bacteriol., 49: 517-526.

Kiss, G.B. and Kalman, Z., 1982. Transformation of Rhizobium meliloti with plasmid DNA. J. Bacteriol., 150: 465-470.

Kucey, R.M.N. and Hynes, M.F., 1989. Populations of Rhizobium leguminosarum biovars phaseoli and viciae in fields after bean or pea rotation with nonlegumes. Can. J. Microbiol., 35: 661-667.

Kuykendall, L.D., Devine, T.E. and Cregan, P.B., 1982. Positive role of nodulation on the establishment of Rhizobium japonicum in subsequent crops of soybean. Current Microbiol., 7: 79-81.

Kuykendall, L.D., Roy, M.A., O'Neill, J.J. and Devine, T.E., 1988. Fatty acids, antibiotic resistance, and deoxyribonucleic acid homology groups of Bradyrhizobium japonicum. Int. J. Syst. Bacteriol., 38: 358-361.

Kvien, C.S. and Ham, G.E., 1985. Effect of soil temperature and inoculum rate on the recovery of three introduced strains of Rhizobium japonicum. Agron. J., 77: 484-489.

Lambert, G.R., Harker, A.R., Zuber, D.A., Hanus, F.J., Russell, S.A. and Evans, H.J., 1985. Characterization, significance and transfer of hydrogenase genes from Rhizobium japonicum. In: H.J. Evans, P.J. Bottomley and W.E. Newton (Editors), Nitrogen Fixation Research. Martinus Nijhoff, Dordrecht, pp. 209-215.

Lawrie, A .C., 1983. Relationships among rhizobia from native Australian legumes. Appl. Environ. Microbiol., 45: 1822-1828.

Lawson, K.A., 1987. Environmental factors influencing growth and interaction of Rhizobium leguminosarum biovar trifolii and its phages in soil. Ph.D. Thesis, University of New South Wales.

Lawson, K.A. and Barnet, Y.M., 1984. The effect of soil water matric potential on growth of Rhizobium trifolii and its phage. In: Proceedings of the Seventh Australian Legume Nodulation Conference. A.I.A.S. Occasional Publication No., 12, Australian Institute for Agricultural Science, Melbourne, pp. 37-38.
Lawson, K.A., Barnet, Y.M. and McGilchrist, C.A., 1987. Environmental factors influencing numbers of Rhizobium leguminosarum biovar trifolii and its bacteriophages in two field soils. Appl. Environ. Microbiol., 53: 1125-1131.
Lepo, J.E., Hanus, F.J. and Evans, H.J., 1980. Chemoautotrophic growth of hydrogen-uptake-positive strains of Rhizobium japonicum. J. Bacteriol., 141: 664-670.
Lesley, S.M., 1982. A bacteriophage typing system for Rhizobium meliloti. Can. J. Microbiol., 28: 180-189.
Li, D. and Alexander, M., 1986. Bacterial growth rates and competition affect nodulation and root colonization by Rhizobium meliloti. Appl. Environ. Microbiol., 52: 807-811.
Lie, T.A., 1981. Gene centres, a source for genetic variants in symbiotic nitrogen fixation: host induced ineffectivity in Pisum sativum ecotype fulvum. Plant Soil, 61: 125-134.
Lie, T.A. and Goktan, D., 1984. Gene centres, a source for genetic variants in symbiotic nitrogen fixation : the symbiotic response of the cultivated pea to Rhizobium leguminosarum strains from Europe and the Middle East. Plant Soil, 82:359-369.
Lie, T.A., Goktan, M., Pijnenborg, J. and Anlarsal, E., 1987. Co-evolution of the legume-Rhizobium association. Plant Soil, 100: 171-181.
Lindström, K. and Lehtomäki, S., 1988. Metabolic properties, maximum growth temperature and phage sensitivity of Rhizobium sp. (Galega) compared with other fast-growing rhizobia. FEMS Microbiol. Lett., 50: 277-287.
Lindström, K., Jarvis, B.D.W., Lindström, P.E. and Patel, J.J., 1983. DNA homology, phage-typing, and cross-nodulation studies of rhizobia infecting Galega species. Can. J. Microbiol., 29: 781-789.
Liu, R., Tran, V.M. and Schmidt, E.L., 1989. Nodulating competitiveness of a nonmotile Tn7 mutant of Bradyrhizobium japonicum in nonsterile soil. Appl. Environ. Microbiol., 55: 1895-1900.
Lochner, H.H., Strijdom, B.W. and Law, I.J., 1989. Unaltered nodulation competitiveness of a strain of Bradyrhizobium sp. (Lotus) after a decade in soil. Appl. Environ. Microbiol., 55: 3000-3008.
Lowendorf, H.S., 1980. Factors affecting the survival of Rhizobium in soil. Advances in Microbial Ecology, 4: 87-124.
McDermott, T.R. and Graham, P.H., 1989. Bradyrhizobium japonicum inoculant mobility, nodule occupancy, and acetylene reduction in the soybean root system. Appl. Environ. Microbiol., 55: 2493-2498.
Madsen, E.L. and Alexander, M., 1982. Transport of Rhizobium and Pseudomonas through soil. Soil Sci. Soc. Am. J., 46: 557-560.
Mahler, R.L. and Bezdicek, D.F., 1980. Serogroup distribution of Rhizobium leguminosarum in peas in the Palouse of eastern Washington. Soil Sci. Soc. Am. J., 44: 292-295.
Mahler, R.L. and Wollum, A.G., 1981a. The influence of soil water potential and soil texture on the survival of Rhizobium japonicum and Rhizobium leguminosarum isolates in the soil. Soil Sci. Soc. Am. J., 45: 761-766.
Mahler, R.L. and Wollum, A.G., 1981b. Seasonal variation of Rhizobium trifolii in clover pastures and cultivated fields in North Carolina. Soil Sci., 132: 240-246.
Mahler, R.L. and Wollum, A.G., 1982a. Seasonal fluctuation of Rhizobium japonicum under a variety of field conditions in North Carolina. Soil Sci., 134: 317-324.

Mahler, R.L. and Wollum, A.G., 1982b. Seasonal variation of Rhizobium meliloti in alfalfa hay and cultivated fields in North Carolina. Agron. J., 74: 428-431.
Makker, N.S. and Casida, L.E., 1987. Technique for estimating low numbers of a bacterial strain(s) in soil. Appl. Environ. Microbiol., 53: 887-888.
Mary, P., Ochin, D. and Tailliez, R., 1986. Growth status of rhizobia in relation to their tolerance to low water activities and desiccation stress. Soil Biol. Biochem., 18: 179-184.
Mellor, H.Y., Glenn, A.R., Arwas, R. and Dilworth, M.J., 1987. Symbiotic and competitive properties of motility mutants of Rhizobium trifolii TA1. Arch. Microbiol., 148: 34-39.
Moawad, H.A., Ellis, W.R. and Schmidt, E.L., 1984. Rhizosphere response as a factor in competition among three serogroups of indigenous Rhizobium japonicum for nodulation of field-grown soybeans. Appl. Environ. Microbiol., 47: 607-612.
Morley, S.J. and Jones, D.G., 1980. A note on a highly sensitive modified ELISA (enzyme-linked immunosorbent assay) technique for Rhizobium strain identification. J. Appl. Bacteriol., 49: 103-109.
Munevar, F. and Wollum, A.G., 1981. Growth of Rhizobium japonicum strains at temperature above 27C. Appl. Environ. Microbiol., 42: 272-
Neal, J.L., Allan, G.C., Morse, R.D. and Wolf, D.D., 1983. Anaerobic nitrate-dependent chemolithotrophic growth by Rhizobium japonicum. Can. J. Microbiol., 29: 316-320.
Newman, E.I. and Watson, A., 1977. Microbial abundance in the rhizosphere: a computer model. Plant Soil, 48: 17-56.
Noel, D.K. and Brill, W.J., 1980. Diversity and dynamics of indigenous Rhizobium japonicum populations. Appl. Environ. Microbiol. 40: 931-938.
Norris, D.O., 1965. Acid production by Rhizobium, a unifying concept. Plant Soil, 22: 143-166.
Nutman, P.S., 1969. Symbiotic nitrogen fixation: legume nodule bacteria. In: Rothamsted Report for 1968, Part 2. pp. 179-181.
Nutman, P.S. and Ross, G.J.S., 1970. Rhizobium in the soils of the Rothamsted and Woburn farms. In: Rothamsted Report for 1969, Part 2. pp. 148-167.
Nye, P.H., 1981. Changes of pH across the rhizosphere induced by roots. Plant Soil, 61: 7-26.
Ozawa, T. and Yamuguchi, M., 1986. Fractionation and estimation of particle-attached and unattached Bradyrhizobium japonicum strains in soils. Appl. Environ. Microbiol., 52: 911-914.
Padan, E., 1984. Adaptation of bacteria to external pH. In: M.J. Klug and C.A. Reddy (Editors), Current Perspectives in Microbial Ecology. American Society for Microbiology, Washington, pp. 49-55.
Parke, D. and Ornston, L.N., 1984. Nutritional diversity of Rhizobiaceae revealed by auxanography. J. Gen. Microbiol., 130: 1743-1750.
Parker, C.A., Trinick, M.J. and Chatel, D.L., 1977. Rhizobia as soil and rhizosphere inhabitants. In: R.W.F. Hardy and A.H. Gibson (Editors), A Treatise on Dinitrogen Fixation IV. Agronomy. John Wiley and Sons, New York, pp. 311-352.
Patel, J.J., 1976. Morphology and host range of virulent phages of Lotus rhizobia. Can. J. Microbiol., 22: 204-212.
Pena-Cabriales, J.J. and Alexander, M., 1983. Growth of Rhizobium in soil amended with organic matter. Soil Sci. Soc. Am. J., 47: 241-245.
Pilacinski, W.P. and Schmidt, E.L., 1981. Plasmid transfer within and between serologically distinct strains of Rhizobium japonicum, using antibiotic resistance mutants and auxotrophs. J. Bacteriol., 145: 1025-1030.
Pinero, D., Martinez, E. and Selander, R.K., 1988. Genetic diversity and relationships among isolates of Rhizobium leguminosarum biovar phaseoli. Appl. Environ. Microbiol., 54: 2825-2832.

Pugashetti, B.K., Angle, J.S. and Wagner, G.H., 1982. Soil microorganisms antagonistic towards Rhizobium japonicum. Soil Biol. Biochem., 14: 45-49.
Rice, W.A., Olsen, P.E. and Page, W.J., 1984. ELISA evaluation of the competitive abilities of two Rhizobium meliloti strains. Can. J. Microbiol., 30: 1187-1190.
Richardson, A.E. and Simpson, R.J., 1988. Acid-tolerance and symbiotic effectiveness of Rhizobium trifolii associated with a Trifolium subterraneum L.-based pasture growing in an acid soil. Soil Biol. Biochem., 21: 87-95.
Richaume, A., Angle, J.S. and Sadowsky, M.J., 1989. Influence of soil variables on in situ plasmid transfer from Escherichia coli to Rhizobium fredii. Appl. Environ. Microbiol., 55: 1730-1734.
Robert, F.M. and Schmidt, E.L., 1983. Population changes and persistences of Rhizobium phaseoli in soil and rhizospheres. Appl. Environ. Microbiol., 45: 550-556.
Robson, A.D. and Loneragan, J.F., 1970. Nodulation and growth of Medicago truncatula on acid soils. II. Colonization of acid soils by Rhizobium meliloti. Aust. J. Agric. Res., 21: 435-445.
Rovira, A.D. and Campbell, R., 1974. Scanning electron microscopy of microorganisms on the roots of wheat. Microbial Ecol., 1: 15-23.
Rupela, O.P., Toomsan, B., Mittal, S., Dart, P.J. and Thompson, J.A., 1987. Chickpea Rhizobium populations: survey of influence of season, soil depth and cropping pattern. Soil Biol. Biochem., 19: 247-252.
Schmidt, E.L., 1978. Legume symbiosis. A. Ecology of the legume root-nodule bacteria . In: Y.R. Dommergues and S.V. Krupa (Editors), Interactions Between Non-pathogenic Soil Microorganisms and Plants 4. Elsevier Scientific Publishing Company, Amsterdam, New York, Oxford, pp. 269-303.
Schmidt, E.L., 1979. Initiation of plant root-microbe interactions. Ann. Rev. Microbiol., 33: 355-376.
Schmidt, E.L. and Robert, F.M., 1985. Recent advances in the ecology of Rhizobium. In: H.J. Evans, P.J. Bottomley and W.E. Newton (Editors), Nitrogen Fixation Research Progress. Martinus Nijhoff, Dordrecht, pp. 379-385.
Schmidt, E.L., Zidwick, M.J. and Abebe, H.M., 1986. Bradyrhizobium japonicum serocluster 123 and diversity among member isolates. App. Environ. Microbiol., 51: 1212-1215.
Schofield, P.R., Gibson, A.H., Dudman, W.F. and Watson. J.M., 1987. Evidence for genetic exchange and recombination of Rhizobium symbiotic plasmids in a soil population. Appl. Environ. Microbiol., 53: 2942-2947.
Shoustari, N.H. and Pepper, I.L., 1985. Mesquite rhizobia isolated from the sonoran desert: competitiveness and survival in soil. Soil Biol. Biochem., 17: 803-806.
Sinclair, J.L. and Alexander, M., 1989. Effect of protozoan predation on relative abundance of fast- and slow-growing bacteria. Can. J. Microbiol., 35: 578-582.
Singleton, P.W., El Swaify, S.A. and Bohlool, B.B., 1982. Effect of salinity on Rhizobium growth and survival. Appl. Environ. Microbiol., 44: 884-890.
Slattery, J.F. and Coventry, D.R., 1989. Populations of Rhizobium lupini in soils used for cereal-lupin rotations in north-east Victoria. Soil Biol. Biochem., 21: 1009-1010.
Smith, L. T., Pocard, J., Bernard, T. and Le Rudelier, D., 1988. Osmotic control of glycine betaine biosynthesis and degradation in Rhizobium meliloti. J. Bacteriol., 170: 3142-3149.
Soberón-Chávez, G. and Nájera, R., 1989. Isolation from soil of Rhizobium leguminosarum lacking symbiotic information. Can. J. Microbiol., 35: 464-468.
Soby, S. and Bergman, K., 1983. Motility and chemotaxis of Rhizobium meliloti in soil. Appl. Environ. Microbiol., 46: 995-998.

Staniewski, R., 1970. Typing of Rhizobium by phages. Can. J. Microbiol., 16: 1003-1009.
Stanley, J., Brown, G.G. and Verma, D.P.S., 1985. Slow-growing Rhizobium japonicum comprise two highly divergent symbiotic types. J. Bacteriol., 163: 148-154.
Stowers, M.D., 1985. Carbon metabolism in Rhizobium species. Ann. Rev. Microbiol., 39: 89-108.
Stowers, M.D. and Eaglesham, A.R.J., 1983. A stem-nodulating Rhizobium with physiological characteristics of both fast and slow growers. J. Gen. Microbiol., 129: 3651-3656.
Stowers, M.D. and Elkan, G.H., 1984. Growth and nutritional characteristics of cowpea rhizobia. Plant Soil, 80: 191-200.
Thurman, N.P. and Bromfield, E.S.P., 1988. Effect of variation within and between Medicago and Melilotus species on the composition and dynamics of indigenous populations of Rhizobium meliloti. Soil Biol. Biochem., 20: 31-38.
Trinick, M.J., 1980. Relationships amongst the fast-growing rhizobia of Lablab purpureus, Leucaena leucocephala, Mimosa spp., Acacia farnesiana and Sesbania grandiflora and their affinities with other rhizobial groups. J. Appl. Bacteriol., 49: 39-53.
Trinick, M.J., Parker, C.A. and Palmer, M.J., 1983. Interactions of the microflora from nodulation problem and non-problem soils towards Rhizobium spp. on agar. Soil Biol. Biochem., 15: 295-301.
Turco, R.F. and Bezdicek, D.F., 1987. Diversity within two serogroups of Rhizobium leguminosarum native to soils in the Palouse of eastern Washington. Ann. Appl. Biol., 111:103-114.
Valdivia, B., Dughri, M.H. and Bottomley, P.J., 1988. Antigenic and symbiotic characterization of indigenous Rhizobium leguminosarum bv. trifolii recovered from root nodules of Trifolium pratense L. sown into subterranean clover pasture soils. Soil Biol. Biochem., 20: 267-274.
Van Rensberg, H., and Strijdom, B.W., 1985. Effectiveness of Rhizobium strains used in inoculants after their introduction into soil. Appl. Environ. Microbiol., 49: 127-131.
Vargas, R. and Hattori, T., 1986. Protozoan predation of bacterial cells in soil aggregates. FEMS Microbiol. Ecol., 38: 233-242.
Vincent, J.M., 1970. A Manual for the Practical Study of Root-nodule Bacteria. International Biological Programme Handbook No., 15. Blackwell Scientific Publications, Oxford and Edinburgh.
Viteri, S.E. and Schmidt, E.L., 1987. Ecology of indigenous soil rhizobia: response of Bradyrhizobium japonicum to readily available substrates. Appl. Environ. Microbiol., 53: 1872-1875.
Viteri, S.E. and Schmidt, E.L., 1989. Chemoautotrophy as a strategy in the ecology of indigenous soil bradyrhizobia. Soil Biol. Biochem. 21: 461-463.
Waldon, H.B., Jenkins, M.B., Virginia, R.A. and Harding E.E., 1989. Characteristics of woodland rhizobial populations from surface- and deep-soil environments of the Sonoran desert. Appl. Environ. Microbiol., 55: 3058-3064.
Wedlock, D.N. and Jarvis, B.D.W., 1986. DNA homologies between Rhizobium fredii, rhizobia that nodulate Galega sp., and other Rhizobium and Bradyrhizobium species. Int. J. Syst. Bacteriol., 36: 550-558.
Whipps, J.M. and Lynch, J.M., 1986. The influence of the rhizosphere on crop productivity. Advances in Microbial Ecology, 9: 187-244.
Wood, M. and Cooper, J.E., 1984. Aluminium toxicity and multiplication of Rhizobium trifolii in a defined growth medium. Soil Biol. Biochem., 16: 571-576.
Wood, M. and Cooper, J.E., 1988a. Acidity, aluminium and multiplication of Rhizobium trifolii: effects of initial inoculum density and growth phase. Soil Biol. Biochem., 20: 83-87.

Wood, M. and Cooper, J.E., 1988b. Acidity, aluminium and multiplication of Rhizobium trifolii: possible mechanisms of aluminium toxicity. Soil Biol. Biochem., 20: 95-99.

Wood, M. and Shepherd, G., 1987. Characterization of Rhizobium trifolii isolated from soils of different pH. Soil Biol. Biochem., 19: 317-321.

Wood, M., Cooper, J.E. and Bjourson, A.J., 1988. Response of Lotus rhizobia to acidity and aluminium in liquid culture and in soil. Plant Soil, 107: 227-231.

Woomer, P., Singleton, P.W. and Bohlool, B.B., 1988. Ecological indicators of native rhizobia in tropical soils. Appl. Environ. Microbiol., 54: 1112-1116.

Yap, S.F. and Lim, S.T., 1983. Response of Rhizobium sp. UMKL 20 to sodium chloride stress. Arch. Microbiol., 135: 224-228.

Yelton, M.M., Yang, S.S., Edie, S.A. and Lim, S.T., 1983. Characterization of an effective salt-tolerant, fast-growing strain of Rhizobium japonicum. J. Gen. Microbiol., 129: 1537-1547.

Young, J.P.W., 1985. Rhizobium population genetics: enzyme polymorphism in isolates from peas, clover, beans and lucerne grown at the same site. J. Gen. Microbiol., 131: 2399-2408.

Young, J.P.W., 1989. The population genetics of bacteria. In: D.A. Hopwood and K.F. Chater (Editors), Genetics of Bacterial Diversity. Academic Press, London, San Diego, New York, Berkeley, Boston, Sydney, Tokyo, Toronto, pp. 417-438.

Young, J.P.W. and Wexler, M., 1988. Sym plasmid and chromosomal genotypes are correlated in field populations of Rhizobium leguminosarum. J. Gen. Microbiol., 134: 2731-2739.

Young, J.P.W., Demetriou, L. and Apte, R.G., 1987. Rhizobium population genetics: enzyme polymorphism in Rhizobium leguminosarum from plants and soil in a pea crop. Appl. Environ. Microbiol., 53: 397-402.

Yousef, A.N., Al-Nassiri, A.S., Al-Azawi, S.K. and Abdul-Hussain, A., 1987. Abundance of peanut Rhizobium as affected by environmental conditions in Iraq. Soil Biol. Biochem., 19: 391-396.

Zeph, L.R. and Casida, L.E., 1986. Gram-negative versus gram-positive (actinomycete) nonobligate bacterial predators of bacteria in soil. Appl. Environy. Microbiol., 52: 819-823.

CHAPTER 11

RECOGNITION AND INFECTION IN LEGUME NODULATION

B.W.S. SOBRAL[1,2], R.J. HONEYCUTT[1], A.G. ATHERLY[1] and K.D. NOEL[3]
[1]Department of Genetics, Iowa State University, Ames, 50011;
[2]present address: California Institute of Biological Research, 11099 N. Torrey Pines Rd., La Jolla, CA 92037.

[3]Department of Biology, Marquette University, Milwaukee, WI 53233

1 INTRODUCTION

This chapter concerns the molecular biology of bacterial nodulation (nod) and polysaccharide biosynthesis (exo, lps, ndv) genes involved in the establishment of nitrogen-fixing symbioses between rhizobia and legumes. The root nodules which develop are specialized organs in which rhizobia reduce dinitrogen to ammonia (Chapter 13), in exchange for carbon sources derived from photosynthate (Chapter 12). Such metabolic cooperation requires an exchange of appropriate molecular signals to coordinate recognition and developmental processes. Any attempts at genetic engineering of these economically and ecologically important nitrogen-fixing symbioses will require an understanding of recognition, infection and development.

Following a brief account of the steps of nodule development, attention is focused on the rhizobial genes thus far identified as being involved in the recognition and infection process. Some recent reviews of interest include Debelle et al., 1986b; Rolfe et al., 1986; Djordjevic et al., 1987b; Kondorosi et al., 1988; Long and Cooper, 1988; and Long, 1989.

2 A BRIEF SUMMARY OF NODULE FORMATION

The formation of nodules occurs via a complex series of steps (Newcomb, 1981). Although rhizobia are chemotactic toward plant roots (Currier and Strobel, 1977; Bergman et al., 1988), nonmotile mutants are still able to form nodules (Ames and Bergman, 1981; Mellor et al., 1987), indicating that motility is not required for nodulation, although it may give a competitive advantage. Once near the root surface, rhizobia cause alterations in the growth of epidermal hairs on the root, leading to root hair curling (Fahraeus, 1957; Nutman, 1959; Yao and Vincent, 1969). Attachment of rhizobia to the root hairs may be an essential step early in the recognition process and may be mediated by plant lectins (Dazzo and Gardiol, 1984). As root hair curling occurs, so does initiation of cell division in the root cortex (Newcomb, 1981; Dudley et al., 1987), progressive divisions leading to nodule formation. Induction of cortical cell division therefore occurs before infection of the outer plant cells, which is the next step, at least in Medicago sativa (alfalfa; Dudley et al., 1987) and Glycine max (soybean; Calvert et al., 1984). The infected plant cells are stimulated to produce cell wall sheaths, known as "infection threads" (Callaham and Torrey, 1981), which continue to penetrate the plant cells, the bacteria being released into the cytoplasm of individual target cells. The bacteria are enveloped by a plasma membrane of plant origin, called the "peribacteroid membrane" (Robertson et al., 1978; Newcomb, 1981; Chapter 12). At least four different rhizobial signals leading to peribacteroid membrane formation by the plant have been suggested (Mellor et al., 1989). The first signal would occur when colonization begins, and the other signals would require the presence of the bacteroids inside the plant cells.

Next, the intracellular bacteria differentiate (see Verma and Long, 1983) and are then known as "bacteroids". The control of this process is not understood, although large genetic rearrangements are probably not involved (Sobral, 1989). More likely candidates for regulation of bacteroid differentiation are protein phosphorylation (Karr and Emerich, 1989), K^+ ion concentration (Gober and Kashket, 1987) and low oxygen tension (Gober and Kashket, 1987, 1989). Once differentiation of both partners is complete, symbiotic nitrogen fixation and its attendant metabolite exchanges can begin (Chapters 12 and 13). Table 1 summarizes the phenotypic codes for these developmental stages (Vincent, 1980).

TABLE 1. Developmental stages in nodule formation

Stage	Phenotypic Code
Multiplication in root area, colonization	Roc
Attachment to root	Roa
Root hair curling	Hac
Infection thread formation	Inf
Nodule initiation and development	Noi
Bacterial release	Bar
Bacteroid differentiation	Bad
Nitrogen fixation	Nif
Nodule function and persistence	Nop

After Vincent (1980) and Long (1984)

3 SPECIFICITY IN THE Rhizobium-LEGUME INTERACTION

Genetic and molecular techniques have been applied extensively in studies of the rhizobial genes involved in all steps of the interaction with legumes (see Dénarié et al., 1981; Kondorosi and Johnston, 1981; Long, 1984; Fisher et al., 1988). Most fast-growing rhizobia contain megaplasmids which code for many of the symbiotically important genes and are usually referred to as pSym (see Prakash and Atherly, 1986). In slow-growing Bradyrhizobium, large plasmids are not generally observed (though Masterson et al. (1982) have reported them in B. japonicum) and symbiotically important genes are located on the chromosome. The main genes involved in symbiosis are nodulation genes (nod), polysaccharide biosynthesis genes (exo and lps), nodule development genes (ndv), and nitrogen fixation genes (nif and fix). Nod genes have generally been identified by transposon mutagenesis or sequencing and are defined as those genes that affect the ability of the bacteria to cause normal nodulation on the original host (Table 2), and/or cause root hair deformation, and possibly curling, on atypical hosts (Long, 1989). Exo genes are involved in the synthesis of extracellular or capsular polysaccharides (Leigh et al., 1985, 1987). Exo⁻ mutants typically cannot invade the plant in a normal manner, although cortical cell divisions and root hair curling are induced. Ndv genes are required for the production and secretion of neutral ß-1,2-cyclic glucan (Dylan et al., 1986; Stanfield et al., 1988). Nif genes (Chapter 6) are defined as those required for nitrogen fixation that have homologues in free-living diazotrophs like Klebsiella pneumoniae, while fix genes are those required for nitrogen fixation but not present in K.

TABLE 2. Root nodule bacteria-legume specificity

Bacterial genus/species	Host
Azorhizobium sesbaniae	Sesbania rostrata
Bradyrhizobium spp	Glycine, Vigna
Bradyrhizobium japonicum	Glycine
Rhizobium meliloti	Medicago sativa, Melilotus, Trigonella
Rhizobium leguminosarum	
bv. viciae	Lathyrus, Lens, Pisum sativum, Vicia
bv. trifolii	Trifolium
bv. phaseoli	Phaseolus
Rhizobium loti	Lotus
Rhizobium NGR234	Leucaena, Parasponia
Sinorhizobium fredii[a]	Glycine

[a] Chen et al., 1988b

pneumoniae. Mutations in nif and fix genes do not affect the nodulation process but produce ineffective nodules (Nod^+Fix^-).

Plant genes which are selectively expressed during symbiosis have also been identified and their products termed nodulins (Gloudemans et al., 1988). The function of only a few nodulins has so far been determined (leghaemoglobin, uricase II, glutamine synthetase, choline kinase, and sucrose synthase) and many of the nodulins with unknown function are localized in the peribacteroid membrane (Brewin et al., 1988; Werner et al., 1988).

Specificity is evident in the Rhizobium-legume interaction from the initial stages of root hair curling and the induction of plant cortical cell division. To understand the nature and specificity of the signalling process between legumes and rhizobia, attention has been focused on the role of nod and polysaccharide biosynthesis genes.

4 THE NODULATION (nod) GENES OF RHIZOBIA

Nod genes seem to be involved in the sending and receiving of signals between bacterium and plant partner and are best known from R. meliloti and R. leguminosarum bvs. viciae and trifolii. Nod genes are generally classified in at least two broad groups; characteristics of these genes and their products are summarized in Table 3. One group, known as the "common" nod genes, are functionally interchangeable among all rhizobia (Kondorosi et al., 1984; Egelhoff and Long, 1985; Fisher et al., 1985; Long et al., 1986; Hong et al., 1987a; Marvel et al., 1987; Yelton et al., 1987). The common nod genes, nodDABC, are highly homologous and can functionally complement mutants in any species of rhizobia (Djordjevic et al., 1987b). Since there is a level of specificity determined by nodD, it could be classified as a regulatory gene and placed in a third category (Hong et al., 1987a, 1987b; Spaink et al., 1987b). Some authors consider nodIJ to be part of the common nod genes because nodIJ are not involved in host specificity and, at least in R. leguminosarum bv. viciae, are transcribed as the nodABCIJ operon (Wijffelman et al., 1986; Hong et al., 1987a, 1987b; Spaink et al., 1989a, 1989b; Winsor, 1989).

The second set of nod genes is known as the host-specific nod genes or hsn genes (Debelle and Sharma, 1986; Debelle et al., 1986b; Horvath et al., 1986; Rolfe et al., 1986; Fisher et al., 1987; Marvel et al., 1987; Rodriguez-Quinones et al., 1987; Swanson et al., 1987; Canter-Cremers et al., 1988; Debelle et al., 1988; Surin and Downie, 1988, 1989). Host-specific nod genes include nodFE, nodG, nodH, nodLMN, nodPQ, nodX, and very likely others. Figure 1 shows the general organization of nod genes in three rhizobial species.

TABLE 3. Characterized nod genes and gene products[a].

Gene	MW(KDa)	Localization	Homologies	Putative Role
nodA	21.8	cytoplasmic (interior membrane?)[b]	None	-synthesis of low MW compounds stimulating plant cell division
nodB	28.0	cytoplasmic	None	-stimulation and root hair curling
nodC	46.8	bacterial outer membrane	None	-cell surface receptor and signal transduction
nodD	33.0	bacterial cytoplasmic membrane	transcriptional activators (LysR, AraC, NahR)	-regulate nod gene autoregulation, interact with flavonoid
nodE	41.8	bacterial cytoplasmic membrane	None	-modify nodABC-derived signal
nodF	9.8	cytoplasmic?[b]	acyl carrier protein	-biosynthesis of LPS, fatty acids, or ß-1,2-glucan
nodG	26.1	?[c]	ribitol-, alcohol-, and glucose-dehydrogenases	?
nodH	28.6	?	None	-modification of alfalfa specific signal
nodI	34.3	bacterial membrane?[b]	ATP binding proteins (hisP, malK)	-active transport of low MW molecules
nodJ	27.6	bacterial membrane?[b]	None	-active transport of low MW molecules
nodL	20.1	?	None	?
nodM	65.8	?	amidophosphoribosyl transferases (purF)	-modification of LPS and EPS
nodN	18.0	?	None	?
nodO	50.0	extracellular	haemoylsin (hlyA)	?
nodP	34.7	?	None	?
nodQ	70.5	?	GDP binding protein (EF-Tu)	?
nodX	41.0	?	None	?

[a]Data compiled from Egelhoff and Long (1985), John et al., (1985,1988), Schmidt et al.(1986), Johnson et al. (1989), Djordjevic et al. (1987b), Schlaman et al. (1989), Henikoff et al. (1988), Shearman et al. (1986), Rossen et al. (1985), Fisher et al. (1987), Spaink et al. (1989a,b), Debelle and Sharma (1986), Faucher et al. (1988), Higgins et al. (1986), Evans and Downie (1986), Davis et al. (1988), Surin and Downie (1988), de Maagd et al. (1989c), Schwedock and Long (1989)
[b]Location inferred from gene product characteristics and/or sequence homologies
[c]Unknown

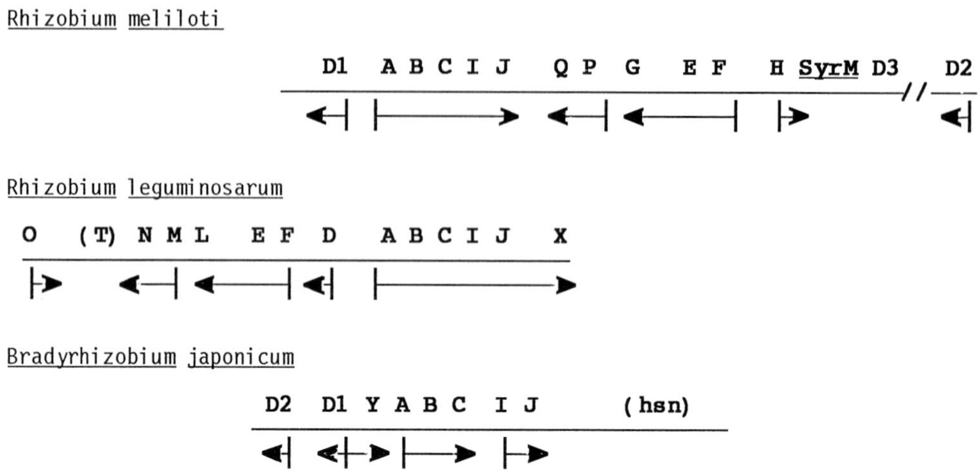

Fig. 1. Organization of *nod* genes in *R. meliloti*, *R. leguminosarum*, and *B. japonicum*. Arrows indicate direction of transcription of genes or operons, (hsn) represents relative location of genes for host-specifictiy in *B. japonicum*, and *SyrM* represents the symbiotic regulator M gene.

4.1 Function and regulation of the common *nod* genes

Mutations in the *nodDABC* genes of *R. leguminosarum* bv. *viciae* (Downie et al., 1985) and bv. *trifolii* (Djordjevic et al., 1987b) lead to the Hac^- Nod^- phenotype (Table 1) (see Long, 1989). These genes are also required for induction of plant cortical cell division (Dudley et al., 1987). In *R. meliloti*, *nodABC* mutants are also Hac^- Nod^-, but *nodD* mutants can still infect, albeit with differences in efficiency, because there are three copies (Fig. 1) of the *nodD* gene (Honma and Ausubel, 1987). Both *Sinorhizobium fredii* strain USDA 191 and *B. japonicum* strain USDA 123 have two copies of *nodD* (Prakash and Atherly, 1984; Kosslak et al., 1987; Appelbaum et al., 1988).

The fact that *nodABC* genes are required for root hair curling and stimulation of cortical cell division suggests a common antecedent for these events (Long, 1989). The action of the *nodABC* products is so central to the nodulation process that they are required (Marvel et al., 1987) for even the nodulation of *Parasponia*, a woody member of the elm family and the only non-legume genus whose members are known to form nodules with rhizobia. It has been suggested that rhizobia expressing *nodABC* export a factor(s) affecting root morphology and root hair growth (Canter-Cremers et al., 1986; van Brussel et al., 1986; Zaat et al., 1987). The *nodABC* products induce the expression of early nodulins such as ENOD2 (Govers et al., 1986). Plant growth substances have also been implicated in the early steps of nodulation. A *nodABC* mutant of *R. meliloti* overproducing zeatin (Long and Cooper, 1988) formed primitive nodule-like structures on M. sativa, although root hair curling was not observed, indicating that cytokinin may imitate a secondary effect of the common nod genes. Hirsch et al. (1989) showed that "anti-auxins" such as naphthylphthalamic acid also cause M. sativa to produce empty nodules which express early nodulins. These data do not define the roles of the nodABC genes, although they clearly must be involved in early signalling events. The development of Lotus as a model for legume genetic transformation by way of Ri plasmids and Agrobacterium rhizogenes (Stougaard et al., 1986; Petit et al., 1987), should allow the use of reporter gene fusions as an assay to purify bacterial signals to the plant (Long, 1989).

DNA sequences are known for the nodABC genes (Torok et al., 1984; Rossen et al., 1984), but no precise biochemical function has yet been assigned to any of these three genes on the basis of sequence homology to known proteins. The NodA (Egelhoff and Long, 1985) proteins are cytoplasmically located (Schmidt et al., 1986; John et al., 1988), although some NodA has also been observed in the membrane fractions (Johnson et al., 1989). NodA and NodB may direct the synthesis of a low molecular weight diffusible factor which stimulates cell division in plant protoplasts (John et al., 1988), although the signalling molecule(s) has not yet been identified. NodC is a hydrophobic, transmembrane protein that may be involved in transmembrane signalling (John et al., 1985, 1988; Djordjevic et al., 1987b), although it is probably produced on free cytoplasmic ribosomes before being assembled in the membrane (Johnson et al., 1989). Cross-linking studies imply that the NodC protein exists in the membrane as a dimer and has similarities to cell surface receptors (John et al., 1988). A truncated form of NodC present in nodules of many legumes is associated with the bacteroids, and increases 2- to 3-fold during nodule development (John et al., 1988), but its function is not known. Since antibodies against NodC reduce the nodulation of M. sativa by R. meliloti approximately 50% (John et al., 1985), NodC may play an important role in transmission of diffusible growth factors from the bacterium to plant (John et al., 1988). Supporting this notion is the finding that NodC is required for the induction or stimulation of expression of RH-42 and RH-44 genes in root hairs after inoculation with rhizobia (Gloudemans et al., 1989).

The regulation of nodABC is under the control of nodD. Unlike most nod genes, which are repressed during growth in culture, nodD is constitutively expressed by rhizobia (Mulligan and Long, 1985; Innes et al., 1985; Rossen et al., 1985). Induction of most nod genes requires both the NodD protein and plant-secreted phenolic compounds, collectively known as flavonoids (Peters et al., 1986; Redmond et al., 1986; Kosslak et al., 1987). Therefore, NodD is a trans-acting protein inducing flavonoid-dependent transcription of other nod genes that are preceded by the consensus, upstream, regulatory region to which NodD binds - the "nod-box" (Hong et al., 1987b). The NodD protein has been localized exclusively to the cytoplasmic membrane in R. leguminosarum bv. viciae (Schlaman et al., 1989). The localization of NodD was independent of the presence of naringenin during growth and it has been calculated that 20-80 NodD molecules are present in each cell (Schlaman et al., 1989). DNA sequence analysis of various nodD genes shows that the predicted proteins have significant homology to several DNA-binding transcriptional activators including LysR (Henikoff et al., 1988), AraC (Shearman et al., 1986), and the NahR regulator of naphthalene degradation (Schlaman et al., 1989). One current model has the amphipathic NodD anchored in the inner monolayer of the cytoplasmic membrane with a substantial hydrophilic domain extending into the cytoplasm (Schlaman et al., 1989), so that the interaction between NodD, flavonoids, and the nod-box occurs in or near the membrane.

Small differences in NodD proteins or in the structure of the flavonoid molecule can lead to either induction or repression of nod genes (Firmin et al., 1986; Wijffelman et al., 1986; Horvath et al., 1987; Spaink et al., 1987a,b, 1989a,b). NodD therefore acts to receive the plant signal, to help determine host specificity, and to activate transcription of other nod genes. Furthermore, NodD can also autoregulate its own level of expression in some rhizobia (Rossen et al., 1985; Fisher et al., 1987). Multiple copies of nodD genes in most rhizobia may suggest even more complex regulatory networks, as in R. meliloti (Honma and Ausubel, 1987; Mulligan and Long, 1989; Honma et al., 1990) or, alternatively, to allow recognition of a broader range of flavonoids (Mulligan and Long, 1989).

The NodD domains that are responsible for its different functions have been studied by engineering nodD genes with point mutations (Burn et al., 1987; Hong et al., 1987b), and by forming interspecies chimeras (Horvath et al., 1987; Spaink et al., 1989a). Unfortunately, the results of these studies have not been conclusive. While studies using chimeras indicate that the C-terminal region and amino acids 90 to 118 are responsible for autoregulation, point mutations in the N-terminal region have also been implicated. Chimera studies have also suggested that the C-terminal half of NodD (amino acids 132 to 270) specifically interacts with flavonoids (Spaink et al., 1989a). Many regions of NodD appear to be involved in its activation by flavonoids, but they are not clearly separated from the regions involved in autoregulation (Spaink et al., 1989a). Much remains to be learned about the common nod genes before nodulation gene function in symbiosis is understood.

4.2 The function and regulation of host-specific nod genes

After the induction of root hair curling and cortical cell division is accomplished by the products of the common nod genes, the high degree of specificity of the subsequent infection process is mediated by the action of the host-specific nod genes (sometimes known as host-specific nodulation, or hsn genes). It now appears that Nod proteins are involved in positive and negative aspects of interaction between plant and bacteria, and some Nod proteins may be involved in both (Long, 1984; Djordjevic et al., 1985, 1987b; Debelle et al., 1986b; Long, 1989).

Although nodIJ is part of the nodABCIJ transcription unit (Fig. 1), these genes are not normally regarded as common nod genes, nor do they appear to be involved in host-specific nodulation (but see Surin and Downie, 1989). NodIJ mutants of R. leguminosarum bv. viciae nodulate Vicia sativa and Vicia hirsuta, though nodulation is slightly delayed (Downie et al., 1985; Canter-Cremers et al., 1988). DNA sequence analysis of nodI shows that the predicted protein has significant homology to a family of ATP-binding proteins usually located in the inner membrane of Escherichia coli or Salmonella typhimurium and involved in the transport of low molecular weight compounds such as histidine, maltose and phosphate (Higgins et al., 1986), and may indeed be a transport protein (Davis et al., 1988). The predicted NodJ protein is very hydrophobic and may be an integral membrane protein (Evans and Downie, 1986). NodIJ gene products may be involved in the uptake of nutrients from the plant (Djordjevic et al., 1987b), or may form a membrane complex with other gene products (NodC?) from the nodABCIJ operon.

Such a membrane complex might include host-specific proteins such as the nodX product, which enables R. leguminosarum bv. viciae strain TOM (Davis et al., 1988) to nodulate P. sativum cv. Afghanistan. NodX appears to be part of an nodABCIJX operon (Fig. 1), and to have no allelic form in strains of R. leguminosarum bv. viciae which cannot nodulate Afghanistan peas (Gotz et al., 1985; Lie et al., 1987; Davis et al., 1988). A homologue of nodX exists in R. leguminosarum bv. trifolii, but not in bv. phaseoli or in R. meliloti (Davis et al., 1988). Furthermore, mutations in the nodFE genes of bv. trifolii result in reduced nodulation of Trifolium repens (white clover) but gain the capacity to nodulate Afghanistan peas, suggesting that the nodX gene can function in bv. trifolii (Davis et al., 1988). The NodX-pea cultivar interaction is reminiscent of the "gene-for-gene" relationships betweem plant pathogens and their hosts (Flor, 1946, 1947, 1958; Gabriel, 1986; Djordjevic et al., 1987b) and indeed the fact that the genetic locus in P. sativum cv. Afghanistan that confers the resistance to nodulation by most strains of R. leguminosarum bv. viciae behaves as a single recessive Mendelian trait (Hall, 1975) is consistent with a gene-for-gene interaction. The nodX gene may therefore be functionally similar to the avirulence (avr) genes of bacterial pathogens.

The nodLMN genes of R. leguminosarum bv. viciae are induced in response to flavonoids, with nodMN forming one operon and nodL part of the nodEFL operon (Fig. 1) (Surin and Downie, 1988). nodLMN homologues occur in R. leguminosarum bv. trifolii, but hybridize only weakly to B. japonicum and R. meliloti DNA even under low stringency conditions (Surin and Downie, 1988). NodL is important for host-specific nodulation of P. sativum, Lens, and Lathyrus, but not for Vicia (Surin and Downie, 1988). Mutants in nodM and nodN produce only slightly fewer nodules than the wild type (Surin and Downie, 1988). The DNA sequence of nodM shows significant homology with 2-amidophosphoribosyl transferases (purF gene products), although nodM did not complement purF mutants in E. coli (Surin and Downie, 1988). NodM may be involved in the transfer of an amido group, possibly to a sugar moiety on the lipopolysaccharide or exopolysaccharide (Surin and Downie, 1988), modifying the rhizobial surface and with possibly affecting the outcome of recognition.

In R. meliloti, the host-specific nod genes nodFEGH are also known as hsnABCD in strain 41 (Horvath et al., 1986). The nodH, nodG and nodQ genes are found only in R. meliloti (Debelle and Sharma, 1986; Horvath et al., 1986; Rodriguez-Quinones et al., 1987; Surin and Downie, 1988). NodH mutants are Hac^-, Inf^- and Nod^- on M. sativa, yet they gain the capacity to nodulate a heterologous host, V. sativa subsp. nigra (common vetch) (Faucher et al., 1988). The NodH protein appears to mediate the production of a specific extracellular signal inducing deformation of root hairs of M. sativa but not V. sativa, i.e. NodH is alfalfa-specific (Faucher et al., 1988). NodH may in fact modify the nodABC-dependent signal to a form specific for M. sativa (Faucher et al., 1988). The predicted NodH protein is very unusual in having a very high proline content (Fisher et al., 1987). The amino acid sequence predicted for the nodG product has homology to the ribitol dehydrogenase of K. pneumoniae (Debelle and Sharma, 1986) and to alcohol dehydrogenase and glucose dehydrogenase (Fisher et al., 1987).

The nodQ gene is thought to play a similar role to the nodH gene in determining host specificity (Faucher et al., 1989), because the phenotype of nodQ and nodH mutants is the same and the presence of both genes in either R. leguminosarum bv. viciae or in R. meliloti prevents the nodulation of V. sativa (Faucher et al., 1989). The symbiotic roles of the two genes are different, however, since nodH is required for the nodulation of M. sativa by either R. meliloti or R. leguminosarum bv. viciae, whereas nodQ is only required for nodulation by R. leguminosarum bv. viciae (Faucher et al., 1989). DNA sequence analysis of nodQ predicts a protein with significant homology to EF-Tu, an E. coli elongation factor in the G protein superfamily (Schwedock and Long, 1989). Transfer of nodH, nodQ, or nodFE from R. meliloti to R. leguminosarum bvs. viciae and trifolii has demonstrated that these genes behave like avr genes from phytopathogenic bacteria (Staskawicz et al., 1984; Gabriel, 1986), inasmuch as they are dominant and cause the suppression or inhibition of "virulence" of the recipient strains on some of their hosts (Faucher et al., 1989).

The nodFE genes of both R. meliloti and R. leguminosarum most likely play a role in infection thread formation within the root hairs, (Debelle et al., 1986a; Horvath et al., 1986; Swanson et al., 1987; Debelle et al., 1988). The phenotype of nodFE mutants is one of exaggerated root hair curling and delayed nodulation; these mutants cannot be complemented by corresponding nod genes from other rhizobial species (Spaink et al., 1987b). The control of root hair curling exercised by the nodFE gene product is negative and prevents excessive root hair curling on heterologous hosts (Debelle et al., 1986a). Despite their involvement with host range functions, the nodFE genes are conserved in R. leguminosarum strains and biovars (Debelle and Sharma, 1986). DNA sequence analyses of nodFE show that the predicted NodF protein is highly homologous to the acyl carrier protein of both E. coli and barley (Shearman et al., 1986; Fisher et al., 1987), leading to suggestions of NodF involvement in lipid or lipopolysaccharide synthesis or modification (Shearman et al., 1986). Since the acyl carrier protein of E. coli is known to be involved in the synthesis of extracellular ß-1,2-glucan (Therisod et al., 1986), it is also plausible that NodF may have a similar function (Fisher et al., 1987). The predicted amino acid sequence for nodE shows no homologies with other known proteins, but its apparent hydrophobic nature suggests a location in the membrane (Fisher et al., 1987)

NodE mutants of R. leguminosarum bv. trifolii acquire the ability to nodulate P. sativum, a heterologous host (Djordjevic et al., 1985). In isogenic strains of R. leguminosarum bv. trifolii, the source of the nodE gene is crucial in determining whether the strain behaves as bv. viciae or bv. trifolii in its host range (Spaink et al., 1989b). Since the nucleic acid sequences of the nodE genes from bvs. trifolii and viciae are highly homologous (Spaink et al., 1989b), there must a particular region of the protein involved in conferring specificity. Hybrid nodE genes show that this region is in the centre of the protein and contains 44 non-conserved amino acids (Spaink et al., 1989b). Using polyclonal antibodies, NodE has been localized to the cytoplasmic membrane (Spaink et al., 1989b), possibly as a transmembrane protein with regions in the cytoplasm and periplasm. NodF and NodE may be involved in the production, modification, or transport of an unidentified, host-specific ß-1,2-linked glucan that may be important in the interaction with plant lectins (Spaink et al., 1989b).

The NodO product of R. leguminosarum bv. viciae is a secreted protein (de Maagd et al., 1989c). The DNA sequence of nodO reveals a poorly conserved nod-box and the predicted NodO protein is homologous to a part of the haemolysin (HylA) from E. coli (de Maagd et al., 1989c). Multiple nodD copies greatly enhance the expression of nodO, but not most other nod genes. Depending on the bacterial chromosomal background, mutations in nodO either do not affect nodulation, or delay and decrease it (de Maagd et al., 1989c). NodO has not been found in other rhizobia and may thus represent a gene specific to R. leguminosarum bv. viciae (de Maagd et al., 1989c). The NodO protein may interact not only with the plant cell surface and thereby mediate communication with the host but also affect expression of the cytoplasmically-located RhiA (de Maagd et al., 1988, 1989c; Economou et al., 1989).

5 THE ROLE OF RHIZOBIAL POLYSACCHARIDES

One idea that has consistently dominated thinking about the Rhizobium-legume symbiosis is that a polysaccharide produced by the bacterium might be critical for recognition by the plant. Just what recognition means and how it is manifested biochemically is not clear, however. One possibility is that a host lectin interacts with a bacterial surface polysaccharide to bring about host-specific attachment of the bacteria to infectible root hairs (Dazzo and Hollingsworth, 1984). Some observations of certain microsymbiont-host combinations are consistent with this notion (Halverson and Stacey, 1986). However, to date it has not been shown that the interaction of a particular lectin with a particular polysaccharide is required for either nodulation or host-specificity. On the other hand, there is now a great deal of evidence from the analysis of bacterial mutants that strongly suggests roles for certain polysaccharides at stages in nodulation beyond host-specific attachment.

The array of polysaccharides produced by rhizobia is quite impressive (Carlson, 1982). Most attention has focused recently on acidic exopolysaccharides (EPS), lipopolysaccharides (LPS) and ß(1,2) glucans, largely because of the isolation of mutants defective in synthesis of these polymers. Such mutants generally elicit incomplete nodule development (Ndv$^-$ phenotype) because infection does not occur or breaks down before the proliferation of bacteroids. However, there are revealing exceptions to this generalization.

5.1 Acidic extracellular polysaccharides (EPS)

Field isolates of Rhizobium and Bradyrhizobium spp. are generally capable of excreting abundant acidic polysaccharides, some representative structures for which are shown in Fig. 2. The acidity is due to uronic acids and/or acyl groups, such as pyruvyl or succinyl, linked to sugar residues.

Mutants lacking EPS production (Exo⁻) have been isolated as non-fluorescent colonies in the presence of Calcofluor (Leigh et al., 1985; Muller et al., 1988; Kim et al., 1989) or as non-mucoid colonies in species whose EPS does not bind Calcofluor. The most extensive account of exo genetics is in R. meliloti SU47 (Table 4).

TABLE 4. Genes for rhizobial polysaccharide synthesis

Polysaccharide	Strain[a]	Genes/Loci[b]	Biochem. function[c]	Sym pheno.[d]	Ref.[e]
EPS I	Rm SU47	exo(ABFLMPQ)*CD'	synthesis	f	1
		exoRS	regulation		1
		exoH*	succinyl-	f	2
EPS II	Rm SU47	exp(ACDEFG)*, exoBC*	synthesis	f	3
		expR, mucR	regulation	Fix⁺	3,4
EPS	NGR234	exoABCD	synthesis	g	5
		exoXY	regulation	g	6
EPS	Rl 8002	pss-1, pss-2	synthesis	g	7
		psi**, psr**	regulation	Ndv⁻	8
	Rl CFN42	exoA1, A2	synthesis	g	9
ß-(1,2)glucan	Rm SU47	ndvAB, exoC	synthesis	Ndv⁻	10
LPS	Rl CFN42	lpsα (lpsABCDEFGHI)	OPS, core	Ndv⁻	11
		lpsß*, Γ[h]	core	Ndv⁻	12,13
	Rl VF39	lpsα (≥5 genes), ß*	i	Ndv⁻	14
	Rl ANU843	lpsα, ß*	i	Ndv⁻	15

[a]Abbreviations: Rm: R. meliloti, Rl: R. leguminosarum
[b]Only a few of the loci listed have been defined as genes. Most are transcriptional units defined by transposon mutagenesis. *: genes found on an indigenous plasmid other than the Sym (nod-nif-fix) plasmid. **: genes on the Sym plasmid.
[c]Synthesis: the genetic locus is required for polysaccharide synthesis, but a role in regulation is not ruled out. OPS: O polysaccharide.
[d]Symbiotic phenotype associated with mutation of the genes indicated.
[e]1: Long et al., 1988; 2: Leigh et al., 1987; 3: Glazebrook and Walker, 1989; 4: Zhan et al., 1989; 5. Chen et al., 1988a; 6: Gray et al., 1990; 7: Borthakur et al., 1986; 8: Borthakur et al., 1987; 9: Diebold and Noel, 1989; 10: Stanfield et al., 1988; 11: Cava et al., 1990; 12: Cava et al., 1989; 13: Brink, 1989; 14: Priefer, 1989; M. Hynes (personal communication); 15: Brink et al., 1990.
[f]Ndv⁻ if EPS I and EPS II are both absent.
[g]Nod⁻, Ndv⁻, or Ndv⁺ Fix⁺ depending on the host.
[h]lps α, ß, and Γ refer to genetic regions cloned on pCOS109.11, pCOS126, and pCOS309, respectively (Cava et al., 1989). Genes in two other chromosomal regions also are required for LPS (Cava et al., 1989; Diebold and Noel, 1989).
[i]Presumed necessary for the same portions of the LPS as the corresponding genetic regions of strain CFN42.

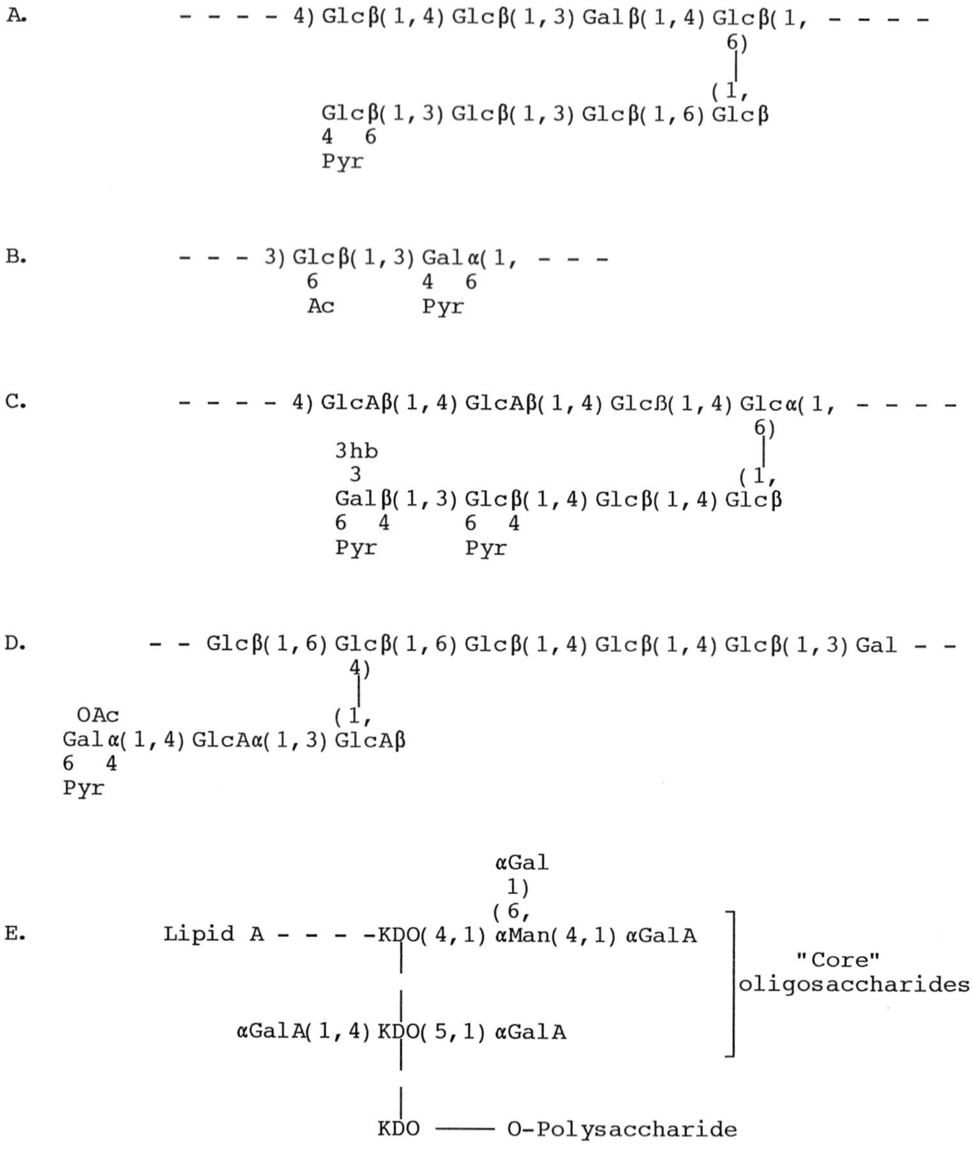

Fig. 2. EPS repeat units and one generalized LPS structure. A) EPS I (succinoglycan) of R. meliloti (Aman et al., 1981), B) EPS II of R. meliloti (Glazebrook and Walker, 1989), C) most common EPS of R. leguminosarum (McNeil et al., 1986), D) EPS of strain NGR234 (Djordjevic et al., 1987a), E) LPS of R. leguminosarum (Hollingsworth et al., 1989; R. Bhat and R. W. Carlson, personal communication). An acetyl and a succinyl group in structure A occur at unknown positions. O-acetyl substitutions occur at variable positions in structure C. Residues to which KDOs are linked at the ketosidic bond are unknown, as indicated by - - - -.

Exo⁻ mutants of R. meliloti SU47 induce nodules that lack infection threads or have very attenuated infections (Finan et al., 1985). An important finding to emerge from analysis of exo genetics in R. meliloti SU47 is that this strain has the cryptic ability to synthesize a second EPS (EPS II) (Fig. 2). EPS II synthesis can be activated by mutation of loci designated as expR and mucR (Glazebrook and Walker, 1989; Zhan et al., 1989). Perhaps most interesting is that EPS II synthesis suppresses the Ndv^- phenotype of mutants unable to make EPS I. Any hypothesis to explain the role(s) of EPS in alfalfa nodulation must accommodate this observation. Other than acidity, the only feature common to EPS I and EPS II is the presence of a Glc ß(1,3) Gal linkage (Fig. 2). One indication that acidity is the key structural feature is that mutants producing EPS I without the acidic succinyl groups are Ndv^- (Leigh et al., 1987).

R. meliloti 41, another wild-type isolate, synthesizes both EPS I and EPS II. A mutant, strain AK631, synthesizes neither, but is Ndv^+ Fix^+ (Putnoky et al., 1988; Williams et al., 1990). Transfer of gene lpsZ from R. meliloti 41 partially suppresses the Ndv^- phenotype of Exo^- mutants of R. meliloti SU47. This gene, normally not present in strain SU47, appears to affect lipopolysaccharide structure (Williams et al., 1990). Therefore, in R. meliloti 41, the feature of EPS I and EPS II that is critical for symbiosis may also reside in lipopolysaccharide.

The EPS of Rhizobium NGR234, which effectively nodulates a wide range of legumes, has also been extensively studied and its structure determined (Fig. 2) (Djordjevic et al., 1987a). A number of Exo^- mutants have been isolated and the mutations mapped (Chen et al., 1988a). The host dictates whether or not EPS synthesis is required for symbiosis. On hosts forming determinate nodules, at least some Exo^- mutants of strain NGR234 are Ndv^+ Fix^+, but all Exo^- mutants are Ndv^- on Leucaena leucocephala, which forms indeterminate nodules (Chen et al., 1985). An important finding is that Exo^- mutants elicit Ndv^+ Fix^+ nodules (at low frequency) when purified EPS or the oligosaccharide repeat unit of NGR234 is supplied exogenously before, during, or after inoculation of L. leucocephala (Djordjevic et al., 1987a). The same authors report a similar suppression of the Ndv^- phenotype of Exo^- R. leguminosarum bv. trifolii ANU843.

The reported structures of EPS isolated from various strains of R. leguminosarum share the same backbone of glucose and glucuronic acid residues (Fig. 2). The most common side chain is the one shown in Fig. 2, but the EPS of some strains deviates somewhat in residues and linkages (McNeil et al., 1986). Moveover, many residues are modified by substituents (3-hydroxybutyryl-, pyruvyl-, and acetyl- groups), whose stoichiometry varies with strain and probably with growth conditions. It has been proposed that these structural variations are involved in host specificity of nodulation (McNeil et al., 1986; Philip-Hollingsworth et al., 1989). However, this conclusion is based on quantitative variations in these modifications in very few strains and might not hold up in a wider survey. Moreover, it seems not to account for the symbiotic phenotype of mutants which lack the EPS entirely.

Whether EPS synthesis is required for nodulation depends on the host or the Sym plasmid of an R. leguminosarum strain. Exo^- mutants are Hac^+ Nod^- on peas (curling root hairs and perhaps initiating a short-lived meristem), Ndv^- on clover (with extensive infection threads and some infected nodule cells), and Ndv^+ Fix^+ on bean host (Chakravorty et al., 1982; Borthakur et al., 1986; Diebold and Noel, 1989). Since Exo^- mutants of Sinorhizobium fredii and B. japonicum are Ndv^+ Fix^+ on soybean (Law et al., 1982; Kim et al., 1989), it appears that the Phaseolae legumes somehow differ from the legumes that require Exo^+ microsymbionts. This difference may be related to the determinate nodule development exhibited by the Phaseolae. Siratro and other hosts of strain NGR234 that have determinate nodulation seem not to require Exo^+ microsymbionts, whereas all the plants shown to require Exo^+ microsymbionts have indeterminate nodule development.

The gene psi found on a bv. phaseoli Sym plasmid negatively regulates the exo (pss) genes of R. leguminosarum (Borthakur et al., 1987). This gene is not expressed under traditional laboratory conditions of growth ex planta. In bean nodules psi is transcribed but exo genes are not (A.W.B. Johnston, personal communication). Psi^- mutants are Ndv^- (Borthakur et al., 1987; A.W.B. Johnston, personal communication). Thus, it may be that bean cannot tolerate EPS synthesis during infection and/or in bacteroids. A gene (exoX) that regulates exo negatively, and is homologous to psi in a short expanse of its sequence, has been identified in the wide host-range strain NGR234 (Gray et al., 1990). The symbiotic phenotype of mutants defective in this gene has not been reported.

5.2 ß-glucans

Cyclic ß(1,2) glucans appear to be produced by all species of Rhizobium and Agrobacterium. Bradyrhizobium spp. have cyclic glucans with ß(1,3) and ß(1,6) linkages (Miller et al., 1990), whose function may be similar. In R. meliloti 102F34, mutations eliminating production of ß(1,2) glucan have been assigned to two genes, ndvA and ndvB (Stanfield et al., 1985). Despite being deficient in root hair attachment and intiation of infection threads, ndv mutants elicit more nodules per plant than normal, but the nodules are Ndv^- (incompletely developed and almost devoid of bacteria) (Dylan et al., 1986, 1990).

It is possible to select mutations that override an ndv mutation such that normally developed nodules are elicited on alfalfa (Dylan et al., 1990). "Pseudorevertants" carrying an ndv mutation and an extragenic suppressor mutation of this type appear not to produce ß(1,2) glucans; therefore, ß(1,2) glucans *per se* seem not to be required for normal nodule development (Ndv$^+$). The targets of the suppressor mutations are not yet known. The pseudorevertants are still deficient in what may be the main function of ß(1,2) glucans during growth *ex planta*, adaptation to low osmolarity, and are only marginally less deficient in root-hair attachment and in infection thread formation than the ndv mutant parents. Therefore, in the laboratory, *R. meliloti* has excess capacity for attachment and infection thread initiation; ndv mutations prevent infection by causing a problem in some other aspect of nodule development.

Rhizobia also produce ß(1,4) glucans (cellulose). Although *R. leguminosarum* mutants that cannot synthesize cellulose fibrils are incapable of forming large aggregates of bacteria on pea root tips, such mutants nodulate pea in a normal manner under laboratory conditions (Smit et al., 1987). These mutants have not been tested for whether cellulose synthesis allows more frequent nodulation in the field.

5.3 Lipopolysaccharide (LPS)

Among rhizobia, lipopolysaccharides have been characterized most extensively in *R. leguminosarum*. As in other Gram-negative eubacteria, the LPS of this species is composed of a hydrophobic lipid A portion that anchors it in the outer membrane and a polysaccharide portion that extends into the bacterial environment (Fig. 2). The polysaccharide portion is composed of a core more or less conserved among all strains of the species and an O-polysaccharide portion that varies from strain to strain. The lipid A of *R. leguminosarum* has two features not previously observed in other bacteria. The disaccharide to which the fatty acids are linked is composed of glucosamine and galacturonic acid (R. Bhat, R.W. Carlson and R.I. Hollingsworth, unpublished). One of the most abundant fatty acids in the lipid A is a C_{28} hydroxy fatty acid (Hollingsworth and Carlson, 1989). The unusual structural features of the polysaccharide and lipid portions may be important in symbiosis or may be simply a reflection of the phylogeny of rhizobia.

In *R. leguminosarum* bv. *phaseoli* CFN42, genes required for synthesis of the polysaccharide portion of the LPS (designated as lps genes) have been grouped into five genetic regions by cosmid cloning (Table 4). A short region necessary for LPS core synthesis is carried on an indigenous plasmid other than the Sym (nod-nif-fix) plasmid (Brink, 1989). A more extensive region located on the chromosome is required for both O-polysaccharide and core synthesis (Carlson et al., 1989; Cava et al., 1990). Two corresponding genetic regions have been identified in _R. leguminosarum_ bv. _viciae_ VF39 and _R. leguminosarum_ bv. _trifolii_ ANU843 (Table 4).

Mutants lacking LPS O-polysaccharide are Ndv⁻ on bean, pea, vetch (_Vicia sativa_ or _V. hirsuta_) and clover. On bean, infection thread development becomes arrested within root hairs and bacteria are not released from the threads (Noel et al., 1986). On clover and _V. hirsuta_, a few nodule cells are infected with bacteroids resulting in low-level nitrogenase activity (de Maagd et al., 1989a; Priefer, 1989; Brink et al., 1990). In order to be Ndv⁺, the bacteria apparently must have O-polysaccharide in near-normal abundance. A mutant which has about one-third the normal amount, but whose LPS otherwise seems normal, has the same degree of deficiency in infection on bean as mutants completely lacking O-polysaccharide (Cava et al., 1989). This result argues against the infection defect having arisen because LPS serves as a cell-bound receptor (Cava et al., 1989). The (free-living) O-polysaccharide structure can be changed by genetic manipulation and still serves its role in infection (Brink et al., 1990).

Because of the unique environment of bacteroids, it has been suggested that bacteroid LPS might differ from the LPS of free-living _R. leguminosarum_. Although the bacteroid LPS generally seems substantially unchanged (de Maagd et al., 1989b), bacteroid-specific LPS epitopes (detected by monoclonal antibodies) and bacteroid-specific bands on gel electrophoresis have been observed from pea and bean nodules (Kannenberg and Brewin, 1989; H. Tao and K.D. Noel, unpublished). The actual structural changes have not been determined. These subtle changes could indicate a requirement for O-polysaccharide in infection, a second role for LPS in symbiosis, or merely a response to the nodule environment. The epitope changes can be mimicked by growth of the bacteria _ex planta_ either at low oxygen concentration or at low pH. Until mutants incapable of switching to the bacteroid LPS form are isolated and tested for symbiotic properties, the importance of the LPS changes cannot be assessed.

Peribacteroid membrane preparations isolated from pea nodules have been reported to contain substantial amounts of LPS (Brewin et al., 1985). LPS and other outer membrane components are known to become detached from the outer membranes of other bacteria under various conditions (Mayrand and Grenier, 1989). In this connection it is intriguing that _R. leguminosarum_ Exo⁻ mutants excrete significant amounts of LPS into liquid growth media, whereas Exo⁺ bacteria do not (Carlson and Lee, 1983; Diebold and Noel, 1989). Repression of EPS synthesis within the nodule could trigger the release of LPS from the bacteria, making LPS a potential signal molecule.

A B. japonicum mutant lacking O-polysaccharide is Nod$^-$ on soybean (Puvanesarajah et al., 1987). The significance of this observation awaits the isolation of additional mutants, but it may indicate that LPS plays an essential role earlier in this symbiosis than in those between R. leguminosarum and its hosts. LPS and Lps$^-$ mutants have not been studied sufficiently in other Rhizobium spp. to comment on the possible function of LPS in their symbioses.

5.4 Lectins, host-specificity and rhizobial attachment

The involvement of host lectins in infection has been studied most intensively in soybean (Halverson and Stacey, 1986), pea (Kijne et al., 1988), and clover (Dazzo and Hollingsworth, 1984). Recent experiments involving the transfer of recombinant DNA bolster the view that lectins are involved in host-specific nodulation of clover and pea by R. leguminosarum. When the pea microsymbiont R. leguminosarum 300 received the nod genes specifying the host range of clover microsymbiont R. leguminosarum ANU843, its binding of trifoliin A (clover lectin) in an in situ assay increased dramatically, from nil to a level comparable to that of clover symbiont ANU843 (Philip-Hollingsworth et al., 1989). Furthermore, when white clover was transformed with pea lectin genes, its roots became susceptible to nodulation by the pea microsymbiont R. leguminosarum 248 (Diaz et al., 1989); most nodules were of the Ndv$^-$ type, but some were Fix$^+$.

While these results are exciting, the biochemical role of the lectins, and the symbiotic events in which they participate, remain subject to debate. A great deal of effort has been expended to elucidate these roles. The reader is referred to extensive reviews on the subject (Dazzo and Hollingsworth, 1984; Halverson and Stacey, 1986). In this chapter it suffices to note that the lectins are assumed to bind to particular rhizobial polysaccharides, although direct assay of purified lectin binding to purified polysaccharide is almost never carried out. Involvement of lectin is often inferred from hapten inhibition of a process (e.g., Kijne et al., 1988). Interaction of a lectin with a particular polysaccharide is often inferred from inhibition by the polysaccharide of a process for which there is evidence of lectin involvement or inhibition of lectin binding to bacterial aggregates (e.g., Dazzo et al., 1984). Assayed in a more direct manner, pea lectin is reported to bind to both EPS (of several Rhizobium species) and LPS (of the only R. leguminosarum bv. viciae strain tested) (Kamberger, 1979). Clover lectin has been inferred to bind capsular polysaccharide (very similar structurally to EPS) and LPS only of R. leguminosarum bv. trifolii strains (Dazzo and Hollingsworth, 1984). The LPS is active only at a certain stage of growth ex planta. The capsule of R. leguminosarum bv. trifolii appears to undergo degradation in a topologically specific manner in the presence of clover roots. Dazzo et al. (1984) speculated that this degradation is important in host-specific attachment to root hairs. Whether the major EPS and LPS interact with lectin in vivo is still an important question. Mutants constructed with available exo and lps mutations provide new tools for answering this question.

Although adopted as a central theme of the lectin hypothesis, the process whereby rhizobia attach to root hairs is in principle a separate issue. Much controversy has centred on whether attachment is host-specific (Dazzo et al., 1984) or nonspecific (Smit et al., 1989b). The answer may well depend on the particular symbiosis. A question of higher priority is whether the several assays of attachment measure events and properties actually required in nodule development. A prototypic approach is being followed by Kijne and colleagues. As mentioned before, they have devised a microscopic assay of root-hair attachment very similar to the one of Dazzo et al. (1984). A "calcium-dependent" adhesin, cellulose fibrils, and, when cells are grown under certain conditions, pea lectin are involved in formation of the bacterial aggregation on root hairs that they assay (Kijne et al., 1988; Smit et al., 1989b). Using Lps mutants, they have shown that LPS O-polysaccharide is not required for attachment in this assay (Smit et al., 1989a). Mutants that lack the adhesin should be very informative, for this protein seems to be required for any attachment at all in their assay. Will such mutants be incapable of infecting or initiating a nodule meristem? This will be an important test of this widely used type of attachment assay.

5.5 Conclusions

Since other nonsymbiotic bacteria have EPS, LPS, and glucan polymers analogous to those of rhizobia, it is clear that these polysaccharides did not evolve only for symbiosis. As the symbioses evolved, however, these polymers were probably adapted and altered to improve the compatibility of the partners. As a consequence, it is difficult to separate the symbiosis-specific from their more general roles, and indeed a given polysaccharide may have several roles in symbiosis. Such roles may vary from one Rhizobium-legume combination to another. Therefore, discerning the roles is a complex problem.

In cases where a mutant totally lacks a given polysaccharide, the symbiotic phenotype indicates the first step in nodule development at which the polysaccharide is required. The insurmountable defect of most of these mutants is in infection rather than nodule meristem formation. Hypothetical roles of surface polysaccharides that have been proposed to explain this phenotype include protecting against host defenses, acting as hormones (oligosaccharins), activating host enzymes that polymerize or hydrolyze polysaccharides, serving as part of the infection thread matrix or wall, delivering bacterial compounds or enzymes, and modulating the osmotic pressure of the infection thread. Complete understanding will require much more information regarding the cell biology and biochemistry of infection thread formation.

There is little evidence from analyzing these mutants that EPS, LPS, or ß-glucans are essential for host-specific nodule formation; polysaccharide-deficient mutants generally induce nodules on the proper hosts. However, since Exo⁻ R. leguminosarum bv. viciae and Lps⁻ B. japonicum display Nod⁻ phenotypes, EPS and LPS may be host-specificity factors for nodulation on peas and soybean. If lectins and polysaccharides are involved in host-specificity in the other symbioses, perhaps the lectins interact with minor EPS or LPS molecules or polysaccharides induced by the plant.

On the other hand, the requirement for a given type of polysaccharide depends on the host. Exo^- mutants are Ndv^+ Fix^+ on bean and soybean. Pea and clover require both the EPS and normal LPS of an R. leguminosarum microsymbiont but the stage in nodule development at which EPS is required is quite different in these two hosts.

The work of Djordjevic et al., (1987a) in suppressing the Ndv^- phenotype of Exo^- mutants by adding purified EPS should be pursued and, if possible, extended to other Rhizobium-legume combinations and other types of polysaccharides. Djordjevic et al. (1987a) observed that only the EPS of NGR234 could restore infection on its indeterminate host and only the EPS of R.l. bv. trifolii could enhance infection on clover. This is the only striking demonstration of structural specificity thus far indicated in studies of polysaccharide-deficient mutants. Indeed, results of genetic suppression of Exo^- R. meliloti and Lps^- R.l. bv. trifolii mutants have been consistent with a lack of structural specificity. In these latter cases, it may be that the EPS need have only appreciable acidity and the LPS merely abundant O-polysaccharide of at least a certain length.

6 REFERENCES

Aman, P., McNeil, M., Franzen, L., Darvill, A.G. and Albersheim, P., 1981. Structural elucidation, using HPLC-MS and GLC-MS, of the acidic polysaccharide secreted by Rhizobium meliloti strain 1021. Carbohydr. Res., 95: 263-287.

Ames, P. and Bergman, K., 1981. Competitive advantage provided by bacterial motility in the formation of nodules by Rhizobium meliloti. J. Bacteriol., 148: 728-729.

Appelbaum, E.R., Thompson, D.V., Idler, K. and Chartrain, N., 1988. Rhizobium japonicum USDA 191 has two nodD genes that differ in primary structure and function. J. Bacteriol., 170: 12-20.

Bergman, H., Gulash-Hofee, M., Hovestadt, R.E., Larosiliere, R.C., Ronco, P.G. and Su, L., 1988. Physiology of behavioral mutants of Rhizobium meliloti: evidence for a dual chemotaxis pathway. J. Bacteriol., 170: 3249-3254.
Borthakur, D. and Johnston, A.W.B., 1987. Sequence of psi, a gene on the symbiotic plasmid of Rhizobium phaseoli which inhibits exopolysaccharide synthesis and nodulation, and demonstration that its transcription is inhibited by psr, another gene on the symbiotic plasmid. Mol. Gen. Genet., 207: 149-154.
Borthakur, D., Barber, C.E., Lamb, J.W., Daniels, M.J., Downie, J.A. and Johnston, A.W.B., 1986. A mutation that blocks exopolysaccharide synthesis prevents nodulation of peas by Rhizobium leguminosarum but not of beans by Rhizobium phaseoli and is corrected by cloned DNA from Rhizobium or the phytopathogen Xanthomonas. Mol. Gen. Genet., 203: 320-323.
Brewin, N.J., Wood, E.A., Bradley, D.J., Harding, S.C., Sindhu, S.S., Perotto, S., Kannenberg, E.L. and vandenBosch, K.A., (1988). The use of monoclonal antibodies to study plant-microbe interactions in the pea nodule. In: H. Bothe, F.J. de Bruijn and W.E. Newton (Editors), Nitrogen Fixation: Hundred Years After. Gustav Fischer, Stuttgart, pp. 517-521.
Brewin, N.J., Robertson, J.G., Wood, E.A., Wells, B., Larkins, A.P., Galfre, G. and Butcher, G.W., 1985. Monoclonal antibodies to antigens in the peribacteroid membrane from Rhizobium-induced root nodules of pea cross-react with plasma membranes and Golgi bodies. EMBO J., 4: 605-611.
Brink, B.A., 1989. Genetic and functional comparison of LPS biosynthetic loci in Rhizobium leguminosarum strains of different host ranges. Ph.D. dissertation. Marquette University, Milwaukee, Wisconsin, USA.
Brink, B.A., Miller, J., Carlson, R.W. and Noel, K.D., 1990. Expression of Rhizobium leguminosarum CFN42 genes for lipopolysaccharide in strains derived from different R. leguminosarum soil isolates. J. Bacteriol., 172: 548-555.
Burn, J., Rossen, L. and Johnston, A.W.B., 1987. Four classes of mutations in the nodD gene of Rhizobium leguminosarum biovar viciae that affect its ability to autoregulate and/or activate other nod genes in the presence of flavonoid inducers. Genes Dev., 1: 456-464.
Callaham, D.A. and Torrey, J.G., 1981. The structural basis for infection of root hairs of Trifolium repens by Rhizobium. Can. J. Bot., 59: 1647-1664.
Calvert, H.E., Pence, M.K., Pierce, M., Malik, N.S.A. and Bauer, W.D., 1984. Anatomical analysis of the development and distribution of Rhizobium infections in soybean roots. Can. J. Bot., 62: 2375-2384.
Canter-Cremers, H.C.J., van Brussel, A.A.N., Plazinski, J. and Rolfe, B.G., 1986. Sym plasmid and chromosomal gene products of Rhizobium trifolii elicit developmental responses on various legume roots. J. Plant Physiol., 122: 25-40.
Canter-Cremers, H.C.J., Wijffelman, C.A., Pees, E., Rolfe, B.G., Djordjevic, M.A. and Lugtenberg, B.J.J., 1988. Host specific nodulation of plants of the pea cross-inoculation group is influenced by genes in fast growing Rhizobium downstream nodC. J. Plant Physiol., 132: 398-404.
Carlson, R.W., 1982. Surface chemistry. In: W.J. Broughton (Editor), Nitrogen Fixation: Rhizobium (Vol. 2). Clarendon, Oxford, pp. 199-234.
Carlson, R.W. and Lee, R.P., 1983. A comparison of the surface polysaccharides from R. leguminosarum 128C53 str^{r} rif^{r} with the surface polysaccharides from its Exo^{-} mutant. Plant Physiol., 71: 223-228.
Carlson, R.W., Garcia, F., Noel, D. and Hollingsworth, R., 1989. The structures of the lipopolysaccharide core components from Rhizobium leguminosarum biovar phaseoli CE3 and two of its symbiotic mutants, CE109 and CE309. Carbohydr. Res., 195: 101-110.
Cava, J.R., Elias, P.M., Turowski, D.A. and Noel, K.D., 1989. Rhizobium leguminosarum CFN42 genetic regions encoding lipopolysaccharide structures essential for complete nodule development on bean plants. J. Bacteriol., 171: 8-15.

Cava J. R., Tao, H. and Noel, K.D., 1990. Mapping of complementartion groups within a Rhizobium leguminosarum CFN42 chromosomal region required for lipopolysaccharide synthesis. Mol. Gen. Genet., 221: 125-128.
Chakravorty, A.K., Zurkowski, W., Shine, J. and Rolfe, B.G., 1982. Symbiotic nitrogen fixation: molecular cloning of Rhizobium genes involved in exopolysaccharide synthesis and effective nodulation. J. Mol. Appl. Genet., 1: 585-596.
Chen, H., Batley, M., Redmond, J. and Rolfe, B.G., 1985. Alteration of the effective nodulation properties of a fast-growing broad host range Rhizobium due to changes in exopolysaccharide synthesis. J. Plant Physiol., 120: 331-349.
Chen, H., Gray, J.X., Nayudu, M., Djordjevic, M.A. Batley, M., Redmond, J.W. and Rolfe, B.G., 1988a. Five genetic loci involved in the symbiosis of acidic exopolysaccharides are closely linked in the genome of Rhizobium sp., strain NGR234. Mol. Gen. Genet., 212: 310-316.
Chen, W.X., Yan, G.H. and Li, J.L., 1988b. Numerical taxonomic study of fast-growing soybean rhizobia and a proposal that Rhizobium fredii be assigned to Sinorhizobium gen. nov. Int. J. Syst. Bacteriol., 38: 392-397.
Currier, W.M. and Strobel, G.A., 1977. Chemotaxis of Rhizobium spp. to a glycoprotein produced by birdsfoot trefoil roots. Science, 196: 434-436.
Davis, E.O., Evans, I.J. and Johnston, A.W.B., 1988. Identification of nodX, a gene that allows Rhizobium leguminosarum biovar viciae strain TOM to nodulate Afghanistan peas. Mol. Gen. Genet., 212: 531-535.
Dazzo, F.B. and Gardiol, A., 1984. Host specificity in Rhizobium-legume interactions. In: D.P.S. Verma and T. Hohn (Editors), Genes Used in Plant-Microbe Interactions. Springer Publishers, New York, pp. 3-31.
Dazzo, F.B. and Hollingsworth, R.E., 1984. Trifoliin A and carbohydrate receptors as mediators of cellular recognition in the Rhizobium trifolii-clover symbiosis. Biol. Cell., 51: 267-274.
Dazzo, F.B., Truchet, G.L., Sherwood, J.E., Hravak, E.R., Abe, M. and Pankratz, S.H., 1984. Specific phases of root hair attachment in the Rhizobium trifolii-clover symbiosis. Appl. Environ. Microbiol., 48: 1140-1150.
Debelle, F. and Sharma, S.B., 1986. Nucleotide sequence of Rhizobium meliloti RCR2011 genes involved in host specificity of nodulation. Nucleic Acids Res., 14: 7453-7472.
Debelle, F., Rosenberg, C., Vasse, J., Maillet, F., Martinez, E., Dénarié, J. and Truchet, G., 1986a. Assignment of symbiotic developmental phenotypes to common and specific nodulation (nod) genetic loci of Rhizobium meliloti. J. Bacteriol., 168: 1075-1086.
Debelle, F., Sharma, S.B., Rosenberg, C., Vasse, J., Maillet, F., Truchet, G. and Dénarié, J., 1986b. Respective roles of common and specific Rhizobium meliloti nod genes in the control of lucerne infection. In: B. Lugentenberg (Editor), Recognition in Microbe-Plant Symbiotic and Pathogenic Interactions. (NATO ASI Series, Vol. H4) Springer-Verlag, Berlin, pp. 17-28.
Debelle, F., Maillet, F., Vasse, J., Rosenberg, C., de Billy, F., Truchet, G., Dénarié, J. and Ausubel, F.M., 1988. Interference between Rhizobium meliloti and Rhizobium trifolii nodulation genes: genetic basis of R. meliloti dominance. J. Bacteriol., 170: 5718-5727.
deMaagd, R.A., Wijffelman, C.A., Pees, E. Lugtenberg, B.J.J., 1988. Detection and subcellular localization of two Sym plasmid-dependent proteins of Rhizobium leguminosarum biovar viciae. J. Bacteriol., 170: 4424-4427.
deMaagd, R.A., Rao, A.S., Mulders, I.H.M., Leentje, G.R., Loosdrecht, M.C.M., Wijffelman, C.A. and Lugtenberg, G.J.J., 1989a. Isolation and characterization of mutants of Rhizobium leguminosarum bv. viciae 248 with altered lipopolysaccharides: possible role of surface charge or hydrophobicity in bacterial release from the infection thread. J. Bacteriol., 171: 1143-1150.

deMaagd, R.A., de Rijk, R., Mulders, I.H.M. and Lugtenberg, G.J.J., 1989b. Immunological characterization of Rhizobium leguminosarum outer membrane antigens by use of polyclonal and monoclonal antibodies. J. Bacteriol., 171: 1136-1142.

deMaagd, R.A., Wijfjes, A.H.M., Spaink, H.P., Ruiz-Sainz, J.E., Wijffelman, C.A., Okker, R.J.H. and Lugtenberg, B.J.J., 1989c. nodO, a new nod gene of the Rhizobium leguminosarum biovar viciae sym plasmid pRL1JI, encodes a secreted protein. J. Bacteriol., 171: 6764-6770.

Dénarié, J., Boistard, P., Casse-Delbart, F., Atherly, A.G., Berry, J.O. and Russell, P., 1981. Indigenous plasmids of Rhizobium. In: K.L. Giles and A.G. Atherly (Editors), Int. Rev. Cytol. Suppl. 13, Biology of the Rhizobiaceae. Academic Press, New York, pp. 225-246.

Diaz, C.L., Melchers, L.S., Hooykaas, P.J.J., Lugtenberg, B.J.J. and Kijne. J.W., 1989. Root lectin as a determinant of host-plant specificity in the Rhizobium legume symbiosis. Nature, 338: 579-581.

Diebold, R. and Noel, K.D., 1989. Rhizobium leguminosarum exopolysaccharide mutants: biochemical and genetic analyses and symbiotic behavior on three hosts. J. Bacteriol., 171: 4821-4830.

Djordjevic, M.A., Schofield, P.R. and Rolfe, B.G., 1985. Tn5 mutagenesis of Rhizobium trifolii host-specific nodulation genes result in mutants with altered host-range ability. Mol. Gen. Genet., 200: 463-471.

Djordjevic, S.P., Chen, H., Batley, M., Redmond, J.W. and Rolfe, G.G., 1987a. Nitrogen fixation ability of exopolysaccharide synthesis mutants of Rhizobium sp. strain NGR234 and Rhizobium trifolii is restored by the addition of homologous exopolysaccharides. J. Bacteriol., 169: 53-60.

Djordjevic, M.A., Gabriel, D.W. and Rolfe, B.G., 1987b. Rhizobium - the refined parasite of legumes. Ann. Rev. Phytopathol., 25: 145-168.

Downie, J.A., Knight, C.D., Johnston, A.W.B. and Rossen, L., 1985. Identification of genes and gene products involved in the nodulation of peas by Rhizobium leguminosarum. Mol. Gen. Genet., 198: 255-262.

Dudley, M.E., Jacobs, T.W. and Long, S.R., 1987. Microscopic studies of cell divisions induced in alfalfa roots by Rhizobium meliloti. Planta, 171: 289-301.

Dylan, T., Telpi, L., Stanfield, S., Kashyap, L., Douglas, C., Yanofsky, M, Nester, E.W., Helinski, D.R. and Ditta, G., 1986. Rhizobium meliloti genes required for nodule development are related to chromosomal virulence genes in Agrobacterium tumefaciens. Proc. Natl. Acad. Sci. USA, 83: 4403-4407.

Dylan, D., Nagpal, P., Helinski, D.R. and Ditta, G.S., 1990. Symbiotic pseudorevertants of Rhizobium meliloti ndv mutants. J. Bacteriol., 172: 1409-1417.

Economou, A., Hawkins, F.K.L., Downie, J.A. and Johnston, A.W.B., 1989. Transcription of rhiA, a gene on Rhizobium leguminosarum bv. viciae Sym plasmid, requires rhiR and is repressed by flavonoids that induce nod genes. Mol. Microbiol., 3: 87-93.

Egelhoff, T.T. and Long, S.R., 1985. Rhizobium meliloti nodulation genes: identification of nodDABC gene products, purification of nodA protein, and expression of nodA in Rhizobium meliloti. J. Bacteriol., 164: 591-599.

Evans, I.J. and Downie, J.A., 1986. The nodI gene product of Rhizobium leguminosarum is closely related to ATP-binding bacterial transport proteins: nucleotide sequence analysis of the nodI and nodJ genes. Gene, 43: 95-101.

Fåhraeus, G., 1957. The infection of clover root hairs by nodule bacteria studied by a simple glass slide technique. J. Gen. Microbiol., 16: 374-381.

Faucher, C., Maillet, F., Vasse, J., Rosenberg, C., van Brussel, A.A.N., Truchet, G. and Dénarié, J., 1988. Rhizobium meliloti host range nodH gene determines production of an alfalfa-specific extracellular signal. J. Bacteriol., 170: 5489-5499.

Faucher, C., Camut, S., Dénarié, J. and Truchet, G., 1989. The nodH and nodQ host range genes of Rhizobium meliloti behave as avirulence genes in R. leguminosarum bv. viceae and determine changes in the production of plant-specific extracellular signals. Mol. Plant-Microbe Interact., 2: 291-300.
Finan, T.M., Hirsch, A.M., Leith, J.A., Johansen, E., Kuldau, G.A., Deegan, S., Walker, G.C. and Signer, E.R., 1985. Symbiotic mutants of Rhizobium meliloti that uncouple plant from bacterial differentiation. Cell, 40: 869-877.
Firmin, J.L., Wilson, K.E., Rossen, L. and Johnston, A.W.B., 1986. Flavonoid activation of nodulation genes in Rhizobium reversed by other compounds present in plants. Nature, 324: 90-92.
Fisher, R.F., Tu, J.K. and Long, S.R., 1985. Conserved nodulation genes in R. meliloti and R. trifolii. Appl. Environ. Microbiol., 49: 1432-1435.
Fisher, R.F., Swanson, J. A., Mulligan, J.T. and Long, S.R., 1987. Extended region of nodulation genes in Rhizobium meliloti 1021. II. Nucleotide sequence, transcription start sites and protein products. Genetics, 117: 191-201.
Fisher, R.F., Egelhoff, T.T., Mulligan, J.T., Yelton, M.M. and Long, S.R., 1988. In: H. Bothe, F.J. de Bruijn and W.E. Newton (Editors), Nitrogen Fixation: Hundred Years After. Gustav Fischer, Stuttgart, pp. 391-398.
Flor, H.H., 1946. Genetics of pathogenicity of Melampsora lini. J. Agric. Res., 73: 335-357.
Flor, H.H., 1947. Inheritance of reaction to rust in flax. J. Agric. Res., 74: 241-262.
Flor, H.H., 1958. Mutation to wider virulence in Melampsora lini. Phytopathology, 48: 297-301.
Gabriel, D.W., 1986. Specificity and gene function in plant-pathogen interactions. Amer. Soc. Microbiol. News, 52: 19-25.
Glazebrook, J. and Walker, G.C., 1989. A novel exopolysaccharide can function in place of the calcofluor-binding exoplysaccharide in nodulation of alfalfa by Rhizobium meliloti. Cell, 56: 661-672.
Gloudemans, T., Moerman, M., vanBeckum, J., Gundersen, J., van Kammen, A. and Bisseling, T., 1988. Identification of plant genes involved in the Rhizobium leguminosarum-pea root hair interaction. In: H. Bothe, F.J. de Bruijn and W.E. Newton (Editors), Nitrogen Fixation: Hundred Years After. Gustav Fischer, Stuttgart, pp. 611-616.
Gloudemans, T., Bhuvaneswari, T.V., Moerman, M., van Brussel, T., van Kammen, A. and Bisseling, T., 1989. Involvement of Rhizobium leguminosarum nodulation genes in gene expression in pea root hairs. Plant Mol. Biol., 12: 157-167.
Gober, J.W. and Kashket, E.R., 1987. K^+ regulates bacteroid-associated functions of Bradyrhizobium. Proc. Natl. Acad. Sci. USA, 84:4650-4654.
Gober, J.W. and Kashket, E.R., 1989. Role of DNA superhelicity in regulation of bacteroid-associated functions of Bradyrhizobium sp. strain 32H1. Appl. Environ. Microbiol., 55: 1420-1425.
Gotz, R., Evans, I.J., Downie, J.A. and Johnston, A.W.B., 1985. Identification of the host-range DNA which allows Rhizobium leguminosarum strain TOM to nodulate cv. Afghanistan peas. Mol. Gen. Genet., 201: 296-300.
Govers, F., Moerman, M., Downie, J.A., Hooykaas, P., Franssen, H.J., Louwerse, J., van Kammen, A. and Bisseling, T., 1986. Rhizobium nod genes are involved in inducing an early nodulin gene. Nature, 323:564-566.
Gray, J.X., Djordjevic, M.A. and Rolfe, B.G., 1990. Two genes that regulate exopolysaccharide production in Rhizobium sp. strain NGR234: DNA sequences and resultant phenotypes. J. Bacteriol., 172: 193-204.
Hall, F.B., 1975. Host plant control of the inheritance of dinitrogen fixation in the Pisum-Rhizobium symbiosis. Euphytica, 24: 767-770.
Halverson, L.J. and Stacey, G., 1986. Signal exchange in plant-microbe interactions. Microbiol. Rev., 50:193-205.
Henikoff, S., Haughn, G.W., Calvo, J.M. and Wallace, J.C., 1988. A large family of bacterial activator proteins. J. Bacteriol., 170: 6602-6606.

Higgins, C.F., Hiles, I.D., Salmond, G.P.C., Gill, D.R., Downie, J.A., 1986. A family of related ATP-binding subunits coupled to distinct biological processes in bacteria. Nature, 323: 448-450.
Hirsch, A.M., Bhuvaneswari, T.V., Torrey, J.G. and Bisseling, T., 1989. Early nodulin genes are induced in alfalfa root outgrowths elicited by auxin transport inhibitors. Proc. Natl. Acad. Sci. USA, 86: 1244-1248.
Hollingsworth, R.I. and Carlson, R.W., 1989. 27-Hydroxyoctacosanoic acid is a major structural fatty acyl component of the lipopolysaccharide of Rhizobium trifolii ANU843. J. Biol. Chem., 246: 9300-9303.
Hollingsworth, R.I., Carlson, R.W., Garcia, F. and Gage, D.A., 1989. A new core tetrasaccharide component from the lipopolysaccharide of Rhizobium trifolii ANU843. J. Biol. Chem., 246: 9294-9299.
Hong, G.-F., Burn, J.E. and Johnston, A.W.B., 1987a. Analysis ofthe mechanism of nod gene regulation in Rhizobium leguminosarum. In: D. von Wettstein and N.-H. Chua (Editors), Plant Molecular Biology, (NATO ASI Series, Series A, Life Sciences) 140: 523-530.
Hong, G.-F., Burn, J.E. and Johnston, A.W.B., 1987b. Evidence that DNA involved in the expression of nodulation (nod) genes in Rhizobium binds to the product of the regulatory gene nodD. Nucleic Acids Res., 15: 9677-9690.
Honma, M.A. and Ausubel, F.M., 1987. Rhizobium meliloti has three functional copies of the nodD symbiotic regulatory gene. Proc. Natl. Acad. Sci. USA, 84: 8558-8562.
Honma, M.A., Asomaning, M. and Ausubel, F.M., 1990. Rhizobium meliloti nodD genes mediate host-specific activation of nodABC. J. Bacteriol., 172: 901-911.
Horvath, B., Kondorosi, E., John, M., Schmidt, J., Torok, I., Gyorgypal, Z., Barabas, I., Wieneke, U., Schell, J. and Kondorosi, A., 1986. Organization, structure and symbiotic function of Rhizobium meliloti nodulation genes determining host specificity for alfalfa. Cell, 46: 335-343.
Horvath, B., Bachem, C.W.B., Schell, J. and Kondorosi, A., 1987. Host-specific regulation of nodulation genes in Rhizobium is mediated by a plant-signal, interacting with the nodD gene product. EMBO J., 6: 841-848.
Innes, R.W., Kuempel, P.L., Plazinski, J., Canter-Cremers, H., Rolfe, B.G. and Djordjevic, M.A., 1985. Plant factors induce expression of nodulation and host-range genes in Rhizobium trifolii. Mol. Gen. Genet., 201: 426-432.
John, M., Schmidt, J., Wieneke, U., Kondorosi, E., Kondorosi, A. and Schell, J., 1985. Expression of the nodulation gene nodC of Rhizobium meliloti in Escherichia coli: the role of the nodC gene product in nodulation. EMBO J., 4: 2425-2430.
John, M., Schmidt, J., Wieneke, U., Krussman, H.-D. and Schell, J., 1988. Transmembrane orientation and receptor-like structure of the Rhizobium meliloti common nodulation protein nodC. EMBO J., 7: 583-588.
Johnson, D., Roth, L.E. and Stacey, G., 1989. Immunogold localization of the NodC and NodA proteins of Rhizobium meliloti. J. Bacteriol., 171: 4583-4588.
Kamberger, W., 1979. An Ouchterlony double diffusion study on the interaction between legume lectins and rhizobial cell surface antigens. Arch. Microbiol., 121: 83-90.
Kannenberg, E.L. and Brewin, N.J., 1989. Expression of a cell surface antigen from Rhizobium leguminosarum 3841 is regulated by oxygen and pH. J. Bacteriol., 171: 4543-4548.
Karr, D.B. and Emerich, D.W., 1989. Protein phosphorylation in Bradyrhizobium japonicum bacteroids and cultures. J. Bacteriol., 171: 3420-3426.
Kijne, J.W., Smit, G., Diaz, C.L. and Lugtenberg, B.J.J., 1988. Lectin-enhanced accumulation of manganese-limited Rhizobium leguminosarum cells of pea root hair tips. J. Bacteriol., 170: 2994-3000.
Kim, C-H., Tully, R.E. and Keister, D.L., 1989. Exopolysaccharide-deficient mutants of Rhizobium fredii HH303 which are symbiotically effective. Appl. Environ. Microbiol., 55: 1852-1854.

Kondorosi, A. and Johnston, A.W.B., 1981. The genetics of Rhizobium. In: K.L. Giles and A.G. Atherly (Editors), Int. Rev. Cytol. Suppl. 13, Biology of the Rhizobiaceae. Academic Press, New York, pp. 191-224.
Kondorosi, E., Banfalvi, Z. and Kondorosi, A., 1984. Physical and genetic analysis of a symbiotic region of Rhizobium meliloti: identification of nodulation genes. Mol. Gen. Genet., 193: 445-452.
Kondorosi, A., Kondorosi, E., Gyorgypal, Z., Banfalvi, Z., Gyuris, J., Putnoky, P., Grosskopf, E., John, M., Schmidt, J., Cam Ha, D.T., Lados, M., Slaska-Kiss, K. and Schell, J., 1988. Rhizobium meliloti nod and fix genes controlling the initiation and development of root nodules. In: H. Bothe, F.J. de Bruijn and W.E. Newton (Editors) Nitrogen Fixation: Hundred Years After. Gustav Fischer, Stuggart, pp. 399-403.
Kosslak, R.M., Bookland, R., Barkei, J., Paaren, H.E. and Appelbaum, E.R., 1987. Induction of Bradyrhizobium japonicum common nod genes by isoflavones isolated from Glycine max. Proc. Natl. Acad. Sci. USA, 84:7428-7432.
Law, I.J., Yamanoto, A.J., Mort, A.J. and Bauer, W.D., 1982. Nodulation of soybean by Rhizobium japonicum mutants with altered capsule synthesis. Planta, 154: 100-109.
Leigh, J.A., Signer, E.R. and Walker, G.C., 1985. Exopolysaccharide-deficient mutants of Rhizobium meliloti that form ineffective nodules. Proc. Natl. Acad. Sci. USA, 82: 6231-6235.
Leigh, J.A., Reed, J.W., Hanks, J.F., Hirsch, A.M. and Walker, G.C., 1987. Rhizobium meliloti mutants that fail to succinylate their calcofluor-binding exopolysaccharide are defective in nodule invasion. Cell, 51: 579-587.
Lie, T. A., Goktan, D., Engin, M., Pijnenborg, J. and Anlarsal, E., 1987. Co-evolution of the legume-Rhizobium association. Plant and Soil, 100: 171-181.
Long, S.R., 1984. Genetics of Rhizobium nodulation. In: T. Kosuge and E.W. Nester (Editors), Plant-Microbe Interactions. (Volume 1 Molecular and Genetic Perspectives) Macmillan Publishing Co., New York, pp. 265-306.
Long, S.R., 1989. Rhizobium-legume nodulation: life together in the underground. Cell, 56: 203-214.
Long, S.R. and Cooper, J., 1988. Overview of Symbiosis. In: D.P.S. Verma and R. Palacios (Editors), Molecular Genetics of Plant-Microbe Interactions. APS Press, St. Paul, Minnesota, pp. 163-178.
Long, S.R., Peters, N.K., Mulligan, J.T., Dudley, M.E. and Fisher, R.F., 1986. Genetic analysis of Rhizobium-plant interactions. In: B. Lugtenberg (Editor), Recognition in Microbe-Plant Interactions (NATO ASI Series, Series H), 4: 1-15.
Long, S.R., Reed, J.W., Himawan, J. and Walker, G.C., 1988. Genetic analysis of a cluster of genes required for synthesis of the calcofluor-binding exopolysaccharide of Rhizobium meliloti. J. Bacteriol., 170: 4239-4248.
Marvel, D.J., Torrey, J.G. and Ausubel, F.M., 1987. Rhizobium symbiotic genes required for nodulation of legume and nonlegume hosts. Proc. Natl. Acad. Sci. USA, 84: 1319-1323.
Masterson, R.V., Russell, P.R. and Atherly, A.G., 1982. Nitrogen fixation (nif) genes and large plasmids of Rhizobium japonicum. J. Bacteriol., 152: 928-931.
Mayrand, D. and Grenier, D., 1989. Biological activities of outer membrane vesicles. Canad. J. Microbiol., 35: 607-613.
McNeil, M., Darvill, J., Darvill, A.G. and Albersheim, P., 1986. The discernible structrual features of the acidic polysaccharides secreted by different Rhizobium species are the same. Carbohyr. Res., 126: 307-326.
Mellor, H.Y., Glenn, A.R., Arwas, R. and Dilworth, M.J., 1987. Symbiotic and competitive properties of motility mutants of Rhizobium trifolii TA1. Arch. Microbiol., 148: 34-39.
Mellor, R.B., Garbers, C., and Werner, D., 1989. Peribacteroid membrane nodulin gene induction by Bradyrhizobium japonicum mutants. Plant Mol. Biol., 12: 307-315.

Miller, K.J., Gore, R.S., Johnson, R., Benesi, A.J. and Reinhold, V.N., 1990. Cell-associated oligosaccharides of Bradyrhizobium spp. J. Bacteriol., 172: 136-142.
Müller, P., Hynes, M., Kapp, D., Neihaus, K. and Pühler, A., 1988. Two classes of Rhizobium meliloti infection mutants differ in exopolysaccharide production and in coinoculation properties with nodulation mutatns. Mol. Gen. Genet., 211: 17-26.
Mulligan, J.T. and Long, S.R., 1985. Induction of Rhizobium meliloti nodC expression by plant exudate requires nodD. Proc. Natl. Acad. Sci. USA, 82: 6609-6613.
Mulligan, J.T. and Long, S.R., 1989. A family of activator genes regulates expression of Rhizobium meliloti nodulation genes. Genetics, 122: 7-18.
Newcomb, W., 1981. Nodule morphogenesis. In: G.H. Bourne and J.F. Danielli (Editors), International Review of Cytology, Supplement 13, G, Academic Press, New York pp. 246-298.
Noel, K.D., Vanden Bosch, K.A. and Kulpaca, B., 1986. Mutations in Rhizobium phaseoli that lead to arrested development in infection threads. J. Bacteriol., 168: 1392-1401.
Nutman, P.S., 1959. Sources of incompatibility affecting nitrogen fixation in legume symbiosis. Symp. Soc. Exp. Biol., 13: 42-58.
Peters, N.K., Frost, J.W. and Long, S.R., 1986. A plant flavone, luteolin, induces expression of Rhizobium meliloti nodulation genes. Science, 233: 977-980.
Petit, A., Stougaard, J., Kuhle, A, Marcker, K.A. and Tempe, J., 1987. Transformation and regeneration of the legume Lotus corniculatus: A system for molecular studies of symbiotic nitrogen fixation. Mol. Gen. Genet., 207: 245-250.
Philip-Hollingsworth, S., Hollingsworth, R.I., Dazzo, F.B., Djordjevic, M.A. and Rolfe, B.G., 1989. The effect of interspecies transfer of Rhizobium host-specific nodulation genes on acidic polysaccharide structure and in situ binding by host lectin. J. Biol. Chem., 264: 5710-5714.
Prakash, R.K. and Atherly, A.G., 1984. Reiteration of genes involved in symbiotic nitrogen fixation by fast-growing Rhizobium japonicum. J. Bacteriol., 160: 785-787.
Prakash, R.K. and Atherly, A.G., 1986. Plasmids of Rhizobium and their role in symbiotic nitrogen fixation. Intl. Rev. Cytol., 104: 1-24.
Priefer, U.B., 1989. Genes involved in lipopolysaccharide production and symbiosis are clustered on the chromosome of Rhizobium leguminosarum biovar viciae VF39. J. Bacteriol., 171: 6161-6168.
Putnoky, P., Grosskopf, E., Camlta, D.T., Kiss, G.B. and Kondorosi, A., 1988. Rhizobium fix genes mediate at least two communication steps in symbiotic nodule development. J. Cell Biol., 106:597-607.
Puvanesarajah, V., Schell, F.M., Gerhold, D. and Stacey, G., 1987. Cell surface polysaccharides from Bradyrhizobium japonicum and a non-nodulating mutant. J. Bacteriol., 169: 137-141.
Redmond, J.W., Batley, M., Innes, R.W., Kuempel, P.L., Djordjevic, M.A. and Rolfe, B.G., 1986. Flavones induce expression of the nodulation genes in Rhizobium. In: B. Lugtenberg (Editor), Recognition in Microbe-Plant Symbiotic and Pathogenic Interactions (NATO ASI Series, Series H), 4: 115-121.
Robertson, J.G., Lyttleton, P., Bullivant, S. and Grayston, G.F., 1978. Membranes in lupin root nodules. I. The role of Golgi bodies in the biogenesis of infection threads and peribacteroid membranes. J. Cell Sci., 30: 129-149.
Rodriguez-Quinones, F., Banfalvi, Z., Murphy, P. and Kondorosi, A., 1987. Interspecies homology of nodulation genes in Rhizobium. Plant Mol. Biol., 8: 61-75.

Rolfe, B.G., Redmond, J.W., Batley, M., Chen, H., Djordjevic, S.P., Ridge, R.W., Bassam, B.J., Sargent, C.L., Dazzo, F.B. and Djordjevic, M.A., 1986. Intercellular communication and recognition in the Rhizobium-legume symbiosis. In: B. Lugtenberg (Editor), Recognition in Microbe-Plant Symbiotic and Pathogenic Interactions (NATO ASI Series, H), 4: 39-54.
Rossen, L., Johnston, A.W.B. and Downie, J.A., 1984. DNA sequence of the Rhizobium leguminosarum nodulation genes nodA, B and C required for root hair curling. Nucleic Acids Res., 12: 9497-9508.
Rossen, L., Shearman, C.A., Johnston, A.W.B. and Downie, J.A., 1985. The nodD gene of Rhizobium leguminosarum is autoregulatory and in the presence of plant exudate induces expression of nodABC genes. EMBO J., 4: 3369-3373.
Schlaman, H.R.M., Spaink, H.P., Okker, R.J.H. and Lugtenberg, B.J.J., 1989. Subcellular localization of the nodD gene product in Rhizobium leguminosarum. J. Bacteriol., 171: 4686-4693.
Schmidt, J., John, M., Wieneke, U., Krussmann, H.-D. and Schell, J., 1986. Expression of the nodulation gene nodA in Rhizobium meliloti and localization of the gene product in the cytosol. Proc. Natl. Acad. Sci. USA, 83: 9581-9585.
Schwedock, J. and Long, S.R., 1989. Nucleotide sequence and protein products of two new nodulation genes of Rhizobium meliloti, nodP and nodQ. Mol. Plant-Microbe Interactions, 2: 181-194.
Shearman, C.A., Rossen, L., Johnston, A.W.B. and Downie, J.A., 1986. The Rhizobium leguminosarum nodulation gene nodF encodes a polypeptide similar to acyl-carrier protein and is regulated by nodD plus a factor in pea root exudate. EMBO J., 5: 647-652.
Smit, G., Kijne, J.W. and Lugtenberg, B.J.J., 1987. Involvement of both cellulose fibrils and a Ca^{2+}-dependent adhesin in the attachment of Rhizobium leguminosarum to pea root hair tips. J. Bacteriol., 169: 4294-4301.
Smit, G., Kijne, J.W. and Lugtenberg, B.J.J., 1989a. Roles of flagella, lipopolysaccharide, and a Ca^{2+}-dependent cell surface protein in attachment of Rhizobium leguminosarum biovar viciae to pea root hair tips. J. Bacteriol., 171: 569-572.
Smit, G, Logman, T.J.J., Boerrigter, M.E.T.I., Kijne, J.W. and Lugtenberg, B.J.J., 1989b. Purification and partial characterization of the Rhizobium leguminosarum biovar viciae Ca^{2+}-dependent adhesin, which mediates the first step in attachment of cells of the family Rhizobiaceae to plant root hair tips. J. Bacteriol., 171: 4054-4062.
Sobral, B.W.S., 1989. Pulsed-Field Gel Electrophoresis Studies of Rhizobiaceae. Ph.D. dissertation, Iowa State University, Ames.
Spaink, H.P., Okker, R.J.H., Wijffelman, C.A., Pees, E. and Lugtenberg, B.J.J., 1987a. Promoters in the nodulation region of the Rhizobium leguminosarum Sym plasmid pRL1JI. Plant Mol. Biol., 9: 27-39.
Spaink, H.P., Wijffelman, C.A., Pees, E., Okker, R.J.H. and Lugtenberg, B.J.J., 1987b. Rhizobium nodulation gene nodD as a determinant of host specificity. Nature, 328: 337-340.
Spaink, H.P., Wijffelman, C.A., Okker, R.J.H. and Lugtenberg, B.E.J., 1989a. Localization of functional regions of the Rhizobium nodD product using hybrid nodD genes. Plant Mol. Biol., 12: 59-73.
Spaink, H.P., Weinman, J., Djordjevic, M.A., Wijffelman, C.A., Okker, R.J.H. and Lugtenberg, B.J.J., 1989b. Genetic analysis and cellular localization of the Rhizobium host specificity-determining NodE protein. EMBO J., 8: 2811-2818.
Stanfield, S., Telpi, L., O'Brochta, D., Helinski, D.R. and Ditta, G.S., 1988. The ndvA gene product of Rhizobium meliloti is required for ß-(1-2)glucan production and has homology to the ATP binding export protein HlyB. J. Bacteriol., 170: 3523-3530.
Staskawicz, B.J., Dahlbeck, D. and Keen, N.T., 1984. Cloned avirulence gene of Pseudomonas syringae pv. glycinea determines race-specific incompatibility on Glycine max (L.) Merr. Proc. Natl. Acad. Sci. USA, 81: 6024-6028.

Stougaard, J., Marcker, K.A., Otten, L. and Schell, J., 1986. Nodule-specific expression of a chimeric soybean leghaemoglobin gene in transgenic Lotus corniculatus. Nature 321: 669-674.
Surin, B.P. and Downie, J.A., 1988. Characterization of the Rhizobium leguminosarum genes nodLMN involved in efficient host-specific nodulation. Mol. Microbiol., 2: 173-183.
Surin, B.P. and Downie, J.A., 1989. Rhizobium leguminosarum genes required for expression and transfer of host specific nodulation. Plant Mol. Biol., 12: 19-29.
Swanson, J.A., Tu, J.K., Ogawa, J., Samga, R., Fisher, R.F. and Long, S.R., 1987. Extended region of nodulation genes in Rhizobium meliloti 1021. I. Phenotypes of Tn5 insertion mutants. Genetics, 117: 181-189.
Therisod, H., Weissborn, A.C. and Kennedy, E.P., 1986. An essential function for acyl carrier protein in the biosynthesis of membrane-derived oligosaccharides of E. coli. Proc. Natl. Acad. Sci. USA, 83: 7236-7240.
Torok, I., Kondorosi, E., Stepkowski, T., Posfai, J. and Kondorosi, A., 1984. Nucleotide sequence of Rhizobium meliloti nodulation genes. Nucleic Acids Res., 12: 9509-9524.
van Brussel, A.A.N., Zaat, S.A.J., Canter-Cremers, H.C.J., Wijffelman, C.A., Pees, E., Tak, T. and Lugtenberg, B.J.J., 1986. Role of plant exudate and sym plasmid-localized nodulation genes in the synthesis by Rhizobium leguminosarum of TSR factor which causes thick and short roots on common vetch. J. Bacteriol., 165: 517-522.
Verma, D.P.S. and Long, S., 1983. The molecular biology of Rhizobium-legume symbiosis. Internatl. Rev. Cytol., 14(Suppl.): 211-245.
Vincent, J.M., 1980. Factors controlling the legume-Rhizobium symbiosis. In: W.E. Newton and W.H. Orme-Johnson (Editors) Nitrogen Fixation, volume II. University Park Press, Baltimore, pp. 103-129.
Werner, D., Bassarab, S., Humbeck, C., Kape, R., innback, A., Mellor, R.B., Mõrschel,Parniske, M., Pausch, G., Rõhm, M., Schenk, S., Thierfelder, H., Thynn, M., Wetzel, A. and Wolff, A., 1988. Nodule proteins and compartments. In: H. Bothe, F.J. de Bruijn and W.E. Newton (Editors), Nitrogen Fixation: Hundred Years After. Gustav Fischer, Stuttgart, pp. 507-515.
Wijffelman, C., Zaat, B., Spaink, H., Mulders, I., van Brussel, T., Okker, R., Pees, E., de Maagd, R. and Lugtenberg, B., 1986. Induction of Rhizobium nod genes by flavonoids: differential adaptation of promoter, nodD gene and inducers for various cross-inoculation groups. In: B. Lugtenberg (Editor), Recognition in Microbe-Plant Symbiotic and Pathogenic Interactions (NATO ASI Series, Series H), 4: 123-135.
Williams, M.N.V., Hollingsworth, R.I., Klein, S. and Signer, E.R., 1990. The symbiotic defect of Rhizobium meliloti exopolysaccharide mutants is suppressed by lpsZ, a gene involved in lipopolysaccharide biosynthesis. J. Bacteriol., 172: 2622-2632.
Winsor, B.A.T., 1989. A nod at differentiation: the nodD gene product and initiation of Rhizobium nodulation. TIG, vol. 5.
Yao, P.Y. and Vincent, J.M., 1969. Host specificity in the root hair "curling factor" of Rhizobium sp. Aust. J. Biol. Sci., 22: 413-423.
Yelton, M.M, Mulligan, J.T. and Long, S.R., 1987. Expression of Rhizobium meliloti nod genes in Rhizobium and Agrobacterium backgrounds. J. Bacteriol., 169: 3094-3098.
Zaat, S.A.J., van Brussel, A.A.N., Tak, T., Pees, E. and Lugtenberg, B.J.J., 1987.Flavonoids induce Rhizobium leguminosarum to produce nodDABC gene-related factors that cause thick, short roots and root hair responses on common vetch. J. Bacteriol., 169: 3388-3391.
Zhan, H., Levery, S.B., Lee, C.C. and Leigh, J.A., 1989. A second exopolysaccharide of Rhizobium meliloti strain SU47 that can function in root nodule invasion. Proc. Natl. Acad. Sci. USA, 86: 3055-3059.

CHAPTER 12

ORGANIC AND INORGANIC INPUTS INTO LEGUME ROOT NODULE NITROGEN FIXATION

LIS ROSENDAHL*, ANDREW R. GLENN and MICHAEL J. DILWORTH
Nitrogen Fixation Research Group, School of Biological and Environmental Sciences, Murdoch University, Murdoch, Western Australia 6150

*Current address: Agricultural Research Department, Riso National Laboratory, DK-4000 Roskilde, Denmark

1 INTRODUCTION

In the mature legume nodule, the bacteroid is completely enclosed by the plant cell, and is therefore totally dependent on the plant for the nutrients required to maintain its cellular functions as well as to fix N_2. Indeed, this is true from the moment the infecting rhizobial cell enters the root tissue, whether the route of infection is via root hairs, cracks created by lateral roots or epidermal (Sprent et al., 1987). The ability of the plant to supply or withhold a variety of inorganic and organic nutrients is therefore an important factor in the overall success of the symbiosis. This chapter is concerned with what is supplied, how it is delivered, whether the supply is controlled and the site of any such controls.

2 NODULE CELL ARCHITECTURE

Infected nodule cells contain up to 20,000 bacteroids per cell in soybean (Bergersen and Goodchild, 1973). The cell vacuole is generally lost with the close-packing of the rhizobia, but remains in pea nodules. Bacteroids are surrounded, singly or in packets, by a peribacteroid membrane (Bergersen and Briggs, 1958; Robertson et al., 1978a,b). The plant membrane enclosing the microsymbiont retains the same orientation as the plasmalemma i.e., the surface facing the cell wall of the host cell now faces the bacteroid. The detailed cytology of the peribacteroid membrane is considered in Section 5.

The consequence of this arrangement is that all exchange of materials between bacteroid and host must occur across both the PBM and the bacteroid membrane. Since all inorganic and organic inputs to the bacteroid must cross the PBM, it constitutes one site where the plant could potentially control microbial activity. The ability of the plant to ration supplies to the bacteroid apparently varies with the nature of the nutrient - the relative impermeability of the PBM to sugars (McKay et al., 1985; Saroso et al., 1986; Udvardi et al., 1988a) appears to limit their access to the bacteroid, whereas C_4-dicarboxylates can be readily transported (Udvardi et al., 1988a).

Controls on access to various compounds could be 'accidental' or 'deliberate'. While legumes have no major metabolic requirement for cobalt, there is a requirement for it for bacteroid development and function. However, 90% or more of the cobalt in acutely deficient lupin nodules occurs as non-cobalamin forms apparently unavailable to the bacteroids for conversion to cobalamins (Dilworth et al., 1979). Thus, there appears to be a restriction on cobalt access to the bacteroids which may well reside in the PBM; since the plant would appear to gain no benefit from such a control it might be reagrded as 'accidental.'

Where the PBM does readily transmit an inorganic nutrient like phosphorus, the question of 'deliberate' control is relevant. Can the plant actually limit the amount it is losing to the effectively external bacteroid compartment? Establishing whether a control exists requires careful experiments if the bacteroids show symptoms of nutrient deficiency in deficient plants but not in adequately-supplied plants. For most inorganic nutrients such experiments have not been done. A 'deliberate' control is ruled out where bacteroids have normal nutrient levels even though the host plants are deficient - thus, even in severely phosphorus-deficient snake beans the bacteroids appear to be phosphorus-adequate, since they contain polyphosphate (Smart et al., 1984).

The bacterial genotype can affect the nutrient concentrations in the plant (and presumably then the bacteroids). Thus, cobalt deficiency in Lupinus angustifolius on some soils can be alleviated by using a specific inoculant strain (Chatel et al., 1978), resulting in higher nodule cobalt concentrations (Dilworth et al., 1979). Peanut nodule concentrations of P, K, Ca, Mg, B, Mn, Zn, Fe, Cu and Na are also affected by rhizobial strain (Howell, 1987). The mechanisms by which the internal root nodule bacteria can exert such effects on external plant uptake or internal partitioning may well be important both theoretically and practically. Whether root nodule bacteria with higher affinity uptake systems could be more effective at fixing N_2 in the nodules of legumes growing under nutrient deficiencies remains to be established.

3 INORGANIC NUTRIENTS

3.1 Nutrition of root nodule bacteria

Inorganic nutrients are required by the root nodule bacteria for a variety of purposes extending through all phases of infection and nodule establishment to steady-state nitrogen fixation. It is functionally difficult to separate off those important only to the steady-state. Many inorganic nutrtients are involved in the metabolism necessary to maintain the bacteroid without being directly involved in the N_2 fixation or closely related machinery. Only a limited range of nutrients has therefore been considered. For reviews of the neglected field of the mineral nutrition of root nodule bacteria, see Munns (1977), Robson (1978) and O'Hara et al. (1988a).

To determine whether an inorganic nutrient is reaching the bacteroid in adequate amounts, it is necessary to develop criteria for the response of the bacteria to deficiency of that nutrient. Such criteria may be indicators of excess (synthesis of reserves like polyphosphate), or limitation (derepressed transport systems, intra- or extra-cellular enzymes), identified with nutrient-limited continuous cultures in the laboratory. Measurements of such parameters in the bacteroid may help to define its nutrient status in the nodule; this approach is only valid if the parameters are regulated in the bacteroid in the same way as they are in laboratory cultures. This 'bioassay' approach has given useful results with phosphorus (Smart et al., 1984) and sulphur (O'Hara et al., 1987).

3.2 Iron

Iron is required by bacteroids for a variety of proteins related to N_2 fixation - nitrogenase, hydrogenase, non-haem electron carriers - and, at least in some systems, for the haem of leghaemoglobin. Despite these important functions, little is known of the mechanisms of iron transfer to bacteroids.

Iron deficiency is common in nodulated legumes in alkaline soils (see e.g., chick pea, Rai et al., 1982; lentil, Rai et al., 1984; French bean, Hemantaranjan and Garg, 1986; peanut, O'Hara et al., 1988b). The most iron-sensitive step may well be different in different symbioses. In L. angustifolius, nodules develop normally under iron-deficiency once nodule initiation has occurred (Tang et al., 1990a). Foliar iron applications to iron-deficient peanuts allow normal development of nodules already initiated but arrested by lack of iron (O'Hara et al., 1988b). By contrast, iron is not mobile in lupins (Tang et al., 1990b) and foliar iron has no effect on nodulation. Rhizobial strain affects nodule formation and function under iron deficiency; cowpea Bradyrhizobium strain NC92 produces more nodules on peanut under iron-deficient conditions than other strains (O'Hara et al., 1988c). The mechanisms behind such effects are unknown.

In other Gram-negative bacteria, iron deficiency results in synthesis of iron-chelating compounds (siderophores) and the transport systems for the iron-siderophore complexes (see Neilands, 1982; Neilands et al., 1987). Root-nodule bacteria produce a variety of iron-chelating compounds (Guerinot, 1990), including citric acid from Bradyrhizobium japonicum 61A152 (Bosch et al., 1988), anthranilic acid from Rhizobium leguminosarum GF160 (Rioux et al., 1986a, b), rhizobactin from Rhizobium meliloti DM14 (Smith and Neilands, 1984; Smith et al., 1985), another rhizobactin-like compound from R. meliloti 1021 (Schwyn and Neilands, 1987) and catechols from R. leguminosarum IARI102 (Patel et al., 1988), a cowpea Rhizobium RA-1 (Modi et al., 1985) and Bradyrhizobium strains NC92 and 5a/70 (Nambiar and Sivaramakrishnan, 1987). The rhizobactins are unusual among bacterial siderophores in carrying neither catechol nor hydroxamate functions (Schwyn and Neilands, 1987).

A mutant of R. leguminosarum which appears to have a defect in its iron acquisition system, possibly in its ability to process iron-anthranilate, is ineffective on peas (Nadler et al., 1990).

While some of these iron chelators accelerate iron uptake into iron-deficient rhizobial cells in the laboratory, their signficance to the nodulated legume is unknown. O'Hara et al. (1988c) reported catechol-like compounds in solutions around peanuts inoculated with strain NC92 but not with other strains. Proof that the improvement in nodulation with NC92 was due to the siderophore requires a control with a siderophore-negative mutant. However, two clones of NC92 from different sources do not produce catechol-type siderophores under the conditions given by Nambiar and Siveramakrishnan (1987) or under conditions where strains of Escherichia coli produce enterochelin (K.C. Carson, unpublished).

3.3 Molybdenum

Molybdenum is directly required for nitrogenase synthesis, both as a component of FeMoco (see Chapter 2) and as a regulator of the synthesis of nitrogenases (see Chapter 3). Molybdate uptake in diazotrophs has been studied in C. pasteurianum, K. pneumoniae, A. vinelandii and more recently in B. japonicum. In C. pasteurianum, molybdate uptake depends on the presence of fermentable substrate and results in accumulation of molybdenum bound to protein; the accumulation is largely prevented by ammonia (Elliott and Mortenson, 1975, 1976). In Klebsiella, Mo accumulation was prevented by O_2 and NH_4^+, conditions which also repressed nitrogenase synthesis (Pienkos and Brill, 1981). Azotobacter is well known as an accumulator of Mo, the metal being stored complexed to a storage protein (Pienkos and Brill, 1981) synthesized irrespective of the presence of ammonia. The occurrence of storage proteins makes it difficult to assess whether these uptake systems are active; if the molybdenum concentration in equilibrium with the storage protein complex is lower than the free concentration in the environment, a facilitated diffusion system would concentrate the metal. Indeed, if the chemistry of the molybdenum is altered by reduction before binding to storage proteins, uptake of Mo could be driven by an inward-directed concentration gradient of molybdate.

For root-nodule bacteria, molybdenum uptake has been studied only in B. japonicum. In bacteroids isolated from soybean nodules, molybdate uptake was followed using filtration assays where cells were washed with solutions containing 1 mM molybdate (Graham and Maier, 1987). It is not clear how much exchange-driven loss of radioactive molybdate might have resulted. Bacteroids from different strains of B. japonicum differed in absolute rate of uptake and in apparent K_m (from 45 nM for USDA136 to 360 nM for USDA140). It cannot be concluded that these differences will reflect differences in actual uptake in the nodule in the absence of data on cytosolic concentrations of Mo, PBM permeability and peribacteroid space concentration. Uptake was unaffected by respiratory inhibitors but was sensitive to the ionophores CCCP and FCCP (Maier and Graham, 1988). Transport was reported to depend on ΔpH; this was interpreted to mean that the species transported was H_2MoO_4. Whether the PBM affects Mo transfer to the bacteroids is unknown; the lack of functional nitrogenase in Mo-deficient nodules is prima facie evidence that under deficiency conditions inadequate Mo reaches the bacteroids.

3.4 Vanadium

If root-nodule bacteria were to carry the vnf nitrogenase system, vanadium uptake into legumes and transfer to bacteroids might be important. It now appears unlikely (see Chapters 3 and 7) that the vnf genes are present in those root nodule bacteria examined. When Medicago truncatula (R. meliloti WSM419), Ornithopus sativus (Bradyrhizobium lupini WU 425) and Trifolium subterraneum (R. trifolii WU95) were grown under severely Mo-deficient conditions in solution cultures, and supplemented with 0.1 μM Mo, or V, or both, no ethane was produced during acetylene reduction assays (M.J. Dilworth and J.F. Loneragan, unpublished). This implies the absence of a V nitrogenase or nitrogenase-3 system (see chapter 3). The question of V uptake and transport therefore appears to be irrelevant.

3.5 Nickel

Nickel is a constituent of a variety of uptake hydrogenases (see Evans et al., 1987), including that from B. japonicum (Harker et al., 1984; Stults et al., 1984; Arp, 1985); deficiency in laboratory culture reduces hydrogenase activity (Klucas et al., 1983). The hup cluster in Alcaligenes eutrophus contains a gene for a nickel uptake system (Eberz et al., 1989). The restoration of hydrogenase activity in some mutants of E. coli by elevated nickel concentrations also implies a transport system mutation (Waugh and Boxer, 1986; Chaudhuri and Krasna, 1987). A mutant of B. japonicum constitutive for uptake hydrogenase also absorbs more nickel than the wild type, implying that nickel concentration in the cell may be involved with the regulation of hydrogenase synthesis (Stults et al., 1987). There appear to be two Ni-uptake systems in A. eutrophus, one carrying Ni or Mg, and the other specific for Ni and hup-associated (Eberz et al., 1989); accumulated nickel was rapidly exchangeable with that in the medium. Nickel uptake in B. japonicum is insensitive to respiratory inhibitors and ionophores and accumulated Ni does not exchange with extracellular metal (Stults et al., 1987). These observations raise questions whether the system genuinely active transport or driven by a complexing reaction. Indeed, an intracellular Ni-complexing system may be essentially pulling Ni into the cell down a concentration gradient. How freely nickel crosses the PBM is yet to be determined. In soybeans in solution culture, nickel deficiency can decrease hydrogenase activity, implying that indequate Ni reaches the bacteroid under such extreme conditions (Klucas et al., 1983). Nickel deficiency seems unlikely to be agronomically important (Dalton et al., 1985).

3.6 Cobalt

Cobalt is vital to bacteroid multiplication because of its effects (as cobalamin derivatives) on ribonucleotide reductase for DNA synthesis (Cowles and Evans, 1968; Cowles et al., 1969), methylmalonyl-CoA mutase for propionate metabolism (De Hertogh et al., 1964) and methionine synthetase (Sato et al., 1974, 1982; Dilworth and Bisseling, 1984). Its transfer from the plant to the bacteroid can be inferred to be slow (see Section 2).

3.7 Sulphur

Sulphur-limited continuous cultures of cowpea or soybean bradyrhizobia induce an alkaline sulphatase and a high affinity (apparent K_m 3 - 20 μM) sulphate transport system (O'Hara et al., 1987). Neither of these traits is evident in bacteroids of sulphur-deficient soybean or peanut, suggesting that bacteroids are adequately supplied through the PBM even when the legume is deficient (O'Hara et al., 1987). The sulphatase is not apparently present in fast-growing root-nodule bacteria (O'Hara et al., 1987), limiting this approach to the bradyrhizobia.

3.8 Selenium

Selenium has been implicated as a component of hydrogenase in B. japonicum (Boursier et al., 1988); the lack of codons for selenocysteine in the published sequence (Sayavendra-Soto et al., 1988) raises the question of whether Se is incorporated into Fe-S or Fe-Ni-S centres in place of sulphur. While Se deficiency lowered hydrogenase activity in bacteroids of Vicia faba, no such effect was seen for soybean or white lupin, and a field effect seems unlikely (Boursier et al., 1988; Evans et al., 1988).

3.9 Oxygen supply

Oxygen is required for bacteroid respiration and may also be a powerful irreversible inhibitor of nitrogenase (Robson and Postgate, 1980). This apparent conflict is solved by leghaemologobin within the bacteroid-infected cells, which facilitates diffusion of O_2 into the bacteroids to support synthesis of ATP (Appleby et al., 1975) in such a way that the equilibrium concentration of free O_2 at the bacteroid surface is very low (Dilworth and Appleby, 1979; see Chapter 5). A number of studies (e.g. Witty et al., 1984; Hunt et al., 1987) have suggested that treatments like stem girdling, continuous darkness or nitrate addition, which lower N_2 fixation, do so by changing the diffusion barrier of the nodule inner cortex. This affects the supply of O_2 to the bacteroid. Recent work by Layzell et al. (1990) using pulse-modulated spectroscopy provides direct evidence that nitrogenase activity is limited by O_2. This raises some intriguing possibilities of developing plants with changes in the diffusion barrier and perhaps increased rates of N_2 fixation.

With the discovery that leghaemoglobin is also found in uninfected cells within the nodule, this protein may also have a role in facilitation of O_2 supply for urate oxidation in ureide-producing legumes (Van den Bosch and Newcomb, 1988; see Chapter 13).

4 ORGANIC INPUTS TO THE BACTEROID

4.1 Compounds available in nodules

The growth and differentiation of bacteroids and N_2 fixation require organic materials from the host. A range of auxotrophic (Scherrer and Dénarié, 1971) and gluconeogenic mutants (McKay et al., 1985; Finan et al., 1988) show that the plant must be able to supply at least small amounts of a wide range of substrates like amino acids, vitamins and sugars.

Substrates that are present in the plant cytosol in high concentrations (i.e., high pool size) might be regarded as suitable candidates for supporting bacteroid metabolism. However, some compounds which are present in large amounts in nodules turn over relatively slowly, and are thus unlikely to contribute significantly to the carbon economy of the bacteroid. In contrast, compounds that are present at relatively low concentrations in the plant cytosol, but which have a very rapid turnover, might be important to bacteroid physiology. It is useful to know something about the variety of carbon sources in the nodule, their rate of turover, and how they vary in concentration with time and bacterial strain.

Nodule cytosol contains a range of carbohydrates, with sucrose being the major ^{14}C-labelled product when leaves are fed $^{14}CO_2$ (Antoniw and Sprent, 1978). Glucose, fructose, maltose, trehalose and cyclitols like myo- and chiro-inositol, D-pinitol and ononitol are also present (Kouchi and Yoneyama, 1984; Streeter, 1987; Lafontaine et al., 1989). Using ^{13}C-labelling techniques, Kouchi and Yoneyama (1984) showed that cyclitols are unlikely to be involved in nodule metabolism because of their slow turnover rate. The role of these compounds is still unknown.

Nodules contain a wide range of organic acids (Stumpf and Burris, 1981; Streeter, 1987a, Rosendahl et al., 1990), predominantly fumarate, malate, succinate, 2-oxoglutarate, citrate, malonate, trans-aconitate and benzoate. There are significant differences in the organic acid composition of nodules from different legumes (see Kouchi and Yoneyama, 1986; Streeter, 1987a; Lafontaine et al., 1989; Rosendahl et al., 1990). Similarly, the concentrations of some organic acids, such as urocanate, vary in the plant cytosol according to the rhizobial strain used as inoculant (Streeter, 1987b). This may indicate that such organic acids are consumed by some strains but not others, or that some rhizobia stimulate the accumulation of these acids and others do not. The use of mutants lacking the ability to transport or metabolize specific organic acids may help to determine if these compounds are likely to be important in nodule bacteroid physiology (see Lafontaine et al., 1989; Rosendahl et al., 1990). Such experiments have shown considerably higher concentrations of malate (and carbohydrates) in French bean nodules containing a C_4-dicarboxylate utilization mutant of R. leguminosarum bv. phaseoli (Lafontaine et al., 1989), whereas pea nodules formed by a dicarboxylate transport mutant of R. leguminosarum bv. viciae contained lower concentrations of organic acids than nodules induced by the wild type (Rosendahl et al., 1990). The reason for this apparent discrepancy is not immediately obvious, though it is intriguing that malate appeared to be the only C_4-dicarboxylate found in significant quantities in the nodules of the French bean.

The legume nodule also contains a range of amino acids, especially aspartate, glutamate, asparagine, arginine, serine, 4-aminobutyrate, glutamine, alanine and proline (see, e.g., Streeter, 1987; Jin et al., 1990a; Rosendahl et al., 1990). A rhizopine (3-O-scyllo-inosamine, Tempe and Petit, 1983), a nitrogenous compound produced by the plant specifically in response to the presence of rhizobia, is present in nodules containing R. meliloti L5-30 but not other strains (Murphy et al., 1987). Since most strains of R. meliloti manage to nodulate and fix N_2 without inducing any known rhizopine, it seems unlikely that these compounds play an essential part in nodule metabolism. Any other role for them, e.g. in survival and persistence in senescing nodules, remains to be evaluated in field experiments.

For any, or all, of these carbon sources to be of value in fuelling the bacteroid inside the PBM two criteria must be satisfied. Firstly, the compound must be transported across both the peribacteroid and bacteroid cytoplasmic membranes, and, secondly, the rhizobial strains must have the capacity to degrade the particular substrate.

4.2 Nutritional capabilities of rhizobia

(Brady)rhizobia utilize a wide range of carbon substrates for growth (Graham, 1964; Dilworth and Glenn, 1981). Disaccharides are utilized by Rhizobium spp., but not by Bradyrhizobium spp., since the latter lack both disaccharide uptake systems and disaccharidases (Glenn and Dilworth, 1981), though there are occasional reports (e.g., Meyer and Pueppke, 1980) of individual strains of bradyrhizobia utilizing C_{12}-sugars.

In most rhizobia, hexoses are catabolized via the Entner-Doudoroff (ED) and the pentose phosphate pathways (Ronson and Primrose, 1979; Glenn et al., 1984; McKay et al., 1989). Most reports of the Entner-Doudoroff pathway in rhizobia have failed to demonstrate the enzymes required for the further degradation of the pyruvate and glyceraldehyde 3-phosphate generated by 6-phosphogluconate dehydratase and 2-keto-3-deoxy-6-phosphogluconate aldolase (the "Entner-Doudoroff enzyme"). However, Saroso et al. (1986) and McKay et al. (1989) showed that in cowpea Rhizobium NGR234 and R. leguminosarum MNF300, respectively, the enzymes of the terminal part of the Embden-Meyerhof-Parnas (EMP) pathway are present. The latter pathway is incomplete in most rhizobia, because of the lack of phosphofructokinase (Ronson and Primrose, 1979; Reibach and Streeter, 1983; Glenn et al., 1984; Saroso et al., 1986; McKay et al., 1989). There are occasional report of low activities of phosphofructokinase (see, e.g., Stowers and Elkan, 1985; Kouchi et al., 1988), but their biological significance is difficult to assess. Claims that both the EMP and ED pathways exist in a wide range of rhizobia (Chakrabarty et al., 1987) are inconsistent with data from other laboratories.

Evidence for the pentose phosphate pathway in rhizobia has usually been based on the presence of 6-phosphogluconate dehydrogenase, the first enzyme of the pathway (see, e.g., Ronson and Primrose, 1979; Glenn et al., 1984). However, the presence of the first enzyme of a pathway, while suggestive, is far from unequivocal proof that the whole pathway is present. Saroso et al. (1986) and McKay et al. (1989) demonstrated all the enzymes of the pentose phosphate pathway in cowpea Rhizobium NGR234 and R. leguminosarum MNF300, respectively.

Catabolism of pentoses like L-arabinose is common in rhizobia and bradyrhizobia, leading to the formation of L-2-oxo-3-deoxyarabinonate (Duncan, 1979; Dilworth et al., 1986). In rhizobia, this is then converted to 2-oxoglutarate (Duncan, 1979; Dilworth et al., 1986), while in bradyrhizobia it is cleaved to pyruvate and glycolaldehyde (Pedrosa and Zancan, 1974; Duncan, 1979).

Aromatic acids like mandelate, 4-hydroxymandelate (Chen et al., 1989), benzoate or 4-hydroxybenzoate (Chen et al., 1984a, b; 1985; Parke and Ornston, 1985) are metabolized by ortho-cleavage of the aromatic ring via the catechol and protocatechuate branches of the 3-oxoadipate pathway (see Stanier and Ornston, 1973). Some strains of root nodule bacteria have both branches e.g., R. trifolii TA1 (Chen et al., 1984; Sinorhizobium fredii USDA201 (D. Parke and A.R. Glenn, unpublished), whereas others e.g., R. leguminosarum MNF300 (Chen et al., 1984a) and R. trifolii WU95 (F. Rynne and D. Parke, unpublished) have only the protocatechuate branch. In addition, Chen et al. (1989) measured the meta-cleavage enzyme, catechol 2,3-dioxygenase, in cell-free extracts of R. trifolii TA1 grown on mandelate or benzoate (Chen et al., 1989), suggesting that some rhizobia may possess the alternative pathway for aromatic ring degradation (see Stanier and Ornston, 1973). However, in the absence of measurements of the remaining meta-cleavage pathway enzymes, its presence remains speculative.

C_4-Dicarboxylic acids like succinate, fumarate and malate are metabolized via the tricarboxylic acid cycle (TCA), involving malic enzyme and pyruvate dehydrogenase (McKay et al., 1988; Copeland et al., 1989). In R. leguminosarum MNF300, malic enzyme appears to be poorly regulated and is present in cells grown on a sugar or an organic acid (McKay et al., 1989). The presence of specific TCA cycle enzymes in rhizobia has been reported in a number of studies (see e.g., Kurz and LaRue, 1977; Rawsthorne et al., 1980), though such studies have usually not measured all of the enzymes and so do not provide sufficient evidence for assessing the importance of the cycle in rhizobial physiology. However, McKay et al. (1989) have shown that the complete TCA cycle exists in R. leguminosarum MNF3841 and that the activities of the enzymes do not vary significantly with the growth conditions.

Amino acids are suitable growth substrates for rhizobia (e.g., glutamate, Poole et al., 1986; Jin et al., 1990a; histidine, Dilworth et al., 1983; 4-aminobutyrate, Jin et al., 1990b; proline, M.J. Dilworth and A.R. Glenn, unpublished). Knowledge of the pathways for the catabolism of amino acids in rhizobia is sketchy and is usually restricted to measurements of the first enzyme of a pathway (e.g., Dilworth et al., 1983). Similarly, little attention has been paid to the mechanism whereby the ammonia is actually released from the amino acid. 4-Aminobutyrate catabolism in cowpea Rhizobium NGR234 involves a series of amino-group transfers involving 4-aminobutyrate:2-oxoglutarate transaminase, glutamate:oxaloacetate aminotransferase and alanine dehydrogenase (Jin et al., 1990b).

Rhizobia growing on substrates like C_4-dicarboxylic acids, or amino acids like glutamate, proline or 4-aminobutyrate that lead directly into the TCA cycle have to overcome the twin problems of maintaining the TCA cycle when intermediates are withdrawn for biosynthesis, and also of synthesizing sugars for cell wall and nucleotide biosynthesis. In some rhizobia, the TCA cycle is maintained by the combined activity of malic enzyme and pyruvate dehydrogenase (McKay et al., 1988; Copeland et al., 1989). Gluconeogenesis is carried out via phosphoenolpyruvate carboxykinase (McKay et al., 1987; Finan et al., 1988).

4.3 Physiology of bacteroids

Naked bacteroids isolated from a range of legumes commonly oxidize organic acids like succinate, fumarate, malate, pyruvate and acetate, but not sugars (see e.g., Tuzimura and Meguro, 1960; Stoval and Cole, 1978). Similarly, such organic acids usually stimulate N_2-fixation by isolated soybean bacteroids (e.g. Bergersen and Turner, 1967) but sugars do not. This orthodoxy was challenged by the work of Rigaud and his collaborators (Trinchant et al., 1981; Herrada et al., 1989), who pointed out that most studies with isolated bacteroids have been carried out at atmospheric pO_2, a condition entirely different to that in the nodule. Some studies (e.g. Tajima et al., 1985, 1986) have used bacteroids suspended in leghaemoglobin solutions as a way of reducing the O_2 concentration to levels that may more closely approximate those in the nodule. Since in some obligately aerobic bacteria a number of transport systems are influenced by pO_2 (Pelliccione et al., 1979) and some rhizobial cell surface antigens are also regulated in part by pO_2 (Kannenberg and Brewin, 1989), the objections of Rigaud and his co-workers has some validity (see also section 5).

The techniques used in the measurement of bacteroid uptake and respiration have not always been appropriate. Claims that isolated lupin bacteroids take up ^{14}C-glucose and convert it into sucrose and a range of TCA cycle intermediates (Kretovich et al., 1985) are difficult to assess because the study failed to give the actual concentration of glucose used. Similarly, when materials are forced into bacteroids by using high external concentrations of substrate (Hooymans and Logman, 1984), the data are impossible to interpret unequivocally. Although in the latter study a high external concentration of sugar increased O_2 uptake by isolated pea bacteroids it is impossible to say if this was due to actual metabolism of the carbohydrate, sugar stimulation of endogenous respiration or to an osmotic effect on the cell.

In addition, bacteroids may be damaged during isolation, perhaps by the loss of periplasmic solute-binding proteins. Thus pea bacteroids fail to transport significant amounts of glucose, fructose or sucrose, even though free-living cells of R. leguminosarum bv. viciae transport these sugars constitutively (Dilworth and Glenn, 1981). Herrada et al. (1989) have reported active uptake of glucose by bacteroids of R. leguminosarum bv. phaseoli. Isolated bacteroids of R. leguminosarum bv. viciae lose a substantial amount of the periplasmic RP1-encoded β-lactamase and constitutive alkaline phosphatase (A.R. Glenn and M.J. Dilworth, unpublished) and San Francisco and Jacobson (1985) have suggested a similar loss of periplasmic proteins during isolation of soybean bacteroids. Although soybean bacteroids take up sugars only very slowly, they accumulate α,α-trehalose (Streeter, 1985) and synthesize trehalose synthetase (Salminen and Streeter, 1986). The trehalose is not metabolized and its function in the bacteroid remains obscure. However, trehalose is accumulated by a wide variety of eukaryotes as a preparation for dormancy (Crowe et al., 1984) and McBride and Ensign (1987) showed that spores of the soil bacterium Streptomyces griseus with high concentrations of trehalose survive better than those with low. By analogy, the trehalose in bacteroids may protect the rhizobia at nodule senescence.

In contrast, isolated bacteroids transport C_4-dicarboxylic acids (Glenn et al., 1980; Ronson et al., 1981; Finan et al., 1983; McAlister and Lepo, 1983; Saroso et al., 1984; San Francisco and Jacobson, 1985; Humbeck and Werner, 1989). In most strains of root nodule bacteria the dicarboxylate transport system (dct) is inducible; its presence in isolated bacteroids therefore suggests that they receive C_4-dicarboxylates from the plant cytosol. The elegant work of Ronson and his collaborators has established the regulatory role of the dctB and dctD genes and the requirement for the ntrA product for expression of the dicarboxylate transport protein gene dctA (Ronson et al., 1985, 1987). Also, mutants defective in the ntrA gene in cowpea Rhizobium NGR234 were unable to transport C_4-dicarboxylates or fix N_2. In R. meliloti, the dct genes are located on the exo-megaplasmid (Watson et al., 1988). Introducing the dct genes from R. meliloti into B. japonicum resulted in increased rates of succinate uptake and in vitro acetylene reduction (Birkenhead et al., 1988), but the effects of such genetic manipulations on yield under field conditions have yet to be established (see Chapter 17).

The rates of C_4-dicarboxylate transport in most bacteroid preparations are slower than in their free-living counterparts (see, in R. leguminosarum WU235,

Glenn et al., 1980; Glenn and Dilworth, 1981; and in cowpea Rhizobium NGR234, Saroso et al., 1984). In some cases (e.g., San Francisco and Jacobson, 1985), reported C_4-dicarboxylate uptake activities in isolated bacteroids were only 5% of those in free-living cells. Salminen and Streeter (1987) have also reported low rates of succinate and malate transport (0.3-0.5 nmol min^{-1} mg $protein^{-1}$) by isolated soybean bacteroids. There is some indication from inhibition (San Francisco and Jacobson, 1985) or kinetic studies (Humbeck and Werner, 1989) that B. japonicum may have more than one C_4-dicarboxylate uptake system. The K_m for succinate uptake by purified soybean bacteroids was similar to that of free-living cells (10 μM), but purified bacteroids failed to take up malate, even though the laboratory cells had a system with a K_m of 6.7 μM. It is not clear if this implies a change in the properties of the permease or the synthesis of a bacteroid-specific transport system. Humbeck and Werner (1987) have reported that B. japonicum 61A101 may have two dct systems with apparent K_m's of 2.4 μM and 172 μM; the use of concentrations of succinate 50 times the apparent K_m of the low affinity system may give misleading data. Succinate uptake in this strain was relatively insensitive to malate, in contrast to the findings of San Francisco and Jacobson (1985).

Amino acids are reported to be transported by isolated bacteroids (see Glenn and Dilworth, 1981; Dilworth and Glenn, 1981, 1985; Salminen and Streeter, 1987; Udvardi et al., 1989; Jin et al., 1990b). A role for glutamate in bacteroid metabolism has been suggested by Kahn and his collaborators, in which glutamate is transported from the plant to the bacteroid and ammonia is released from the bacteroid to the plant. A number of Tn5-induced glutamate catabolic mutants were apparently unable to fix N_2. In contrast, all of seven mutants of R. leguminosarum unable to grow on glutamate as a sole source of C and N nodulated and fixed N_2 (P.S. Poole, unpublished). Some organisms utilize L-glutamate only poorly as a carbon source (Jin et al., 1990a) and the synthesis of the L-glutamate transport system in some rhizobia is sensitive to ammonia repression (Jin et al., 1990a). However, Kahn also suggested a shuttle system whereby the transport of one solute is linked to the conjoint movement of another; in particular, Kahn et al. (1985) raised the possibility of a malate-aspartate shuttle and Streeter and Salminen (1988) have presented some data suggesting a malate-aspartate shuttle in soybean bacteroids. The observed rates of uptake of amino acids by isolated bacteroids can be very slow (<0.08 nmol min^{-1} mg $protein^{-1}$; Salminen and Streeter, 1987), making the significance of such rates difficult to assess. Studies using peri-bacteroid units (see section 5) are likely to be of more use in providing answers about the importance of shuttle mechanisms.

Transport studies with isolated bacteroids have given useful, if somewhat limited, information. The in vitro assay conditions have, in most cases, been quite different from those in the nodule and it is therefore difficult to guage the significance of extremely slow transport rates. Given that the rate of solute transport is frequently the rate-limiting step in the utilization of a carbon source and that the surface area to volume ratio of bacteroids is lower than for free-living cells, it is not unexpected that bacteroids should have lower rates of transport than their free-living counterparts. In addition, the mature bacteroids are not actively growing cells and are likely to be more akin to resting cells than the logarithmic-phase cells with which they are usually compared. In theory, substrates which contribute significantly to the fuelling of nitrogenase activity should have rates of transport consistent with the observed rates of acetylene reduction. When a transport rate represents only a minute fraction of the necessry carbon flow required to support nitrogenase it is not possible to draw firm conclusions about its significance. Because of these limitations on the information that can be gained from transport studies, alternative indices of the metabolic capacities of bacteroids are required, such as studies with defined catabolic mutants or measurements of the activities of intracellular catabolic enzymes.

4.4 Studies with mutants

Several studies have examined the nodulation and fixation characteristics of mutants unable to catabolize specific carbon sources. To be of value, such mutants must be well-characterized with respect to growth properties, not be metabolically crippled, demonstrated to be due to a specific change in DNA base sequence or the unique insertion of a transposon (by transductional analysis or DNA hybridization), and to be stable; these criteria have not always been satisfied. Mutants defective in the ability to degrade C_{12}, C_6 or C_5 sugars (see e.g., Ronson and Primrose, 1979; Glenn et al., 1984; Arwas et al., 1986; Dilworth et al., 1986) are able to nodulate and fix N_2 as well as the wild type. However, many reports (see e.g., Glenn and Brewin, 1981; Ronson et al., 1981; Finan et al., 1983; Arwas et al., 1985; Finan et al., 1988; Humbeck and Werner, 1989) have shown that defects in the ability to transport C_4-dicarboxylates (succinate, fumarate and malate) lead to a Fix^- phenotype. This suggests that a supply of these organic acids is essential for N_2 fixation. The route whereby these organic acids is provided is not clear; it is unlikely that the mitochondria would be able to provide the required amounts as they possess low activities of malic enzyme (Day and Mannix, 1988) and because the low pO_2 in nodules is likely to restrict mitochondrial activity (Rawsthorne and LaRue, 1986). Other anaplerotic mechanisms may be involved in the supply of organic acid to bacteroids (Rosendahl et al., 1990). The reason why C_4-dicarboxylic acids should be such an essential element in N_2 fixation is not obvious, but may be involved with the thermodynamic efficiency of utilization of these compounds (Westerhoff et al., 1983). Since bacteroids of many rhizobia contain significant carbon reserves in the form of poly-β-hydroxybutyrate, they should theoretically be able to generate C_4-dicarboxylates in situ, which should then be available to support N_2 fixation.

Since dct-defective mutants nodulate and their bacteroids differentiate, it is clear that they must be able to obtain sufficient carbon from the host plant to fuel such processes. Despite attempts to determine the identity of the carbon sources being used under such circumstances, they remain unknown. Double mutants of R. leguminosarum lacking the dct system and the ability to catabolize a range of other carbon sources including C_5, C_6 and C_{12} sugars and glutamate (Arwas et al., 1986) still nodulated as well as the dct^- parent strain.

4.5 Studies of intracellular enzyme levels

Although surface properties like membrane transport may be damaged during bacteroid isolation, intracellular enzymes should not be affected. If the regulation of metabolic pathways in the free-living form is known, the presence or absence of particular enzymes in the bacteroid may be a useful indication about metabolites that pass from the plant to the bacteroid and induce particular enzymes. This approach is predicated on the assumption that the genetic regulation of catabolic pathways is the same in bacteroids and free-living cells. In evaluating data of this type, it is important to consider the likely biological significance of the activities being measured, as well as the stability of the enzymes. The method used to isolate the bacteroids from the nodule tissue is of some importance. Pea bacteroids of Tn5-induced mutants defective in certain catabolic enzymes were contaminated with plant proteins when isolated by differential centrifugation (McKay et al., 1989). To overcome this problem of contamination bacteroids have been isolated using self-generating Percoll (Reibach et al., 1981; McKay et al., 1985, 1989; Herrada et al., 1989) or sucrose gradients (Karr et al., 1984).

A reasonable amount of data is now available on the activities of particular enzymes in bacteroids, ranging from those of sugar metabolism (Ronson and Primrose, 1979; Glenn et al., 1984; Streeter and Salminen, 1985; Kouchi et al., 1988; McKay et al., 1989), acetate metabolism (Preston et al., 1989), succinate metabolism (McKay et al., 1988; Copeland et al., 1989), poly-β-hydroxybutyrate metabolism (Karr et al., 1984) to the enzymes of the TCA cycle (McKay et al., 1989). Bacteroids appear to oxidize C_4-dicarboxylates via the TCA cycle, malic enzyme and pyruvate dehydrogenase (McKay et al., 1988; Copeland et al., 1989). Antoun et al. (1984) claimed that there was a link between the capacity of the free-living form of R. meliloti to utilize TCA cycle intermediates and the effectiveness of the bacteroids. These experiments were, however, based on measuring bacterial growth on plates, which can be misleading, especially if there is significant polysaccharide production, and there was no statistical evidence provided for improved plant dry matter production.

In bacteroids of R. leguminosarum MNF300 McKay et al. (1989) reported an elevation of most of the TCA cycle enzymes (up to 15-fold) compared to the activities found in free-living cells. Only isocitrate dehydrogenase appeared not to increase in specific activity, and it was already present at a high activity in free-living cells. Other studies (DeVries et al., 1980; Waters et al., 1985; Kouchi et al., 1988) have also reported high activities of malate dehydrogenase in pea and soybean bacteroids. It is perhaps noteworthy that most of these studies have been carried out with bacteroids isolated from plants of one age. In a time course experiments with soybean bacteroids, Karr et al. (1984) reported that the specific activity of fumarase and pyruvate dehydrogenase increased, while that of malate dehydrogenase remained constant and that of isocitrate dehydrogenase decreased.

In contrast, sugar catabolic enzymes in bacteroids are either at basal levels (e.g. in snake bean bacteroids of Rhizobium NGR234, Saroso et al., 1986) or present at low activities (e.g. in French bean bacteroids of R. phaseoli, Salminen and Streeter, 1987 and pea bacteroids of R. leguminosarum MNF300, McKay et al., 1989).

A notable omission in some studies of bacteroid physiology has been the lack of comparable data for the free-living form grown on a variety of carbon sources. The significance of the presence or absence of an enzyme in bacteroids is likely to be interpretable only if something is known of the regulation of the enzyme. Thus, snake bean bacteroids of cowpea Rhizobium NGR234 do not induce sugar catabolic enzymes (Saroso et al., 1986) or those of 4-aminobutyrate degradation (Jin et al., 1990b), even though snake bean nodules contain significant concentrations of these substrates and the enzyme systems are inducible in the free-living form. The discovery of aromatic organic acids like benzoate in some root nodules raised the possibility that they were being used by bacteroids as a source of carbon. The suggestion has been made (Werner et al., 1975; Rohm and Werner, 1985) that aromatic compounds might be used by bacteroids, since in long (25 h) incubations radioactivity in phenolic compounds was very slowly converted into $^{14}CO_2$ (at about 0.1 nmol min^{-1} 10^{-9} cells) and some of the enzymes of aromatic degradation could be detected. In a more thorough investigation of the enzymes of the 3-oxoadipate pathway, Parke and Ornston (1986) showed that bradyrhizobia (unlike the rhizobia) express the enzymes of this pathway (apart from the dioxygenases) constitutively; their presence cannot therefore be used as evidence for the utilization of aromatic compounds in nodule bacteroids.

These data suggest that although free-living rhizobia are capable of utilizing a range of substrates, bacteroids may have access to only a limited range of compounds in sufficient concentration to support N_2 fixation and bacteroid development. The indirect evidence based on the physiology of the bacteroid suggests that the PBM may act as a significant barrier to the movement of materials from the plant cytosol to the bacteroid.

5 THE PERIBACTEROID MEMBRANE (PBM)

5.1 Origin of the PBM

Upon infection of the legume root cell by Rhizobium/Bradyrhizobium, the wall of the infection thread is degraded by cellulolytic enzymes of rhizobial origin (Verma et al., 1978; Chalifour and Benhamou, 1989). Following this degradation of the infection thread wall, numerous vesicles produced by Golgi bodies fuse with the membrane near the tip of the infection thread (Robertson and Lyttleton, 1982). The microsymbiotic bacteria start to bud off by endocytosis of the infection thread membrane. The subsequent proliferation proceeds by membrane flow from the endoplasmic reticulum (ER) and Golgi bodies (Robertson et al., 1978a; Verma et al., 1978; Brewin et al., 1985; Mellor and Werner, 1987). A substantial synthesis of PBM occurs in this way and infected soybean cells have been recorded to contain up to about 40 times more PBM than plasma membrane (Verma et al., 1978). The PBM is of plant origin, but a recent report by Mellor et al. (1989) suggests that at least four different signals from the microsymbiont are required for the formation of PBM, and four new fix loci influencing PBM development have been identified in R. meliloti (Putnoky et al., 1988). The PBM is a necessary feature of an effective symbiosis. Free bacteroids elicit a phytoalexin response (Werner et al., 1985) and ineffective symbioses often disintegrate or lack PBM (Werner et al., 1980; Arwas et al.,1985). The bacteria within the PBM differentiate into the metabolically and morphologically distinct bacteroids normally capable of nitrogen fixation. The differentiation is believed to be stimulated by the combined action of specific signals from the host plant and microenvironmental factors as low pO_2, high K^+ concentrations and low pH (Bradley et al., 1988; Kannenberg and Brewin, 1989). Enclosed in the PBM may be a single (e.g. in lupin and alfalfa) or several (e.g. in soybean and Phaseolus) bacteroids (Robertson et al., 1978a; Verma et al.,1978; Miller et al., 1988; Pladys and Rigaud, 1988). The different numbers of bacteroids enclosed by the PBM is under control of the host plant and may arise due to different degrees of adhesion between the PBM and the bacterial outer membrane (Robertson and Lyttleton, 1984). The PBM and enclosed bacteroids have been termed peribacteroid units (PBU) or "symbiosomes"; the latter term has been proposed by Roth et al. (1988) to describe discrete endosymbiotic organelles in general. Objections to such common terminology for endosymbioses have been raised by Smith and Smith (1990) because of the lack of functional similarity between them.

5.2 Properties of the PBM

The PBM resembles the plasma membrane in staining with phosphotungstic acid and silver but not with zinc iodide-osmium tetroxide, as opposed to ER and Golgi (Robertson et al., 1978a). Furthermore, the PBM shares antigens with plasma- and other plant membranes (Brewin et al., 1985; Bradley et al., 1988). However, several characteristcs of the PBM are unique, and it must be considered a true symbiotic structure combining gene functions from both symbiotic partners. The particle density on freeze-fracture replicas is half as dense in PBM as in plasma membranes of lupin nodules (Robertson et al., 1978a). A recent report by Werner et al. (1988) demonstrates that at one stage of development the particle density of the PBM in soybean is affected by the genotype of the microsymbiont. As the work did not include a time course, it is possible that the recorded differences arose from unsynchronized nodule ontogeny.

A high lipid:protein ratio (6:1) has been recorded in the PBM (Robertson et al. 1978b), possibly implying that a restricted complement of proteins is required to carry out the metabolic functions of the PBM. Furthermore, the high lipid:protein ratio may result in a lower rate of simple diffusion of hydrophilic compounds (e.g. sugars) across the PBM than across other membranes. The recorded decrease in the density of the PBM during development of nitrogen fixation may reflect changes in the lipid:protein ratio related to changes in bacteroid metabolism, or loss from the PBM of proteins associated with cell wall synthesis (Robertson et al., 1978b). The phospholipid (Mellor et al., 1985) and fatty acid (Bassarab et al., 1989) composition of the PBM resembles that of ER.

The biosynthesis of proteins associated with the PBM is controlled by the host cell as well as the microsymbiont (Verma et al., 1978; Brewin et al., 1985; Werner et al., 1988). The polypeptide composition of the PBM shows no obvious homology to the plasma membrane or other major plant membranes like ER, Golgi or mitochondria (Mellor and Werner, 1986). The PBM a characteristic Ca^{2+}-dependent protein kinase (Bassarab and Werner, 1987), along with several nodule-specific proteins (nodulins) (Katinakis and Verma, 1985; Fortin et al., 1985, 1987; Verma et al., 1986; Mellor et al., 1989). The PBM acquires strongly acidic phospholipids (Mellor et al., 1985) and the interaction of PBM nodulins with these may provide one of the bases for their specific incorporation into the PBM. Nodulin-26 is a transmembrane protein of the PBM, which is already being synthesised at maximal rate when infection threads are formed; its expression is therefore independent of PBM formation. The expression of a surface PBM nodulin (Nodulin-24) depends on the release of bacteria from the infection threads. This temporal separation suggests that these two PBM nodulins are independently regulated (Fontin et al., 1987). The roles of the PBM nodulins still remain to be elucidated, but it is tempting to suggest that some of them may be related to the control of metabolite exchange between the symbiotic partners.

The PBM of soybean and lupin possesses ATPase activity (Blumwald et al., 1985; Domigan et al., 1988; Udvardi and Day, 1989). The properties of the enzyme in soybean PBM are similar to those of plasma membrane ATPase with respect to pH optimum, vanadate inhibition and stimulation by cations. The existence of an additional tonoplast-like ATPase in soybean PBM was suggested by Bassarab et al. (1986) on the basis of inhibition of ATPase activity by nitrate. However, minor contamination of the PBM preparations with mitochondria may have resulted in some nitrate inhibition of ATPase activity, and definitive proof from immunological studies or isolation of the ATPase polypeptides is still lacking. Udvardi and Day (1989) have challenged the existence of a tonoplast-like ATPase in the PBM, as they found no vanadate-insensitive peak of PBM ATPase activity between pH 7-8, a characteristic of tonoplast ATPase. The properties of ATPase from lupin PBM differ from those of soybean PBM by showing a) a more acidic pH optimum, b) only weak stimulation by potassium ions, c) sensitivity to molybdate, and d) lack of stimulation by magnesium and manganese (Domigan et al., 1988). The plasma membrane ATPase of lupin has not yet been characterized and it is therefore not known if lupin PBM ATPase is unique or resemples lupin plasma membrane ATPase. The ATPase activity in the PBM translocates protons and creates a positive membrane potential across the membrane, which may result in some acidification of the peribacteroid space (PBS). The bacteroids may counteract acidification by changing their surface membrane or surface charge (Kannenberg and Brewin, 1989), and uptake and metabolism of dicarboxylic acids and excretion of ammonia would also counteract acidification of the PBS. However, direct recordings of the pH of the PBS, and its fluctuations in response to bacteroid metabolism are still lacking, and the occurrence of pH changes in the PBS remain speculative.

5.3 The peribacteroid space

A unique set of proteins is recorded in the PBS (Katinakis et al., 1988), including α-mannisodase isoenzyme II (Kinnback et al.,1987), acid proteases (Mellor et al., 1984), acid trehalase (Mellor, 1988) and a low molecular weight protease (thermolysin) inhibitor (Garbers et al., 1988). The protease inhibitor may protect the bacteroids from digestion by the host (Garbers et al., 1988). The existence of an internal lytic compartment of the PBU has led to the suggestion that PBUs are nodule-specific forms of the plant lysosome (see Mellor, 1989).

The apparent absence of malonate from the PBS despite its abundance in the plant fraction of soybean nodules was interpreted as an exclusion of malonate by the PBM (Humbeck and Werner, 1987). However, the possibility of PBS malonate being metabolised during the procedure of separation cannot be excluded. Furthermore, malonate is a competitive inhibitor of succinate dehydrogenase and is likely to be sequestered in the vacuole of the plant cell (Beevers et al., 1966). Spatially separated from the plant cytoplasm, malonate may not have ready access to the PBM.

5.4 Transport across the PBM

Information on the transport of compounds across the PBM is valuable because it is likely to explain how the bacteroids retain a mutualistic relationship with the host plant. Furthermore, it is likely that the PBM is essential for regulation of nitrogen fixation by controlling the exchange of compounds between the symbiotic partners. However, little direct information is available on the permeability and transport properties of the PBM because of the difficulties in obtaining large quantities of pure and intact PBUs for in vitro assays, and in distinguishing between the properties of the PBM and those of the bacteroid membranes in the assays performed. It is important for in vitro assays to approximate to in vivo conditions (e.g. anaerobic isolation of pure, intact PBUs, inclusion of an O_2 carrier, and appropriate pO_2, pH and ionic environment), in order to obtain an accurate reflection of the in vivo exchange of compounds. Few reports have observed all these precautions.

Studies on substrate-supported bacteroid respiration indicate that the membrane system enclosing the bacteroids in soybean (Price et al., 1987) and in alfalfa (Miller et al., 1988) is selectively permeable. In soybean, the permeability of the PBM was distinguished from that of the bacteroid membranes, by comparing substrate-supported bacteroid respiration in intact PBUs with that of free bacteroids released from isolated PBUs (Price et al., 1987). The results reveal that C_4-dicarboxylic acids are readily utilized by the PBUs, whereas glutamate, aspartate, pyruvate and sugars supported respiration of free bacteroids but not that of intact PBUs. Although studies on substrate-supported respiration provide valuable information on which compounds are readily respired, the permeability of compounds not readily respired cannot be assessed by this method.

Further studies on the permeability of the PBM have been conducted by incubating PBUs or free bacteroids with a ^{14}C-labelled compound to determine the rate of ^{14}C accumulation in the bacteroids (Udvardi et al., 1988a,b; Herrada et al., 1989; McRae et al., 1989). Based on inhibitor studies Udvardi et al. (1988a) interpret their results as demonstrating carrier-mediated transport of succinate and malate across the PBM as well as the bacteroid membrane. Ultimate proof of an active process, demonstrated by a concentration gradient across the PBM, has not yet been provided. The presence of a dicarboxylate carrier on the PBM would provide an additional feature to distinguish the PBM from the plant plasma membrane. A PBM dicarboxylate carrier may be one of the nodulins associated with it. The reported PBM carrier has a lower affinity for malate and succinate (K_m=156 and 391 μM, respectively) than does the bacteroid carrier (K_m=9 and 13 μM, respectively), while the response of malate uptake to external pH may indicate that the monovalent dicarboxylate anion is the true substrate for the PBM carrier (Udvardi et al., 1988a). The transport of malate into soybean PBUs is inhibited by KCN (Day et al., 1989), which may imply that dicarboyxlate transport across the PBM is also affected by uptake and metabolism of the substrate in the bacteroids. Glutamate passes the PBM by simple diffusion, but is actively transported across the bacteroid membranes (Udvardi et al., 1988b). Sugars do not appear to cross the PBM readily, and they are most likely to pass both membrane systems by simple diffusion (Day et al., 1989).

When interpreting ^{14}C uptake studies as ^{14}C accumulation there is a risk of underestimating compounds which are readily metabolised and/or excreted from the PBUs. Thus, precautions should be taken not to classify a particular compound as unimportant or not transferrable _in vivo_ solely on the basis of ^{14}C accumulation in ^{14}C uptake studies. Valuable information on the physiological significance of various substrates can be obtained by detecting the fate of the compounds supplied to the bacteroids. Using soybean bacteroids, Salminen and Streeter (1987) have demonstrated that an increase in the rate of $^{14}CO_2$ formation from various labelled substrates coincided with a maximum accumulation of label in the glutamate pool of the bacteroids. This indicates some involvement of glutamate in the respiratory metabolism of bacteroids, despite the lack of active transport of glutamate across the PBM. A function for glutamate may be supported by the large pools of glutamate recorded in bacteroids directly isolated from nodules (Streeter, 1987; Rosendahl et al., 1990).

The possibility of a shuttle to exchange metabolites between the symbiotic partners has been proposed (Akkermans et al., 1981; Kahn et al., 1985). Shuttle activity in the interface between the symbiotic partners has the potential to transfer NADH from the cytoplasm of the host plant to the microsymbiont without nett C transfer. A malate/aspartate shuttle in the membrane of mammalian mitochondria performs such a transfer of NADH (LaNoue and Tischler, 1974). The possibility of a malate/aspartate shuttle between the host and the microsymbiont in legume nodules has been supported by a "dual-substrate" experiment, which demonstrates that a second unlabelled substrate has an effect on the return of a labelled compound originally supplied to the bacteroids (Streeter and Salminen, 1988). Additional support for the hypothesis is provided by labelling studies with $^{14}CO_2$ and ^{14}C-aspartate (Snapp and Vance, 1986; Rosendahl et al., 1990), and by the high activities of aspartate aminotransferase (AAT) and malate dehydrogenase (MDH) in root nodules. Furthermore, there are reports of a nodule-specific form of AAT (Griffith and Vance, 1989) and a nodule-stimulated form of MDH (Appels and Haaker, 1988). Association of a malate/aspartate shuttle with the PBM has not yet been confirmed, and the subcellular location of the required enzymes still needs to be established.

6 CONCLUSION

The picture of the metabolic relationship between bacteroid and plant host tha emerges from this review is one in which the PBM exercises a significant control on the exchange of materials. It is not known if the root nodule bacteria themselves influence the permeability of the plant PBM. Since modification of the PBM could influence the characteristics of the metabolic exchange in the symbiiosis, it represents one potential site for genetic manipulation of the plant side of the association.

Despite the volume of work which has addressed the question of nutrient inputs to bacteroids, much remains to be discovered. There is now a growing appreciation of the inadequacies in the techniques which have been used by most investigators. The development of techniques for isolating and using highly purified PBUs in experiments which attempt as far as possible to simulate the nodule environment should yield valuable data. However, there is a need for a more integrated interdisciplinary approach than has usually been evident. Much research on nodule function has been carried out with rhizobial strains of undefined physiology or genetic background. Few studies of plant function have taken advantage of the wide range of well-characterized rhizobial mutants. Similarly, microbial physiologists have usually given insufficient attention to the in fluence of the plant on the performance of the bacteroids in the symbiosis. The time has surely come for more interaction between those who work primarily with the bacteria and those who work mainly with the host plant.

7 REFERENCES

Akkermans, A.D.L., Huss-Danell, K. and Roelofsen, W., 1981. Enzymes of the tricarboxylic acid cycle and the malate-aspartate shuttle in the N_2-fixing endophyte of Alnus glutinosa. Physiol. Plant., 53: 289-294.

Antoniw, L.D. and Sprent, J.I., 1976. Primary metabolites of Phaseolus vulgaris nodules. Phytochem., 17: 675-678.

Antoun, H., Bordeleau, L.M. and Sauvageau, R., 1984. Utilization of the tricarboxylic acid cycle intermediates and symbiotic effectiveness in Rhizobium meliloti. Plant Soil, 77: 29-38.

Appels, M.A. and Haaker, H., 1988. Identification of cytoplasmic nodule-associated forms of malate dehydrogenase involved in the symbiosis between Rhizobium leguminosarum and Pisum sativum. Eur. J. Biochem., 171: 515-522.

Applcby, C.A., Turner, G.L. and MacNicol, P.K., 1975. Involvement of oxyleghaemoglobin and cytochrome P450 in an efficient oxidative phosphorylation pathway which supports nitrogen fixtion in Rhizobium. Biochim. Biophys. Acta, 387: 461-474.

Arp, D.J., 1985. Rhizobium japonicum hydrogenase: purification to homogeneity from soybean nodules, and molecular characterization. Arch. Biochem. Biophys., 237: 504-512.

Arwas, R., McKay, I.A., Rowney, F.R.P., Dilworth, M.J. and Glenn, A.R., 1985. Properties of organic acid utilization mutants of Rhizobium leguminosarum strain 300. J. Gen. Microbiol., 131: 2059-2066.

Bassarab, S., Mellor, R.B. and Werner, D., 1986. Evidence for two types of Mg^{++}-ATPase in the peribacteroid membrane from Glycine max root nodules. Endocyt. Cell Res., 3: 189-196.

Bassarab, S., Schenk, S.U. and Werner, D., 1989. Fatty acid composition of the peribacteroid membrane and the ER in nodules of Glycine max varies after infection by different strains of the microsymbiont Bradyrhizobium japonicum. Botanica Acta, 102: 196-202.

Bassarab, S. and Werner, D., 1987. A calcium-dependent protein kinase on the peribacteroid membrane from nodules of Glycine max. J. Plant Physiol., 130: 233-241.

Beevers, H., Stiller, M.L. and Butt, V.S., 1966. Metabolism of the organic acids. In: F.C. Steward (Editor), Plant Physiology: A Treatise. Vol 4B. Academic Press, New York, pp. 119-262.

Bergersen, F.J. and Briggs, M.J., 1958. Studies on the bacterial component of soybean root nodules: cytology and organization in the host tissue. J. Gen. Microbiol., 19: 482-490.

Bergersen, F.J. and Goodchild, D.J., 1973. Cellular location and concentration of leghaemoglobin in soybean root nodules. Aust. J. Biol. Sci., 26: 741-756.

Bergersen, F.J. and Trinchant, J.C., 1985. Flux of O_2 affects apparent optimal concentrations of O_2 in suspensions of bacteria respiring microaerobically. J. Theoret. Biol., 115: 93-102.

Bergersen, F.J. and Turner, G.L., 1967. Nitrogen fixation by the bacteroid fraction of breis of soybean root nodules. Biochim. Biophys. Acta, 141: 507-515.

Birkenhead, K., Manian, S.S. and O'Gara, F., 1988. Dicarboxylic acid transport in Bradyrhizobium japonicum: use of Rhizobium meliloti dct gene(s) to enhance nitrogen fixation. J. Bacteriol., 170: 184-189.

Blumwald, E., Fortin, M.G., Rea, P.A., Verma, D.P.S. and Poole, R.J., 1985. Presence of host-plasma membrane type H^+-ATPase in the membrane envelope enclosing the bacteroids in soybean root nodules. Plant Physiol., 78: 665-672.

Bosch, I., Meidle, E.J., Hoult, M., Plessner, O and Guerinot, M.L., 1988. Iron uptake and metabolism in the Bradyrhizobium/soybean symbiosis. In: H. Bothe, F.J. de Bruijn and W.E. Newton (Editors), Nitrogen Fixation: Hundred Years After. Gustav Fischer, Stuttgart, p. 652.

Boursier, P., Hanus, F.J., Papen, H., Becker, M.M., Russell, S.A. and Evans, H.J., 1988. Selenium increases hydrogenase expression in autotrophically cultured Bradyrhizobium japonicum and is a constituent of the purified enzyme. J. Bacteriol., 170: 5594-5560.

Bradley, D.J., Wood, E.A., Larkins, A.P., Galfre, G., Butcher, G.W. and Brewin, N.J., 1988. Isolation of monoclonal antibodies reacting with peribacteroid membranes and other components of pea root nodules containing Rhizobium leguminosarum. Planta, 173: 149-160.

Brewin, N.J., Robertson, J.G., Wood, E.A., Wells, B., Larkins, A.P., Galfre, G. and Butcher, G.W., 1985. Monoclonal antibodies to antigens in the peribacteroid membrane from Rhizobium-induced root nodules of pea cross-react with plasma membranes and Golgi bodies. EMBO J., 4: 605-611.

Chakrabarty, S.K., Mishra, A.K. and Chakrabarty, T.T.Y., 1987. Metabolism of glucose and gluconate in fast- and slow-growing rhizobia. Phytochem., 26: 85-87.

Chalifour, F.-P. and Benhamou, N., 1989. Indirect evidence for cellulase production by Rhizobium in pea root nodules during bacteroid differentiation: cytochemical aspects of cellulose breakdown in rhizobial droplets. Can. J. Microbiol., 35: 821-829.

Chatel, D.L., Robson, A.D., Gartrell, J.W. and Dilworth, M.J., 1978. The effect of inoculation and cobalt application on the growth of and nitrogen fixation by sweet lupins. Aust. J. Agric. Res., 29: 1191-1202.

Chaudhuri, A. and Krasna, A.I., 1987. Isolation of genes required for hydrogenase synthesis in Escherichia coli. J. Gen. Microbiol., 133: 3289-3298.

Chen, Y.P., Dilworth, M.J. and Glenn, A.R., 1984b. Aromatic metabolism in Rhizobium trifolii - protocatechuate 3,4-dioxygenase. Arch. Microbiol., 138 187-190.

Chen, Y.P., Dilworth, M.J. and Glenn, A.R., 1989. Degradation of mandelate and 4-hydroxymandelate by Rhizobium leguminosarum biovar trifolii TA1. Arch. Microbiol., 151: 520-523.

Chen, Y.P., Glenn, A.R. and Dilworth, M.J., 1984a. Uptake and oxidation of aromatic substrates by Rhizobium leguminosarum MNF3841 and Rhizobium trifolii TA1. FEMS Microbiol. Lett., 21: 201-205.

Chen, Y.P., Glenn, A.R. and Dilworth, M.J., 1985. Aromatic metabolism in Rhizobium trifolii - catechol 1,2-dioxygenase. Arch. Microbiol., 141: 225-228.

Copeland, L., Quinnell, R.G. and Day, D.A., 1989. Malic enzyme activity in bacteroids from soybean nodules. J. Gen. Microbiol., 135: 2005-2011.

Copeland, L., Vella, J. and Hong, Z., 1988. Enzymes of carbohydrate metabolism in soybean nodules. Phytochem., 28: 57-61.

Cowles, J.R. and Evans, H.J., 1968. Some properties of the ribonucleotide reductase of Rhizobium meliloti. Arch. Biochem. Biophys., 127: 770-778.

Cowles, J.R., Evans, H.J. and Russell, S.A., 1969. B_{12}-coenzyme-dependent ribonucleotide reductase in Rhizobium species and the effect of cobalt deficiency on the activity of the enzyme. J. Bacteriol., 97: 1460-1465.

Crowe, J.H., Crowe, L.M. and Chapman, D., 1984. Preservation of membranes in anhydrobiotic organisms: the role of trehalose. Science, 223: 701-703.

Dalton, D.A., Evans, H.J. and Hanus, F.J., 1985. Stimulation by nickel of soil microbial urease activity and urease and hydrogenase activities in soybeans grown in a low-nickel soil. Plant Soil., 88: 245-258.

Day, D.A. and Mannix, M., 1988. Malate oxidation by soybean nodule mitochondria and the possible consequences for nitrogen fixation. Plant Physiol. Biochem., 26: 567-573.

Day, D.A., Price, G.D. and Udvardi, M.K., 1989. Membrane interface of the Bradyrhizobium japonicum-Glycine max symbiosis: Peribacteroid units from soybean nodules. Aust. J. Plant Physiol., 16: 69-84.
De Hertogh, A.A., Mayeux, P.A. and Evans, H.J., 1963. The relationshiop of cobalt requirements to propionate metabolism in Rhizobium. J. Biol. Chem., 239: 2446-2455.
DeVries, G.E., in T'Veld, P. and Kijne, J.W., 1980. Production of organic acids in Pisum sativum root nodules as a result of oxygen stress. Plant Sci. Lett., 20: 115-123.
Dilworth, M.J. and Appleby, C.A., 1979. Leghemoglobin and Rhizobium hemoproteins. In: R.W.F. Hardy, F. Bottomley and R.C. Burns (Editors) A Treatise on Dinitrogen Fixation. John Wiley, New York, pp. 691-764.
Dilworth, M.J., Arwas, R., McKay, I.A.. Saroso, S. and Glenn, A.R., 1986. Pentose metabolism in Rhizobium leguminosarum MNF300 and in cowpea Rhizobium NGR234. J. Gen. Microbiol., 132: 1733-1742.
Dilworth, M.J. and Bisseling, T., 1984. Cobalt and nitrogen fixation in lupinus angustifolius. III. DNA and methionine in bacteroids. New Phytol., 98: 311-316.
Dilworth, M.J. and Glenn, A.R., 1981. Control of carbon substrate utilization by rhizobia. In: A.H. Gibson and W.E. Newton (Editors), Current Perspectives in Nitrogen Fixation. Australian Academy of Sciences, Canberra, pp. 244-251.
Dilworth, M.J. and Glenn, A.R., 1984. How does a legume nodule work? Trends Biochem. Sci., 9: 519-522.
Dilworth, M.J. and Glenn, A.R., 1985. Transport in Rhizobium and its significance to the legume symbiosis. In: P.W. Ludden and J.E. Burris (Editors), Nitrogen Fixation and Photosynthesis. Elsevier, New York, pp. 53-61.
Dilworth, M.J., McKay, I.A.. Franklin, M. and Glenn, A.R., 1983. Catabolite effects on enzyme induction and substrate utilization in Rhizobium leguminosarum. J. Gen. Microbiol., 129:359-366.
Dilworth, M.J., Robson, A.D. and Chatel, D.L., 1979. Cobalt and nitrogen fixation in Lupinus angustifolius. II. Nodule formation and function. New Phytol., 83: 63-79.
Domigan, N.M., Farnden, K.J.F., Robertson, J.G. and Monk, B.C., 1988. Characterization of the peribacteroid membrane ATPase of lupin root nodules. Arch. Biochem. Biophys., 264: 564-573.
Duncan, M.J., 1979. Arabinose metabolism in rhizobia. J. Gen. Microbiol., 113 415-419.
Eberz, G., Eitinger, T. and Friedrich, B., 1989. Genetic determinants of a nickel-specific transport system are part of the plasmid-encoded hydrogenase gene cluster in Alcaligenes eutrophus. J. Bacteriol., 171: 1340-1345.
Elliott, B.B. and Mortenson, L.E., 1975. Transport of molybdate by Clostridium pasteurianum. J. Bacteriol., 124: 1295-1301.
Elliott, B.B. and Mortenson, L.E., 1976. Regulation of molybdate transport by Clostridium pasteurianum. J. Bacteriol., 127: 770-779.
Evans, H.J., Harker, A.R., Papen, H., Russell, S.A., Hanus, F.J. and Zuber, M., 1987. Physiology, biochemistry and genetics of the uptake hydrogenase in Rhizobium. Ann. Rev. Microbiol., 41: 335-361.
Evans, H.J., Russell, S.A., Hanus, F.J., Papen, H., Sayavendra-Soto, L., Zuber, M. and Boursier, P., 1988. Hydrogenase and nitrogenase relationships in Rhizobium: some recent developments. In:H. Bothe, F.J. de Bruijn and W.E. Newton (Editors), Nitrogen Fixation: Hundred Years After. Gustav Fischer, Stuttgart, pp. 577-582.
Finan, T.M., Oresnik, I. and Bottacin, A., 1988. Mutants of Rhizobium meliloti defective in succinate metabolism. J. Bacteriol., 170: 3396-3403.
Finan, T.M., Wood, J.M. and Jordan, D.C., 1983. Symbiotic properties of C_4-dicarboxylic acid transport mutants of Rhizobium leguminosarum. J. Bacteriol., 154: 1403-1413.

Fortin, M.G., Morrison, N.A. and Verma, D.P.S., 1987. Nodulin-26, a peribacteroid membrane nodulin is expressed independently of the development of the peribacteroid compartment. Nucl. Acids Res., 15: 813-824.
Fortin, M.G., Zelechowska, M. and Verma, D.P.S., 1985. Specific targetting of membrane nodulins to the bacteroid-enclosing compartment in soybean nodules. EMBO J., 4: 3041-3046.
Garbers, C., Meckbach, R., Mellor, R.B. and Werner, D., 1988. Protease (thermolysin) inhibition activity in the peribacteroid space of Glycine max root nodules. J. Plant Physiol., 132: 442-445.
Glenn, A.R. and Brewin, N.J., 1981. Succinate-resistant mutants of Rhizobium leguminosarum. J. Gen. Microbiol., 126: 237-241.
Glenn, A.R. and Dilworth, M.J., 1981. Oxidation of substrates by isolated bacteroids and free-living cells of Rhizobium leguminosarum. J. Gen. Microbiol., 126: 243-247.
Glenn, A.R., McKay, I.A., Arwas, R. and Dilworth, M.J., 1984. Sugar metabolism and the symbiotic properties of carbohydrate mutants of Rhizobium leguminosarum. J. Gen. Microbiol., 130: 239-245.
Glenn, A.R., Poole, P.S. and Hudman, J.F., 1980. Succinate uptake by free-living and bacteroid forms of Rhizobium leguminosarum. J. Gen. Microbiol., 119: 267-271.
Graham, L. and Maier, R.J., 1987. Variability in molybdenum uptake activity in Bradyrhizobium japonicum strains. J. Bacteriol., 169: 2555-2560.
Graham, P.H., 1964. Studies on the utilization of carbohydrates and Krebs cycle intermediates by rhizobia using an agar plate method. Antonie van Leeuwenhoek, 30: 68-72.
Griffith, S.M. and Vance, C.P., 1989. Aspartate aminotransferase in alfalfa root nodules. 1. Purification and partial characterization. Plant Physiol., 90: 1622-1629.
Guerinot, M.L., 1990. Iron uptake and metabolism in the rhizobia/legume symbiosis. J. Plant Nutr., in press.
Harker, .A.R., Xu, L.-S., Hanus, F.J. and Evans, H.J., 1984. Some properties of the nickel-containing hydrogenase of chemolithotrophically grown Rhizobium japonicum. J. Bacteriol., 159: 850-856.
Hemantaranjan, A. and Garg, O.K., 1986. Introduction of nitrogen-fixing nodules through iron and zinc fertilization in non-nodule forming French bean (Phaseolus vulgaris L.). J. Plant Nutr., 9: 281-288.
Herrada, G., Puppo, A and Rigaud, J., 1989. Uptake of metabolites by bacteroid-containing vesicles and by free bacteroids from French bean nodules. J. Gen. Microbiol., 135: 3165-3171.
Hooymans, J.J.M. and Logman, G.J.J., 1984. The effect of high concentrations of glucose and fructose on the respiration rate of isolated Rhizobium leguminosarum bacteroids. J. Plant Physiol., 116: 273-277.
Howell, R.K., 1987. Rhizobium-induced mineral uptake in peanut tissues. J. Plant Nutr., 10: 1297-1305.
Humbeck, C and Werner, D., 1987a. Separation of malate and malonate pools by the peribacteroid membrane in soybean nodules. Endocyt. Cell Res., 4: 185-196.
Humbeck, C and Werner, D., 1987b. Two succinate uptake systems in Bradyrhizobium japonicum. Current Microbiol., 14: 259-262.
Humbeck, C and Werner, D., 1989. Delayed nodule development in a succinate transport mutant of Bradyrhizobium japonicum. J. Plant Physiol., 116: 273-277.
Hunt, S., King, B.J., Canvin, D.T. and Layzell, D.B., 1987. Steady and non-steady state gas exchange characteristics of soybean nodules in relation to the O_2 diffusion barrier. Plant Physiol., 84: 164-172.
Jin, H.N., Dilworth, M.J. and Glenn, A.R., 1990a. 4-Aminobutyrate is not available to bacteroids of cowpea Rhizobium MNF2030 in snake bean nodules. Arch. Microbiol., 153: 455-462.

Jin, H.N., Glenn, A.R. and Dilworth, M.J., 1990b. How does L-glutamate transport relate to selection of mixed nitrogen sources in Rhizobium leguminosarum biovar trifolii and cowpea Rhizobium MNF2030? Arch. Microbiol., 153: 448-454.
Kahn, M.L., Kraus, J. and Somerville, J.E., 1985. A model of nutrient exchange in the Rhizobium-legume symbiosis. In: H.J. Evans, P.J. Bottomley and W.E. Newton (Editors), Nitrogen Fixation Research Progress. Martinus Nijhoff, Dordrecht, pp. 193-199.
Kannenberg, E.L. and Brewin, N.J., 1989. Expression of a cell surface antigen from Rhizobium leguminosarum 3841 is regulated by oxygen and pH. J. Bacteriol., 171: 4543-4548.
Karr, D.B., Waters, J.K., Suzuki, F. and Emerich, D.W., 1984. Enzymes of the poly-β-hydroxybutyrate and citric acid cycles in Rhizobium japonicum bacteroids. Plant Physiol., 75: 1158-1162.
Katinakis, P. and Verma, D.P.S., 1985. Nodulin 24 gene of soybean codes for a peptide of the peribacteroid membrane and was generated by tandem duplication of a sequence resembling an insertion element. Proc. Natl. Acad. Sci. USA, 82: 4157-4161.
Katinakis, P., Lankhorst, R.M.K., Louwerse, J., Van Kammen, A. and Van den Bos, R.C., 1988. Bacteroid-encoded proteins are secreted into the peribacteroid space by Rhizobium leguminosarum. Plant Mol. Biol. 11: 183-190.
Kinnback, A., Mellor, R.B. and Werner, D., 1987. Alpha-mannosidase isoenzymeII in the peribacteroid space of Glycine max root nodules. J. Exp. Bot.,38: 1373-1377.
Klucas, R.V., Hanus, F.J., Russell, S.A. and Evans, H.J., 1983. Nickel a micronutrient element for hydrogen-dependent growth of Rhizobium japonicum and for expression of urease activity in soybean leaves. Proc. Natl. Acad. Sci. USA, 80: 2253-2257.
Kouchi, H., Fukai, K., Katagiri, H., Minamisawa, K. and Tajima, S., 1988. Isolation and enzymological characterization of infected and uninfected protoplasts from root nodules of Glycine max. Physiol. Plant., 73: 327-334.
Kouchi, H. and Yoneyama, T., 1984. Dynamics of carbon photosynthetically assimilated in nodulated soya bean plants grown under steady state conditions. 2. The incorporation of ^{13}C into carbohydrate, organic acids, amino acids and some storage compounds. Anal. Bot., 53: 883-896.
Kouchi, H. and Yoneyama, T., 1986. Metabolism of [^{13}C]-labelled photosynthate in plant cytosol and bacteroids of root nodules of Glycine max. Physiol. Plant., 68: 238-244.
Kretovich, W.L., Romanov, V.I., Abdullaeva, B.R., Ivanov, B.F., Chermenskaya, I.E. and Fedulova, N.G., 1985. [^{14}C]Glucose utilization by Rhizobium lupini bacteroids. Plant Soil, 85: 211-217.
Kurz, W.G.W. and LaRue, T.A., 1977. Citric acid cycle enzymes and nitrogenase in nodules of Pisum sativum. Can. J. Microbiol., 23: 1197-1200.
Lafontaine, P.J., Lafreniere, C., Chalifour, F.P., Dion, P. and Antoun, H., 1989. Carbohydrate and organic acid composition of effective and ineffective nodules of Phaseolus vulgaris. Physiol. Plant., 76: 507-513.
LaNoul, K.F. and Tischler, M.E., 1974. Electrogenic characteristics of the mitochondrial glutamate-aspartate antiporter. J. Biol. Chem., 249: 7522-7528.
Layzell, D.B., Hunt, S. and Palmer, G.R., 1990. Mechanisms of nitrogenase inhibition in soybean nodules. Pulse modulated spectroscopy indicates that nitrogenase activity is limited by O_2. Plant Physiol., 92: 1101-1107.
McAllister, C.F. and Lepo, J.E., 1983. Succinate transport by free-living forms of Rhizobium japonicum. J. Bacteriol., 153: 1155-1162.
McBride, M.J. and Ensign, J.C., 1987. Effects of intracellular trehalose content on Streptomyces griseus spores. J. Bacteriol., 169: 4995-5001.
McKay, I.A., Dilworth, M.J. and Glenn, A.R., 1988. C_4-Dicarboxylate metabolism in free-living and bacteroid forms of Rhizobium leguminosarum MNF3841. J. Gen. Microbiol., 134: 1433-1440.

McKay, I.A., Dilworth, M.J. and Glenn, A.R., 1989. Carbon metabolism in continuous cultures and bacteroid forms of Rhizobium leguminosarum MNF3841. Arch. Microbiol., 152: 606-610.
McKay, I.A., Glenn, A.R. and Dilworth, M.J., 1985. Gluconeogenesis in Rhizobium leguminosarum MNF3841. J. Gen. Microbiol., 131: 2067-2073.
McRae, D.G., Miller, R.W., Berndt, W.B. and Joy, K., 1989. Transport of C_4-dicarboxylates and amino acids by Rhizobium meliloti bacteroids. Mol. Plant-Microbe Interact., 5: 273-278.
Maier, R.J. and Graham, L., 1988. Molybdate transport by Bradyrhizobium japonicum bacteroids. J. Bacteriol., 170: 5613-5619.
Mellor, R.B., 1988. Distribution of trehalase in soybean root nodule cells: Implications for trehalose metabolism. J. Plant Physiol., 133: 173-177.
Mellor, R.B., 1989. Bacteroids in the Rhizobium-legume symbiosis inhabit a plant internal lytic compartment: implications for other microbial endosymbioses. J. Exp. Bot.,40: 831-839.
Mellor, R.B., Christensen, T.M.I.E., Bassarab, S. and Werner, D., 1985. Phospholipid transfer from ER to the peribacteroid membrane in soybean nodules. Z. Naturforsch., 40: 73-79.
Mellor, R.B., Garbers, C. and Werner, D., 1989. Peribacteroid membrane nodulin gene induction by Bradyrhizobium japonicum mutants. Plant Mol. Biol., 12: 307-315.
Mellor, R.B., Morschel, E. and Werner, D., 1984. Legume root response to symbiotic infection: enzymes of the peribacteroid space. Z. Naturforsch., 39: 123-125.
Mellor, R.B. and Werner, D., 1986. The fractionation of Glycine max root nodule cells: a methodological overview. Endocyt. Cell Res., 3: 317-336.
Mellor, R.B. and Werner, D., 1987. Peribacteroid membrane biogenesis in mature legume root nodules. Symbiosis, 3: 75-100.
Meyer, M.C. and Pueppke, S.G., 1980. Differentiation of Rhizobium japonicum strain derivatives by antibiotic sensitivity patterns, lectin binding and utilization of biochemicals. Can. J. Microbiol., 26: 606-612.
Miller, R.W., McRae, D.G., Al-Jobore, A. and Berndt, W.B., 1988. Respiration supported nitrogenase activity of isolated Rhizobium meliloti bacteroids. J. Cell. Biochem., 38: 35-49.
Modi, M., Shah, K.S. and Modi, V.V., 1985. Isolation and characterization of catechol-like siderophore from cowpea Rhizobium RA-1. Arch. Microbiol., 141: 156-158.
Munns, D.N., 1977. Mineral nutrition and the legume symbiosis. In: R.W.F. Hardy and A.H. Gibson (Editors), A Treatise on Dinitrogen Fixation, Section IV. Agronomy and Ecology. John Wiley, New York, pp. 353-391.
Murphy, P.J., Heycke, N., Treuz, S., Ratet, P., de Bruijn, F.J. and Schell, J., 1988. Synthesis of an opine-like compound - a rhizopine - in alfalfa nodules is symbiotically regulated. Proc. Natl. Acad. Sci. USA, 85: 9133-9137.
Nadler, K.D., Johnstone, A.W.B., Chen, J.-W. and John, T.R., 1990. A Rhizobium leguminosarum mutant defective in symbiotic iron acquisition. J. Bacteriol., 172: 670-677.
Nambiar, P.T.C. and Sivaramakrishnan, S., 1987. Detection and assay of siderophores in cowpea rhizobia (Bradyrhizobium) using radioactive Fe (^{59}Fe). Applied Microbiol. Lett., 4: 37-40.
Neilands, J.B., 1982. Microbial envelope proteins related to iron. Ann. Rev. Microbiol., 36: 285-309.
Neilands, J.B., Konopka, K., Schwyn, B., Coy, M., Francis, R.T., Paw, B. and Bagg, S., 1987. Comparative biochemistry of microbial iron assimilation. In: G. Winkelmann, D. van der Helm and J.B. Neilands (Editors), Iron Transport in Microbes, Plants and Animals. VCH Publishers, Weinheim, pp. 3-34.
O'Hara, G.W., Boonkerd, N. and Dilworth, M.J., 1988a. Mineral constraints to nitrogen fixation. Plant Soil, 108: 93-110.

O'Hara, G.W., Dilworth, M.J., Boonkerd, N. and Parkpian, P., 1988b. Iron deficiency specifically limits nodule development in peanut inoculated with Bradyrhizobium spp. New Phytol., 108: 51-57.
O'Hara, G.W., Franklin, M. and Dilworth, M.J., 1987. Effect of sulfur supply on sulfate uptake, and alkaline sulfatase activity in free-lving and symbiotic bradyrhizobia. Arch. Microbiol., 149: 163-167.
O'Hara, G.W., Hartzook, A., Bell, R.W. and Loneragan, J.F., 1988c. Response to Bradyrhizobium strain of peanut cultivars grown under iron stress. J. Plant Nutr., 11: 843-852.
Parke, D. and Ornston, L.N., 1984. Nutritional diversity of Rhizobium revealed by auxanography. J. Gen. Microbiol., 130: 1743-1750.
Parke, D. and Ornston, L.N., 1986. Enzymes of the β-ketoadipate pathway are inducible in Rhizobium and Agrobacterium spp. and constitutive in Bradyrhizobium spp. J. Bacteriol., 165: 288-292.
Patel, H.N., Chakraborty, R.N. and Desai, S.B., 1988. Isolation and characterization of a phenolate siderophore from Rhizobium leguminosarum IARI102. FEMS Microbiol. Lett., 56: 131-134.
Pedrosa, F.O. and Zancan, G.T., 1974. L-Arabinose metabolism in Rhizobium japonicum. J. Bacteriol., 119: 336-338.
Pelliccione, N., Jaffin, B., Sobel, M.E., and Krulwich, T.A., 1979. Induction of the phosphoenolpyruvate:hexose phosphotransferase system associated with relative anaerobiosis in an obligate aerobe. Eur. J. Biochem., 95: 69-75.
Pienkos, P.T. and Brill, W.J., 1981. Molybdenum accumulation and storage in Klebsiella pneumoniae and Azotobacter vinelandii. J. Bacteriol., 145: 743-751.
Pladys, D. and Rigaud, J., 1988. Lysis of bacteroids in vitro and during the senescence in Phaseolus vulgaris nodules. Plant Physiol. Biochem., 26: 179-186.
Poole, P.S., Franklin, M., Glenn, A.R. and Dilworth, M.J., 1986. The transport of L-glutamate by Rhizobium leguminosarum involves a common amino acid carrier. J. Gen. Microbiol., 131: 1441-1448.
Preston, G.G., Zeiher, C., Wall, J. and Emerich, D.W., 1989. Acetate-activating enzymes of Bradyrhizobium japonicum bacteroids. Appl. Environ. Microbiol., 55:165-170.
Price, G.D., Day, D.A. and Gresshoff, P.M., 1987. Rapid isolation of intact peribacteroid envelopes from soybean nodules and demonstration of selective permeability to metabolites. J. Plant Physiol., 130: 157-164.
Putnoky, P., Grosskopf, E, Cam Ha, D.T., Kiss, G.B. and Kondorosi, A., 1988. Rhizobium fix genes mediate at least two communication steps in symbiotic nodule development. J. Cell Biol., 106: 597-607.
Rai, R., Prasad, V., Choudhury, S.K. and Sinha, N.P., 1984. Iron nutrition and symbiotic N_2 fixation of lentil (Lens culinaris) genotypes in calcareous soil. J. Plant Nutr., 7: 399-405.
Rai, R., Singh, S.N. and Prasad, V., 1982. Effect of press-mud amend pyrite on symbiotic N_2 fixation - active iron contents of nodules, grain yield and quality of chick pea (Cicer arietinum Linn.) genotypes in calcareous soil. J. Plant Nutr., 5: 905-913.
Rawsthorne, S. and LaRue, T.A., 1986. Metabolism under microaerobic conditions of mitochondria from cowpea nodules. Plant Physiol., 81: 1097-1102.
Rawsthorne, S., Minchin, F.R., Summerfield, R.J., Cookson, C. and Coombs, J., 1980. Carbon and nitrogen metabolism in legume root nodules. Phytochem., 19: 341-355.
Reibach, P.H. and Streeter, J.G., 1983. Metabolism of ^{14}C-labeled photosynthate and the distribution of enzymes of glucose metabolism in soybean nodules. Plant Physiol., 72: 643-646.
Reibach, P.H., Mask, P.L. and Streeter, J.G., 1981. A rapid one-step method for the isolation of bacteroids from root nodules of soybean plants, utilizing self-generating Percoll gradients. Can. J. Microbiol., 27: 491-495.

Rioux, C.R., Jordan, D.C. and Rattray, J.B.M., 1986a. Iron requirement of Rhizobium leguminosarum and secretion of anthranilic acid during growth on an iron-deficient medium. Arch. Biochem. Biophys., 248: 175-182.
Rioux, C.R., Jordan, D.C. and Rattray, J.B.M., 1986b. Anthranilate-promoted iron uptake in Rhizobium leguminosarum. Arch. Biochem. Biophys., 248: 183-189.
Robertson, J.G. and Lyttleton, P., 1982. Coated and smooth vesicles in the biogenesis of cell walls, plasma membrane, infection thread and peribacteroid membranes in root hairs and nodules of white clover. J. Cell Sci., 58: 63-78.
Robertson, J.G. and Lyttleton, P., 1984. Division of peribacteroid membranes in root nodules of white clover. J. Cell Sci., 69: 147-157.
Robertson, J.G., Lyttleton, P., Bullivant, S. and Grayston, G.F., 1978a. Membranes in lupin root nodules.I. The role of Golgi bodies in the biogenesis of infection threads and peribacteroid membranes. J. Cell Sci. 30: 129-149.
Robertson, J.G., Warburton, M.P., Lyttleton, P., Fordyce, A.M. and Bullivant, S., 1978b. Membranes in lupin root nodules. II. Preparation and properties of peribacteroid membranes and bacteroid envelope inner membranes from developing lupin nodules. J. Cell Sci., 30: 151-174.
Robson, A.D., 1978. Mineral nutrients limiting nitrogen fixation in legume. In: C.S. Andrew and E.J. Kamprath (Editors), The Mineral Nutrition of Legumes in Tropical and Subtropical Soils. CSIRO, Melbourne, pp. 277-293.
Robson, R.L. and Postgate, J.R., 1980. Oxygen and hydrogen in biological nitrogen fixation. Ann. Rev. Microbiol., 34: 183-207.
Rohm, M. and Werner, D., 1985. Regulation of the β-ketoadipate pathway in Rhizobium japonicum bacteroids by succinate. Arch. Microbiol., 140: 375-379.
Ronson, C.W. and Astwood, P.M., 1985. Genes involved in the carbon metabolism of bacteroids. In: H.J. Evans, P.J. Bottomley and W.E. Newton (Editors), Nitrogen Fixation Research Progress. Martinus Nijhoff, Dordrecht, pp. 201-207.
Ronson, C.W., Astwood, P.M., Nixon, B.T. and Ausubel, F.M., 1987. Deduced products of C_4-dicarboxylate transport regulatory genes of Rhizobium leguminosarum are homologous to nitrogen regulatory gene products. Nucleic Acids Res., 15: 7921-7934.
Ronson, C.W., Lyttleton, P. and Robertson, J.G., 1981. C_4-Dicarboxylate transport mutants of Rhizobium trifolii form ineffective nodules on Trifolium repens. Proc. Natl. Acad. Sci. USA, 78: 4284-4288.
Ronson, C.W. and Primrose, S.B., 1979. Carbohydrate metabolism in Rhizobium trifolii: identification and symbiotic properties of mutants. J. Gen. Microbiol., 112: 77-88.
Rosendahl, L., Vance, C.P. and Pedersen, W.B., 1990. Products of dark CO_2 fixation in pea root nodules support bacteroid metabolism. Plant Physiol., 92: (in press).
Roth, E., Jeon, K. and Stacey, G., 1988. Homology in endosymbiotic systems: The term "symbiosome". In: R. Palacios and D.P.S. Verma (Editors), Molecular Genetics of Plant Microbe Interactions. APS, St Paul, pp. 220-225.
Salminen, S.O. and Streeter, J.G., 1986. Enzymes of α,α-trehalose metabolism in soybean nodules. Plant Physiol., 81: 538-541.
Salminen, S.O. and Streeter, J.G., 1987a. Involvement of glutamate in the respiratory metabolism of Bradyrhizobium japonicum bacteroids. J. Bacteriol., 169: 495-499.
Salminen, S.O. and Streeter, J.G., 1987b. Uptake and metabolism of carbohydrates by Bradyrhizobium japonicum bacteroids. Plant Physiol., 83: 535-540.
San Francisco, M.J.D. and Jacobson, G.R., 1985. Uptake of succinate and malate in cultured cells and bacteroids of two slow-growing species of Rhizobium. J. Gen. Microbiol., 131: 765-773.
Saroso, S., Dilworth, M.J. and Glenn, A.R., 1986. The use of activities of carbon catabolic enzymes as a probe for the carbon nutrition of snake bean nodule bacteroids. J. Gen. Microbiol., 132: 243-249.

Saroso, S., Glenn, A.R. and Dilworth, M.J., 1984. Carbon utilization by free-living and bacteroid forms of cowpea Rhizobium NGR234. J. Gen. Microbiol., 130: 1809-1814.
Sato, K., Inukai, S. and Shimizu, S., 1974. Vitamin B_{12}-dependent methionine synthesis in Rhizobium meliloti. Biochem. Biophys. Res. Commun., 60: 723-728.
Sato, K., Inukai, S. and Shimizu, S., 1982. Effects of vitamin B_{12}-dependent enzymes and folate compounds on morphological changes of Rhizobium meliloti. Agric. Biol. Chem., 46: 501-505.
Sayavendra-Soto, L.A., Powell, G.K., Evans, H.J. and Morris, R.O., 1988. Nucleotide sequence of the genetic loci encoding subunits of Bradyrhizobium japonicum uptake hydrogenase. In: H. Bothe, F.J. de Bruijn and W.E. Newton (Editors), Nitrogen Fixation: Hundred Years After. Gustav Fischer, Stuttgart, p. 589.
Scherrer, A. and Dénarié, J., 1971. Symbiotic properties of some auxotrophic mutants of Rhizobium meliloti and of their prototrophic revertants. Plant Soil, special volume, 39-45.
Schwyn, B. and Neilands, J.B., 1987. Siderophores from agronomically important species of the Rhizobiaceae. Comments Agric. Food Chem., 1: 95-113.
Smart, J.B., Robson, A.D. and Dilworth, M.J., 1984. Effect of phosphorus supply on phosphate uptake and alkaline phosphatase activity in rhizobia. Arch. Microbiol., 140: 281-286.
Smith, M.J. and Neilands, J.B., 1984. Rhizobactin, a siderophore from Rhizobium meliloti. J. Plant Nutr., 7: 449-458.
Smith, M.J., Shoolery, J.N., Schwyn, B., Holden, I. and Neilands, J.B., 1985. Rhizobactin, a structurally novel siderophore from Rhizobium meliloti. J. Amer. Chem. Soc., 107: 1739-1743.
Smith, S.E. and Smith, F.A., 1990. Structure and function of the interfaces in biotropic symbioses as they relate to nutrient transport. New Phytol., 114: 1-38.
Snapp, S.S. and Vance, C.P., 1986. Asparagine biosynthesis in alfalfa (Medicago sativa L.) root nodules. Plant Physiol., 82: 390-395.
Sprent, J.I., Sutherland, J.M. and de Faria, S.M., 1987. Some aspects of the biology of nitrogen-fixing organisms. Phil. Trans. Roy. Soc. London, B 317: 111-129.
Stanier, R.Y. and Ornston, L.N., 1973. The β-ketoadipate pathway. Adv. Microbial Physiol., 9: 89-151.
Stanley, J., Van Slooten, J., Dowling, D., Finan, T. and Broughton, W.J., 1989. Molecular cloning of the ntrA gene of the broad host range Rhizobium NGR234, and phenotypes of a site-directed mutant. Mol. Gen. Genet., 217: 528-532.
Stovall, I. and Cole, M., 1978. Organic acid metabolism by isolated Rhizobium japonicum bacteroids. Plant Physiol., 61: 787-790.
Stowers, M.D. and Elkan, G.H., 1985. Regulation of hexose catabolism in Rhizobium sp. 32H1. FEMS Microbiol. Lett., 26: 45-48.
Streeter, J.G., 1987a. Carbohydrate, organic acid, and amino acid composition of bacteroids and cytosol from soybean nodules. Plant Physiol., 85: 768-773.
Streeter, J.G., 1987b. Effect of nitrate on the organic acid and amino acid composition of legume nodules. Plant Physiol., 85: 774-779.
Streeter, J.G. and Salminen, S.O., 1985. Carbon metabolism in legume nodules. In: H.J. Evans, P.J. Bottomley and W.E. Newton (Editors), Nitrogen Fixation Research Progress. Martinus Nijhoff, Dordrecht. pp. 277-283.
Streeter, J.G. and Salminen, S.O., 1988. Carbon metabolism and the exchange of metabolites between symbionts in legume nodules. In: F. O'Gara, S. Manian and J.J. Drevon (Editors), Physiological Limitations and the Genetic Improvement of Symbiotic Nitrogen Fixation. Kluwer Academic, Dordrecht, pp. 11-20.
Stults, L., Mallick, S. and Maier, R.J., 1987. Nickel uptake in Bradyrhizobium japonicum. J. Bacteriol., 169: 1398-1402.

Stults, L., O'Hara, E. and Maier, R.J., 1984. Nickel is a component of hydrogenase in Rhizobium japonicum. J. Bacteriol., 159: 153-158.
Stumpf, D.K. and Burris, R.H., 1981. Organic acid contents of soybean: age and source of nitrogen. Plant Physiol., 68: 989-991.
Tajima, S. and Kouzai, K., 1989. Nucleotide pools in soybean nodule tissues: a survey of NAD(P)/NAD(P)H ratios and energy charge. Plant Cell Physiol., 30: 589-593.
Tajima, S., Kimura, I. and Sasahara, H., 1986. Succinate metabolism of isolated soybean nodule bacteroids at low oxygen concentration. Agric. Biol. Chem., 50: 1009-1014.
Tajima, S., Sasahara, H., Kouchi, H., Yoneyama, T. and Ishizuka, J., 1985. Effects of oxygen concentration and leghemoglobin on organic acid degradation by isolated soybean nodule bacteroids. Agric. Biol. Chem., 49: 3473-3479.
Tang, C.-X., Robson, A.D. and Dilworth, M.J., 1990a. The role of iron in nodulation and nitrogen fixation in Lupinus angustifolius L. New Phytol., 114: 173-182.
Tang, C.-X., Robson, A.D. and Dilworth, M.J., 1990b. A split-root experiment shows that iron is required for nodule initiation in Lupinus angustifolius L. New Phytol., in press.
Tempé, J. and Petit, A., 1983. La piste des opines. In: A. Pühler (Editor), Molecular Genetics of the Bacteria Plant Interaction. Springer Verlag, Berlin, pp. 14-32.
Trinchant, J.C., Birot, A.M. and Rigaud, J., 1981. Oxygen supply and energy-yielding substrates for nitrogen fixation (acetylene reduction) by bacteroid preparations. J. Gen. Microbiol., 125: 159-165.
Tu, J.C., 1974. Relationship between membrane envelopes of rhizobial bacteroids and the plasma membrane of the host cell as demonstrated by histochemical localization of adenyl cyclase. J. Bacteriol., 119: 986-991.
Tu, J.C., 1975. Structural similarity of the membrane envelopes of rhizobial bacteroids and the plasma membrane of the host cell as revealed by freeze-fracturing. J. Bacteriol., 122: 691-694.
Tuzimura, K. and Meguro, H., 1960. Respiration substrates for Rhizobium in the nodules. J. Biochem., 47: 391-397.
Udvardi, M.K. and Day, D., 1989. Electrogenic ATPase activity on the peribacteroid membrane of soybean (Glycine max L.) root nodules. Plant Physiol., 90: 982-987.
Udvardi, M.K., Price, G.D., Gresshoff, P.M. and Day, D.A., 1988a. A dicarboxylate transporter on the peribacteroid membrane of soybean nodules. FEBS Lett., 231: 36-40.
Udvardi, M.K., Salom, C.L. and Day, D.A., 1988b. Transport of L-glutamate across the bacteroid membrane but not the peribacteroid membrane from soybean root nodules. Mol. Plant-Microbe Interact., 1: 250-254.
Van den Bosch, K.A. and Newcomb, E.H., 1988. The occurrence of leghaemoglobin protein in the uninfected interstitial cells of soybean root nodules. Planta, 175: 442-451.
Verma, D.P.S., Fortin, M.G., Stanley, J., Mauro, V.P., Purohit, S. and Morrison, N., 1986. Nodulins and nodulin genes of Glycine max: a perspective. Plant Mol. Biol., 7: 51-61.
Verma, D.P.S., Kazazian, V., Zogbi, V. and Bal, A.K., 1978. Isolation and characterization of the membrane envelope enclosing the bacteroids in soybean root nodules. J. Cell Biol., 78: 919-936.
Waters, J.K., Karr, D.B. and Emerich, D.W., 1985. Malate dehydrogenase from Rhizobium japonicum 311b-143 bacteroids and Glycine max root nodule mitochondria. Biochem., 24: 6479-6486.
Watson, R.J., Chan, Y.K., Wheatcroft, R., Yang, A.P. and Han, S., 1988. Rhizobium meliloti genes required for C_4-dicarboxylate transport and symbiotic nitrogen fixation are located on a megaplasmid. J. Bacteriol., 170: 678-688.

Waugh, R. and Boxer, D.H., 1986. Pleiotropic hydrogenase mutants of Escherichia coli K-12: growth in the presence of nickel can restore hydrogenase activity. Biochim., 68: 157-166.

Werner, D., Mellor, R.B., Hahn, M.G. and Grisebach, H., 1985. Soybean root response to symbiotic infection: Glyceollin I accumulation in an ineffective type of soybean nodule with an early loss of the peribacteroid membrane. Z. Naturforsch., 40: 179-181.

Werner, D., Morschel, E., Garbers, C., Bassarab, S. and Mellor, R.B., 1988. Particle density and protein composition of the peribacteroid membrane from soybean root nodules is affected by mutation in the microsymbiont Bradyrhizobium japonicum. Planta, 174: 263-270.

Werner, D., Morschel, E., Stripf, R. and Winchenbach, B., 1980. Development of nodules of Glycine max infected with an ineffective strain of Rhizobium japonicum. Planta, 147: 320-329.

Werner, D., Wilcockson, J. and Kalkowski, B., 1975. Nitrogen-fixing activities in Rhizobium japonicum separated from plant cell cultures. Z. Naturforsch., 30: 678-688.

Westerhoff, H.V., Hellingwerf, K.J. and van Dam, K., 1983. Thermodynamic efficiency of microbial growth is low but optimal for maximal growth rate. Proc. Natl. Acad. Sci. USA, 80: 305-309.

Wilcockson, J. and Werner, D., 1979. Organic acids and prolonged nitrogenase activity by non-growing free-living Rhizobium japonicum. Arch. Microbiol., 122: 153-159.

Witty, J.F., Minchin, F.R., Sheehy, J.E. and Minguez, M.I., 1984. Acetylene-induced changes in the oxygen diffusion reistance and nitrogenase activity of legume root nodules. Ann. Bot., 53: 13-20.

CHAPTER 13

AMMONIA ASSIMILATION AND EXPORT OF NITROGEN FROM THE LEGUME NODULE

C. A. ATKINS
Botany Department, University of Western Australia, Nedlands, W. Australia 6009.

1 AMMONIA ASSIMILATION

The first stable product of N_2 fixation in legume nodules is ammonia (Kennedy 1966a, b; Bergersen and Turner, 1967; Ta et al., 1986) which is transferred from the bacterium through the surrounding plant-derived peribacteroid membrane (PBM) to the host cell cytoplasm (Bergersen and Turner, 1967). Here it is assimilated to form organic solutes of N which are largely transported out of the nodule in xylem (Pate and Atkins, 1983). The extent to which neutral ammonia or NH_4^+ traverses the peribacteroid space and PBM has not been established. Similarly, any requirement for active transport, or a role for PBM-localised ATP-ases (Blumwald et al., 1985) in ammonia translocation, has not been demonstrated. In fact, recent measurements of free ammonia in nodules of soybean (Streeter, 1989) indicate levels considerably lower than those recorded previously (Kennedy 1966 a, b; Becana et al.,1985), suggesting that rapid movement from bacteroids to host cytosol could be accomplished simply by passive diffusion.

Earlier labelling studies (Aprison et al., 1954; Bergersen, 1965; Kennedy, 1966a, b) with $^{15}N_2$ concluded that reductive amination to form glutamate was the initial and principal reaction of ammonia assimilation in legume nodules. More recent studies with $^{13}N_2$ (Meeks et al., 1978; Schubert and Coker, 1982), or $^{15}N_2$ (Ohyama and Kumazawa, 1978; Minamisawa et al., 1986; Ta et al., 1986; Atkins et al., 1988) indicate glutamine to be the initial organic product of fixation with subsequent transfer of amide-N to glutamate. In nodules, as in other plant tissues (Miflin and Lea, 1977; Joy, 1988), these reactions appear to be accomplished by the coupled activities of glutamine synthetase (GS) and glutamate synthase (GOGAT). Although isolated bacteroids show activities of both (Brown and Dilworth, 1975), each enzyme is expressed at a much higher level in the plant tissue of the nodule, especially in the central infected medulla (Atkins et al., 1980). Indeed, Miflin and Cullimore (1984) indicate that over 95% of GS in nodules of <u>Phaseolus vulgaris</u> is outside the bacteroids where the enzyme may account for as much as 2% of total soluble protein. The ammonia assimilating pathway is induced during nodule development (Robertson et al., 1975a, b; Boland et al., 1978; Atkins et al., 1980; Reynolds et al., 1982; Atkins et al., 1984 b) and in some symbioses (e.g. Medicago sativa, P. vulgaris), but not others (e.g. Pisum sativum, Glycine max), involves expression of nodule-specific isozymes of GS (Cullimore and Bennett, 1988) and, in the case of P. vulgaris, GOGAT as well (Chen and Cullimore, 1988). In P. vulgaris four GS genes appear to be transcribed (Cullimore and Bennett, 1988), one of which (GS g gene) is nodule-specific. The subunits are assembled to form three octameric GS isozymes (Cullimore and Bennett, 1988), two of which are expressed in the cytosol and one in the plastid. Although activities of both cytosolic isozymes (GSN1, GSN2) increased in P. vulgaris nodules during early development, most of the increase was due to the nodule-specific form (GSN1) (Lara et al., 1983).

Three forms of GOGAT have been identified in plants, at least two of which occur in legume nodules (Suzuki et al., 1986). The ferredoxin-dependent form, which is most abundant in photosynthetic tissues and is localised in chloroplasts (Wallsgrove et al., 1979), has been found in soybean nodules and appears to be very similar to the leaf isozyme (Suzuki et al., 1986). Although the source of reductant for ferredoxin-dependant enzymes in plastids of non-green tissues has been problematic (Lee, 1980), functional ferredoxin-like electron carriers which apparently transfer reductant from NADPH generated in the pentose phosphate pathway to enzymes like Fd-GOGAT, have now been detected in root tissues (Ninomiya and Sato, 1984; Suzuki et al., 1985).

An NADH-GOGAT has been purified from lupin nodules (Boland and Benny, 1977) and fractionation of nodules of a range of symbioses indicates a plastid localisation (Awonaike et al., 1981; Boland et al., 1982; Shelp et al., 1983; Shelp and Atkins, 1984). Generally, the activity of NADH-GOGAT in nodules is considerably greater than that of the ferredoxin-dependent enzyme and Chen and Cullimore (1988) have shown two NADH-GOGAT isozymes in P. vulgaris. One is nodule-specific (NADH-GOGAT II) and accounts for most of the induced activity.

In addition to the GS/GOGAT mechanism for NH_4^+ assimilation, nodules contain activities of other enzymes which could potentially serve a similar function. These include NADH-dependent alanine and glutamate dehydrogenases in bacteroids (Atkins et al., 1982; Müller and Werner, 1982) and a plant mitochondrial NADH-glutamate dehydrogenase (GDH) (Shelp et al., 1983; Shelp and Atkins, 1984). There is no convincing evidence of a role for any of these in ammonia assimilation by nodules under normal conditions. Each requires concentrations of ammonia (Miflin and Lea, 1977; Müller and Werner, 1982) higher than those recorded in nodules (Streeter, 1989), except perhaps during senescence (Klucas, 1974; Sutton, 1983) when a role for GDH might be envisaged.

Nevertheless, very high activities of GDH in nodules remain to be explained. In cowpea nodules exposed to 80% Ar : 20% O_2 for periods of up to 6 days, GOGAT levels declined sharply while those of GS and GDH remained high (Atkins et al., 1984b,c). Transferring these plants back to air resulted in prompt renewal of fixation with glutamine, but not glutamate, being exported to the xylem. In the absence of the normal activity of GOGAT, there is a strong implication that glutamate synthesis was via GDH. Lupin nodules also show an NADPH-dependent GDH in addition to their mitochondrial GDH, and the former increases in activity as nodules develop (Ratajczak, 1986a). Furthermore, at least one of the eight isoforms of GDH appears to be nodule specific (Ratajczak, 1986b). Studies of $^{13}NH_4^+$ assimilation by nodules of Alnus glutinosa suggest that both GDH and GS/GOGAT function in non-legume symbioses (Schubert and Coker, 1982).

Since the GS/GOGAT pathway is present at high activity in nodules (Robertson et al., 1975 a,b; Atkins, 1982), induction may be associated with the onset of ammonia production. Two approaches have been used to study this question. The first uses Fix$^-$ mutants of Rhizobium. Soybean nodules formed with Tn5-induced nitrogenase (nifD) mutants, lacked nitrogenase but contained normal levels of the component polypeptides of GS (Sengupta-Gopalan and Pitas, 1986). This was not the case for other mutants in which bacteria did not leave the infection threads, or where the PBM disintegrated early in development (Werner et al., 1980). Recent studies by Dunn and Ausubel (cited in Cullimore and Bennett, 1988) found that mRNA for nodule-specific GS is present in nodules formed by a number of Fix$^-$ mutants which allow relatively normal bacteroid differentiation, confirming the view (Cullimore and Bennett, 1988) that release of bacteria from the infection thread is required for induction of the enzyme. Ammonia production would not then seem to be required for induction of GS polypeptides.

The second involves inhibiting the formation of ammonia, either by replacing N_2 with Ar or by exposing nodulated plants to NO_3^-. Cowpea nodules formed in 80% Ar:20% O_2 develop normally and, for 16-18 days at least, contain levels of nitrogenase and leghaemoglobin comparable with those cultured in air (Atkins et al., 1984a,b). Though both GS and GOGAT activities were reduced in cowpea (Atkins et al., 1984b) and soybean nodules (Hirel et al., 1987) grown in this way, their presence suggests that while ammonia production may influence the extent of gene expression it is probably not a primary effector. Similarly, both nodule-specific and constitutive GS appear to be expressed in P. vulgaris cultured in the absence of N_2 (Cullimore and Giller, cited in Cullimore and Bennett, 1988). Application of NO_3^- to nodulated plants results in a rapid decline of nodule functioning and N_2 fixation (Pate and Atkins, 1983). However, for up to 7 days after application of NO_3^- to soybean, activity of the GS/GOGAT pathway was unaffected despite nitrogenase activity falling to close to zero (Schuller et al., 1986). In contrast, exposing nodulated root systems of cowpea or lupin to Ar:O_2 resulted in a rapid decline in GOGAT activity but little or no change in that of GS for periods up to 6 days (Atkins et al., 1984c). Returning plants to air restored GOGAT activity within 3 days. In both symbioses xylem N was predominantly glutamine following transfer from Ar:O_2 to air. These data are consistent with a role for ammonia production in regulating the level of expression of some of the enzymes required for its assimilation. The failure of nitrate treatment to lower GOGAT may be due to continued ammonia production as a result of localised NO_3^- reduction in nodules (Pate and Atkins, 1983).

2 EXPORT OF FIXED N FROM NODULES

The organic products of N_2 fixation exported from legume nodules have been identified mainly by analysis of xylem fluid exuded from the stumps of decapitated plants or of tracheal sap vacuum-extracted from shoots (Pate, 1977, 1980). Clearly such "xylem saps", even from plants wholly dependent for N on fixation, could contain N solutes from sources other than nodules. These include phloem to xylem exchanges, metabolic activities in roots or, in the case of tracheal sap, N metabolism in leaves (Pate and Atkins, 1983). As expected a greater proportion of N in tracheal sap is derived from sources other than current nodule export than is the case in root bleeding exudate (Peoples et al., 1985; Rerkasem et al., 1988; Peoples et al., 1989). Exported products of fixation are best defined by analyses of xylem exudates collected above and below nodulated root zones (Atkins, 1982; Peoples et al., 1986; Peoples et al., 1989), directly from detached nodules (see Table 1) or following fixation of $^{15}N_2$ (Herridge et al., 1978). Compilations of species (Atkins, 1982; Pate and Atkins, 1983; Schubert, 1986) whose xylem contents have been analysed by any, or all, of these techniques, suggest that nodules of most legumes export either amides, (principally asparagine and to a lesser extent glutamine), or ureides (allantoin and allantoic acid) (Table 1). Although glutamine may comprise a substantial fraction of the transported N in either class of legume it rarely, if ever, constitutes the major N solute of xylem in nodulated plants dependent on N_2. Reactions which transfer the amide-N of glutamine to other products therefore constitute major metabolic routes for N in the nodule.

In the Subfamily Papilionoideae, members of the tribes Aeschynomeneae, Loteae, Viceae, Cicereae, Trifolieae, Mirbelieae, Bossiaeeae and Genisteae have principally asparagine in their xylem sap, while members of six of the eight sub-tribes of the Phaseoleae (Erythrininae, Diocleinae, Glycininae, Kennediinae, Phaseolinae and Cajaninae) contain ureides. Two other tribes of the

Papillionoideae (Indigoforeae and Desmodieae) transport ureides while in the Robinieae ureide transport occurs in Sesbania but not Glyricidia or Robinia (Table 1). These two groups may also be largely of "temperate" (amide-forming) or "tropical" (ureide-forming) origin. Although such a simple division may not be justified, it has been used to speculate about why legumes differ in the nature of their translocated solutes (Sprent, 1980; Pate and Atkins, 1983; Corby et al., 1983). Ureides are much less soluble in water than asparagine and on this basis alone their translocation in tropical species seems more likely than in temperate species. Moreover, nodules of these two types of legume also differ in that those forming ureides show a determinate growth pattern and calcium oxalate crystals in their outer cortical layers, while those forming asparagine are indeterminate and show no oxalate deposits (Sprent, 1980; Sutherland and Sprent, 1984). Initially the two types were also thought to differ in their vasculature, with the vein network in determinate nodules being closed, thus aiding xylem flow and export of the less soluble ureides (Sprent, 1980; Pate and Atkins, 1983). However, the vasculature of the determinate nodules of soybean is not continuous (Walsh et al., 1989) and not all ureide-forming species have determinate nodules; those of Erythrina, Kennedya and Hardenbergia show well-defined meristems.

Among the sub family Mimosoideae, the xylem translocates of only Leucaena, Acacia and Albizia have been examined. In each case asparagine was predominant, though Bollard (1957) has reported substantial levels of citrulline (as well as ureides) in tracheal sap of Albizia lophantha and van Kessel et al. (1988) that one of the acacias (A. mearnsii, Table 1) contained 81.5% xylem N as ureides. van Kessel et al. (1988) examined a number of other acacias, as did Hansen and Pate (1987), but in none of these were ureides predominant. Some of the acacias of SW Western Australia contain the sulphur amino acid, djenkolic acid, as a predominant solute of xylem (Hansen and Pate, 1987). Djenkolate is unlikely to be a product of fixation as its levels increase in xylem when plants are cultured with NO_3^- (Hansen and Pate, 1987). Bollard's (1957) report is the only one which documents citrulline as the major N solute exported from a legume nodule. However, in a number of Frankia/non legume symbioses (Bollard, 1957; Leaf et al., 1958; Walsh et al., 1984) and in the xylem of some Nostoc/cycad symbioses (Pate et al., 1988) this ureide is predominant. Interestingly, the cycad symbioses seem to fall into two groups, one which exports citrulline in xylem and a second which contains largely glutamine and no citrulline (Pate et al., 1988). Van Kessel et al.,(1988) have analyzed root bleeding sap from a number of other mimosoid species establishing that none translocated a high proportion of xylem N as ureides. No analyses of xylem from nodulated species of the subfamily

TABLE 1. Principal nitrogenous solutes translocated in xylem by nodulated members of Leguminosae[a]

Subfamily/Tribe	Species	Analysis[b]	Solutes[c]	References[d]
CAESALPINOIDEAE			no reports	-
MIMOSOIDEAE				
Mimoseae	Leucaena leucocephala	R	Asn	4
Acacieae	Acacia redolans	R, T	Asn	6
	Acacia alata	R, T	Asn, Gln	32
	Acacia extensa	R, T	Asn, Gln	32
	Acacia pulchella	R, T	Asn, Gln	32
	Acacia mearnsii	T	Urd	33
Ingeae	Albizia distachya	T, N	Gln, Asn	1, 26
	Albizia lophantha	T	Cit, Urd	3
PAPILIONOIDEAE				
Robinieae	Gliricidia spp.	T	Asn	1
	Robinia pseudoacacia	R, T	Asn	1
	Sesbania grandiflora	T	Urd	33
Indigoforae	Cyanopsis tetragonoloba	R	Urd, Gln	14
Desmodieae	Desmodium dillenii	U	Urd	34
	Lespedeza thunbergia	U	Urd	34
Phaseoleae	Cajanus cajan	R, N, T	Urd	20
	Canavalia ensiformis	R	Urd	24
	Erythrina crystagalli	T	Asn, Urd	26
	Glycine max	R, N, T	Urd, Asn	11, 12, 14, 27
	Hardenbergia comptoniana	R	Urd, Asn	9
	Kennedia coccinea	R	Urd, Asn	9
	Macrotyloma geocarpum	R	Urd	7
	Macrotyloma uniflorum	R	Urd, Asn	14, 7
	Phaseolus vulgaris	R	Urd	5
	Psophocarpus tetragonolobus	R, N	Urd	14
	Pueraria spp.	R, T	Urd	1
	Vigna angularis	R	Urd	13
	Vigna mungo	R	Urd	14
	Vigna radiata	R	Urd	14
	Vigna subterranea	R	Urd	14
	Vigna triloba	R	Urd	13
	Vigna umbellata	R, T	Urd	14, 25
	Vigna unguiculata	R, N, T	Urd	13, 14, 22
Aeschynomeneae	Arachis hypogea	R, N, T	Asn	23
Loteae	Lotus pedunculatus	R	Asn	30
	Lotus rectum	U	Amd	34

Viciae	Lathyrus cicera	R	Asn, Gln	13
	Lathyrus sativus	R	Asn, Gln	13
	Lens culinaris	N	Asn, Gln	21
	Lens esculenta	R	Asn, Gln	13
	Pisum arvense	R	Asn, Gln	18, 19
	Pisum sativum	R, N	Asn, Gln	10, 21, 28
	Vicia calcarata	R	Asn, Gln	13
	Vicia ervillea	R	Asn, Gln	13
	Vicia faba	R, N	Asn, Gln	16, 21
	Vicia sativa	R	Asn, Gln	13
Cicereae	Cicer arietinum	R, N	Asn	21, 24
Trifolieae	Medicago sativa	R	Asn	29
	Ononis repens	U	Amd	34
	Parochetus communis	U	Amd	34
	Trifolium repens	R	Asn, Asp	8
Mirbelieae	Viminaria juncea	R	Asn, Gln	31
Bossieae	Bossiaea aquifolium	T	Asn	32
Genisteae	Lupinus albus	R, N, T	Asn	15, 17
	Lupinus angustifolius	R, T	Asn	2
	Lupinus cosentinii	R	Asn, Asp	13
	Lupinus mutabilis	R	Asn, Gln	13
	Ornithopus sativus	R	Asn, Gln	1

[a]: classification after Polhill and Raven (1981); a principal nitrogenous solute generally accounts for 50% or more of total xylem-borne N

[b]: R= root bleeding exudate; N= nodule bleeding exudate; T= tracheal sap; U= unspecified

[c]: Asn= asparagine; Asp= aspartate; Urd= ureide (allantoin + allantoic acid); Amd= amides (Asn and/or Gln, not defined)

[d]: **1** C.A. Atkins (unpublished); **2** C.A. Atkins and J.S. Pate (unpublished); **3** Bollard, 1957; **4** Avery and Rhodes, 1985; **5** Cookson et al., 1980; **6** G.F. Craig and C.A. Atkins (unpublished); **7** F.D. Dakora and C.A. Atkins (unpublished); **8** Copeland and Pate, 1970; **9** R. Flugge, C.A. Atkins and J.S. Pate (unpublished); **10** Gunning et al., 1974; **11** Herridge et al., 1988; **12** McClure and Israel, 1979; **13** Pate and Atkins, 1983; **14** Pate et al., 1980; **15** Pate et al., 1979; **16** Pate et al., 1969; **17** Pate et al., 1975; **18** Pate et al., 1965; **19** Pate and Wallace, 1962; **20** Peoples et al., 1989; **21** Peoples et al., 1987; **22** Peoples et al., 1985; **23** Peoples et al., 1986; **24** R.M.R. Rainbird (unpublished); **25** Rerkasem et al., 1988; **26** P.J. Storer and C.A. Atkins (unpublished); **27** Streeter, 1979; **28** van Beusichem, 1983; **29** Vance et al., 1985; **30** Vance et al., 1987; **31** Walker et al., 1983; **32** Hansen and Pate, 1987; **33** van Kessel et al., 1988; **34** Sutherland and Sprent, 1984.

Caesalpinioideae have been reported, but Goi and Neves (1982) reported that nodules of Cassia fistula contain levels of ureides similar to those of soybean.

One exception to the "ureide or asparagine" division in the Papilionoideae has been noted previously (Pate and Atkins, 1983; Schubert, 1986). Arachis hypogea xylem exudate contains substantial methylene glutamine (Me-Gln) which, in young seedlings, is a major N solute (Fowden,1954), accounting for as much as 90% of xylem-mobile N (Peoples et al., 1986). However, as plants mature and fixation of N_2 increases, asparagine becomes the predominant N solute of xylem (Peoples et al., 1986). Little or no net synthesis of Me-Gln occurs after germination and it is not a product of $^{15}N_2$ fixation (Peoples et al., 1986) rendering the suggestions that Me-Gln serves as a N carrier in groundnut (Powell and Dekker, 1983) unlikely.

Other solutes, especially non-protein amino acids, have been suggested to function in N transport in legumes. Canavanine (a guanidinoxy analogue of arginine) is a prominent translocated solute in Canavalia ensiformis (Rosenthal, 1972), but any significant function for this compound seems to be restricted to early seedling development before the onset of fixation (Rosenthal and Rhodes, 1984). Similarly although homoserine (an aspartate derivative) is a major component of the soluble N of peas (Pate, 1977) and has been detected in xylem exudate from nodules (Peoples et al., 1987), there is no evidence for its synthesis following N_2 fixation. In plants exposed to increased levels of salt or subjected to water stress, xylem fluid frequently carries a large proportion of N as proline. For nodulated Acacia redolans, cultured under N-free conditions with increasing salt, proline accounted for as much as 70% of N in xylem, replacing asparagine as the major solute (G.F. Craig and C.A. Atkins, unpublished). An enzymic mechanism for proline synthesis in nodules has been described (Kohl et al., 1988), but utilisation of fixed N has not been established. Le Redulier et al. (1982) have also noted increased levels of proline in nodules of legumes growing in dry coastal habitats. They also contained a large proportion of their soluble N as unidentified methylated onium compounds which were apparently translocated.

$^{15}N_2$ feeding studies show that other compounds besides the major translocated products of fixation are formed from fixed N in nodules; some of which may be exported in xylem. For example, alanine is a prominent product of fixation in soybean (Schubert and Coker, 1982), while nodule xylem exudate of Lupinus albus (Pate et al., 1979), and other species (Atkins, 1982; Pate and Atkins, 1983; Peoples et al., 1987), contain a range of amino acids, the N of which is probably derived glutamate (Ta et al., 1988a).

Differences in xylem composition may also be ascribed to variation in strain of Bradyrhizobium (Neves et al., 1985). In the soybean symbioses studied, the proportion of xylem-N as ureides was closely correlated with the rate of fixation. Unusual amino acids formed by bacteroids of Rhizobium may be transferred to the host (Shaw et al., 1986). Although their synthesis from fixed N has not been proven, such N-containing compounds could well contribute to a sustained symbiosis, though contributing little to the N nutrition of the plant.

3 SYNTHESIS AND EXPORT OF ASPARAGINE

Labelling studies using nodules of Ornithopus sativus with $^{15}N_2$ identified asparagine as a major end product of fixation (Kennedy et al., 1966; Kennedy, 1966 a,b). More recently, Ta et al. (1986) have demonstrated rapid labelling and turnover of the amide group of asparagine in $^{15}N_2/^{14}N_2$ pulse-chase experiments. Changes in asparagine labelling following application of inhibitors of GS, GOGAT or aminotransferase reactions (methionine sulfoximine, azaserine and amino-oxyacetate, respectively) to nodules were consistent with synthesis of the amide, principally though glutamine-dependent amidation of aspartate (Ta et al., 1986). An indication of limited, direct incorporation of ammonia into the amide group of asparagine was subsequently confirmed in studies involving infiltration of nodules with $^{15}NH_4^+$ (Ta et al., 1988b) .

The enzymes involved in asparagine synthesis in nodules appear to be similar to those found elsewhere in the plant (Joy, 1988). Activities of both aspartate aminotransferase and asparagine synthetase increased with nodule development and the onset of N_2 fixation in lupin (Scott et al., 1976; Boland et al., 1978; Reynolds and Farnden, 1979), and asparagine synthetase behaved similarly in nodules of M. sativa (Ta et al., 1988b). The activities of asparagine synthetase initially detected in vitro were extremely low possibly as a consequence of both bacteroid (Lees and Blakeney, 1970; Streeter, 1977) and plant (Scott et al., 1976) asparaginases in nodule extracts, the instability of the enzyme itself (Lea and Miflin, 1980; Huber and Streeter, 1985) and the possible presence of inhibitors (Joy et al., 1983). Improved conditions of extraction, together with the use of protease inhibitors like phenylmethylsulfonyl fluoride and calcium ion chelators yielded in vitro rates close to one third those of asparagine synthesis measured in vivo with ^{15}N (Ta et al., 1988b).

Asparagine synthetase has been partially purified from lupin (Lea and Fowden, 1975; Rognes, 1975) and soybean (Huber and Streeter, 1985). Both enzymes utilised the amide-group of glutamine (K_m = 0.16 mM), but NH_4^+ could partially replace glutamine giving, in the case of crude extracts of M. sativa, rates 40% of those with the amide (Ta et al., 1988 b). The elevated levels of asparagine synthetase in nodules led Vance et al. (1988) to suggest that the enzyme is a nodulin (nodule specific protein). In lupin nodules, the enzyme appears to be cytosolic (Shelp and Atkins, 1984) with neglible activity associated with organelles, though plastids from soybean nodules show a low level of activity (Boland et al., 1982).

Two isozymes of aspartate aminotransferase (AAT) occur in the plant cell fraction of lupin (Reynolds and Farnden, 1979) and soybean (Ryan et al., 1972) nodules. In lupin, the physical and kinetic properties of the two forms differ

(Reynolds et al., 1981) and, although the distinction is not so clear for the two enzymes from soybean (Ryan et al., 1972), in both species one form is cytosolic while the other is localized in plastids (Boland et al., 1982). In soybean, the cytosolic form appears to be restricted to the uninfected cells of the nodule (Hanks et al., 1983). The plastid isozyme in lupin (AAT-P2) increases during nodule development (Reynolds and Farnden, 1979), and it may be nodule-specific (Vance et al., 1988), while the other (AAT-P1) is constitutive in roots.

Although soybean transports ureides as the main product of N_2 fixation (Table 1), asparagine synthesis has also been studied in its nodules. The amide is relatively weakly labelled in nodules following exposure to $^{13}N_2$ (Schubert and Coker, 1982) or $^{15}N_2$ (Ohyama and Kumazama, 1978; Fujihara and Yamaguchi, 1980). During early nodule development asparagine synthetase activity measured in vitro increased, but as ureide export began to predominate, activity declined (Werner et al., 1980; Reynolds et al., 1982). The pool of asparagine in mature soybean nodules is substantial, frequently exceeding that of ureides and other amino compounds by a factor of three or more (Ohyama and Kumazawa, 1978; McClure and Israel, 1979). This asparagine may arise from synthesis in the nodule, or be imported via the phloem following synthesis from ureides in the tops of both soybean and cowpea (Pate et al., 1984). Nodules receive as much as half of their N for growth from phloem import (Layzell et al., 1979; Layzell et al., 1981) and it seems likely that in ureide, as well as in amide-exporting species, a substantial amount is asparagine. Interestingly, the plant asparaginase in lupin nodules is only present during early nodule development when asparagine-N is presumably utilised for growth (Scott et al., 1976). Cowpea nodules also contain a significant level of asparagine, but in $^{15}N_2$ feeding studies with intact plants negligible label was recovered in asparagine even after labelling of glutamine and glutamate had reached that of $^{15}N_2$ (Atkins et al., 1988). Asparagine synthetase activity could not be detected in extracts of cowpea nodules even when precautions were taken against proteolysis, implying that asparagine synthesis was specifically repressed (Atkins et al., 1988). The mechanism of control is unknown. Purified nodule asparagine synthetase from soybean shows competitive inhibition by asparagine (Huber and Streeter, 1985) so that excessive accumulation of phloem-delivered amide could regulate activity.

Continuing synthesis and export of asparagine from nodules requires a ready supply of C_4 precursors of aspartate (oxalacetate or malate), most of which is likely to come from sugar oxidation. However, nodule extracts from several species show relatively high activities of PEP carboxylase (Lawrie and Wheeler, 1975; Christeller et al., 1977; Layzell et al., 1979; Cookson et al., 1980; Vance and Stade, 1984), suggesting (Christeller et al., 1977; Schubert and Coker, 1982;

Gadal, 1983; Schubert, 1986) that an anaplerotic pathway of CO_2 fixation provides part of the C skeletons required for export of N as asparagine. Studies with the amide-exporting species lupin (Christeller et al., 1971; Laing et al., 1979), M. sativa (Vance et al., 1983; Maxwell et al., 1984), birdsfoot trefoil (Maxwell et al., 1984) and Vicia faba (Lawrie and Wheeler, 1975), have all concluded that CO_2 fixation by PEP carboxylase is linked to amide synthesis. The rate of $^{14}CO_2$ fixation by soybean nodules has been estimated to be equivalent to 14% of nett nodule respiration (King et al., 1986), indicating adequate potential for such a role. However, 75-92% of the fixed ^{14}C was rapidly released again as CO_2 (Coker and Schubert, 1982; King et al., 1986), implying that most of the organic acids formed by PEP carboxylase provided reductant and ATP for nitrogenase, rather than C skeletons specifically for asparagine export (King et al., 1986). PEP carboxylase may also provide some of the carboxylate (malate) required to maintain charge balance in xylem (Israel and Jackson, 1982). Since soybean translocates predominantly ureides, the role of PEP carboxylase in asparagine synthesis can hardly be judged from the results of Schubert and Coker (1982) and King et al. (1986) and more detailed study of amide-exporting species is needed.

4 SYNTHESIS AND EXPORT OF UREIDES

Allantoin and allantoic acid are formed in plants through oxidation of purine nucleotides, derived either from de novo synthesis or nucleic acid hydrolysis. Consistent with their role as translocated products of N_2 fixation, ureides are rapidly labelled, both in nodules and xylem of cowpea (Herridge et al., 1978; Atkins et al., 1988) and soybean (Matsumoto et al., 1977; Minamisawa et al., 1986) with $^{15}N_2$ and in soybean (allantoic acid only) with $^{13}N_2$ (Schubert and Coker, 1982). In cowpea, $^{14}CO_2$ supplied to leaves, also rapidly labelled allantoin and allantoic acid translocated from nodules in xylem (Herridge et al ., 1978), indicating their formation from sugar recently delivered in phloem. A pathway involving purine oxidation in nodules was indicated by (1) inhibition of ureide synthesis and export by allopurinol (an inhibitor of xanthine-oxidizing enzymes) applied to nodulated root systems (Fujihara and Yamaguchi, 1978; Atkins et al., 1980), (2) high activities of urate oxidase, xanthine dehydrogenase and allantoinase activities in extracts of nodules from ureide-producers but not from those of amide-forming species (Atkins et al., 1980; Tajima et al., 1977; Herridge et al., 1978) and (3) the ready conversion of a range of purines (bases, nucleosides, nucleotides) to allantoin by cell-free extracts from cowpea nodules (Woo et al., 1980, 1981; Atkins, 1981a).

That ureides were derived from de novo synthesis of purines involved the incorporation of ^{14}C-glycine into the glyoxylate moiety of allantoic acid by slices of cowpea nodules (Atkins et al., 1980). Cell-free preparations from both cowpea and soybean nodules (Atkins et al., 1982; Boland et al., 1982; Boland and Schubert, 1983; Shelp and Atkins,1983) were shown to form purines from glutamine, aspartic acid, glycine, HCO_3^-, a source of activated "C1" units (formyl tetrahydrofolate, THFA), Mg-ATP and phosphoribose pyrophosphate (PRPP) at rates commensurate with those of N_2 fixation (Atkins et al., 1982). Although precursors of the purine ring structure are required (Atkins et al., 1982), detailed labelling studies have not been carried out.

Purine synthesis in nodule extracts has generally been measured as incorporation of ^{14}C from glycine, serine or CO_2 into nucleotides or purine bases (Atkins et al., 1982). Only formyl glycinamide ribonucleotide (FGAR) and aminoimidazole carboxamide ribotide (AICAR) have been identified as intermediates (Atkins et al., 1982; Shelp and Atkins, 1983). Enzymes of the pathway have now been demonstrated in, and partially purified from, extracts of soybean or cowpea nodules include, PRPP synthetase (Schubert, 1981), PRPP amidotransferase (Reynolds et al., 1984), glycinamide ribonucleotide (GAR) synthetase and GAR transformylase (D. Hall and C.A. Atkins, unpublished) as well as an adenylosuccinate-AMP lyase, which also shows succinyl-AICAR lyase activity (D.Hall, C.A. Atkins and Y. Zhou, unpublished). PRPP amidotransferase from soybean has been recovered as a large (8×10^6 kD), highly labile complex containing activities of other enzymes of purine synthesis (Reynolds et al., 1984) suggesting that the pathway in nodules occurs as a soluble multi-enzyme complex. In avian liver, the existence of a postulated multi-enzyme complex (Rowe and Wyngaarden, 1968) has not been confirmed (Rowe et al., 1978). The nodule PRPP synthetase from soybean has a molecular size of $1\text{-}2 \times 10^4$ kD and occurs in at least two isoforms (Schubert, 1987).

Preliminary studies with GAR synthetase activites in cowpea nodules have shown that the bacteroid and plant enzymes are readily distinguishable, and that each plant enzyme occurs as at least two isozymes (D.Hall and C.A. Atkins, unpublished). The relative levels of the isozymes change during nodule development but no "nodule specific" form has been identified. Furthermore, the synthetase and transformylase activities have been clearly separated by ion exchange chromatography and electrophoresis. Although they may still occur in vivo as components of a multi-enzyme complex, the two activities in plants cannot occur as a single multifunctional protein as in some animal tissues (Young et al., 1984; Daubner et al., 1985). In avian liver, for example, purine synthesis involves four multifunctional proteins, which together account for nine of the fourteen enzymic steps (including synthesis of formyl THFA) from PRPP to IMP (Daubner et al., 1985). Further study of the pathway in higher plants will be required to detect any multi-functional proteins.

Plastids (or proplastids), from cowpea or soybean nodules synthesize purines de novo (Boland et al., 1982; Shelp et al., 1983; Boland and Schubert, 1983), and contain several of the enzymes involved in providing substrates for purine synthesis. These include phosphoglycerate dehydrogenase (Boland and Schubert, 1983), phosphoserine aminotransferase (Reynolds and Blevins, 1986) and serine hydroxymethyltransferase (Mitchell et al., 1986). Together with a nodule phosphoserine-specific phosphatase, (Boland and Schubert, 1982), these enzymes would provide both glycine and a source of "C1" units from sugar oxidation. Plastids from cowpea nodules also contain methylene THFA dehydrogenase (Shelp et al., 1983) and, as mentioned earlier, GOGAT, aspartate aminotransferase and possibly malate dehydrogenase. With cytosolic GS and a plastid source of ATP, the ancilliary enzymes would provide all the precursors for purine synthesis from PRPP within the organelle. The location of PRPP synthetase has not been clearly resolved although in soybean nodules it may also be associated with plastids (Boland et al., 1982). Infected cells contain much more (84%) of the purine synthesis enzymes than the uninfected cells (16%; Shelp et al., 1983), but it is not known whether the pathway is located in plastids in each case, or indeed whether plastids are involved in tissues other than nodules.

In cell-free preparations of cowpea nodules ammonia can partially replace glutamine (Atkins et al., 1982) in the synthesis of purines, but high concentrations (40 mM) were required. Much more FGAR accumulated with ammonia than with glutamine as substrate, indicating that the second amidotransferase (FGAR amidotranserase) used ammonia less readily than the first (PRPP amidotransferase). Given the low ammonia concentrations in soybean nodule cytosol (Streeter, 1989), it seems unlikely that ammonia could replace glutamine in ureide biosynthesis. Contrary to previous studies (Kapoor and Waygood, 1962) asparagine was not an effective amide donor for purine biosynthesis in cowpea nodules (Atkins et al., 1982).

De novo purine synthesis increases sharply with nodule development (Atkins et al., 1982) and appears to require NH_4^+ export from bacteroids (Atkins et al., 1984b). While the activities of enzymes of NH_4^+ assimilation, or of purine oxidation (xanthine dehydrogenase, urate oxidase and allantoinase), were reduced two- or three-fold in nodules cultured in 80% Ar:20% O_2, the de novo purine pathway under these conditions was then less than 1% of that in nodules reducing N_2 (Atkins et al., 1984b). It seems clear that a fully functional symbiosis is a prerequisite for ureide formation in nodules. Kohl et al. (1988) have hypothesised that purine biosynthesis in nodules might be regulated by the supply of ribose 5-phosphate (R5P), but it is not known whether the supply of R5P limits ureide formation.

Models of soybean nodule metabolism (Boland et al., 1982; Reynolds et al., 1982; Schubert, 1986) envisage the conversion of IMP to xanthosine or inosine within the plastids of infected cells, but rely on the very slight metabolism of IMP in plastid preparations recovered from sucrose density gradients (Boland and Schubert, 1983). However, IMP dehydrogenase in cowpea nodules appears to be wholly cytosolic (Shelp and Atkins, 1983; Shelp et al., 1985), as is xanthosine-specific purine nucleosidase (Atkins et al ., 1989). These enzymes, together with 5'-nucleotidase (Doremus and Blevins, 1988), appear to provide a high affinity pathway for IMP utilisation outside the plastid (Shelp and Atkins, 1983). The location of IMP dehydrogenase may be critical for feedback control of purine synthesis, as PRPP amidotranferase appears to be regulated by IMP (Reynolds et al., 1984). Although IMP dehydrogenase in nodules was initially overlooked because of its extreme lability (Atkins, 1981a,b), its presence explains why allopurinol, applied to intact soybean or cowpea plants, results in accumulation of xanthine (Atkins et al., 1980; Boland and Schubert, 1982) rather than hypoxanthine. In cell-free extracts (Atkins, 1981a), where IMP dehydrogenase was not protected against inactivation, allopurinol caused accumulation of hypoxanthine. Nodules exposed to allopurinol and supplied $^{15}N_2$ produced ^{15}N-xanthine but not ^{15}N-hypoxanthine (Atkins et al., 1988) confirming the pathway as well as demonstrating, for the first time, recovery of currently fixed N in a purine.

Xanthine oxidation in ureide-forming nodules is catalysed by an NAD-specific xanthine dehydrogenase (Atkins et al., 1980; Triplett et al., 1980). Early reports of xanthine oxidase in soybean (Tajima and Yamamato, 1975; Fujihara and Yamaguchi, 1978) have not been sustained (Triplett et al., 1980). The enzyme from soybean or cowpea is extremely sensitive to allopurinol (Atkins et al., 1980) and its hydroxylation product, oxypurinol (Atkins et al., 1988). While this would explain complete suppression of ureide synthesis and accumulation of xanthine (Boland and Schubert, 1982; Atkins et al., 1988), nitrogenase activity in cowpea is also inhibited by allopurinol (Atkins et al., 1988). How a block in ureide synthesis decreased nitrogenase activity is not known, but neither simple feedback inhibition of N_2 fixation by xanthine, nor a direct effect of allopurinol on bacteroid functioning, seems to be involved. Suppression of nitrogenase activity by allopurinol could be reversed by increasing O_2 from 20 to 40% (Atkins et al., 1988), indicating that diffusional resistance in the organ (Dakora and Atkins, 1989) was increased as a consequence of allopurinol treatment.

Xanthine dehydrogenase from nodules occurs only in the cytosol of infected cells (Triplett, 1985; Nguyan et al., 1986), while the next enzyme in the pathway, urate oxidase (or uricase), is largely confined to uninfected cells. Isolation and separation of the two types of cell (Shelp et al., 1983) or their protoplasts (Hanks et al., 1983; Kouchi et al., 1988) has led to some controversy (see Schubert, 1986) as to the proportion of each enzyme in each cell type. However, immunocytochemical localisation in soybean (Bergmann et al., 1983; Nguyen et al., 1985; Van den Bosch and Newcomb, 1986) and cowpea (Webb and Newcomb, 1987) has shown that in the medulla uricase is virtually restricted to the enlarged microbodies of uninfected cells. The three layers of cells comprising the inner cortex of soybean nodules also contain enlarged microbodies which, like those of the interstitial uninfected cells in the medulla, are enriched in uricase protein (Newcomb et al., 1989). This explains why Shelp et al. (1983) found 40% of the uricase activity of cowpea nodules in the cortex. It is interesting to note that the cells of the inner cortex of determinate legume nodules have been implicated in diffusion adjustment (see Dakora and Atkins , 1989) and control of nitrogenase. If it is not fortuitous that these cells also contain uricase, their functioning in ureide synthesis may be related to a role in diffusional control of O_2 supply.

Purified urate oxidase from both cowpea (Rainbird and Atkins, 1981) and soybean nodules (Lucas et al., 1983) has an extremely low affinity for O_2 (K_m = approx. 30 μM), which is consistent with its absence from infected cells, where free O_2 levels are extremely low (3-10 nM; Appleby, 1984). It seems likely that even in uninfected cells the level of free O_2 would be restricted, because of the diffusional resistance of the tissue, O_2-binding by leghaemoglobin and bacteroid respiration in surrounding infected cells. With as much as 25% of the leghaemoglobin in soybean nodules in the uricase-containing interstitial cells of the medulla, the pigment might also facilitate O_2 diffusion to host organelles of both cell types (Van den Bosch and Newcomb, 1988). The fact that uricase is localised within a microbody which also contains catalase (Hanks et al., 1981) should not be overlooked particularly because the kinetic properties of many enzymes are markedly altered when they are confined within lipophillic micellar structures (Martinek et al., 1987).

The only enzyme of ureide synthesis to receive attention at the genetic level is urate oxidase. Expression of the 35kD subunit of the tetrameric native enzyme is nodule-specific (nodulin 35) and Nguyen et al. (1985) isolated the gene following immunoprecipitation of the nascent polypeptide. The structure of the gene has been determined and potential regulatory sequences identified. The gene for the nodule-specific uricase from P. vulgaris appears similar (Sanchez et al., 1987). While the specific localisation of uricase indicates that gene expression

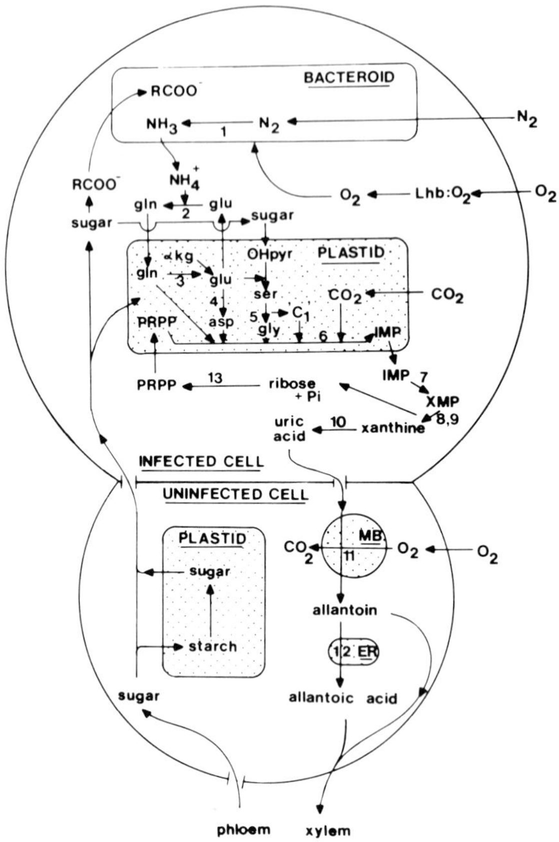

Fig. 1. Intercellular and intracellular localisation of the component reactions of assimilation of fixed N in ureide-forming nodules (cowpea, soybean). The numbers refer to the following enzymes: **1.** nitrogenase; **2.** glutamine synthetase; **3.** glutamate synthase; **4.** aspartate aminotransferase; **5.** serine hydroxymethyltransferase; **6.** component enzymes of de novo purine synthesis; **7.** NAD:IMP dehydrogenase; **8.** nucleotidase or phosphatase; **9.** xanthosine nucleosidase; **10.** NAD:xanthine dehydrogenase; **11.** urate oxidase; **12.** allantoinase; **13.** PRPP synthetase. MB = microbody, ER = endoplasmic reticulum.

is related to the differentiation of microbodies in uninfected cells, the mechanism involved is not known. Production of ammonia due to nitrogenase is not required as nodules from plants cultured in 80% Ar:20% O_2 contain reduced, though still significant, levels of the enzyme (Atkins et al., 1984b). In cultured soybean callus uricase may be regulated by pO_2 (Larsen and Jochimsen, 1987), but in cowpea the enzyme was present in nodules formed in from 1 to 80% O_2 (F.D.Dakora and C.A. Atkins, unpublished). Furthermore, a nodule-specific uricase is not restricted to ureide-forming species. Western blotting with polyclonal antisera to the purified 35kD subunit of cowpea uricase has shown the presence of an homologous protein of similar size in nodules in a range of asparagine-forming symbioses (C.A.Atkins, unpublished). These species also have a uricase activity which shows a correlation with the frequency of uninfected interstitial cells in the central medulla. Thus, while nodules of lupin, with essentially no uninfected interstitial cells, have no uricase activity and no 35kD subunit, those with a high proportion of uninfected cells (eg Robinia pseudoacacia) show levels of activity and subunit approaching those of ureide-formers. Differentiation of uninfected cells in the medulla of cowpea nodules is markedly altered by pO_2 (F.D.Dakora and C.A.Atkins, unpublished), so the level of uricase may be regulated through an effect of pO_2 on the frequency of enlarged microbodies.

Allantoinase, which catalyses the hydrolysis of allantoin to allantoic acid, occurs in the microsomal fraction of sucrose density gradients (Hanks et al., 1981), suggesting that in vivo it is attached to the endoplasmic reticulum, although in other plant tissues (Parish, 1972) it appears to be in microbodies. As the location of uricase might imply allantoinase is found in uninfected cells which also show a very prominent and highly folded endoplasmic reticulum (Newcomb et al., 1985). It is worth noting that these ultrastructural features are also characteristic of the cells of the inner cortex of soybean nodules (Newcomb et al., 1989) and that, in cowpea at least, as much as half of the allantoinase of the nodule is recovered in the cortex (Atkins et al., 1980; Shelp et al., 1983). Although the complex intra- and intercellular compartmentalisation of the steps of ureide synthesis in nodules (see Fig. 1) has been deduced from isolation of cellular or subcellular fragments, the actual transfer of intermediates from one organelle or cell to another remains to be demonstrated.

5 COSTS AND EFFICIENCIES RELATED TO EXPORT OF N FROM NODULES

Nodules consume as much as 30% of the photosynthate produced by the legume plant (Atkins et al., 1979; Layzell et al., 1979; Rainbird et al., 1983) with, on average, around half of this being respired as CO_2. Theoretical estimates of the costs of nodule functioning (Pate et al., 1981; Atkins, 1984), indicate that, although the greatest expense relates to nitrogenase (60%), assimilation of ammonia and ancillary C metabolism consume as much as 20% (Pate et al., 1981; Atkins, 1984). Experimental determinations of the C and N budgets of ureide-forming (Vigna unguiculata, V. radiata) and amide-forming (Lupinus albus,

Pisum sativum) symbioses indicate that those based on ureides are generally more economical of C, requiring as little as 1.4 g $C.g^{-1}$ N fixed in cowpea compared to a minimum of 3.9 g $C.g^{-1}$ N fixed in lupin (Atkins, 1984). These symbioses differ in a range of characters (temperate or tropical, H_2 evolution, nodule morphology and anatomy, activity of PEP carboxylase, etc.) as well as in the nature of the N solute exported from their nodules. Thus, it is not possible to associate economy of C use with the form of N translocated. Ureides, with a C:N ratio of 1 compared to 2 for asparagine, theoretically require less C for their transport, while the energy required for their synthesis is about the same (3-3.5 ATP equivalents per N; Atkins, 1982). Schubert (1986) has pointed out that "costs" associated with the synthesis and transport of compounds formed in N_2 fixation need to also consider eventual utilisation of both the N and C they contain. Allantoin and allantoic acid would then appear to be the least expensive of the major N solutes formed in nodules, and their synthesis an element in the apparently favourable C economy of the "ureide-formers".

Clearly it would be of interest to compare economies based on ureides or amides using the same legume. Triplett (1986) has proposed two procedures (one involving allopurinol) which might be used to screen plants for their synthesis of ureides. While allopurinol appears not to be as specific as initially thought, its use does nevertheless discriminate between symbioses forming ureides and those forming amides (Atkins et al., 1988). In addition to selecting lines or mutants with altered nodule biochemistry, a second approach would be to manipulate symbioses so as to alter their exported solutes. Both asparagine- and ureide-forming nodules produce and export glutamine, following their short term exposure to 80% Ar:20% O_2 (Atkins et al., 1984b,c). Allopurinol treatment results in continued export of N from cowpea nodules as xanthine and glutamine (Atkins et al., 1988). Preliminary experiments indicate that methionine sulphoximine (MSX) applied topically to intact nodules on cowpea plants causes a rapid increase in NH_4^+ export in xylem (C.A. Atkins and B.J.Shelp, unpublished), while a second GS inhibitor, tabtoxinine-β-lactam (TβL), also resulted in substantial changes in translocated N solutes in both soybean and M. sativa (Langston-Unkefer et al., 1984; Knight et al., 1988). Significantly, in the case of TβL formed as a consequence of infection of roots by Pseudomonas syringae pv tabaci (Knight et al ., 1988), the altered spectrum of translocated N solutes was accompanied by significant yield increases, greater nodulation and elevated N_2 fixation. If NH_4^+ or simpler organic solutes of N (urea, glycine etc) could be translocated to the tops of nodulated plants, metabolism might be more directly linked to photosynthesis, thus enhancing yield by conserving fixed C in the plant and reducing the costs associated with assimilation in the nodules.

6 REFERENCES

Appleby, C.A., 1984. Leghaemoglobin and Rhizobium respiration. Ann. Rev. Plant Physiol., 35: 443-478.

Aprison, M.H., Magee, W.E. and Burris, R.H., 1954. Nitrogen fixation by excised soybean root nodules. J. Biol. Chem., 208: 29-39.

Atkins, C.A., 1981a. Metabolism of purine nucleotides to form ureides in nitrogen fixing nodules of cowpea (Vigna unguiculata L. Walp.). FEBS. Lett., 125: 89-93.

Atkins, C.A., 1981b. The legume-Rhizobium symbiosis: ureide biosynthesis. In: A.H. Gibson and W.E. Newton (Editors), Current Perspectives in Nitrogen Fixation. Aust. Acad. Sci., Canberra, pp. 271-272.

Atkins, C.A., 1982. Ureide metabolism and the significance of ureides in legumes. In: N.S.S. Rao (Editor), Advances in Agricultural Microbiology. Oxford and IBH, New Delhi, pp. 53-88.

Atkins, C.A., 1984. Efficiencies and inefficiencies in the legume/Rhizobium symbiosis - a review. Plant and Soil, 82: 273-284.

Atkins, C.A., Herridge, D.F. and Pate, J.S., 1978. The economy of carbon and nitrogen in nitrogen-fixing annual legumes - experimental observation and theoretical consideration. In: C.N. Welsh (Editor), Proc. I.A.E.A. Meeting on N_2 Fixation, IAEA Press Vienna, pp. 221-242.

Atkins, C.A., Pate, J.S. and Shelp, B.J., 1984c. Effects of short-term N deficiency on N metabolism in legume nodules. Plant Physiol., 76: 705-710.

Atkins, C.A., Rainbird, R.M. and Pate, J.S., 1980. Evidence for a purine pathway of ureide synthesis in N_2-fixing nodules of cowpea (Vigna unguiculata L. Walp.). Z. Pflanzenphysiol., 97: 249-260.

Atkins, C.A., Ritchie, A., Rowe, P.B., McCairns, E. and Sauer, D., 1982. De novo purine synthesis in nitrogen-fixing nodules of cowpea (Vigna unguiculata L. Walp) and soybean (Glycine max L. Merr). Plant Physiol., 70: 55-60.

Atkins, C.A., Shelp, B.J., Storer, P.J. and Pate, J.S., 1984a. Nitrogen nutrition and the development of biochemical functions associated with nitrogen fixation and ammonia assimilation of nodules on cowpea seedlings. Planta, 162: 327-333.

Atkins, C.A., Sanford, P.J., Storer, P.J. and Pate, J.S., 1988. Inhibition of nodule functioning in cowpea by a xanthine oxidoreductase inhibitor, Allopurinol. Plant Physiol., 88: 1229-1234.

Atkins, C.A., Shelp, B.J., Kuo, J., Peoples, M.B. and Pate, J.S., 1984b. Nitrogen nutrition and the development and senescence of nodules on cowpea seedlings. Planta, 162: 316-326.

Atkins, C.A., Storer, P.J. and Pate, J.S., 1988. Pathways of nitrogen assimilation in cowpea nodules studied using $^{15}N_2$ and allopurinol. Plant Physiol., 86: 204-207.

Atkins, C.A., Storer, P.J. and Shelp, B.J., 1989. Purification and properties of purine nucleosidase from N_2-fixing nodules of cowpea (Vigna unguiculata). J. Plant Physiol., 134: 447-452.

Avery, M.E. and Rhodes, D., 1988. Nitrogenous solutes of Leucaena leucocephala. Plant Physiol., 86: Suppl. p. 125.

Awonaike, K.O., Lea, P.J. and Miflin, B.J., 1981. The location of the enzymes of ammonia assimilation in root nodules of Phaseolus vulgaris L. Plant Sci. Lett., 23: 189-195.

Becana, M., Aparicio-Tejo, P.M. and Sanchez-Diaz, M., 1985. Levels of ammonia, nitrite and nitrate in alfalfa root nodules supplied with nitrate. J. Plant Physiol., 119: 359-367.

Bergersen, F.J., 1965. Ammonia - an early stable product of nitrogen fixation by soybean root nodules. Aust. J. Biol. Sci., 18: 1-9.

Bergersen, F.J. and Turner, G.L., 1967. Nitrogen fixation by the bacteroid fraction of breis of soybean root nodules. Biochim. Biophys. Acta, 141: 507-515.

Bergmann, H., Preddie, E. and Verma, D.P.S., 1983. Nodulin-35: a subunit of specific uricase (uricase II) induced and localised in the uninfected cells of soybean nodules. EMBO. J., 2: 2333-2339.

Blumwald, E., Fortin, M.G., Rea, P.A., Verma, D.P.S. and Poole, R.J., 1985. Presence of host-plama membrane type H^+-ATPase in the membrane envelope enclosing the bacteroids in soybean root nodules. Plant Physiol., 78: 665-672.

Boland, M.J. and Benny, A.G., 1977. Enzymes of nitrogen metabolism in legume nodules. Purification and properties of NADH-dependent glutamate synthase from lupin nodules. Eur. J. Biochem., 79: 355-362.

Boland, M.J., Fordyce, A.M. and Greenwood, R.M., 1978. Enzymes of nitrogen metabolism in legume nodules - a comparative study. Aust. J. Plant Physiol., 5: 553-559.

Boland, M.J. and Schubert, K.R., 1982. Purine biosynthesis and catabolism in soybean root nodules. Incorporation of ^{14}C from $^{14}CO_2$ into xanthine. Arch. Biochem. Biophys., 213: 486-491.

Boland, M.J. and Schubert, K.R., 1983. Biosynthesis of purines by a proplastid fraction from soybean nodules. Arch. Biochem. Biophys., 220: 179-187.

Boland, M.J., Hanks, J.F., Reynolds, P.H.S., Blevins, D.G., Tolbert, N.E. and Schubert, K.R., 1982. Subcellular organization of ureide biogenesis from glycolytic intermediates and ammonium in nitrogen-fixing soybean nodules. Planta, 155: 45-51.

Bollard, E.G., 1957. Translocation of organic nitrogen in the xylem. Aust. J. Biol. Sci., 10: 292-301.

Brown, C.M. and Dilworth, M.J., 1975. Ammonium assimilation by Rhizobium cultures and bacteroids. J. Gen. Microbiol., 86: 39-48.

Chen, F-L. and Cullimore, J.V., 1988. Two isoenzymes of NADH-dependent glutamate synthase in root nodules of Phaseolus vulgaris L. Purification, properties and activity changes during nodule development. Plant Physiol., 88: 1411-1417.

Christeller, J.T., Laing, W.A. and Sutton, W.D., 1977. Carbon dioxide fixation by lupin root nodules. I. Characterization, association with phosphoenolpyruvate carboxylase, and correlation with nitrogen fixation during nodule development. Plant Physiol., 60: 47-50.

Cookson, C., Hughes, H. and Coombs, J., 1980. Effects of combined nitrogen on anapleuratic carbon assimilation and bleeding sap composition in Phaseolus vulgaris L. Planta, 148: 338-345.

Copeland, R. and Pate, J.S., 1970. In: White clover research, Brit. Grassland Soc., 6: 71-79

Corby, H.D.L., Polhill, R.M. and Sprent, J.I., 1983. Taxonomy. In: W.J. Broughton (Editor), Nitrogen Fixation Vol. 3, Legumes. Clarendon Press, Oxford pp 1-35.

Cullimore, J.V. and Bennett, M.J., 1988. The molecular biology and biochemistry of plant glutamine synthetase from root nodules of Phaseolus vulgaris L. and other legumes. J. Plant Physiol., 132: 387-393.

Dakora, F.D. and Atkins, C.A., 1989. Diffusion of oxygen in relation to structure and function in legume root nodules. Aust. J. Plant Physiol., 16: 131-140.

Daubner, S.C., Schrimsher, J.L., Schendel, F.J., Young, M., Henikoff, S., Patterson, D., Stubbe, J. and Benkovic, S.J., 1985. A multifunctional protein possessing glycinamide ribonucleotide synthetase, glycinamide ribonucleotide transformylase and aminoimidazole ribonucleotide synthetase in de novo purine biosynthesis. Biochem., 24: 7059-7062.

Doremus, H.D. and Blevins, D.G., 1988. Nucleoside diphosphatase and 5'-nucleotidase activities of soybean root nodules and other tissues. Plant Physiol. 87: 36-40.

Fowden, L., 1954. The nitrogen metabolism of groundnut plants: The role of methylene glutamine and methylene glutamic acid. Ann. Bot., 18: 417-429.

Fujihara, S. and Yamaguchi, M., 1978. Effects of allopurinol [4-Hydroxypyrazolo (3,4-d) Pyrimidine] on the metabolism of allantoin in soybean plants. Plant Physiol., 62: 134-138.
Fujihara, S. and Yamaguchi, M., 1980. Asparagine formation in soybean nodules. Plant Physiol., 66: 139-141.
Gadal, P., 1983. Phosphoenolpyruvate carboxylase and nitrogen fixation. Physiol. Veg., 21: 1069-1074.
Goi, S.R. and Neves, M.C.P., 1982. Ureide content, nodule type and nitrogenase activity of grain, forage and forest tree legume species. Desq. Agropec. bras., Brasilia, 17: 43-50.
Gunning, B.E.S., Pate, J.S., Minchin, F.R. and Marks, I., 1974. Quantitative aspects of transfer cell structure in relation to vein loading in leaves and solute transport in legume nodules. Symp. Soc. Exp. Biol., 28: 87-126.
Hanks, J.F., Schubert, K. R. and Tolbert, N.E., 1983. Isolation and characterization of infected and uninfected cells from soybean nodules. Role of uninfected cells in ureide synthesis. Plant Physiol., 71: 869-873.
Hanks, J.F., Tolbert, N.E. and Schubert, K.R., 1981. Localization of enzymes of ureide biosynthesis in peroxisomes and microsomes of nodules. Plant Physiol., 68: 65-69.
Hansen, A.P. and Pate, J.S., 1987. Evaluation of ^{15}N natural abundance method and xylem sap analysis for assessing N_2 fixation of understorey legumes in Jarrah (Eucalyptus marginata Donn ex Sm.) forest in S.W. Australia. J. Expt. Bot., 38: 1446-1458.
Henson, I.E. and Wheeler, G.T., 1977. Hormones in plants bearing nitrogen-fixing root nodules: Cytokinin transport from the root nodules of Alnus glutinosa (L.) Gaertn. J. Expt. Bot., 28: 1099-1110.
Herridge, D.F., Atkins, C.A., Pate, J.S. and Rainbird, R.M., 1978. Allantoin and allantoic acid in the nitrogen economy of the cowpea (Vigna unguiculata) [L.] Walp.). Plant Physiol., 62: 495-498.
Herridge, D.F., O'Connel, P. and Donnelly, K., 1988. The xylem ureide assay of nitrogen fixation: Sampling procedures and sources of error. J. Expt. Bot., 29: 12-22.
Hirel, B., Bouet, C., King, B., Layzell, D., Jacobs, F. and Verma, D.P.S., 1987. Glutamine synthetase genes are regulated by ammonia provided externally or by symbiotic nitrogen fixation. EMBO. J., 6: 1167-1171.
Huber, T.A. and Streeter, J.G., 1985. Purification and properties of asparagine synthetase from soybean root nodules. Plant Sci., 42: 9-17.
Israel, D.W. and Jackson, W.A., 1982. Ion balance, uptake, and transport processes in N_2 fixing and nitrate and urea dependent soybean plants. Plant Physiol., 69: 171-178.
Joy, K.W., 1988. Ammonia, glutamine and asparagine: a carbon-nitrogen interface. Can. J. Bot., 66: 2103-2109.
Joy, K.W., Ireland, R.J. and Lea, P.J., 1983. Asparagine synthesis in pea leaves, and the occurrence of an asparagine sythetase inhibitor. Plant Physiol., 73: 165-168.
Kapoor, M. and Waygood, E.R., 1962. Initial steps of purine biosynthesis in wheat germ. Biochem. Biophys. Res. Commun., 9: 7-10.
Kennedy, I.R., 1966a. Primary products of symbiotic nitrogen fixation. I. Short-term exposures of serradella nodules to $^{15}N_2$. Biochim. Biophys. Acta 130: 285-294.
Kennedy, I.R., 1966b. Primary products of symbiotic nitrogen fixation. II. Pulse labelling of serradella nodules with $^{15}N_2$. Biochim. Biophys. Acta, 120: 295-303.
Kennedy, I.R., Parker, C.A. and Kidby, D.K., 1966. The probable site of nitrogen fixation in root nodules of Ornithopus sativus. Biochim. Biophys. Acta, 130: 517-519.
King, B.J., Layzell, D.B. and Canvin, D.T., 1986. The role of dark carbon dioxide fixation in root nodules of soybean. Plant Physiol., 81: 200-205.

Klucas, R.V.,1974. Studies on soybean nodule senescence. Plant Physiol., 54: 612-616.
Knight, T.J. and Langston-Unkefer, P.J., 1988. Enchancement of symbiotic dinitrogen fixation by a toxin-releasing plant pathogen. Science, 241: 951-954.
Kohl, D.H., Schubert, K.R., Carter, M.B., Hagedorn, C.H. and Shearer, G., 1988. Proline metabolism in N_2-fixing root nodules: Energy transfer and regulation of purine synthesis. Proc. Natl. Acad. Sci. USA., 85: 2036-2040.
Kouchi, H., Fukai, K., Katagiri, H., Minamisawa, K. and Tajima, S., 1988. Isolation and enzymological characterization of infected and uninfected cell protoplasts from root nodules of Glycine max. Physiol. Plant., 73: 327-334.
Laing, W.A., Christeller, J.T. and Sutton, W.D., 1979. Carbon dioxide fixation by lupin root nodules. II. Studies with ^{14}C-labeled glucose, the pathway of glucose catabolism, and the effects of some treatments that inhibit nitrogen fixation. Plant Physiol., 63: 450-454.
Langston-Unkefer, P.L., Macy, P.A. and Durbin, R.D., 1984. Inactivation of glutamine synthetase by tabtoxinine-β-lactam. Effects of substrates and pH. Plant Physiol., 76: 71-74.
Lara, M., Cullimore, J.V., Lea, P.J., Miflin, B.J., Johnstone, A.W.B. and Lamb, J.W., 1983. Appearance of a novel form of plant glutamine synthetase during nodule development of Phaseolus vulgaris. Planta, 157: 254-258.
Larsen, K. and Jochimsen, B.V., 1987. Appearance of purine-catabolizing enzymes in Fix^+ and Fix^- root nodules on soybean and effect of oxygen on the expression of the enzymes in callus tissue. Plant Physiol., 85: 452-456.
Lawrie, A.C. and Wheeler, C.T., 1975. Nitrogen fixation in the root nodules of Vicia faba L. in relation to the assimilation of carbon. II. The dark fixation of carbon dioxide. New Phytol., 74: 429-436.
Layzell, D.B., Pate, J.S., Atkins, C.A. and Canvin, D.T., 1981. Partitioning of carbon and nitrogen and nutrition of root and shoot apex in a nodulated legume. Plant Physiol., 67: 30-36.
Layzell, D.B., Rainbird, R.M., Atkins, C.A. and Pate, J.S., 1979. Economy of photosynthate use in N-fixing legume nodules. Observation on two contrasting symbioses. Plant Physiol., 64: 888-891.
Lea, P.J. and Fowden, L., 1975. The purification and properties of glutamine-dependent asparagine synthase isolated from Lupinus albus. Proc. Roy. Soc. London, Series B., 192: 13-26.
Lea, P.J. and Miflin, B.J., 1980. Transport and metabolism of asparagine and other nitrogen compounds within the plant. In: B.J. Miflin (Editor), The Biochemistry of Plants. Vol. 5. Academic Press, N.Y., pp. 569-607.
Leaf, G., Gardner, I.C. and Bond, G., 1958. Observations on the composition and metabolism of the nitrogen-fixing root nodules of Alnus. J. Expt. Bot., 9: 320-331.
Lee, B., 1980. Sources of reductant for nitrate assimilation in non-photosynthetic tissue. A review. Plant Cell Environ., 3: 65-90.
LeRedulier, D., Goas, G. and Larher, F., 1982. Onium compounds, amides and amino acids level in nodules and other organs of nitrogen fixing plants. Z. Pflanzenphysiol., 105: 417-426.
Lees, E.M. and Blakeney, A.B., 1970. The distribution of asparaginase activity in plants. Biochim. Biophys. Acta, 215: 145-151.
Legocki, R.P. and Verma, D.P.S., 1979. A nodule-specific plant protein (Nodulin-35) from soybean. Science, 205: 190-193.
Lucas, K., Boland, M.J., and Schubert, K.R., 1983. Uricase from soybean nodules: Purification, properties, and comparison with the enzyme from cowpea. Arch. Biochem. Biophys., 226: 190-197.
Matsumoto, T., Yatazawa, M. and Yamamoto, Y., 1977. Incorporation of ^{15}N into allantoin in nodulated soybean plants supplied with $^{15}N_2$. Plant and Cell Physiol, 18: 459-462.

McClure, P.R. and Israel, D.W., 1979. Transport of nitrogen in the xylem of soybean plants. Plant Physiol., 64: 411-416.
Martinek, K., Berezin, I.V., Knmelnitski, V.L., Klyachko, N.L. and Levashov, A.V., 1987. Micellar enzymology: potentialities in applied areas (biotechnology). Coll. Czech. Chem. Commun., 52: 2589-2602.
Maxwell, C.A., Vance, C.P., Heichel, G.H. and Stade, S., 1984. CO_2 fixation in alfalfa and birdsfoot trefoil root nodules and partitioning of ^{14}C to the plant. Crop Sci., 24: 257-264.
Meeks, J.C., Wolk, C.P., Schilling, N., Shaffer, P.W., Avissar, V. and Chien, W.S., 1978. Initial organic products of fixation of [^{13}N] dinitrogen by root nodules of soybean (Glycine max). Plant Physiol., 61: 980-983.
Miflin, B.J. and Cullimore, J.V., 1984. Nitrogen assimilation in the legume-Rhizobium symbiosis: a joint endeavour. In: D.P.S. Verma and T. Honn (Editors), Genes involved in Microbe - Plant Interactions. Springer-Verlag, Vienna, pp. 129-196.
Miflin, B.J. and Lea, P.J., 1977. Amino acid metabolism. Ann. Rev. Plant Physiol., 28: 299-329.
Minamisawa, K., Arima, Y. and Kumazawa, K., 1986. Characteristics of asparagine pool in soybean nodules in comparison with ureide pool. Soil Sci. Plant Nutr., 32: 1-14.
Mitchell, M.K., Reynolds, P.H.S. and Blevins, D.G., 1986. Serine hydroxymethylase from soybean root nodules. Purification and kinetic properties. Plant Physiol., 81: 553-557.
Müller, P. and Werner, D., 1982. Alanine dehydrogenase from bacteroids and free living cells of Rhizobium japonicum. Z. Naturforsch., 37c: 927-936.
Neves, M.C.P., Didonet, A.D., Duque, F.F. and Döbereiner, J., 1985. Rhizobium strain effects on nitrogen transport and distribution in soybeans. J. Expt. Bot., 36: 1179-1192.
Newcomb, E.H., Kaneko, Y. and Van den Bosch, K.A., 1989. Specialization of the inner cortex for ureide production in soybean root nodules. Protoplasma, 150: 150-159.
Newcomb, E.H., Tandon, S.R. and Kowal, R.R., 1985. Ultrastructural specialization for ureide production in uninfected cells of soybean root nodules. Protoplasma, 125: 1-12.
Nguyen, J., Machal, L., Vidal, J., Perrot-Rechenmann, C. and Gadal, P., 1986. Immunochemical studies on xanthine dehydrogenase of soybean root nodules. Ontogenic changes in the level of enzyme and immunocytochemical localization. Planta, 167: 190-195.
Nguyen, T., Zelechowska, M., Foster, V., Bergmann, H. and Verma, D.P.S., 1985. Primary structure of the soybean nodulin-35 gene encoding uricase II localized in the peroxisomes of uninfected cells of nodules. Proc. Natl. Acad. Sci. USA., 82: 5040-5044.
Ninomiya, Y. and Sato, S., 1984. A ferredoxin like electron carrier from non-green cultured tobacco cells. Plant Cell Physiol., 25: 453-458.
Ohyama, T. and Kumazawa, K., 1978. Incorporation of ^{15}N into various nitrogenous compounds in intact soybean nodules after exposure to $^{15}N_2$ gas. Soil. Sci. Plant Nutr., 24: 525-533.
Parish, R.W., 1972. Peroxisomes from the Arum italicum appendix. Z. Pflanzenphysiol., 67: 430-442.
Pate, J.S., 1977. Nodulation and nitrogen metabolism. In: J.F. Sutcliffe and J.S. Pate (Editors), The Physiology of the Garden Pea. Academic Press, London, pp. 349-383.
Pate, J.S., 1980. Transport and partitioning of nitrogenous solutes. Ann. Rev. Plant Physiol., 31: 313-340.
Pate, J.S. and Atkins, C.A., 1983. Nitrogen uptake, transport and utilization in nitrogen fixation. In: W.J. Broughton (Editor), Ecology of Nitrogen Fixation, Vol. 3. Clarendon Press, Oxford, pp. 245-298.

Pate, J.S., Atkins, C.A., Hamel, K., McNeil, D.L. and Layzell, D.B., 1979. Transport of organic solutes in phloem and xylem of a nodulated legume. Plant Physiol., 63: 1076-1081.
Pate, J.S., Atkins, C.A. and Rainbird, R.M., 1981. Theoretical and experimental costing of nitrogen fixation and related processes in nodules of legumes. In: A.H. Gibson and W.E. Newton (Editors), Current Perspectives in Nitrogen Fixation. Australian Academy of Sciences, Canberra, pp. 105-116.
Pate, J.S., Atkins, C.A., White, S.T., Rainbird, R.M. and Woo, K.C., 1980. Nitrogen nutrition and xylem transport of nitrogen in a ureide-producing grain legume. Plant Physiol., 65: 961-965.
Pate, J.S., Gunning, B.E.S. and Briarty, L.G., 1969. Ultrastructure and functioning of the transport system of the legume root nodule. Planta, 85: 11-34.
Pate, J.S., Lindblad, P. and Atkins, C.A., 1988. Pathways of assimilation and transfer of fixed nitrogen in coralloid roots of cycad - Nostoc symbioses. Planta, 176: 461-471.
Pate, J.S., Peoples, M.B. and Atkins, C.A., 1984. Spontaneous phloem bleeding from cryopunctured fruits of a ureide-producing legume. Plant Physiol., 74: 499-505.
Pate, J.S., Sharkey, P.J. and Lewis, O.A.M., 1975. Xylem to phloem transfer of solutes in fruiting shoots of legumes, studied by a phloem bleeding technique. Planta, 122: 11-26.
Pate, J.S., Walker, J. and Wallace, W., 1965. Nitrogen-containing compounds in the shoot system of Pisum arvense L. II. The significance of amino acids and amides released from nodulated roots. Ann. Bot., 29: 475-493.
Pate, J.S. and Wallace, W., 1964. Movement of assimilated nitrogen from the root system of the field pea (Pisum arvense L.). Ann. Bot., 28: 80-99.
Peoples, M.B., Hebb, D.M., Gibson, A.H. and Herridge, D.F., 1989. Development of the xylem ureide assay for the measurement of nitrogen fixation by pigeon pea (Cajanus cajan (L.) Millsp.). J. Expt. Bot., 40: 535-542.
Peoples, M.B., Pate, J.S. and Atkins, C.A., 1985. The effect of nitrogen source on transport and metabolism of nitrogen in fruiting plants of cowpea (Vigna unguiculata (L.) Walp.). J. Exp. Bot., 36: 567-582.
Peoples, M.B., Pate, J.S., Atkins, C.A. and Bergersen, F.J., 1986. Effect of nitrogen nutrition on xylem composition of peanut (Arachis hypogea L. cv. Virginia Bunch). Plant Physiol., 82: 946-951.
Peoples, M.B., Sudin, M-N. and Herridge, D.F., 1987. Translocation of nitrogenous compounds in symbiotic and nitrate-fed amide-exporting legumes. J. Expt. Bot., 38: 567-579.
Polhill, R.M. and Raven, P.H., 1981. Advances in legume systematics, part 1. Royal Botanic Gardens, Kew.
Powell, G.K. and Dekker, E.E., 1983. Purification and properties of a 4-methylene- L-glutamine amidohydrolase from peanut leaves. J. Biol. Chem., 28: 8677-8683.
Rainbird, R.M. and Atkins, C.A., 1981. Purification and some properties of urate oxidase from nitrogen-fixing nodules of cowpea (Vigna unguiculata L. Walp.). Biochim. Biophys. Acta., 659: 132-140.
Rainbird, R.M., Atkins, C.A. and Pate, J.S., 1983. Diurnal variation in the functioning of cowpea nodules. Plant Physiol, 72: 308-312.
Ratajczak, L., Koroniak, D., Mazurowa, H., Ratajczak, W. and Prus-Glowacki, W., 1986a. Glutamate dehydrogenase isoforms in lupine roots and root nodules. Immunological studies. Physiol. Plant., 67: 685-689.
Ratajczuk, W., Ratajczak, L. and Mazurowa, H., 1986b. Activity changes of ammonia assimilating enzymes in developing root nodules of Lupinus luteus L. Acta. Physiol. Plant., 8: 33-41.
Rerkasem, B., Rerkasem, K., Peoples, M.B., Herridge, D.F. and Bergersen, F.J., 1988. Measurement of N_2 fixation in maize (Zea mays L.) - ricebean (Vigna umbellata [Thunb.] Ohmi and Ohashi) intercrops. Plant and Soil 180: 125-135.

Reynolds, P.H.S. and Blevins, D.G., 1986. Phosphoserine aminotransferase in soybean root nodules. Demonstration and localization. Plant Physiol., 81: 293-296.
Reynolds, P.H.S., Blevins, D.G. and Randall, D.D., 1984. 5-Phosphoribosylpyrophosphate amidotransferase from soybean root nodules: kinetic and regulatory properties. Arch. Biochem. Biophys., 229: 623-632.
Reynolds, P.H.S., Boland, M.J., Blevins, D.G., Randall, D.D. and Schubert, K.R., 1982. Ureide biogenesis in leguminous plants. Trends Biochem. Sci., 7: 366-368.
Reynolds, P.H.S., Boland, M.J. and Farnden, K.J.F., 1981. Enzymes of nitrogen metabolism in legume nodules: partial purification and properties of the aspartate aminotransferases from lupin nodules. Arch. Biochem. Biophys., 209: 524-533.
Reynolds, P.H.S. and Farnden, K.J.F., 1979. The involvement of aspartate aminotransferases in ammonium assimilation in lupin nodules. Phytochem., 18: 1625-1630.
Robertson, J.G., Farnden, K.J.F., Warburton, M.P. and Banks, J.A.M., 1975a. Induction of glutamine synthetase during nodule development in lupin. Aust. J. Plant Physiol., 2: 265-272.
Robertson, J.G., Warburton, M.P. and Farnden, K.J.F., 1975b. Induction of glutamate synthase during nodule development of lupin. FEBS. Lett., 55: 33-37.
Rognes, S.E., 1975. Glutamine-dependent asparagine synthetase from Lupinus luteus. Phytochem., 14: 1977-1982.
Rosenthal, G.A., 1972. Investigations of canavanine biochemistry in the Jack bean plant Canavalia ensiformis (L.) D.C. Plant Physiol., 50: 328-331.
Rosenthal, G.A. and Rhodes, D., 1984. L-Canavanine transport and utilization in developing Jack Bean (Canavalia ensiformis (L) D.C. (Leguminosae)). Plant Physiol., 76: 541-544.
Rowe, P.B. and Wyngaarden, J.B., 1968. Glutamine, phosphoribosylpyrophosphate amido-transferase. Purification, substructure, amino acid composition, and absorption spectra. J. Biol. Chem., 243: 6373-6383.
Rowe, P.B., McCairns, E., Madsen, G., Sauer, D. and Elliott, H.,1978. De novo purine synthesis in avian liver. Co-purification of the enzymes and properties of the pathway. J. Biol. Chem., 253: 7711-7721.
Ryan, E., Bodley, F. and Fottrell, P.F., 1972. Purification and characterization of aspartate aminotransferases from soybean root nodules and Rhizobium japonicum. Phytochem., 11: 957-963.
Sanchez, F., Campos, F., Padilla, J., Bohneville, J-M., Enriquez, C. and Caput, D., 1987. Purification, cDNA cloning, and developmental expression of the nodule-specific uricase from Phaseolus vulgaris L. Plant Physiol, 84: 1143-1147.
Schubert, K.R., 1981. Enzymes of purine biosynthesis and catabolism in Glycine max: Comparison of activities with N_2 fixation and composition of xylem exudate during nodule development. Plant Physiol., 68: 1115-1122.
Schubert, K.R., 1986. Products of biological nitrogen fixation in higher plants: synthesis, transport and metabolism. Ann. Rev. Plant Physiol., 37: 539-574.
Schubert, K.R., 1987. Purification of glutamine-dependant phosphoribosylpyrophosphate amidotransferase from soybean root nodules. Plant Physiol., 83: Suppl. p. 20.
Schubert, K.R. and Coker, G.T. III., 1982. Studies of nitrogen and carbon assimi-lation in N_2-fixing plants: Short-term studies using [^{13}N] and [^{11}C]. In: J.W. Root and K.A. Krohn (Editors), Recent Developments in Biological and Chemical Research with Short-lived Radioisotopes. Adv. Chem. Ser. 197: 317-339.
Schuller, K.A., Day, D.A., Gibson, A.H. and Gresshoff, P.M., 1986. Enzymes of ammonia assimilation and ureide biosynthesis in soybean nodules: Effect of nitrate. Plant Physiol., 80: 646-650.

Scott, D.B., Farnden, K.J.F. and Robertson, J.G., 1976. Ammonia assimilation in lupin nodules. Nature, 263: 705.
Sengupta-Gopolan, C. and Pitas, J.W., 1986. Expression of nodule-specific glutamine synthetase genes during nodule development in soybeans. Plant Mol. Biol., 7: 189-199.
Shaw, G.J., Scott, D.B. and Ellingham, P.J., 1986. Interaction of Rhizobium loti strain and host on the biosynthesis of unusual amino acids in leguminous plants. Experientia, 42: 42-43.
Shelp, B.J. and Atkins, C.A., 1983. Role of inosine monophosphate oxidoreductase in the formation of ureides in nitrogen-fixing nodules of cowpea (Vigna unguiculata L. Walp). Plant Physiol., 72: 1029-1034.
Shelp, B.J. and Atkins, C.A., 1984. Subcellular location of enzymes of ammonia assimilation and asparagine synthesis in root nodules of Lupinus albus L. Plant Sci. Lett., 36: 225-230.
Shelp, B.J., Atkins, C.A. and Storer, P.J., 1985. Purification and properties of inosine monophosphate oxidoreductase from nodules of cowpea (Vigna unguiculata L. Walp). Arch. Biochem. Biophys., 236: 807-814.
Shelp, B.J., Atkins, C.A., Storer, P.J. and Canvin, D.T., 1983. Cellular and subcellular organization of pathways of ammonia assimilation and ureide synthesis in nodules of cowpea (Vigna unguiculata L. Walp). Arch. Biochem. Biophys., 224: 429-441.
Sprent, J.I., 1980. Root nodule anatomy, type of export product and evolutionary origin in some Leguminosae. Plant Cell Environ., 3: 35-43.
Streeter, J.G., 1977. Asparaginase and asparagine transaminase in soybean leaves and root nodules. Plant Physiol., 60: 235-239.
Streeter, J.G., 1979. Allantoin and allantoic acid in tissues and stem exudate from field-grown soybean plants. Plant Physiol., 63: 478-480.
Streeter, J.G., 1989. Estimation of ammonium concentration in the cytosol of soybean nodules. Plant Physiol., 90: 779-782.
Sutherland, J.M. and Sprent, J.I., 1984. Calcium-oxalate crystals and crystal cells in determinate root nodules of legumes. Planta, 161: 193-200.
Sutton, W.D., 1983. Nodule development and senescence. In: W.J. Broughton (Editor), Nitrogen Fixation, Vol. 3: Legumes. Clarendon Press, Oxford, pp. 144-212.
Suzuki, A., Oaks, A., Jacquot, J.P., Vidal, J. and Gadal, P., 1985. An electron transport system in maize roots for reactions of glutamate synthase and nitrite reductase. Plant Physiol., 78: 374-378.
Suzuki, A., Vidal, J., Nguyen, J. and Gadal, P., 1986. Occurrence of ferredoxin dependent glutamate synthase in plant cell fraction of soybean root nodules. FEBS. Lett., 173: 204-208.
Ta, T.C., Faris, M.A. and Macdowall, F.D.H., 1986. Pathways of nitrogen metabolism in nodules of alfalfa (Medicago sativa L.). Plant Physiol., 80: 1002-1005.
Ta, T.C., Macdowall, F.D.H., Faris, M.A. and Joy, K.W., 1988a. Metabolism of nitrogen fixed by root nodules of alfalfa (Medicago sativa L.): I. Utilization of [^{14}C, ^{15}N] glutamate and [^{14}C, ^{15}N] glutamine in the synthesis of aminobutyrate. Biochem. Cell Biol, 66: 1342-1348.
Ta, T.C., Macdowall, F.D.H., Faris, M.A. and Joy, K.W., 1988b. Metabolism of nitrogen fixed by nodules of alfalfa (Medicago sativa L.). II. Asparagine synthesis. Biochem. Cell Biol., 66: 1349-1354.
Tajima, S. and Yamamoto, Y., 1975. Enzymes of purine catabolism in soybean plants. Plant Cell Physiol, 16: 271-282.
Tajima, S., Yatazawa, M. and Yamamoto, Y. 1977. Allantoin production and its utilization in relation to nodule formation in soybeans. Enzymatic studies. Soil. Sci. Plant Nutr., 23: 225-235.
Triplett, E.W., 1985. Intercellular nodule localization and nodule specificity of xanthine dehydrogenase in soybean. Plant Physiol., 77: 1004-1009.

Triplett, E.W., 1986. Two indirect methods for detecting ureide synthesis by nodulated legumes. Plant Physiol, 81: 566-571.
Triplett, E.W., Blevins, D.G. and Randall, D.D., 1980. Allantoic acid synthesis in soybean root nodule cytosol via xanthine dehydrogenase. Plant Physiol., 65: 1203-1206.
van Beuschem, M.L., 1983. Xylary charge distribution and nitrogen transport in Pisum sativum L. during dinitrogen fixation or nitrate nutrition. Z. Pflanzenphysiol., 109: 449-458.
Vance, C.P., Boylan, K.L.M., Maxwell, C.A., Heichel, G.H. and Hardman, L.L., 1985. Transport and partitioning of CO_2 fixed by root nodules of ureide and amide producing legumes. Plant Physiol., 78: 774-778.
Vance, C.P., Egli, M.A., Griffith, S.M. and Miller, S.S., 1988. Plant regulated aspects of nodulation and N_2 fixation. Plant Cell Environ., 11: 413-427.
Vance, C.P., Reibach, P.H. and Pankhurst, C.E., 1987. Symbiotic properties of Lotus pedunculatus root nodules induced by Rhizobium loti and Bradyrhizobium sp. (Lotus). Physiol. Plant., 69: 435-442.
Vance, C.P. and Stade, S., 1984. Alfalfa root nodule carbon dioxide fixation. II. Partial purification and characterization of root nodule phosphoenolpyruvate carboxylase. Plant Physiol., 75: 261-264.
Vance, C.P., Stade, J. and Maxwell, C.A., 1983. Alfalfa root nodule carbon dioxide fixation. I. Association with nitrogen fixation and incorporation into amino acids. Plant Physiol., 72: 469-473.
Van den Bosch, K.A. and Newcomb, E.H., 1986. Immunogold localization of nodule-specific uricase in developing soybean root nodules. Planta, 167: 425-436.
Van den Bosch, K.A. and Newcomb, E.H., 1988. The occurrence of leghaemoglobin protein in the uninfected interstitial cells of soybean root nodules. Planta, 175: 442-451.
Van Kessel, C., Roskoski, J.P. and Keane, K., 1988. Ureide production by N_2-fixing and non-N_2-fixing leguminous trees. Soil Biol. Biochem., 20: 891-897.
Walker, B.A., Pate, J.S. and Kuo, J., 1983. Nitrogen fixation by nodulated roots of Viminaria juncea (Schrad. & Wendl.) Hoffmans. (Fabaceae) when submerged in water. Aust. J. Plant Physiol., 10: 409-421.
Wallsgrove, R.M., Lea, P.J. and Miflin, B.J., 1979. The distribution of the enzymes of nitrogen assimilation within the pea leaf. Plant Physiol., 63: 232-236.
Walsh, K.B., McCully, M.E. and Canny, M.J., 1989. Vascular transport and soybean nodule function: nodule xylem is a blind alley, not a throughway. Plant Cell Environ., 12: 395-405.
Walsh, K.B., Ng, B.H. and Chandler, G.E., 1984. Effects of nitrogen nutrition on xylem sap composition of Casuarinaceae. Plant and Soil, 81: 291-293.
Webb, M.A. and Newcomb, E.H., 1987. Cellular compartmentation of ureide biogenesis in root nodules of cowpea (Vigna unguiculata (L.) Walp.). Planta, 172: 162-175.
Werner, D., Morschel, E., Stripf, R. and Winchenbach, B., 1980. Development of nodules of Glycine max infected with an ineffective strain of Rhizobium japonicum. Planta, 147: 320-329.
Woo, K.C., Atkins, C.A. and Pate, J.S., 1980. Biosynthesis of ureides from purines in a cell-free system from nodule extracts of cowpea (Vigna unguiculata L. Walp). Plant Physiol., 66: 735-739.
Woo, K.C., Atkins, C.A. and Pate, J.S., 1981. Ureide synthesis in a cell-free system from cowpea (Vigna unguiculata [L.] Walp.) nodules. Studies with O_2, pH and purine metabolites. Plant Physol., 67: 1156-1160.
Young, M., Sammons, R.D., Mueller, W.T. and Benkovic, S.J., 1984. An antibody probe to determine the native species of glycinamide ribonucleotide transformylase in chicken liver. Biochem., 23: 3979-3986.

CHAPTER 14

LIMITATIONS IN THE USE OF LEGUMES IN AGRICULTURE AND FORESTRY

A.D. ROBSON[1] and P.J. BOTTOMLEY[2]
[1]Soil Science and Plant Nutrition, School of Agriculture, The University of Western Australia, Nedlands, Western Australia, 6009.
[2]Department of Microbiology and Soil Science, Oregon State University, Corvallis, Oregon, USA.

1 INTRODUCTION

A wide diversity of legumes is used in agriculture and forestry for human consumption (pulses, green vegetables, sprouts), for feeding animals (grain legumes, grazed pasture legumes, legumes cut for hay, browse shrubs), for firewood, for rehabilitation of degraded lands and to provide nitrogen inputs because of their ability to fix atmospheric N_2. After the legume has senesced, this fixed nitrogen can be used by either companion plants or succeeding crops. In addition to the direct effect of nitrogen inputs into crop growth, the relief of nitrogen limitation in the soil may result in organic matter accumulation and changes in soil structure and water infiltration under legumes. Periods of pasture legumes and grain legumes between successive cereal and oilseed crops may permit the control of pathogens and insect pests. The beneficial effects of pasture and grain legumes on soil fertility may be substantial (e.g. see Robson and Abbott, 1989; Robson, 1989). However, growth of legumes may lead to soil acidification because of increased nitrogen levels leading to accelerated nitrification and leaching of nitrate, a higher content of acidic organic materials and a greater uptake of cations than of anions (see Helyar and Porter, 1989).

Legumes generally form nitrogen-fixing symbioses with species of Rhizobium, Bradyrhizobium and Azorhizobium (see Sprent, 1989). Several legume species can be nodulated by both Rhizobium and Bradyrhizobium (Trinick 1980; Lin et al., 1987; Barnet, 1988). The majority of the members of the Caesalpiniae are non-nodulating (Chapter 13).

The legumes as a group differ markedly in their mechanisms of infection (root hairs, epidermal cells, or sites of lateral root emergence), the character of their nodules (determinate or indeterminate), and the forms of fixed nitrogen exported from their nodules (Sprent, 1980; Chapter 13). Despite this enormous diversity among legumes, most generalizations regarding the response of legumes to stress have been made after study of either a single or only a very limited number of species. Variation in the response of the host legume and the root nodule bacteria, as well as their symbiotic association, has allowed the selection of genotypes and associations tolerant of many adverse environments. However, to progress most rapidly towards increasing the productivity of legumes in these environments it is important to emphasize selection pressure of the component which is most sensitive.

For annual legumes, there is a period when the root nodule bacteria must persist in the soil in the absence of the host. Additionally, during the growth of both annual and perennial legumes there may be periods when roots of the host die due to extremes of temperature and water supply. The root nodule bacteria must then be saprophytically competent within the soil if an effective symbiosis is to be re-established. In this account we will consider some general rhizobial limitations to legume production and the effect of five adverse environments (low temperatures, salinity, low phosphorus supply, acidity, and seasonal drought) on the nitrogen-fixing symbiosis in legumes in agriculture and forestry. Because these topics have been reviewed by Gibson (1977), Munns (1977), Robson (1983, 1988), Sprent and Minchin (1985) and O'Hara et al. (1988), this review will focus on recent advances and on differences among legumes and their root nodule bacteria in their responses to limiting environmental conditions.

2 RHIZOBIAL LIMITATIONS TO LEGUME PRODUCTION

Legumes may fail to produce profitable yields because of the absence of symbiotically effective rhizobia, particularly for the first planting of legumes with highly specific rhizobial requirements (e.g. Leucaena, Glycine, Cajanus, Medicago and Cicer) in soil having no prior history of the crop (Arsac and Cleyet-Marel, 1986; Brockwell et al.,1987, 1988; Herridge et al., 1987; Owiredu and Danso, 1988; Somaseragan et al., 1988; Sanginga et al., 1989). In most situations rhizobia persist in soils after their introduction and responses to further inoculation occur only occasionally due to low numbers of effective rhizobia. However, in other situations introduced rhizobia fail to persist, leading to poor re-establishment of annual pasture legumes (Chatel et al., 1968, 1973; Chatel and Greenwood, 1973a,b; Chatel and Parker, 1973a,b; Howieson and Ewing, 1986) and poor nodulation of annual grain legumes (Rupela et al., 1987; Dudeja and Khurana, 1989).

There is also concern that legumes may fail to reach their yield potential in post-establishment years because of a decline in either numbers or effectiveness of the soil population of rhizobia (see Brockwell et al., 1977, 1982).

However, there is little evidence to justify this concern except from south-western Australia (Chatel et al., 1968; Howieson and Ewing, 1986). A decline in productivity of legume pastures in south-eastern Australia (Carter et al., 1982; Coventry et al., 1987; Evans et al., 1988) appears to involve other factors besides persistence and effectiveness.

Legume production can be limited by the presence of indigenous populations of rhizobia which are symbiotically ineffective, but sufficiently competitive to prevent nodulation by inoculant strains (Ireland and Vincent, 1968; Holland, 1970; Jones et al., 1978). There are many reports that indigenous rhizobial populations in soil are less effective than inoculant strains for both forage legumes (Bottomley and Jenkins, 1983) and soybeans (Damirgi et al., 1967). Current inoculation technologies are incapable of displacing less effective serogroups with more effective strains (Bohlool and Schmidt, 1973). Whether or not soil rhizobia of moderate effectiveness restrict the nitrogen fixation of legumes in the field has not been resolved.

There are several potential approaches to removing limitations to legume production associated with a failure to form effective associations with rhizobia. First, strains can be screened for tolerance to soil and climatic stresses (see below) and for competitive ability (see Robert and Schmidt, 1983), but this approach has only led to one success in the field (Howieson and Ewing, 1986). This lack of success has often been because the rhizobial component of the symbiosis is not the most sensitive one. A second approach may be to improve inoculant technology by better understanding the physiology of rhizobia in peat carriers. A "curing" period for peat inoculants enhances the survival of rhizobia on the seed (Burton, 1975; Materon and Weaver, 1985) but might slow their response to signals from the plant. In *R. trifolii*, *nod* genes were not inducible in cells in the lag and stationary phase (Djordjevic et al., 1987). The induction of *nod* genes by flavonoids is known to be influenced by the composition of the growth medium (Peters and Long, 1988; Dusha et al., 1989) and it is possible that the presence of such compounds in the peat might have effects on the rhizobia.

Another strategy may be to produce effective strains which inhibit nodulation by less effective strains. Thus, the transfer to a symbiotically effective strain of *Rhizobium leguminosarum* bv. *trifolii* of the genes for trifolitoxin production and resistance (Triplett, 1988; Triplett et al., 1989), which allowed the ineffective strain T24 to inhibit growth and nodulation of other strains (Schwinghamer and Belkengren, 1968; Triplett and Barta, 1984), resulted in enhanced competition in the recipient (Triplett, 1990).

Any long term approach to relieving rhizobial limitations requires an understanding of competition for nodulation. Most studies of mixtures of strains of similar effectiveness show a relationship between nodulation success and the relative numbers of strains, either in the inoculum or in the rhizosphere (for recent examples, see Beattie et al., 1989; Fuhrmann and Wollum, 1989; Mullen and Wollum, 1989). However, in non-sterile soil with indigenous rhizobial populations, the outcome of competition was not related to the relative numbers of different serogroups in rhizospheres as determined by immunofluorescence (Ellis et al., 1984; Moawad and Bohlool, 1984; Robert and Schmid, 1985; Demezas and Bottomley, 1986a,b). The reasons for this discrepancy require further investigation and may involve (i) the relation between the numbers of root hairs infected and sub-population size, (ii) environmental effects on root development, (iii) the isolation and use of strains representative of indigenous sub-populations competitive under field conditions, (iv) the physiological state of soil and inoculant rhizobia and (v) nodulation patterns and N_2-fixing characteristics. Success in this area will depend upon refining methods for studying sub-populations in the rhizosphere and will require a closer liaison between rhizobial ecologists and molecular biologists.

3 SALINITY

Despite the widespread occurrence of saline soils, the effects of salinity on nodulation and nitrogen fixation by legumes are little studied.

The growth and survival of root nodule bacteria appear to be more tolerant of salinity than either nodule initiation or host legume growth. High concentrations of NaCl show litle effect on the growth and survival of root nodule bacteria in both soil and culture (Singleton et al., 1982; Hafeez et al., 1988), but considerable variation exists both among rhizobia (Singleton et al., 1982) and legumes (Russell, 1976; Keating and Fisher, 1985). For the root nodule bacteria, adding 100 mM NaCl (equivalent to 1.3 S m^{-1}) increased mean doubling times by only 1.7 times in some strains but completely prevented growth of other strains (Singleton et al., 1982). Within the temperate legumes, for example, a soil salinity of 0.88 S m^{-1} was required to decrease plant yield of M. sativa to half, whereas only 0.42 S m^{-1} was required for Trifolium semipilosum (Russell, 1976). Increasing concentrations of NaCl decrease nodule weight and nitrogen fixation in several legumes (Vigna radiata, Hafeez et al., 1988; Glycine max, Singleton and Bohlool, 1984; Medicago sativa, Bernstein and Ogata, 1966; Cicer arietinum, Lauter et al., 1981; Vicia faba, Yousef and Sprent, 1983) as well as in non-legumes (Casuarina equisetifolia, Ng, 1987). In some cases, effects of salinity on nodule weight and shoot weight were closely correlated (for example

Ng, 1987; Hafeez et al., 1988), suggesting that effects of NaCl on nodule weight were due to indirect effects on host plant growth, but this finding may not be generally applicable. For some legumes, however, three lines of evidence indicate that effects of salinity on nodulation and nitrogen fixation are more marked than those on host legume growth: (1) the growth of legumes reliant on symbiotic nitrogen fixation was depressed to a greater extent by salinity than was growth with combined nitrogen (Glycine wightii, Wilson 1970; C. arietinum, Lauter et al., 1981), (2) increasing concentrations of NaCl decreased nitrogen concentrations in shoots of legumes reliant on N_2, but not in those supplied with mineral nitrogen (C. arietinum, Lauter et al., 1981; V. faba, Yousef and Sprent, 1983) and (3) in split root experiments, effects of salinity on nodule weight in G. max occurred in the absence of effects on shoot dry weight (Singleton and Bohlool, 1984).

Although the number of studies is limited, the initiation of nodules appears to be particularly sensitive to salinity for legumes nodulated by both bradyrhizobia and rhizobia. Nodulation of soybean was decreased by NaCl without effects either on the number of bradyrhizobia in the rooting medium or on colonization of roots (Singleton and Bohlool, 1984). In V. faba, adding 100 mM NaCl did not decrease rhizobial colonization of roots, but lowered the number of root hairs curling or containing infection threads (Zahran and Sprent, 1986). In soybean, delaying the addition of NaCl until 96 hours after inoculation markedly decreased the effects of NaCl on nodule number and weight (Singleton and Bohlool, 1984).

Zahran and Sprent (1986) suggested that salinity may depress the infection of root hairs by decreasing the uptake and intracellular concentrations of calcium. Indeed, in Leucaena leucocephala increasing the concentration of $CaSO_4$ from 0.5 to 2.5 mM increased nodule weight at 25, 50 and 100 mM NaCl, but did not affect nodule weight at 1 mM NaCl (Hansen and Munns, 1988). However, there was no similar interaction between NaCl and $CaSO_4$ affecting nodule number, indicating that effects on nodule weight may be associated with alleviation by $CaSO_4$ of the effects of salinity on host legume growth.

4 PHOSPHORUS SUPPLY

Phosphorus deficiency is widespread in legumes, although as a group they do not appear to have greater internal or external requirements for growth than non-legumes (Munns, 1977). There has been considerable research on the role of phosphorus in nodulation and nitrogen fixation including the involvement of vesicular-arbuscular (VA) mycorrhizas.

A greater external requirement for phosphorus for symbiotic N_2 fixation than for growth of legume shoots has frequently been claimed (see Israel, 1987). An alternative view is that increased phosphate supply increases nodulation and nitrogen fixation by increasing host legume growth (see Robson, 1983; Jakobsen, 1985). The key evidence relating to the involvement of phosphorus in symbiotic nitrogen fixation comes from three areas: (i) comparison of the external and internal requirements for phosphorus between legumes supplied with mineral nitrogen with those reliant on symbiotic N_2, (ii) effect of phosphorus supply on nodule weight and number and (iii) the effect of phosphorus supply on shoot nitrogen concentrations.

4.1 Phosphorus requirements for nitrogen fixation

Legumes reliant on N_2 do not always appear to have a greater internal requirement (assessed by critical phosphorus concentrations in shoots) than those supplied with mineral nitrogen (*Lablab purpureus*, Zaroug and Munns, 1979; *G. max* 40 days after transplanting, Israel, 1987). However, 97 days after transplanting the internal phosphate requirement for soybeans reliant on N_2 was greater than those supplied with mineral nitrogen (Israel, 1987).

The interaction between nitrogen and phosphorus is usually positive, indicating that correcting phosphorus deficiency does not increase nitrogen supply to the legume. Adding nitrogen after the onset of nitrogen fixation does not generally appear to increase growth to a greater extent in phosphate-deficient than in phosphate-adequate plants (Table 1; subterranean clover, Robson et al., 1981; field peas, Jakobsen, 1985; soybeans 40 days after transplanting, Israel, 1987). However, in soybeans 97 days after transplanting (Israel, 1987), the growth of those reliant on symbiotic nitrogen fixation was increased by raising the phosphate supply, but the growth of soybeans supplied with mineral nitrogen was not. This negative interaction between nitrogen and phosphorus on the growth of soybeans occurred only at levels of applied phosphate sufficient for maximum growth of plants supplied with nitrate and sufficient for approximately 80% of maximum growth for plants reliant on N_2.

4.2 Phosphorus and nodule weight

Adding phosphorus, which increases the growth of shoots, generally increases both nodule number and weight (see Robson, 1983, 1988). Indeed, nodule weight is closely correlated with shoot weight (Fig. 1; Graham and Rosas, 1979; Cassman et al., 1981b; Robson et al., 1981; Israel, 1987) over a wide range of phosphorus supply, providing little support for the frequent comment (see Israel, 1987) that adding phosphorus increases nodule weight more than shoot weight. This comment, which is based largely on effects of applied phosphorus in increasing the ratio of nodule weight to shoot weight, occurs only because there must be a certain shoot weight before nodules are formed.

TABLE 1 Nature of the interaction between phosphorus and nitrogen on legume growth.

Legume	Timing of N application	Nature of interaction	Reference
Glycine max	20 d after transplanting (after shoots had re-greened)	+ve, 40 d after transplanting +ve at severe deficiency -ve at moderate deficiency 97 d after transplanting	Israel, 1987
Glycine max	17 and 23 d after planting	+ve, 33 d after planting	Singleton et al., 1985
Glycine max	3 d after sowing	+ve	Cassman et al., 1981
Lablab purpureus	At emergence	+ve, 6 weeks after emergence	Zaroug and Munns, 1979
Stylosanthes humilis	Before sowing	+ve	Gates and Wilson, 1974
Lotus pedunculatus	Before sowing	+ve	Hart et al., 1981
Pisum sativum	9 d after emergence (approx. 30 pink nodules on main root	+ve, 22 d after emergence	Jakobsen, 1985
Trifolium subterraneum	12 d after sowing (after commencement of nitrogen fixation)	+ve, 32 d after sowing	Robson et al., 1981
Trifolium repens	Before sowing	+ve	Hart et al., 1981

The strong positive correlation between nodule weight and shoot weight with increasing phosphorus supply is in marked contrast to the effects of deficiencies of nutrients that are directly involved in nodule function. Thus, nodule weight may frequently be greater in plants deficient in calcium, molybdenum and cobalt and adding the nutrient may decrease nodule weight but increase shoot weight (Robson, 1983).

Why nodule weight and shoot weight are strongly correlated is not clear. In soybean, fully expanded and expanding leaves of phosphorus-deficient plants had higher starch, but lower sugar concentrations than in phosphorus-adequate plants (Fredeen et al., 1989; Israel and Rufty, 1989). In contrast, fibrous roots of

P-deficient soybean plants had higher concentrations of both starch and sucrose than those with an adequate supply (Fredeen et al., 1989). The high concentrations of starch in phosphate-deficient plants indicate ample photosynthate. Indeed, the concentration of soluble carbohydrates in roots of subterranean clover was much higher in phosphate-deficient than phosphate-adequate plants (Jasper et al., 1979; Same et al., 1983; Thomson et al., 1986). The higher concentrations of soluble carbohydrates in phosphate-deficient plants result from effects of phosphorus deficiency on the production of new leaves. Although phosphorus deficiency in soybeans did not decrease rates of photosynthesis per unit leaf area (Fredeen and Terry, 1988), it markedly decreased total leaf area, mainly by decreasing the expansion of individual leaves (Fredeen and Terry, 1988; Fredeen et al., 1989).

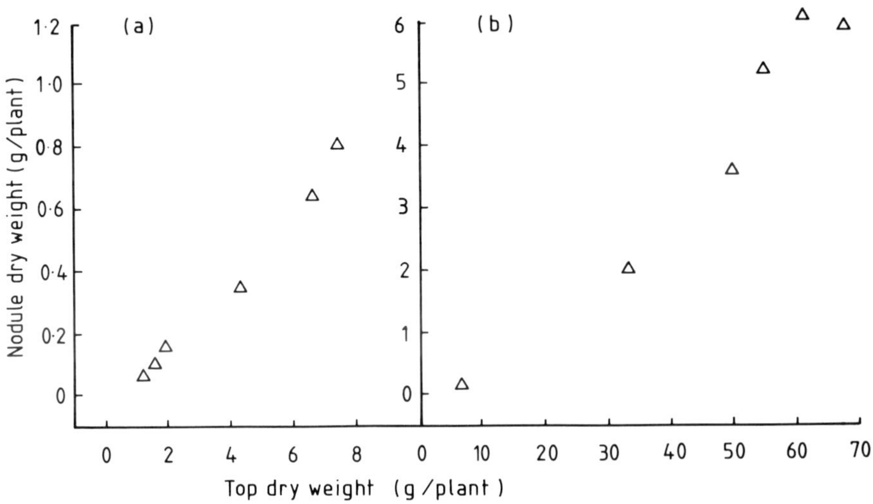

Fig. 1. The relationship between nodule weight and shoot weight for soybeans supplied with increasing phosphate. (a) data of Cassman et al. (1981b), (b) data of Israel (1987).

Increasing CO_2 concentration around shoots (Hardy and Havelka, 1975), increasing light (Lawn and Brun, 1974) and increasing leaf area (Streeter, 1974) all elevate nitrogen fixation on both a plant and nodule weight basis, indicating a close link between the supply of photosynthate and nitrogen fixation. However, the high concentrations of soluble carbohydrates within roots of P-deficient legumes do not suggest that nodule weight is limited by the supply of carbon. In subterranean clover reliant on symbiotic N_2 fixation, concentrations of free amino-N in root extracts decreased more than ten-fold on alleviating phosphorus deficiency (Thomson et al., 1986), probably because of decreased leaf production in phosphate-limited plants. These high concentrations of free amino-nitrogen may inhibit nodule development.

4.3 Nitrogen concentrations in shoots

Alleviating phosphorus deficiency in legumes reliant on N_2 generally increases nitrogen concentrations in shoots (subterranean clover, McLachlan and Norman, 1961; Robson et al., 1981; Stylosanthes humilis, Shaw et al., 1966; Gates et al., 1966; Gates and Wilson, 1974; several tropical pasture legumes, Andrew and Robins, 1969; field peas, Jakobsen, 1985; soybean, Singleton et al., 1985; Israel, 1987). For legumes supplied with mineral nitrogen, there was no effect of phosphorus supply on nitrogen concentrations in shoots. The effects of phosphate deficiency on nitrogen concentrations in shoots of nitrogen-fixing legumes cannot be due to effects of phosphorus on nitrogen metabolism within the plant, or on plant structure. Jakobsen (1985) has suggested that increased nitrogen concentra-tions in shoots may result from a relatively larger increase in specific nodule activity than in shoot dry weight.

4.4 Growth of root nodule bacteria

In sterilized soils, the addition of dipotassium phosphate, disodium phosphate and dicalcium phosphate in the absence of a host plant increased the numbers of Rhizobium meliloti (Truesdell, 1917). The phosphate nutrition of root nodule bacteria has recently been extensively studied in both dialysed batch culture (Cassman et al., 1981a; Beck and Munns, 1984, 1985) and in continuous culture (Smart et al., 1984a,b). Rhizobial strains varied widely in tolerance to low solution concentrations of phosphate in equilibrium with an iron oxide system (Cassman et al., 1981a; Beck and Munns, 1984) and the difference within a species was as large as between species. All strains could grow well at phosphate concentrations commonly found in soil solutions, although growth may be limited by the much lower phosphate concentrations in the rhizosphere.

Both R. meliloti and R. trifolii require much higher phosphate concentra-tions to achieve maximum growth at both low calcium (Beck and Munns, 1984) and high hydrogen ion (Khan et al., 1984) concentrations.

Root nodule bacteria store considerable phosphate when grown with luxury supplies (Cassman et al., 1981). Growing Bradyrhizobium japonicum at 0.5 µM phosphorus instead of 1 mM delayed infection of the primary root of soybean in growth pouches without affecting nodule number (Mullen et al., 1988). However, in glasshouse studies, varying the phosphate supply in bacterial growth media had little or no effect on nodule weight or nitrogen content of shoots in 45 day-old soybeans (Mullen et al., 1988).

Phosphorus concentrations in nodules are frequently much greater than those in shoots and roots over a wide range of phosphorus supply (Robson et al., 1981). Moreover, even in snake-beans severely deficient in phosphorus, bacteroids still contained polyphosphate, indicating an adequate supply of phosphate (Smart et al., 1984b).

Adding phosphate may markedly change the strains of root nodule bacteria forming nodules. For example, at 4 sites, adding phosphate markedly decreased nodule occupancy by serogroup 123 in soybeans and increased nodule occupancy by unknown serogroups (Hicks and Loynachan, 1987). For _R. trifolii_, adding KH_2PO_4 to a soil low in extractable phosphate increased occupancy by serogroup 36 but not by serogroup 6 (Almendras and Bottomley, 1987). Parallel studies indicated that serogroups 36 and 6 differed in (i) their critical external phosphate requirements for growth (serogroup 6, 0.51 μM; serogroup 36, 0.89 μM) and (ii) their ability to nodulate _Trifolium subterraneum_ under low external phosphate at pH 5.5, but not at pH 6.5 (Leung and Bottomley, 1987). Serogroups 6 and 36 appeared to have similar internal phosphate requirements. Zaroug and Munns (1979) found no effect of phosphate addition on nodulation of _L. purpureus_ by 2 strains of root nodule bacteria which differed in response to phosphate in culture, perhaps because 10^4 cells were added per seed.

Rhizobia may utilize phosphatases to hydrolyse soil organophosphates, but the extracellular nature of these hydrolytic enzymes will still make phosphorus uptake dependent on their inorganic phosphate uptake system. Soil isolates of _R. trifolii_ express acid phosphatase activity (optimum pH, 6.5) when grown under inorganic phosphate-limited conditions (Leung and Bottomley, 1987). In contrast, alkaline phosphatase (optimum pH 9.0) was detected in all strains of _Rhizobium_, while strains of _Bradyrhizobium_ spp. produced neither alkaline nor acid phosphatases under similar conditions (Smart et al., 1984a). These variations in phosphate metabolism among rhizobia need to be resolved to determine if they have any significance for soil colonizing ability or nodule formation and function in phosphate-deficient soils.

4.5 VA mycorrhizas

Inoculation with VA mycorrhizal fungi increases nodulation and nitrogen fixation by many legumes grown on P-deficient soils (see Munns and Mosse, 1980; Barea and Azcon-Aquilar, 1983), mostly through effects on phosphorus uptake.

Nodulation and nitrogen fixation in several studies were only increased by inoculation with VA mycorrhizal fungi when insufficient phosphorus was added for maximum growth of the non-mycorrhizal plant (G. max, Carling et al., 1978; Bethlenfalvay and Yoder, 1981; Fredeen and Terry, 1988; T. subterraneum, Robson et al., 1981). In many studies nodule weight and shoot weight were positively correlated in both mycorrhizal and non-mycorrhizal plants (for example, Fig. 2; soybeans, Pacovsky et al., 1986; Bayne and Bethlenfalvay, 1987; Bethlenfalvay et al.,1989; subterranean clover, Robson et al., 1981). Indeed, Cluett and Boucher (1983), in an analysis of 20 published data sets, concluded that the effect of mycorrhizal infection on nodulation was almost entirely accounted for by the regression of nodulation on plant size. In one study, where inoculation with a VA mycorrhizal fungus depressed shoot weight of Phaseolus vulgaris, it also decreased nodule weight (Bethlenfalvay et al., 1982), which was closely correlated with shoot weight.

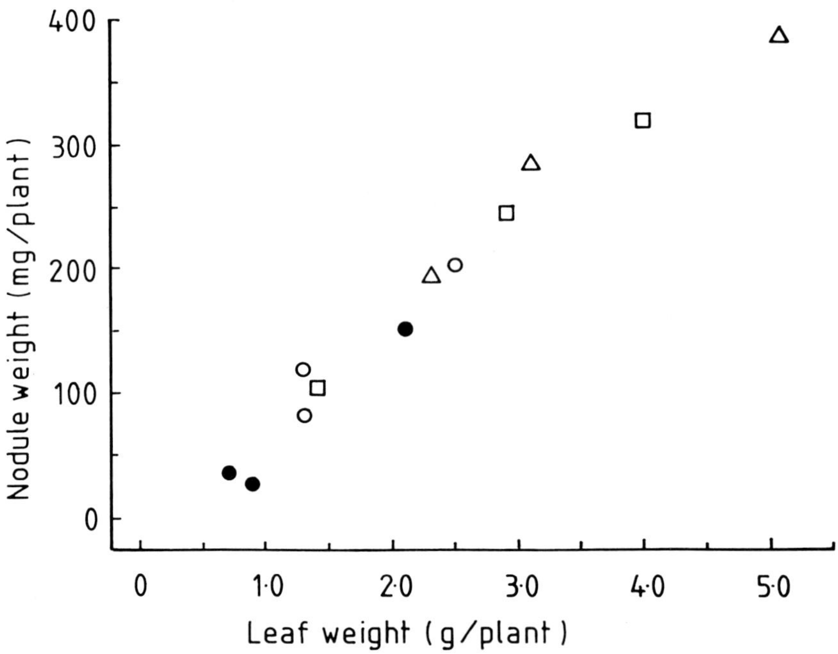

Fig. 2. Relationship between nodule weight and leaf weight for soybeans either uninoculated but supplied with phosphate (△) or inoculated with Glomus versiforme (●), Glomus mosseae (○) or Glomus fasciculatum (□). Data calculated from Bayne and Bethlenfalvay (1987).

However, some effects of VA mycorrhizal infection on nodule formation and functioning are puzzling. Inoculation with a mycorrhizal fungus at planting depressed nodule weight in soybean if inoculation with the bradyrhizobia were delayed until 20 days later (Bethlenfalvay et al., 1985), but not if the bradyrhizobia and the mycorrhizal fungus were both either inoculated at planting or at 20 days afterwards. In a split root experiment with cowpea, inoculation of one compartment with a VA mycorrhizal fungus increased root weight by approximately 25% and doubled acetylene reduction per pot in that compartment (Ames and Bethlenfalvay, 1987). Surprisingly, shoot dry weight was increased by 15% when one compartment was inoculated with VA mycorrhizal fungus, but phosphate addition did not increase shoot weight, while the increased acetylene reduction in the inoculated compartment did not increase nitrogen concentrations in either leaves or roots. In a similar split-root experiment with Acacia saligna, inoculation of the same or different compartments with a VA mycorrhizal fungus and root nodule bacteria did not affect nodule formation (Juniper, 1989), nodule weights being positively correlated with shoot weights whatever the inoculation regime.

Nodulating isolines of soybeans showed a greater response to mycorrhizal inoculation than did non-nodulating isolines (Schenck and Hinson, 1973), which yielded less and contained lower nitrogen concentrations. Differences in response to inoculation with a mycorrhizal fungus may therefore indicate differences in the adequacy of nitrogen supply for plant growth. However, an inability to form nodules may be linked to an inability to form mycorrhizal associations; in peas, 7 of 14 Nod^- genotypes did not form mycorrhizas, whereas all of 12 Nod^+ (both Fix^+ and Fix^-) genotypes formed mycorrhizas (Duc et al., 1989). In this experiment, mineral nitrogen was added to ensure an adequate nitrogen supply for both nodulating and non-nodulating genotypes. The basis for the correlation between the nodule and mycorrhizal associations should be further investigated.

In an interesting study, inoculation with a VA mycorrhizal fungus did not affect the relative competitive abilities of two serogroups of B. japonicum (Hicks and Loynachan, 1987). Unfortunately, these authors presented no plant growth data so that the adequacy of soil phosphate for plant growth cannot be assessed. It would be worthwhile to examine the effect of VA mycorrhizal infection on the competitive ability of serogroups 6 and 36 (Almendras and Bottomley, 1987) of R. trifolii in subterranean clover grown with a wide range of external phosphate supply, so that effects of internal and external phosphate supply on competitive ability could be separated.

5 SOIL ACIDITY

The effects of soil acidity on the growth and nodulation of legumes have been extensively reviewed (for example Munns, 1986; Coventry and Evans, 1989). We have therefore only considered four aspects of the overall area. (Unless otherwise stated, all soil pH values quoted are measured in 1:5 suspensions of soil in 0.01M $CaCl_2$).

5.1 Distribution and abundance of root nodule bacteria

Soil acidity may limit the colonization and survival of root nodule bacteria both in the presence and absence of plants (see Vincent and Waters, 1954; Coventry et al., 1985), but the effects of acidity on soil populations of rhizobia are not easy to determine because of several complicating factors. The distribution of root-nodule bacteria may be uneven both horizontally and vertically (see Chapter 10). Thus, more than 70% of the viable population of *R. trifolii* under a dead subterranean clover stand were located within senesced nodules (Chatel and Greenwood, 1973a). Moreover, more than 80% of the cells of *R. trifolii* in an acid soil were located in the less acid top 2 cm (Richardson and Simpson, 1988). Cultivation of such soils mixes the layers and may markedly decrease nodulation (Richardson et al., 1985).

A second complication arises because some of the effects on numbers of root nodule bacteria may vary seasonally. In one study, lime application to three acid (pH 4.2-4.5) soils increased numbers of *R. trifolii* in autumn and winter, after the start of the growing season, whereas later in the season there was little difference in numbers between limed and unlimed soils (Evans et al., 1988). Considerable care should therefore be taken in sampling soils to examine whether soil acidity is limiting numbers of root nodule bacteria.

A final problem is that soil acidity may limit nodulation by certain rhizobial strains much more than others. In an unlimed soil at pH 4.8, serogroup 6 of *R. trifolii* occupied half the nodules on subterranean clover, whereas in limed soil (pH 6.4) this serogroup occupied less than 10% (Dughri and Bottomley, 1983). In contrast, serogroup 36 of *R. trifolii* formed very few nodules in the acid soil, but 40% of the nodules in the limed soil. Lime application did not affect relative numbers of strains in serogroups 6 and 36 (Almendras and Bottomley, 1987). Liming may enhance nodule occupancy by serogroup 36 by affecting the chemistry of phosphorus in the soil, since it increased organic phosphate content. Effects of lime on nodule occupancy by serogroup 36 were duplicated by the addition of KH_2PO_4.

Studies of indigenous rhizobia in acid soils have not always produced results supporting the concept that they are successful colonizers because of their superior acid tolerance. In a survey of Hawaiian soils of pH 4.9 to pH 7.9 (1/2 water), the most dense populations of *Bradyrhizobium* spp. were recovered in Trophomults of pH <5.0 (Woomer et al., 1988). No difference in acid tolerance was noted among isolates of *R. trifolii* recovered across a soil pH transect (Wood and Shepherd, 1987).

5.2 Selection for acid tolerance in root nodule bacteria and legumes

In research to select root nodule bacteria and host legumes tolerant of soil acidity (see Munns, 1986; Coventry and Evans, 1989), the rhizobia and the host legume have generally been screened separately for their ability to tolerate acidity-related stresses such as high concentrations of hydrogen ions, aluminium and manganese and low concentrations of calcium, magnesium and molybdenum. We have focused here only on recent studies of annual Medicago species and R. meliloti and on P. vulgaris and Rhizobium leguminosarum bv. phaseoli (referred to as R. phaseoli).

The growth and survival of R. meliloti is much more sensitive to soil acidity than nodulation, nodule function or growth of the host legume (Robson and Loneragan, 1970a). Strains of R. meliloti differ considerably in their ability to colonize and persist in acid soils (Munns, 1965; Robson and Loneragan, 1970b; Howieson and Ewing, 1986; Howieson et al., 1988; Hartel and Bouton, 1989). However, for many medic species (for example M. truncatula, Robson and Loneragan, 1970b; M. sativa, Hartel and Bouton, 1989) these differences in acid-tolerance are insufficient to permit adequate nodulation on acid soils. Significantly, there are differences both among and within annual Medicago species in ability to nodulate on acid soils (Howieson and Ewing, 1986, 1989). A combination of a host able to nodulate on acid soils (M. polymorpha, M. murex) with an acid-tolerant rhizobial strain has permitted annual medics to nodulate and persist in soils as acid as pH 4.8 (Ewing and Howieson, 1989).

Why annual medics differ in their ability to nodulate on acid soils is not understood. On a soil of pH 4.9 increasing the concentration of inoculum 100-fold (from 0.02 to 2 g peat m^{-2}) did not change the differences in nodulation among species (Howieson and Ewing, 1989). However, although increasing inoculum to 20 g m^{-2} did not increase nodulation in "tolerant" species (M. polymorpha, M. murex, M. soleirolii), it did so in a "sensitive" species (M. truncatula). When the soil was limed to pH 6.8, all medic species were nodulated at an inoculation rate of 2.0 g peat m^{-2}. These differences between "sensitive" and "tolerant" species were not related to differences in rhizosphere pH (Ewing and Robson, 1990a), but Ewing and Robson (1990b) showed that in solution culture, a "sensitive" species (M. truncatula) required much higher pH and calcium concentrations than did two tolerant species (M. polymorpha, M. murex). Perhaps there are differences among the medic species in their ability to adsorb R. meliloti cells (see below).

The selection of acid tolerant strains of R. meliloti has also demonstrated some important principles. One is that growth on acidic agar is poorly correlated with ability to persist in acid soils (Howieson et al., 1988), whereas growth in acid broth, which appears to reflect an ability to maintain intracellular pH (pH_i) did identify strains which persisted poorly in acid soils (Table 2; O'Hara et al., 1989).

TABLE 2 Relationship between laboratory and field performance of *R. meliloti* in three studies

Strain	pH_i pH 5.6[a]	Growth pH 5.6[b]	Growth pH 5.6 and 5.8[c] (agar)	Field performance[d]	Field performance[e]
WSM419	7.25	+	31	20	36
WSM533	7.21	+	36	14	nd
WSM539	7.27	+	34	15	nd
WSM540	7.20	+	40	40	nd
U45	6.15	+/-	41	nd	16
CC169	6.52	+/-	30	nd	12
WSM244	6.24	-	nd	nd	9

a: Internal pH in cells incubated at pH 5.6 (O'Hara et al., 1989)

b: Growth on laboratory media: +, growth; +/-, poor growth; -, no growth (O'Hara et al., 1989)

c: Growth rating on laboratory media - maximum score 60 (Howieson et al., 1988)

d: Percentage of plants nodulated in the region 11-20 cm from initial point of inoculation 1 year previously in soil of pH 4.8 (Howieson et al., 1988)

e: Percentage plants nodulated in the region 11-20 cm from initial point of inoculation 1 year previously in soil of pH 5.0 (Howieson and Ewing, 1986)

nd: not determined

Acid-sensitive mutants derived from an acid tolerant strain (WSM419) were less able to maintain internal pH when grown with 0.1 mM calcium than with 2.0 mM calcium (O'Hara et al., 1989), but no calcium effect was observed with 4 acid-tolerant strains. Strains which grew well in broth at pH 5.6 and maintained pH_i differed in their ability to persist in acid soils (pH 4.8), indicating that within these strains there are other important differences. Strains collected from acid soils and isolated onto acid media appeared to be more "acid-tolerant" than strains collected at neutral sites and isolated onto neutral media (Howieson et al., 1988). Similarly, the strains of *R. trifolii* with the best symbiotic performance in the acidic soils were those isolated from the most acid soil (Lindström and Myllyniemi, 1987).

Acid-tolerance varies widely among strains of *R. phaseoli* (Graham et al., 1982; Vargas and Graham, 1988). Moreover, cultivars of *P. vulgaris* varied in their ability to nodulate on acid soils (Vargas and Graham, 1988). Again there has been relatively little study of the mechanisms underlying this variation. However, in sand culture at pH 4.5, an acid-tolerant rhizobial strain had higher nodule occupancy than an acid-sensitive strain with an acid-sensitive plant cultivar, but there was equal nodule occupancy in an acid-tolerant plant cultivar (Vargas and Graham, 1989). At pH 5.5 and 6.5, the acid-sensitive rhizobial strain formed approximately 70% of the nodules for both plant cultivars.

5.3 Calcium

Calcium supply could potentially affect the growth of root nodule bacteria, the formation of nodules and the function of nodules as well as host legume growth (see Munns, 1977; Robson, 1983; O'Hara et al., 1988).

Although there is a low calcium requirement for growth of most rhizobia under optimum laboratory conditions (Vincent, 1962), its role in several physiological processes means that a critical role for calcium in the ecology of rhizobia cannot be discounted. Supplemental calcium is essential for growth of R. meliloti, both at low phosphate concentrations (Beck and Munns, 1984, 1985), and at low matric water potential created by a nonpermeating solute (Busse and Bottomley, 1989a). Acid-tolerant strains of R. meliloti required higher levels of calcium for growth than did other R. meliloti strains (O'Hara et al., 1989). Calcium plays a role in the synthesis of a surface adhesin protein important for attachment of R. leguminosarum bv. viciae to pea roots (Smit et al., 1987, 1989) and in the same biovar calcium is involved in stabilizing oligomeric forms of outer membrane proteins which are covalently linked to peptidoglycan (DeMaagd et al., 1989).

Rhizobia and bradyrhizobia differ considerably in the effects of calcium on their growth. In electrodialysed clays, the addition of calcium strongly stimulated the growth of rhizobia, but not of bradyrhizobia (McCalla, 1937; Norris, 1959). Similarly, increasing the calcium concentration to 1 mM alleviated toxic effects of hydrogen ions on the survival of R. meliloti, R. leguminosarum bv viciae (referred to as R. viciae) and R. trifolii without affecting the survival of B. japonicum, cowpea Bradyrhizobium spp. or lupin Bradyrhizobium spp. (Rerkasem, 1973). Moreover, increasing the calcium concentration in solution from 0.03 to 3 mM alleviated toxic effects of aluminium on R. trifolii, whereas it was without effect in lupin Bradyrhizobium spp. (Khan et al., 1984).

Notwithstanding these striking effects of calcium supply on growth of the root nodule bacteria, it is likely that nodule formation has a greater requirement for calcium than growth of the bacteria (Lowther and Loneragan, 1968; Munns,

1968, 1970). Complexing calcium with an ionophore markedly decreased nodule number and weight in soybeans at a low calcium concentration (0.1 mM), but was without effects at higher calcium concentrations (0.5-1 mM) (Blevins et al., 1977). The adsorption of cells of R. meliloti to alfalfa roots increased linearly with increasing concentrations of both calcium and magnesium up to 1.5 mM at pH 7.0 (Caetono-Anolles et al., 1989). The requirement for calcium during adsorption could not be replaced by calcium pretreatment of the roots, but preincubation of the bacteria with calcium permitted adsorption at calcium concentrations otherwise too low for it to occur. The adsorption of R. viciae to pea root hairs required the bacteria to be grown in the presence of calcium (Smit et al., 1986), and it has been proposed that a calcium-dependent adhesin is required for rhizobial adsorption to root hairs (Smit et al., 1987, 1989).

Increasing the number of rhizobia added (Lowther, 1972) decreased the effects of calcium on nodulation in subterranean clover, which were greatest in moderately acid solutions (Loneragan and Dowling, 1958; Andrew, 1976). Both pH and numbers of added rhizobia will affect the number of rhizobia adhering to roots (Caetano-Anolles et al., 1989), so that much higher concentrations of calcium were required for the maximum adsorption of R. meliloti at pH 6.5 than at pH 7.0 (Caetono-Anolles et al., 1989). Magnesium could replace calcium at pH 7.0, but not at pH 6.5. In a similar way, calcium cannot be replaced by magnesium for nodulation in moderately acid soils (Munns, 1970).

Motile cells of R. meliloti (Ames and Bergman, 1981) and R. trifolii (Mellor et al., 1987) produce many more nodules than non-motile mutants when both are placed in the rhizosphere. Although addition of 1 mM $CaCl_2$ increased motility of R. viciae, 1 mM $MgSO_4$ did not (Bowra and Dilworth, 1981). Effects of calcium and pH on motility do not appear to account for effects on adsorption of R. meliloti on roots, because adsorption of a non-flagellate mutant was similarly affected by low pH and low calcium (Caetono-Anolles et al., 1989).

Low pH and low calcium concentrations decreased the ability of exudates from roots of clover (grown in the absence of rhizobia) to induce the expression of nodulation genes in R. trifolii (Richardson et al., 1988a, 1988b), which are maximally inducible during early logarithmic growth (Djordjevic et al., 1987). This may be related to effects on both rhizobial multiplication in the rhizosphere and their adsorption to roots. Since even large numbers of root nodule bacteria may fail to nodulate in acid (Munns, 1968) and low calcium solutions (Lowther and Loneragan, 1968; Munns, 1970), these conditions may prevent bacterial adsorption onto roots or the induction of nodulation genes by plant exudates.

Since adding large numbers of rhizobia can produce nodulation at a pH where rhizobial multiplication is prevented, and this strategy probably fails with low calcium, it follows that the effects of calcium deficiency and low pH are unlikely to be the same.

5.4 Aluminium

For most legumes, nodule initiation is decreased by aluminium concentrations that do not limit either nodule function or host legume growth with combined nitrogen (Stylosanthes species, Carvalho et al., 1981; subterranean clover, Kim et al., 1985; soybean, Brady et al., 1989). Aluminium probably affects nodule initiation by influencing both rhizobial numbers (Wood et al., 1984; Kim et al., 1985; Whelan and Alexander, 1986) and early events in nodulation. In soybeans, even after inoculation with dense bradyrhizobial suspensions, nodule number and nodule weight on the tap root were greatly reduced by a 12-hour exposure to 8 μM aluminium (Brady et al., 1989). There have been no studies of the effect of aluminium on the adsorption of root nodule bacteria to roots. Interference with induction of nod genes of R. trifolii is unlikely (Richardson et al., 1988a), since aluminium concentrations as high as 60 μM did not influence the exudation of compounds from roots of subterranean clover. The mechanism of this aluminium inhibition is unknown, but as little as 2.5 μM aluminium limited root hair formation in soybeans (Hecht-Bucholz et al., 1989) and this may be associated with nodulation failure.

6 LOW SOIL TEMPERATURES

Annual crop and pasture legumes are frequently required to establish effective symbioses at less than optimal soil temperatures, while perennial legumes may also grow symbiotically in soil with low temperatures. The effects of low soil temperatures have been examined on a very limited number of legumes (largely Trifolium species) and the generalizations that follow may not hold for other legumes.

The growth of nodulated plants appears to be decreased more by low temperatures than the growth of plants supplied with mineral nitrogen (T. subterraneum, Meyer and Anderson, 1959; Gibson, 1961; Trifolium spp., Small and Joffe, 1968; soybeans, Matthews and Hayes, 1982) indicating that either multiplication of the bacteria, nodule formation or nodule function is affected to a greater extent by low temperatures than growth of the host legume. Nitrogen concentrations in shoots as well as shoot weight may be increased by raising the soil temperature (soybeans, Matthews and Hayes, 1982; lupins, S.C.Peltzer, unpublished).

Multiplication of the rhizobia and nodule formation appear to be more limited by low soil temperatures than nodule function. Nodules of perennial legumes such as Trifolium repens can persist over winter, enabling new bacteroid production and nitrogen fixation to commence at soil temperatures too low for nodule formation (Masterson and Murphy, 1976).

The effect of low soil temperatures varies considerably for different isolates of root nodule bacteria and legumes. At 10°C growth in broth culture of isolates of R. trifolii from the sub-arctic region of Scandinavia was greater than that of isolates from the warmer southern areas (Ek-Jander and Fahraeus, 1971). Root nodule bacteria isolated from arctic soils also grew better at low temperatures than isolates from temperate soils (Nelson and Parkinson, 1978; Prévost et al., 1987). For example, all of 48 isolates from root nodules of arctic legumes were able to grow at 5°C, but no temperate reference strain could (Prévost et al., 1987). Effects of low temperature on the efficiency of nitrogen fixation by soybean (Layzell et al., 1984) and Lotus (Pankhurst and Layzell, 1984) also varied with the strain of root nodule bacteria.

Further evidence for the variation among strains of root nodule bacteria in response to low temperatures is provided by the large effect of root temperature on nodule occupancy (Rice and Olsen, 1988). The percentage of nodules containing R. meliloti strain NRE195 increased from less than 10%, to more than 70%, when root temperature was raised from 8°C to 26°C. In contrast, nodule occupancy by the Balsac strain was unaffected by root temperature in the range 8 - 22°C and fell when root temperature rose from 22°C to 26°C. Similar strain variation has been observed for B. japonicum (Weber and Miller, 1972) and R. trifolii (Hardarson and Jones, 1979).

The effects of low soil temperatures on the processes of nodule formation have been reviewed by Gibson (1977) and Sprent and Minchin (1983). There has been little advance since those reviews.

7 WATER RELATIONS

Soil populations of Bradyrhizobium spp. are more persistent than Rhizobium spp. in soils which endure seasonal moisture deficits (Chatel and Parker, 1973a, b; Woomer et al., 1988). It has been suggested that cells of Bradyrhizobium enter dormancy more readily than those of Rhizobium, because at low water potentials the former maintain a lower internal water content than the latter (Bushby and Marshall, 1977). Appealing as this hypothesis may be, other studies do not support it (Hartel and Alexander, 1979; Van Rensberg and Strijdom, 1980; Singleton et al., 1982). There is a need to address this controversy by conducting a systematic study of the water relations of well-characterized populations of both genera indigenous to the same soil.

Rhizobia can survive prolonged periods of exposure to extremely low water potential (Fuhrmann et al., 1986). However, few studies have addressed either the efficiency of rhizobial growth at low matric potential (Singleton et al., 1982), or their capacity for growth after rewetting (Salema et al., 1982). When the matric water potential of growth medium was lowered with polyethylene glycol, field isolates of two serotypes of R. meliloti differed both in their ability to grow as matric potential was lowered to -1.5 MPa and in their recovery after 'upshock' into conventional growth medium (-0.1 MPa) (Busse and Bottomley, 1989). After such a transfer from -1.0 MPa growth medium onto alfalfa seedlings, nodulation by the upshock-sensitive strains was delayed relative to the insensitive strains. It is worth noting that although root hair infection of subclover does not occur at matric potentials lower than -0.3 MPa (Worrall and Roughley, 1976) subclover seeds can germinate and undergo primary development at water potentials as low as -1.2 MPa (McWilliam et al., 1969; K. Leung and P.J.Bottomley, unpublished). It remains to be determined whether rhizobia inhabiting dry surface soil can obtain nodulating advantage by regaining metabolic competence at water potentials sufficient for seed germination, yet too low for root hair infection.

8 CONCLUSION

There are many challenges to overcome limitations on the use of legumes in agriculture and forestry. For some problems there is a need to focus on the rhizobial component (for example, low competitiveness of inoculant strains); for others it is the initiation of nodules which is most sensitive (for example, salinity) and for other limitations the focus should be on the legume (for example, low phosphate supply). In overcoming these limitations it is essential that skills in rhizobial ecology and legume physiology are supplemented, rather than replaced, by skills in molecular biology.

Acknowledgments

Dr L.K. Abbott made very helpful comments and suggestions in the development of the review.

9 REFERENCES

Almendras, A.S. and Bottomley, P.J., 1987. Influence of lime and phosphate on nodulation of soil-grown Trifolium subterraneum L. by indigenous Rhizobium trifolii. Appl. Environ. Microbiol., 53: 2090-2097.

Ames, P. and Bergman, K., 1981. Competitive advantage provided by bacterial motility in the formation of nodules by Rhizobium meliloti. J. Bacteriol., 148: 728-729.

Ames, R.N. and Bethlenfalvay, G.J., 1987. Localized increase in nodule activity but no competitive interaction of cowpea rhizobia due to pre-establishment of vesicular-abbuscular mycorrhiza. New Phytol., 106: 207-215.

Andrew, C.S., 1976. Effect of calcium, pH and nitrogen on the growth and chemical composition of some tropical and temperate pasture legumes. I. Nodulation and growth. Aust. J. Agric. Res., 27: 612-623.
Andrew, C.S. and Robins, M.F., 1969. The effect of phosphorus on the growth and chemical composition of some tropical pasture legumes. II. Nitrogen, calcium, magnesium, potassium and sodium contents. Aust. J. Agric. Res., 20: 675-685.
Arsac, J.F. and Cleyet-Marel, J.C., 1986. Serological and ecological studies of Rhizobium spp. (Cicer arietinum L.) by immunofluorescence and ELISA technique: competitive ability for nodule formation between Rhizobium strains. Plant Soil, 94: 411-423.
Barea, J.M. and Azcon-Aquilar, C., 1983. Mycorrhizas and their significance in nodulating nitrogen-fixing plants. Adv. Agron., 36: 1-54.
Barnet, Y.M., 1988. Nitrogen-fixing symbioses with Australian native legumes. In: W.M.G. Murrell and I.R. Kennedy (Editors), Microbiology in Action. John Wiley and Sons, Chichester, pp. 81-92.
Bayne, H.G. and Bethlenfalvay, G.J., 1987. The Glycine-Glomus-Rhizobium symbiosis. IV. Interactions between the mycorrhizal and nitrogen-fixing endopohytes. Plant Cell Environ., 10: 607-612.
Beattie, G.A., Clayton, M.K. and Handelsman, J., 1989. Quantitative comparison of the laboratory and field competitiveness of Rhizobium leguminosarum biovar phaseoli. Appl. Environ. Microbiol., 55: 2755-2761.
Beck, D.P. and Munns, D.N., 1984. Phosphate nutrition of Rhizobium spp. Appl. Environ. Microbiol., 47: 278-282.
Beck, D.P. and Munns, D.N., 1985. Effect of calcium on the phosphorus nutrition of Rhizobium meliloti. Soil Sci. Soc. Am. J., 49: 334-337.
Bernstein, L. and Ogata, G., 1966. Effects of salinity on nodulation, nitrogen fixation and growth of soybean and alfalfa. Agron. J., 58: 201-203.
Bethlenfalvay, G.J. and Yoder, J.F., 1981. The Glycine-Glomus-Rhizobium symbiosis. I. Phosphorus effect on nitrogen fixtion and mycorrhizal infection. Physiol. Plant., 52: 141-145.
Bethlenfalvay, G.J., Brown, M.S. and Stratford, A.E., 1985. The Glycine-Glomus-Rhizobium symbiosis. II. Antagonistic effects between mycorrhizal colonization and nodulation. Plant Physiol., 79: 1045-1058.
Bethlenfalvay, G.J., Franson, R.L., Brown, M.S. and Mihara, K.L., 1989. The Glycine-Glomus-Rhizobium symbiosis. IX. Nutritional, morphological and physiological responses of nodulated soybean to geographic isolates of the mycorrhizal fungus Glomus mosseae. Physiol. Plant., 76: 226-232.
Bethlenfalvay, G.J., Pacovsky, R.S., Brown, M.S. and Fuller, G., 1982. Mycotrophic growth and mutualistic development of host plant and fungal endophyte in an endomycorrhizal symbiosis. Plant Soil., 68: 43-54.
Blevins, D.G., Barnett, N.M. abnd Bottino, P.J., 1977. The effects of calcium deficiency and the ionophore A23187 on nodulation, nitrogen fixation and growth of soybeans. Physiol. Plant., 41: 235-238.
Bohlool, B.B. and Schmidt, E.L., 1973. Persistence and competition aspects of Rhizobium japonicum observed by immunofluorescence microscopy. Soil Sci. Soc. Am. Proc., 37: 561-564.
Bottomley, P.J. and Jenkins, M.B., 1983. Some characteristics of Rhizobium meliloti isolates from alfalfa fields in Oregon. Soil Sci. Soc. Am. J., 47: 1153-1157.
Bowra, B.J. and Dilworth, M.J., 1981. Motility and chemotaxis towards sugars in Rhizobium leguminosarum. J. Gen. Microbiol., 126: 231-235.
Brady, D.J., Hecht-Bucholz, C., Asher, C.J. and Edwards, D.G., 1990. Effects of low activities of aluminium on soybean. 1. Early growth and nodulation of soybean. Plant Soil (in press).

Brockwell, J., Gault, R.R., Zorin, M. and Roberts, M.J., 1982. Effects of environmental variables on the competition between inoculum strains and naturalized populations of Rhizobium trifolii for nodulation of Trifolium subterraneum L. and on rhizobial persistence in the soil. Aust. J. Agric. Res., 33: 803-815.
Brockwell, J., Holliday, R.A., Daoud, D.M. and Materon, L.A., 1988. Symbiotic characteristics of a Rhizobium-specific annual medic, Medicago rigidula (L.) All. Soil Biol. Biochem., 20: 593-600.
Brockwell, J., Roughley, R.J. and Herridge, D.F., 1982. Population dynamics of Rhizobium japonicum strains used to inoculate three successive crops of soybean. Aust. J. Agric. Res., 38: 61-74.
Brockwell, J., Schwinghamer, E.A. and Gault, R.R., 1977. Ecological studies of root-nodule bacteria introduced into field environments. V. A critical examination of the stability of antigenic and streptomycin-resistance markers for identification of strains of Rhizobium trifolii. Soil Biol. Biochem., 9: 19-24.
Burton, J.C., 1975. Methods of inoculating seeds and their effect on survival of rhizobia. In: P.S. Nutman (Editor), Symbiotic Nitrogen Fixation in Plants. Cambridge University Press, London, IBP Vol. 7, pp. 175-189.
Bushby, H.V.A. and Marshall, K.C., 1977. Water status of rhizobia in relation to their susceptibility to desiccation and to their protection by montmorillonite. J. Gen. Microbiol., 99: 19-27.
Busse, M.D. and Bottomley, P.J., 1989. Growth and nodulation responses of Rhizobium meliloti to water stress induced by permeating and nonpermeating solutes. Appl. Environ. Microbiol., 55: 2431-2436.
Caetono-Annollés, G., Lagares, A. and Favelukes, G., 1989. Adsorption of Rhizobium meliloti to alfalfa roots: dependence on divalent cations and pH. Plant Soil, 117: 67-74.
Carling, D.E., Riehle, W.G., Brown, M.F. and Johnson, D.R., 1978. Effects of a vesicular-arbuscular mycorrhizal fungus on nitrate reductase and nitrogenase activities in nodulating and non-nodulating soybeans. Phytopath., 68: 1590-1596.
Carter, E.D., Wolfe, E.C. and Francis, C.M., 1982. Problems of maintaining pastures in the cereal-livestock areas of southern Australia. In: M.J. Norman (Editor), Proc. 2nd Australian Agronomy Conference, Wagga Wagga, pp. 68-87.
Carvalho, de, M.M., Edwards, D.G., Andrew, C.S. and Asher, C.J., 1981. Aluminium toxicity, nodulation and growth of Stylosanthes species. Agron. J., 73: 261-265.
Cassman, K.G., Munns, D.N. and Beck, D.P., 1981a. Growth of Rhizobium strains at low concentrations of phosphate. Soil Sci. Soc. Am. J., 45 520-523.
Cassman, K.G., Whitney, A.S. and Fox, R.L., 1981b. Phosphorus requirements of soybean and cowpea as affected by mode of N nutrition. Agron. J., 73 17-22.
Chatel, D.L. and Greenwood, R.M., 1973a. The location and distribution in soil of rhizobia under senesced annual legume pastures. Soil Biol. Biochem., 5: 799-808.
Chatel, D.L. and Greenwood, R.M., 1973b. Differences between strains of Rhizobium trifolii in ability to colonize soil and plant roots in the absence of their specific host plants. Soil Biol. Biochem., 5: 809-813.
Chatel, D.L. and Parker, C.A., 1973a. Survival of field-grown rhizobia over the dry summer period in Western Australia. Soil Biol. Biochem., 5: 415-423.
Chatel, D.L. and Parker, C.A., 1973b. The colonization of host-root and soil by rhizobia. I. Species and strain differences in the field. Soil Biol. Biochem., 5: 425-432.
Chatel, D.L., Greenwood, R.M. and Parker, C.A., 1968. Saprophytic competence as an important character in the selection of Rhizobium for inoculation. Trans. Ninth Inter. Congr. Soil Sci. Angus and Robertson, Syd., Vol. 2, pp. 65-73.

Chatel, D.L., Shipton, W.A. and Parker, C.A., 1973. Establishment and persistence of Rhizobium trifolii in Western Australian soils. Soil Biol. Biochem., 5: 815-824.
Cluett, H.C. and Boucher, D.H., 1983. Indirect mutualism in the legume-Rhizobium mycorrhizal fungus interaction. Oecologia, 59: 405-408.
Coventry, D.R. and Evans, J., 1989. Symbiotic nitrogen fixation and soil acidity. In: A.D. Robson (Editor), Soil Acidity and Plant Growth. Academic Press, New York, pp. 103-137.
Coventry, D.R., Hirth, J.R. and Fung, K.K.H., 1987. Nutritional restraints on subterranean clover grown on acid soils used for crop-pasture rotation. Aust. J. Agric. Res., 38: 163-176.
Coventry, D.R., Hirth, J.R. Reeves, T.G. and Jones, H.R., 1985. Development of population of Rhizobium trifolii and nodulation of subterranean clover following the cropping phase in crop-pasture rotations in south-east Australia. Soil Biol. Biochem., 17: 17-22.
Damirgi, S.M., Frederick, L.R. and Anderson, I.C., 1967. Serogroups of Rhizobium japonicum in soybean nodules as affected by soil type. Agron. J., 59: 10-12.
De Maagd, R.A., Wientjes, F.B. and Lugtenberg, B.J.J., 1989. Evidence for divalent cation (Ca^{2+})-stabilized oligomeric proteins and covalently bound protein-peptidoglycan complexes in the outer membrane of Rhizobium leguminosarum. J. Bacteriol., 171: 3989-3995.
Demezas, D.H. and Bottomley, P.J.., 1986a. Autecology in rhizospheres and nodulating behaviour of indigenous Rhizobium trifolii. Appl. Environ. Microbiol., 52: 1014-1019.
Demezas, D.H. and Bottomley, P.J.., 1986b. Interstrain competition between representatives of indigenous serotypes of Rhizobium trifolii. Appl. Environ. Microbiol., 52: 1020-1025.
Djordjevic, M.A., Redmond, J.W., Batley, M. and Rolfe, B.G., 1987. Clovers secrete specific phenolic compounds which either stimulate or repress nod gene expression in Rhizobium trifolii. EMBO J., 6: 1173-1179.
Duc, G., Trouvelot, A., Gianinazzi-Pearson, V. and Gianinazzi, S., 1989. First report of non-mycorrhizal plant mutants (Myc^-) obtained in pea (Pisum sativum L.) and faba-bean (Vicia faba L.). Plant Sci., 60: 215-222.
Dudeja, S.S. and Khurana, A.L., 1989. Persistence of Bradyrhizobium sp. (Cajanus) in a sandy loam. Soil Biol. Biochem., 21: 709-713.
Dughri, M.H. and Bottomley, P.J., 1983. Effect of acidity on the composition of an indigenous soil population of Rhizobium trifolii found in nodules of Trifolium subterraneum L. Appl. Environ. Microbiol., 45 1207-1213.
Dusha, T., Bakos, A., Kondorosi, A., de Bruijn, F.J. and Schell, J., 1989. The Rhizobium meliloti early nodulation genes (nodABC) are nitrogen-regulated: isolation of a mutant strain with efficient nodulation capacity on alfalfa in the presence of ammonium. Mol Gen. Genet., 219: 89-96.
Ek-Jander, J. and Fahraeus, G., 1971. Adaptation of Rhizobium to sub-arctic environments in Scandinavia. Plant Soil Special Vol., 127-129.
Ellis, W.R., Ham, G.E. and Schmidt, E.L., 1984. Persistence and recovery of Rhizobium japonicum inoculum in a field soil. Agron. J., 76: 573-576.
Evans, J., Hochman, Z., O'Connor, G.E. and Osborne, G.J., 1988. Soil acidity and Rhizobium: their effects on nodulation of subterranean clover on the slopes of southern New South Wales. Aust. J. Agric. Res., 38: 605-618.
Ewing, M.A. and Howieson, J.G., 1989. The development of Medicago polymorpha L. as an important pasture species for southern Australia. Proc. 16th Grasslands Congr., Nice, pp. 197-198.
Ewing, M.A. and Robson, A.D., 1990a. The effect of nitrogen supply on the early growth and nodulation of several annual Medicago species. Aust J. Agric. Res., in press.

Ewing, M.A. and Robson, A.D., 1990b. The effect of solution pH and external calcium concentration on the early growth and nodulation of several annual Medicago species. Aust J. Agric. Res., in press.
Freeden, A.L. and Terry, N., 1988. Influence of vesicular-arbuscular mycorrhizal infection and soil phosphorus level on growth and carbon metabolism of soybean. Can. J. Bot., 66: 2311-2316.
Freeden, A.L., Rao, I.M. and Terry, N., 1989. Influence of phosphorus nutrition on growth and carbon partitioning in Glycine max. Plant Physiol., 89: 225-230.
Fuhrmann, J. and Wollum, A.G., 1989. Nodulation competition among Bradyrhizobium japonicum strains as influenced by rhizosphere bacteria and iron availability. Biol. Fert. Soils, 7: 108-112.
Fuhrmann, J., Davey, C.B. and Wollum, A.G., 1986. Desiccation tolerance of clover rhizobia in sterile soils. Soil Sci. Soc. Am. J., 50: 639-644.
Gates, C.T. and Wilson, J.R., 1974. The interaction of nitrogen and phosphorus on the growth, nutrient status and nodulation of Stylosanthes humilis HBK (Townsville Stylo). Plant Soil, 41: 325-333.
Gates, C.T., Wilson, J.R. and Shaw, N.H., 1966. Growth and chemical composition of Townsville lucerne (Stylosanthes humilis). 2. Chemical composition with special reference to cations as affected by the principal constituent elements of molybdenized superphosphate. Aust. J. Exp. Agric. Anim. Husb., 6: 266-276.
Gibson, A.H., 1961. Root temperature and symbiotic nitrogen fixation. Nature, 191: 1080-1081.
Gibson, A.H., 1977. The influence of the environment and managerial practices on the legume-Rhizobium symbiosis. In: R.W.F. Hardy and A.H. Gibson (Editors), A Treatise on Dinitrogen Fixation, Vol. IV, Ecology. Wiley, New York, pp. 393-450.
Graham, P.H. and Rosas, J.C., 1979. Phosphorus fertilization and symbiotic nitrogen fixation in common bean. Agron J., 71: 925-926.
Graham, P.H., Viteri, S.E., Mackie, F., Vargas, A.T. and Palacios, A., 1982. Variation in acid tolerance among strains of Rhizobium phaseoli. Field Crops Res., 5: 121-128.
Hafeez, F.Y., Aslam, Z. and Malik, K.A., 1988. Effect of salinity and inoculation on growth, nitrogen fixation and nutrient uptake of Vigna radiata (L.) Wilczek. Plant Soil., 106: 3-8.
Hansen, E.H. and Munns, D.N., 1988. Effects of $CaSO_4$ and NaCl on growth and nitrogen fixation of Leucaena leucocephala. Plant Soil, 107: 95-99.
Hardarson, G. and Jones, D.G., 1979. Effect of temperature on competition amongst strains of Rhizobium trifolii for nodulation of two white clover varieties. Ann. Appl. Biol., 92: 229-236.
Hardarson, G., Golbs, M. and Danso, S.K.A., 1989. Nitrogen fixation in soybean (Glycine max L. Merrill) is affected by nodulation patterns. Soil Biol. Biochem., 21: 783-787.
Hardy, R.W.F. and Havelka, U.D., 1975. Photosynthate as a major factor limiting nitrogen fixation by field-grown plants with emphasis on soybeans. In: P.S. Nutman (Editor), Symbiotic Nitrogen Fixation by Plants. Cambridge University Press, New York, pp. 421-439.
Hart, A.L., Jessop, D.J. and Galpin, J., 1981. The response to phosphorus of white clover and lotus inoculated with rhizobia or given KNO_3. N.Z. J. Agric. Res., 24: 27-32.
Hartel, P.G. and Bouton, J.H., 1989. Rhizobium meliloti inoculation of alfalfa selected for tolerance to acid, aluminium-rich soils. Plant Soil, 116: 283-285.
Hecht-Bucholz, C., Brady, D.J., Asher, C.J. and Edwards, D.G., 1990. Effects of low activities of aluminium on soybean. 2. Root cell structure and root hair development. Plant Soil, in press.

Helyar, K.R. and Porter, W.M., 1989. Soil acidification, its measurement and the processes involved. In: A.D. Robson (Editor), Soil Acidity and Plant Growth. Academic Press, New York, pp. 61-101.
Herridge, D.F., Roughley, R.J. and Brockwell, J., 1987. Low survival of Rhizobium japonicum inoculant leads to reduced nodulation, nitrogen fixation and yield of soybean in the current crop but not in the subsequent crop. Aust. J. Agric. Res., 38: 75-82.
Hicks, P.M. and Loynachan, T.E., 1987. Phosphorus fertilization reduces vesicular-arbuscular mycorrhizal infection and changes nodule occupancy of field-grown soybeans. Agron. J., 79: 841-844.
Holland, A.A., 1970. Competition between soil- and seed-borne Rhizobium trifolii in nodulation of introduced Trifolium subterraneum. Plant Soil, 32: 293-302.
Howieson, J.G. and Ewing, M.A., 1986. Acid tolerance in the Rhizobium meliloti - Medicago symbiosis. Aust. J. Agric. Res., 37: 55-64.
Howieson, J.G. and Ewing, M.A., 1989. Annual species of Medicago differ greatly in their ability to nodulate on acid soils. Aust. J. Agric. Res., 40: 843-850.
Howieson, J.G., Ewing, M.A. and D'Antuono, M.F., 1988. Selection for acid tolerance in Rhizobium meliloti. Plant Soil, 105: 179-188.
Ireland, J.A. and Vincent, J.M., 1968. A quantitative study of competition for nodule formation. Trans. Ninth Int. Congr. Soil Sci. Angus and Robertson, Sydney, Vol. 2, pp. 85-93.
Israel, D.W., 1987. Investigation of the role of phosphorus in symbiotic dinitrogen fixation. Plant Physiol., 84: 835-840.
Israel, D.W. and Rufty, T.W. Jr., 1988. Influence of phosphorus nutrition on phosphorus and nitrogen utilization efficiencies and associated physiological responses in soybean. Crop Sci., 28: 954-960.
Jakobsen, I., 1987. The role of phosphorus in nitrogen fixation by young pea plants. Physiol. Plant., 64: 190-196.
Jasper, D.A., Robson, A.D. and Abbott, L.K., 1979. Phosphorus and the formation of vesicular-arbuscular mycorrhizas. Soil Biol. Biochem., 11: 501-505.
Jones, M.B., Burton, J.C. and Vaughn, C.E., 1978. Role of inoculation in establishing subclover in California annual grasslands. Agron. J., 70: 1081-1085.
Juniper, S., 1989. The interaction between a vesicular-arbuscular mycorrhizal fungus and root nodule bacteria on growth and nitrogen fixation in legumes (Acacia saligna). Undergraduate dissertation, The University of Western Australia.
Keating, B.A. and Fisher, M.J., 1985. Comparative tolerance of tropical grain legumes to salinity. Aust. J. Agric. Res., 36: 373-383.
Khan, S.I., Robson, A.D. and Parker, C.A., 1984. Comparison of effects of acid soil factors on growth of fast and slow growing species of Rhizobium. In: I.R. Kennedy and L. Copeland (Editors), 7th Australian Legume Nodulation Conference, Occasional Publication 12, Australian Institute of Agricultural Science, Melbourne, pp. 151-152.
Kim, M.M., Asher, C.J., Edwards, D.G. and Date, R.A., 1985. Aluminium toxicity: effects on growth and nodulation of subterranean clover. Proc. 15th Internationl Grasslands Congress., Kentucky, pp. 501-503.
Lauter, D.J., Munns, D.N. and Clarkin, K.L., 1981. Salt response of chick pea as influenced by nitrogen supply. Agron. J., 73: 961-966.
Lawn, R.J. and Brun, W.A., 1974. Symbiotic nitrogen fixation in soybeans. 1. Effect of photosynthetic source-sink manipulations. Crop Sci., 14: 11-16.
Layzell, D.B., Rochman, P. and Canvin, D.T., 1984. Low root temperatures and nitrogenase activity in soybean. Can. J. Bot., 62: 965-971.

Leung, K. and Bottomley, P.J., 1987. Influence of phosphate on the growth and nodulation characteristics of Rhizobium trifolii. Appl. Environ. Microbiol., 53: 2098-2105.
Lin, J., Walsh, K.B., Johnson, D.A., Canvin, D.T., Wang, S.-J. and Layzell, D.B., 1987. Characterization of R. fredii QB 1130, a strain effective on commercial soybean cultivars. Plant Soil, 99: 441-446.
Lindström, K. and Myllneimi, H., 1987. Sensitivity of red clover rhizobia to soil acidity factors in pure culture and in symbiosis. Plant Soil, 98: 353-362.
Loneragan, J.F. and Dowling, E.F., 1958. The interaction of calcium and hydrogen ions in the nodulation of subterranean clover. Aust. J. Agric. Res., 9: 464-472.
Lowther, W.L., 1970. Calcium in the nodulation and growth of legumes. Ph. D. Thesis, The University of Western Australia.
Lowther, W.L. and Loneragan, J.F., 1968. Calcium and nodulation in subterranean clover (Trifolium subterraneum L.). Plant Physiol., 43: 1362-1366.
Masterson, C.L. and Murphy, P.M., 1976. Application of the acetylene reduction technique to the study of nitrogen fixation by white clover in the field. In: P.S. Nutman (Editor), Symbiotic Nitrogen Fixation in Plants. Cambridge University Press, London, IBP Vol. 7, pp. 299-316.
Materon, L.A. and Weaver, R.W., 1985. Inoculant maturity influences survival of rhizobia on seed. Appl. Environ. Microbiol., 49: 465-467.
Matthews, D.J. and Hayes, P., 1982. Effect of root zone temperature on early growth, nodulation and nitrogen fixation in soyabeans. J. Agric. Sci., 98: 371-376.
McCalla, T.M., 1937. Behaviour of legume bacteria (Rhizobium) in relation to exchangeable calcium and hydrogen ion concentration of the colloidal fraction of the soil. University of Missouri College of Agriculture, Res. Bull. 250.
McLachlan, K.D. and Norman, B.W., 1961. Phosphorus and symbiotic nitrogen fixation in subterranean clover. J. Aust. Inst. Agric. Sci., 27: 244-245.
McWilliam, J.R., Clements, R.J. and Dowling, P.M., 1969. Some factors influencing the germination and early seedling development of pasture plants. Aust. J. Agric. Res., 21: 19-32.
Mellor, H.Y., Glenn, A.R., Arwas, R. and Dilworth, M.J., 1987. Symbiotic and competitive properties of motility mutants of Rhizobium trifolii TA1. Arch. Microbiol., 148: 34-39.
Meyer, D.R. and Andersen, A.J., 1959. Temperature and symbiotic nitrogen fixation. Nature, 183: 61.
Moawad, H. and Bohlool, B.B., 1984. Competition among Rhizobium spp. for nodulation of Leucaena leucocephala in tropical soils. Appl. Environ. Microbiol., 48: 5-9.
Mullen, M.D. and Wollum, A.G., 1989. Variation among different cultures of Bradyrhizobium japonicum strains USDA 110 and 122. Can. J. Microbiol., 35: 583-588.
Mullen, M.D., Israel, D.W. and Wollum, A.G., 1988. Effects of Bradyrhizobium japonicum and soybean (Glycine max (L.) Merr.) phosphorus nutrition on nodulation and dinitrogen fixation. Appl. Environ. Microbiol., 54: 2387-2392.
Munns, D.N., 1965. Soil acidity and growth of a legume. I. Interactions of lime with nitrogen and phosphate on growth of Medicago sativa L. and Trifolium subterraneum L. Aust. J. Agric. Res., 16: 733-741.
Munns, D.N., 1968. Nodulation of Medicago sativa in solution culture. I. Acid-sensitive steps. Plant Soil, 28: 129-146.
Munns, D.N., 1970. Nodulation of Medicago sativa in solution culture. V. Calcium and pH requirements during infection. Plant Soil, 32: 90-102.
Munns, D.N., 1977. Mineral nutrition and the legume symbiosis. In: R.W.F. Hardy and A.H. Gibson (Editors), A Treatise on Dinitrogen Fixation. Wiley, New York, Vol. IV., pp. 353-391.

Munns, D.N., 1986. Acid tolerance in legumes and rhizobia. Adv. Plant Nutr., 2: 63-91.
Munns, D.N. and Mosse, B., 1980. Mineral nutrition of legume crops. In: R.J. Summerfield and A.H. Bunting (Editors), Advances in Legume Science. HMSO, London, pp. 115-125.
Nelson, L.M. and Parkinson, D., 1978. Growth characteristics of three bacteria isolates from an arctic soil. Can. J. Microbiol., 24: 909-914.
Ng, B.H., 1987. The effects of salinity on growth, nodulation and nitrogen fixation of Casuarina equisetifolia. Plant Soil, 103: 123-125.
Norris, D.O., 1959. The role of calcium and magnesium in the nutrition of Rhizobium. Aust. J. Agric. Res., 10: 651-658.
O'Hara, G.W., Boonkerd, N. and Dilworth, M.J., 1988. Mineral constraints to nitrogen fixation. Plant Soil, 108: 91-110.
O'Hara, G.W., Goss, T.J., Dilworth, M.J. and Glenn, A.R., 1989. Maintenance of intracellular pH and acid tolerance in Rhizobium meliloti. Appl. Environ. Microbiol., 55: 1870-1876.
Owiredu, J.D. and Danso, S.K.A., 1988. Response of soybean (Glycine max (L.) Merrill) to Bradyrhizobium japonicum inoculation in three soils in Ghana. Soil Biol. Biochem., 20: 311-314.
Pacovsky, R.S., Paul, E.A. and Bethlenfalvay, G.J., 1986. Response of mycorrhizal and P-fertilized soybeans to nodulation by Bradyrhizobium or ammonium nitrate. Crop Sci., 26: 145-150.
Pankhurst, C.E. and Layzell, D.B., 1984. The effect of bacterial strain and temperature changes on the nitrogenase activity of Lotus pedunculatus root nodules. Physiol. Plant., 62: 404-409.
Peters, N.K. and Long, S.R., 1988. Alfalfa root exudates and compounds which promote or inhibit induction of Rhizobium meliloti nodulation genes. Plant Physiol., 88: 396-400.
Prēvost, D., Bordeleau, L.M., Candry-Reznick, S., Schulman, H.M. and Antoun, H., 1987. Characteristics of rhizobia isolated from three legumes indigenous to the Canadian high arctic: Astragalus alpinus, Oxytropis inagdelliana and Oxytropis arctobia. Plant Soil, 98: 313-324.
Rerkasem, B., 1977. Differential sensitivity to soil acidity of legume-Rhizobium symbioses. Ph. D. Thesis, The University of Western Australia.
Rice, W.A. and Olsen, P.E., 1988. Root-temperature effects on competition for nodule occupancy between two Rhizobium meliloti. Biol. Fert. Soils, 6: 137-140.
Richardson, A.E. and Simpson, R.J., 1988. Enumeration and distribution of Rhizobium trifolii under a subterranean clover-based pasture growing in an acid soil. Soil Biol. Biochem., 20: 431-438.
Richardson, A.E., Djordjevic, M.A., Rolfe, B.G. and Simpson, R.J., 1988a. Effects of pH, Ca and Al on the exudation from clover seedlings of compounds that induce the expression of nodulation genes in Rhizobium trifolii. Plant Soil, 109: 34-47.
Richardson, A.E., Simpson, R.J., Djordjevic, M.A. and Rolfe, B.G., 1988b. Expression of nodulation genes in Rhizobium leguminosarum bv. trifolii is affected by low pH and by Ca and Al ions. Appl. Environ. Microbiol., 54: 2541-2548.
Robert, F.M. and Schmidt, E.L., 1983. Population changes and persistence of Rhizobium phaseoli in soil and rhizospheres. Appl. Environ. Microbiol., 45: 550-556.
Robert, F.M. and Schmidt, E.L., 1985. Response of three indigenous serogroups of Rhizobium japonicum to the rhizosphere of pre-emergent seedlings of soybean. Soil Biol. Biochem., 17: 579-580.
Robson, A.D., 1983. Mineral nutrition. In: W.J. Broughton (Editor), Nitrogen Fixation, Vol. 3, Legumes. Clarendon Press, Oxford, pp. 36-55.

Robson, A.D., 1988. Nutrient requirements of pulses. In: R.J. Summerfield (Editor), World Crops: Cool Season Food Legumes. Kluwer Academic, London, pp. 869-881.
Robson, A.D., 1989. The role of self regenerating pastures in rotation with cereals in Mediterranean areas. In: A.E. Osman, M.M. Ibrahim and M.A. Jones (Editors), The Role of Legumes in the Farming Systems of the Mediterranean Areas. Kluwer Academic, Dordrecht, pp. 217-236.
Robson, A.D. and Abbott, L.K., 1988. Improving the biological and chemical fertility of temperate area soils. In: J. Loveday (Editor), National Soils Conference Review Papers, Aust. Soc. Soil Sci., pp. 73-86.
Robson, A.D. and Loneragan, J.F., 1970a. Nodulation and growth of Medicago truncatula. I. Effect of calcium carbonate and inoculation level on the nodulation of Medicago truncatula on a moderately acid soil. Aust J. Agric. Res., 21: 427-434.
Robson, A.D. and Loneragan, J.F., 1970b. Nodulation and growth of Medicago truncatula. II. Colonization of acid soils by Rhizobium meliloti. Aust J. Agric. Res., 21: 435-445.
Robson, A.D., O'Hara, G.W. and Abbott, L.K., 1981. Involvement of phosphorus in nitrogen fixation by subterranean clover (Trifolium subterraneum L.). Aust. J. Plant Physiol., 8: 427-436.
Rupela, O.P., Toomsan, B., Mittal, S., Dart, P.J. and Thompson, J.A., 1987. Chickpea Rhizobium populations: survey of influence of season, soil depth and cropping pattern. Soil Biol. Biochem., 19: 247-252.
Russell, J.S., 1976. Comparative salt tolerance of some tropical and temperate legumes and tropical grasses. Aust. J. Exp. Agric. Anim. Husb., 16: 103-109.
Salema, M.P., Parker, C.A., Kidby, D.K., Chatel, D.L. and Armitage, T.M., 1982. Rupture of nodule bacteria on drying and rehydration. Soil Biol. Biochem., 14: 15-22.
Same, B.I., Robson, A.D. and Abbott, L.K., 1983. Phosphorus, soluble carbohydrates and endomycorrhizal infection. Soil Biol. Biochem., 15: 593-597.
Sanginga, N., Mulongoy, K. and Ayanaba, A., 1989. Effectivity of indigenous rhizobia for nodulation and early nitrogen fixation with Leucaena leucocephala grown in Nigerian soils. Soil Biol. Biochem., 21: 231-235.
Schenck, N.C. and Hinson, K., 1973. Response of nodulating and non-nodulating soybeans to a species of Endogone mycorrhiza. Agron. J., 65: 849-850.
Schwinghamer, E.A. and Belkengren, R.P., 1968. Inhibition of rhizobia by a strain of Rhizobium trifolii: some properties of the antibiotic and of the strain. Arch. Mikrobiol., 64: 1301-45.
Shaw, N.H., Gates, C.T. and Wilson, J.R., 1966. Growth and chemical composition of Townsville lucerne (Stylosanthes humilis). 1. Dry matter yield and nitrogen content in response to superphosphate. Aust. J. Exp. Agric. Anim. Husb., 6: 150-156.
Singleton, P.W. and Bohlool, B.B., 1984. Effect of salinity on nodule formation by soybean. Plant Physiol., 74: 72-76.
Singleton, P.W., Abdel-Magi, H.M. and Tavares, J.W., 1985. Effect of phosphorus on the effectiveness of strains of Rhizobium japonicum. Soil Sci. Soc. Am. J., 49: 613-616.
Singleton, P.W., El Swaify, S.A. and Bohlool, B.B., 1982. Effect of salinity on Rhizobium growth and survival. Appl. Environ. Microbiol., 44: 884-890.
Small, J.G.C. and Joffe, A., 1968. Physiological studies on the genus Trifolium with special reference to the South African species. II. Influence of root temperature on growth, nodulation and symbiotic nitrogen fixation. South African J. Agric. Sci., 11: 41-56.
Smart, J.B., Dilworth, M.J. and Robson, A.D., 1984b. Effect of phosphorus supply on phosphate uptake and alkaline phosphatase activity in rhizobia. Arch. Microbiol., 140: 281-286.

Smart, J.B., Robson, A.D. and Dilworth, M.J., 1984a. A continuous culture study of the phosphorus nutrition of Rhizobium trifolii WU95, Rhizobium NGR234 and Bradyrhizobium CB756. Arch. Microbiol., 140: 276-280.
Smit, G., Kijne, J.W. and Lugtenberg, B.J.J., 1987. Involvement of both cellulose fibrils and a Ca^{2+}-dependent adhesin in the attachment of Rhizobium leguminosarum to pea root hair tips. J. Bacteriol., 169: 4294-4301.
Smit, G., Logman, T.J.J., Boerrigter, M.E.T.I., Kijne, J.W. and Lugtenberg, B.J.J., 1989. Purification and partial characterization of the Rhizobium leguminosarum bv. viciae Ca^{2+}-dependent adhesin, which mediates the first step in attachment of cells of the family Rhizobiaceae to plant root hair tips. J. Bacteriol., 171: 4054-4062.
Somasegaran, P., Hoben, H.J. and Gurgun, V., 1988. Effect of inoculation rate, rhizobial strain competition and nitrogen fixation in chickpea. Agron. J., 80: 68-73.
Sprent, J.I., 1980. Root nodule anatomy, type of export product and evolutionary origin in some Leguminosae. Plant Cell Environ., 3: 35-43.
Sprent, J.I., 1989. Which steps are essential for the formation of functional legume nodules? New Phytol., 111: 129-153.
Sprent, J.I. and Minchin, F.R., 1983. Environmental effects on the physiology of nodulation and nitrogen fixation. In: D.G. Jones and D.R. Davies (Editors), Temperate Legumes. Pitman, London, pp. 269-318.
Sprent, J.I. and Minchin, F.R., 1985. Rhizobium, nodulation and nitrogen fixation. In: R.J. Summerfield and E.H. Roberts (Editors), Grain Legume Crops. Collins, London, pp. 115-144.
Streeter, J.G., 1974. Growth of two soybean shoots on a single root. J. Exp. Bot., 25: 189-198.
Thomson, B.D., Robson, A.D. and Abbott, L.K., 1986. Effects of phosphorus on the formation of mycorrhizas by Gigaspora calospora and Glomus fasciculatum in relation to root carbohydrates. New Phytol., 103: 751-765.
Trinick, M.J., 1980. Relationship among the fast-growing rhizobia of Lablab purpureus, Leucaena leucocephala, Mimosa spp., Acacia farnesiana and Sesbania grandiflora and their affinities with other rhizobial groups. J. Appl. Bact., 49: 39-53.
Triplett, E.W., 1988. Isolation of genes involved in nodulation competitiveness from Rhizobium leguminosarum bv. trifolii T24. Proc. Natl. Acad. Sci. USA, 85: 3810-3814.
Triplett, E.W., 1990. Construction of a symbiotically effective strain of Rhizobium leguminosarum bv. trifolii with increased nodulation competitiveness. Appl. Environ. Microbiol., 56: 98-103.
Triplett, E.W. and Barta, T.M., 1987. Trifolitoxin production and nodulation are necessary for the expression of superior nodulation competitiveness by Rhizobium leguminosarum bv. trifolii T24 on clover. Plant Physiol., 85: 335-342.
Triplett, E.W., Schink, M.J. and Noeldner, K.I., 1989. Mapping and subcloning of the trifolitoxin production and resistance genes from Rhizobium leguminosarum bv. trifolii T24. Mol. Plant Microbe Interact., 2: 202-208.
Truesdell, H.W., 1917. The effect of phosphorus on alfalfa and alfalfa bacteria. Soil Sci., 3: 77-98.
Van Rensberg, H.J. and Strijdom, B.W., 1980. Survival of fast- and slow-growing Rhizobium spp. under conditions of relatively mild desiccation. Soil Biol. Biochem., 12: 353-356.
Vargas, A.A.T. and Graham, P.H., 1988. Phaseolus vulgaris cultivar and Rhizobium strain variation in acid-pH tolerance and nodulation under acid conditions. Field Crops Res., 19: 91-101.
Vargas, A.A.T. and Graham, P.H., 1989. Cultivar and pH effects on competition for nodule sites between isolates of Rhizobium in beans. Plant Soil, 117: 195-200.

Vincent, J.M., 1962. Influence of calcium and magnesium on the growth of Rhizobium. J. Gen. Microbiol., 28: 653-663.
Vincent, J.M. and Waters, L.M., 1954. The root-nodule bacteria as factors in clover establishment in the red basaltic soils of the Lismore district, New South Wales. II. Survival and success of inocula in laboratory trials. Aust. J. Agric. Res., 5: 61-76.
Weber, D.F. and Miller, V.L., 1972. Effect of soil temperature on Rhizobium japonicum serogroup distribution in soybean nodules. Agron. J., 64: 796-798.
Whelan, A.M. and Alexander, M., 1986. Effects of low pH and high Al, Mn and Fe levels on the survival of Rhizobium trifolii and nodulation of subterranean clover. Plant Soil, 92: 363-371.
Wilson, J.R., 1970. Response to salinity in Glycine. VI. Some effects of a range of short-term salt stresses on the growth, nodulation and nitrogen fixaqtion of Glycine wightii (formerly javanica). Aust. J. Agric. Res., 21: 571-582.
Wood, M. and Shepherd, G., 1987. Characterization of Rhizobium trifolii isolated from soils of different pH. Soil Biol. Biochem., 19: 317-321.
Wood, M., Cooper, J.E. and Holding, A.J., 1984. Aluminium toxicity and nodulation of Trifolium repens. Plant Soil, 78: 381-391.
Woomer, P., Singleton, P.W. and Bohlool, B.B., 1988. Ecological indicators of native rhizobia in tropical soils. Appl. Environ. Microbiol., 54: 1112-1116.
Worrall, V.S. and Roughley, R.J., 1976. The effect of moisture stress on infection of Trifolium subterraneum L. by Rhizobium trifolii Dang. J. Exp. Bot., 27: 1233-1241.
Yousef, A.N. and Sprent, J.I., 1983. Effects of NaCl on growth, nitrogen incorporation and chemical composition of inoculated and NH_4NO_3 fertilized Vicia faba L. plants. J. Exp. Bot., 34: 941-950.
Zahran, H.H. and Sprent, J.I., 1986. Effects of sodium chloride and polyethyleneglycol on root hair infection and nodulation of Vicia faba L. plants by Rhizobium leguminosarum. Planta, 167: 303-309.
Zaroug, M.G. and Munns, D.N., 1979. Nodulation and growth of Lablab purpureus (Dolichos lablab) in relation to Rhizobium strain, liming and phosphorus. Plant Soil, 53: 329-339.

CHAPTER 15

ACTINORHIZAL PLANTS: FRANKIA-SYMBIOSES

C.R. SCHWINTZER and J.D. TJEPKEMA
Department of Botany & Plant Pathology, University of Maine, Orono, ME 04469, USA

1 INTRODUCTION

Actinorhizal plants rival legumes in the amount of N_2 that they fix on a global basis, yet basic knowledge of many aspects of their biology and physiology is limited or very recent. A major impediment to their study was a long standing inability to isolate the N_2-fixing microsymbiont Frankia (an actinomycete) and grow it in pure culture. This obstacle was finally overcome in 1978 with the publication of the first confirmed isolation (Callaham et al., 1978) and numerous isolations have followed. Studies of actinorhizal symbioses prior to 1978 have been reviewed by Quispel (1990).

Actinorhizal plants, which were referred to as nonleguminous nitrogen-fixing plants until recently, are dicotyledons belonging to eight families and 25 genera (Table 1). Although taxonomically diverse, actinorhizal plants have some common features. All are perennial and all except Datisca, which has herbaceous shoots, are woody shrubs or trees (Tjepkema et al., 1986). Well-known genera include Alnus (alder), Myrica (e.g. bayberry), Hippophaë (sea buckthorn), Elaeagnus (e.g. autumn olive) and Casuarina (e.g. beef wood or she-oak).

Actinorhizal plants are found on every continent except Antarctica (Baker and Schwintzer, 1990). Their modern distribution reflects both their native habitats and regions where they have been introduced for horticultural purposes, land reclamation, or forestry. Although their distribution is world wide, actinorhizal plants are found primarily in cooler climates. They are common in the north and south temperate zones and occur at high altitudes in the tropics. Casuarina and Myrica are exceptions and include subtropical and tropical species as well (Bond 1976; Silvester, 1977). Actinorhizal plants are especially important in high latitude regions such as Scandinavia, Canada and New Zealand where they grow vigorously under conditions that are not favorable for legumes (Silvester, 1977).

Ecologically, the majority of actinorhizal plants are pioneers on nitrogen-poor, open sites. In late glacial and postglacial times they colonized nitrogen-poor glacial deposits and accelerated soil development by adding nitrogen and organic matter (Silvester, 1977). In modern times actinorhizal plants are found primarily on sandy and gravelly sites, shores of lakes and streams, wetlands, and exposed raw mineral soil (Silvester, 1977). Many of the environments occupied by actinorhizal plants are relatively harsh. For example, water stress and temperature extremes are common at open sites with coarse textured soils; anaerobiosis is common in soils at streamsides and in wetlands; and low temperatures and acidic soils prevail at high latitudes. Thus many actinorhizal plants adapt to a variety of stresses in addition to their ability to fix nitrogen.

TABLE 1 Actinorhizal host plants

Family	Nodulated genera
Betulaceae	Alnus
Casuarinaceae	Allocasuarina, Casuarina, Ceuthostoma, Gymnostoma
Coriariaceae	Coriaria
Datiscaceae	Datisca
Elaeagnaceae	Elaeagnus, Hippophaë, Shepherdia
Myricaceae	Comptonia, Myrica
Rhamnaceae	Ceanothus, Colletia, Discaria, Kentrothamnus, Retanilla, Talguenea, Trevoa
Rosaceae	Cercocarpus, Chamaebatia, Cowania, Dryas, Purshia, Rubus

Modified from Baker and Schwintzer (1990)

Economically, actinorhizal plants have potential as timber and fuelwood, in biomass production, land reclamation and amenity planting. In the Pacific Northwest of North America Alnus rubra is used as sawtimber and pulpwood. It also serves as an important alternative crop in areas infected with conifer root diseases (Hibbs and Cromack, 1990). In highland areas of Central and South America, Alnus acuminata is used in pastures to provide shade for livestock while producing fuelwood and timber (Dawson, 1990). In Europe Alnus, Elaeagnus, and Hippophaë are used widely for reclamation of mine spoils and land stabilization. In addition a wide variety of actinorhizal plants is used in horticulture for amenity planting (Wheeler and Miller, 1990). In the tropics and subtropics several species of Casuarinaceae play important roles as sources of timber and fuelwood. They are also used as windbreaks, to stabilize sand dunes and rehabilitate waste lands, and for landscaping where they provide shade and beauty (Diem and Dommergues, 1990).

2 FRANKIAE

2.1 Morphology

Frankiae have three morphological forms: hyphae, vesicles, and sporangia.

All three forms can be observed both in pure culture and symbiotically. The septate hyphae are commonly 0.5-2.0 µm in diameter (Lechevalier and Lechevalier, 1990) and form a mycelium on which the vesicles and sporangia are borne.

Vesicles are specialized nitrogen-fixing structures (Newcomb and Wood, 1987; Huss-Danell, 1990) formed at the tips of short branch hyphae. This use of the term "vesicle" differs from that of electron microscopists who use the same term to refer to intracellular, membrane-bound structures in eukaryotic and prokaryotic cells (Newcomb and Wood, 1987). In culture, vesicles are septate and lightbulb-shaped with a spherical body and a hyphal stalk. In symbiosis (Fig. 1a), they are septate or nonseptate and take a variety of forms ranging from spherical through club-shaped to filamentous. The degree of septation and the shape are largely determined by the host (Newcomb and Wood, 1987). Frankiae from Casuarina and Allocasuarina apparently do not form vesicles within the nodule (Berry and Sunell, 1990) although frankiae isolated from these nodules form spherical vesicles when induced for nitrogen fixation in pure culture (Gauthier et al., 1981; Murry et al., 1985). The vesicles are surrounded by a lipid-rich, laminated wall that probably impedes the diffusion of oxygen to the site of nitrogenase activity (Silvester et al., 1990).

Sporangia and spores are produced both in pure culture and in symbiosis (Fig. 1b). They are borne either terminally or intrahyphally. Individual sporangia contain a few to several hundred spores each and may reach sizes of 60 µm in length (Newcomb and Wood, 1987; Torrey, 1987). Formation of sporangia within nodules requires an appropriate host and the presence of a Frankia strain capable of sporulating in planta. Nodules without sporangia are more common than those with sporangia (Schwintzer, 1990).

Mature spores are released directly into the medium by frankiae in culture or from nodules by decay. They are nonmotile, spherical or angular in shape, and vary in size from 0.5 to 2.0 µm. Ultrastructurally, they have thick walls and dense cytoplasm containing conspicuous electron translucent granules that are likely to be food reserves (Gardner, 1976; Newcomb and Wood, 1987). They probably serve as resistant propagules that persist in the soil and are able to infect further generations of host plants (van Dijk, 1979). More detail on the morphology and fine structure of Frankia is given in a recent review by Newcomb and Wood (1987).

2.2 Isolation

Numerous strains of Frankia have been isolated. Most are from species of Alnus, but isolates have also been obtained from several other genera (Baker, 1989). In spite of this success, strains of Frankia have not yet been isolated

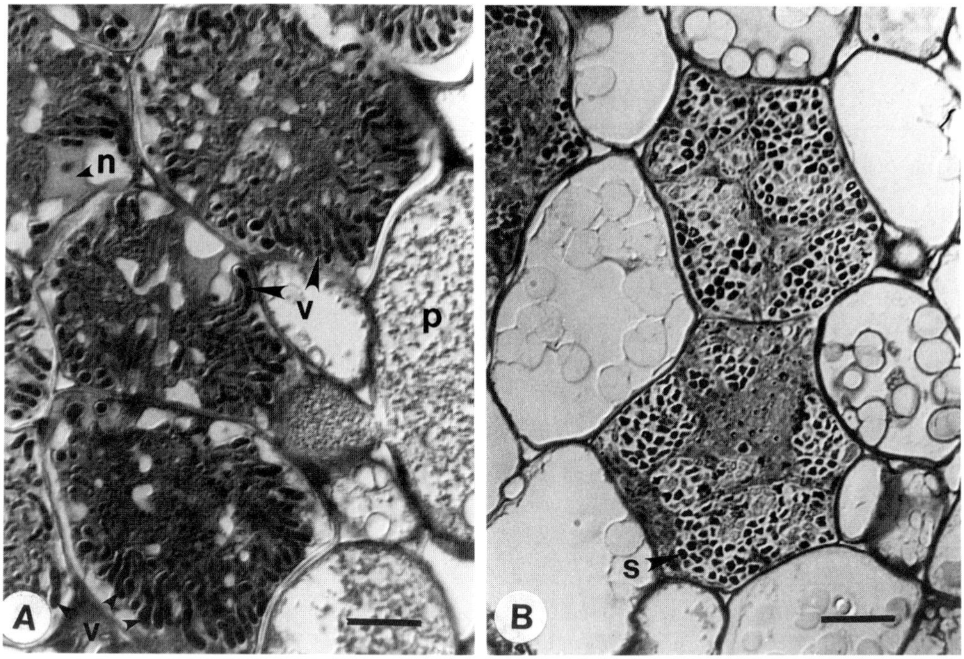

Fig. 1. Frankia within mature cells in Myrica gale stained with toluidine blue. (a) Vesicles (v), the sites of nitrogenase activity, are darkly staining, club-shaped termini of peripheral branch hyphae. Cells containing vesicles are living and contain a nucleus (n). Accumulations of polyphenols (p) are commonly found in uninfected cells. (b) Sporangia containing numerous darkly staining, polyhedral spores (s). The length of the bar is 10 µm. Reproduced with permission from Schwintzer et al. (1982).

from some genera of actinorhizal plants. In particular, strains which are infective on genera of the Rosaceae and Rhamnaceae are lacking (Baker, 1989). Difficulties in isolating frankiae are apparently caused by their slow growth rates and, in some cases, by as yet unidentified nutritional requirements (Baker, 1989; Lechevalier and Lechevalier, 1990).

Most protocols for isolating frankiae involve surface sterilizing the nodule, homogenizing or slicing the nodule aseptically to release the microsymbiont, and incubating vesicle clusters or nodule pieces in agar or liquid growth media. The probability of obtaining an isolate can be increased by using more than one medium for each isolation attempt. Plates or tubes are then examined at intervals of a few days, weeks, or even months, for outgrowth of colonies (Baker, 1989; Lechevalier and Lechevalier, 1990). Although many frankiae can be isolated on relatively simple media, some strains from Alnus glutinosa require inclusion of a lipid found in root extracts. This has recently been identified as dipterocarpol (Quispel et. al., 1989).

Frankiae can be maintained on a variety of defined and complex media. Just as media requirements for isolating frankiae vary with the strain, so do the requirements for maintenance media. The compositions of commonly used isolation and maintenance media and details of various isolation procedures are given by Baker (1989) and Lechevalier and Lechevalier (1990).

2.3 Genetics

The study of the molecular genetics of Frankia has been difficult because frankiae are multicellular, grow very slowly, and to date there is no vector to introduce and replicate DNA within Frankia cells (Mullin and An, 1990). However, substantial progress has been made in some areas. Great genetic diversity has been shown to exist among frankiae by DNA-DNA hybridization of total cellular DNA's. Hybridization values range from 97% to 10% in Frankia strains from 12 different host species located in five different countries (An et al., 1985). Interestingly, a similar range of hybridization values also exists in Frankia strains isolated from a single host species in the state of New Jersey (U.S.A.) (Bloom et al., 1989).

Considerable effort has been devoted to searching for plasmids in Frankia and although plasmids of varying sizes have been found (Normand et al., 1983; Simonet et al., 1985; Bloom et al., 1989), no functions have yet been assigned to them (Mullin and An, 1990).

Substantial progress has been made in identifying and describing the nif genes of Frankia. All three of the structural genes for the Mo nitrogenase complex, nifD, nifK and nifH, are present in Frankia (Ruvkun and Ausubel, 1980) and have been partially characterized (Mullin and An, 1990). In addition nifA and nifB hybridizing regions have been found (Simonet et al., 1988).

Little progress has been made in identifying nod genes in Frankia. Attempts to probe Frankia DNA for sequences that hybridize with known nodulation genes in Rhizobium and Bradyrhizobium have been largely unsuccessful (Mullin and An, 1990). However, use of functional complementation shows some promise. Frankia DNA is able to complement a Nod$^-$ Rhizobium mutant and some of the resulting transconjugants are able to nodulate their host plant (Reddy et al., 1988).

Additional information on the molecular genetics of *Frankia* is given in recent reviews by Simonet et al. (1989) and Mullin and An (1990).

2.4 Taxonomy

Frankia belongs to the family Frankiaceae together with the closely related genus *Geodermatophilus* (Hahn et al., 1989). Like other actinomycetes, frankiae are filamentous, branching, Gram-positive bacteria. Unique among actinomycetes, *Frankia* is characterized by formation, *in submerged liquid culture*, of sporangia filled with spores and by formation of nitrogen-fixing vesicles under appropriate cultural conditions (Lechevalier and Lechevalier, 1990). Thus an organism can be readily identified as belonging to the genus *Frankia*, but identification of species is difficult. Although there has been a recent proposal to assign some frankiae to species (Lalonde et al., 1988), most workers in the field prefer not to do so until more reliable criteria for species definition have been developed (Lechevalier and Lechevalier, 1990). Instead frankiae are designated as *Frankia* sp. and given an acronym of three letters followed by up to 10 numbers, the first four of which represent the plant genus and species from which the strain was isolated (Lechevalier, 1983).

Even though it has not been possible to identify species, frankiae can be separated into two broad groups designated groups A and B. Group A strains are genetically very heterogeneous, relatively aerobic, capable of using carbohydrates, and grow relatively rapidly. Further, they are usually not infective on the host plant from which they were isolated. Group B strains, in contrast, are much more homogeneous and are genetically, serologically, and chemically related. They are also less tolerant of O_2, use carbohydrates poorly, grow relatively slowly, and are generally infective on their host of origin (Lechevalier and Lechevalier, 1990).

2.5 Ecology

Little is known about the ecology of free-living frankiae. It is almost certain that at least some frankiae are able to grow saprophytically in soil because frankiae have been isolated from root zone soil (Baker and O'Keefe, 1984) and many frankiae can be cultured *in vitro* on relatively simple media. Under these conditions they make substantial growth, fix dinitrogen, and sporulate. Moreover, infective *Frankia* particles are present in most soils including soils that have never been known to support host plants or have been devoid of hosts for many years (Baker and Schwintzer, 1990). Birch, a nonhost plant that is closely related to alder (Bousquet and Lalonde, 1990), appears to provide an especially favourable environment for frankiae in the soil because more infective

Frankia particles occur in soils under birch than under alder itself (van Dijk, 1984; Smolander and Sundman, 1987). Moderate soil pH favours survival of frankiae added to soils whereas low pH is detrimental (Smolander et al., 1988). Moreover, the number of infective Frankia particles found at field sites devoid of actinorhizal plants is positively correlated with the pH (Smolander and Sundman, 1987).

3 ACTINORHIZAL ROOT NODULE SYMBIOSES

3.1 Morphology, initiation and development

(i) Morphology and anatomy. Actinorhizal nodules are perennial clusters of modified lateral roots. Each nodule lobe has a central vascular cylinder, and the infected cells are confined to the surrounding cortex. Depending on the host species, the lobes may be loosely clustered (as in Comptonia peregrina) or tightly packed (as in Alnus spp.) (Berry and Sunell, 1990). In most species the apex of the nodule lobe is terminal, but in the Myricaceae and in some Casuarinaceae uninfected roots, termed nodule roots, arise from the apices of the nodule lobes (Newcomb and Wood, 1987).

In nitrogen-fixing cortical tissue, infected cells are interspersed with uninfected cells. The infected cells are living and have a large fraction of their volume occuppied by Frankia hyphae and vesicles. In most cases the Frankia vesicles are peripheral but in Datisca and Coriaria the vesicles are central. Frankia is isolated from the the host cytoplasm by the host plasmalemma and by host-derived encapsulation material that is continuous with the host cell wall (Berry and Sunell, 1990; Huss-Danell, 1990).

Within each nodule lobe the host cells and the microsymbiont are younger near the apex of the lobe than at the base. In growing nodule lobes, new host cells are formed by an apical meristem and are invaded by frankial hyphae when they are partially expanded. The hyphae proliferate within the expanding host cells and vesicles differentiate as the cells mature. In spore-containing nodules, sporangia begin to develop in the region where the vesicles have just matured (van Dijk and Merkus, 1976; VandenBosch and Torrey, 1985). The host cell rapidly senesces following the onset of sporangial differentiation and is dead when the sporangia are mature. Mature sporangia are found in the region of vesicle-containing cells and become increasingly common toward the base of the nodule lobe (VandenBosch and Torrey, 1985).

(ii) Seasonal patterns of development and activity. These patterns are the result of interactions between the local climate, the phenology of the host plant and the Frankia microsymbiont and thus vary with location and host. The best documented case is that of Myrica gale, a deciduous shrub, in Massachusetts

(Schwintzer, 1979; Schwintzer et al., 1982; Schwintzer and Tjepkema, 1983; Schwintzer et al., 1983). There the Frankia microsymbiont overwinters as hyphae in partially expanded cells near the nodule lobe apex. Vesicles first appear in late spring coincident with the onset of nitrogenase activity, occupy the bulk of the current season's infected tissue in summer, and disappear in autumn as nitrogenase activity ceases and the host cell dies (Schwintzer et al., 1982). Sporangia begin to differentiate in June in vesicle-containing cells and become more numerous in late summer (VandenBosch and Torrey, 1985). Although the nodules are perennial, the nitrogen-fixing cells live for only one season.

In Myrica gale, nitrogenase activity (Fig. 2a) is absent in winter, appears in spring when the leaves unfold, reaches a peak in July or early August when the shoots are fully expanded, and then declines until it is no longer measurable in autumn after all leaves have fallen (Sprent et al., 1978; Schwintzer, 1979; Schwintzer et al., 1982). The seasonal pattern of nodule respiration (Fig. 2a) is similar to that for nitrogenase activity except that respiration continues at low rates when nitrogenase activity is absent. The bulk of nodule respiration is apparently associated with nitrogenase activity and related processes, resulting in a relatively constant energy cost of nitrogen fixation throughout the bulk of the growing season (Fig. 2b). The seasonal patterns of Myrica gale nodules are probably typical of winter-deciduous actinorhizal plants in temperate climates (Tjepkema et al., 1986). Further information on seasonal patterns is given in a recent review by Tjepkema et al. (1986).

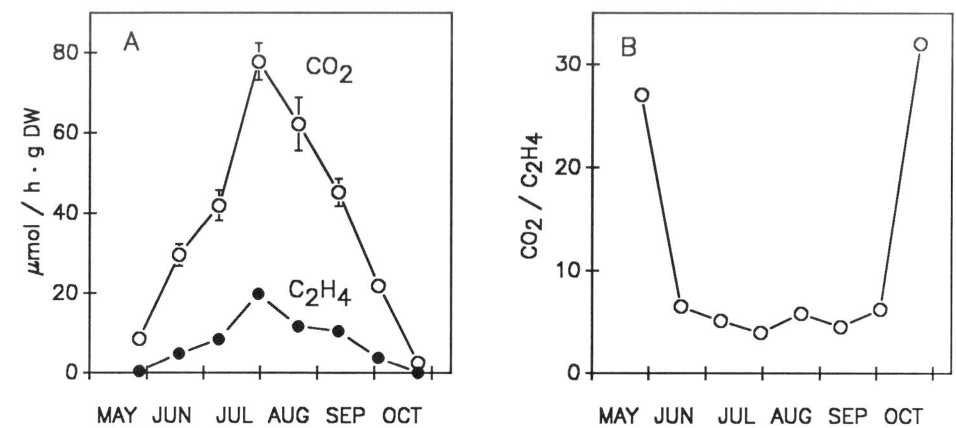

Fig. 2. Seasonal patterns of nitrogenase activity (acetylene reduction), nodule respiration (CO_2 evolution), and the energy cost of nitrogenase activity (CO_2/C_2H_4) in Myrica gale nodules from Tom Swamp in Massachusetts. Modified from Schwintzer and Tjepkema (1983).

(iii) Nodule initiation and development. Nodule formation begins with the invasion of a young host root by *Frankia*. The microsymbiont usually enters the host via deformed root hairs but in some genera of actinorhizal plants it grows between adjacent epidermal cells. Although nodules usually contain only a single *Frankia* strain, more than one strain may be present (Mullin and An, 1990). The exact sequence of events following invasion of the host varies with the host genus (Berry and Sunell, 1990). In all cases the microsymbiont eventually penetrates host cell walls in developing nodule lobes and proliferates within the cells.

Very little is known about the control of infection and development in actinorhizal nodules. Exudates from the host root may help to regulate nodulation (Berry and Sunell, 1990). Plant growth regulators produced by *Frankia* may help to stimulate the meristematic activity that is associated with the early steps of nodule initiation. *Frankia* secretes both cytokinin and auxin (Berry and Sunell, 1990) and elevated levels of cytokinins, auxins, and gibberellins are found in nodules (Wheeler et al., 1979). More information on the infection process and nodule development is given in recent reviews by Newcomb and Wood (1987) and Berry and Sunell (1990).

Both nodule initiation and nodule development are affected by the presence of nitrate or ammonium in the medium, with the size of the effect depending on the host species. In general, low levels of nitrogen (< 10 ppm N) have little if any inhibitory effect on nodule initiation while enhancing nodule development (nodule weight) and plant growth (total plant weight). Combined nitrogen at moderate concentrations often reduces nodule initiation, while development and nodulation is suppressed entirely at sufficiently high concentrations (Bond et al., 1954; MacConnell and Bond, 1957; Kohls and Baker, 1989; Thomas and Berry, 1989). Combined nitrogen apparently affects nodule initiation by more than one mechanism. Stored internal nitrogen in the host plant may help to regulate nodulation (Thomas and Berry, 1989). Moreover, in genera nodulated via root hair deformation, concentrations of nitrate above 20 ppm N probably suppress nodulation by preventing root hair formation in the host plant (Kohls and Baker, 1989).

(iv) Cross-inoculation groups and host specificity. Host-*Frankia* specificity is a complex process that involves the interaction of host and *Frankia* genetics. In addition, our understanding is limited because infective isolates (and many isolates are not infective!) are presently lacking for several host genera and the existing isolates have not been tested on all genera. Many of the existing *Frankia* isolates fit into four host-specificity groups (Baker, 1987). These groups contain strains that nodulate: 1. *Alnus* (Betulaceae) and

Myrica (Myricaceae); 2. Casuarina (Casuarinaceae) and Myrica; 3. Elaeagnus (Elaeagnaceae) and Myrica; and 4. only members of the Elaeagnaceae. Additional groupings may be expected when sufficient infective strains have been isolated from the families that are not part of this scheme.

It is also useful to look at the situation from the point of view of the host. Some hosts are promiscuous, that is they are nodulated effectively by Frankia isolates from many other genera and even families. For example, Elaeagnus angustifolia is nodulated effectively not only by Frankia isolates from Elaeagnus, but also by isolates from six other genera belonging to five famlxies (Baker, 1987). Other promiscuous hosts include Alnus glutinosa, Myrica gale, Myrica cerifera, and Gymnostoma papuanum (Torrey, 1990). It should be noted, however, that promiscuous hosts are not receptive to all Frankia isolates obtained from host species yielding isolates capable of infecting these hosts. For example, Elaeagnus angustifolia is nodulated by only one of four Myrica cerifera isolates tested (Baker, 1987). Other hosts are much more restrictive. For example, Casuarina equisetifolia and Allocasuarina verticillata are nodulated by only two of the Frankia isolates tested to date (Torrey and Racette, 1989). There is also within species variation in host receptivity to nodulation (Bousquet and Lalonde, 1990). More information on cross-inoculation groups and host specificity is given in a recent review by Torrey (1990).

3.2 Carbon metabolism

Relatively little is known about the metabolism of carbon in actinorhizal nodules, either by the host or microsymbiont. It is generally assumed that carbon enters nodules as sucrose (Dixon and Wheeler, 1983; Huss-Danell, 1990), but the form transported into the microsymbiont is uncertain. Group B strains of Frankia use carbohydrates poorly when grown in culture (section 2.4), but whether this is also true in the symbiotic condition is not clear. Vesicle clusters isolated from nodule homogenates can metabolize sugars, but it is still not clear whether sugars are a major carbon source for Frankia in the symbiotic state (Vikman and Huss-Danell, 1987; Huss-Danell, 1990). Although carboxylic acids are another possible carbon source for the microsymbiont, being excellent growth substrates for most pure cultures of Frankia, they are not actively respired by vesicle clusters, with the exception of a preparation from Datisca cannabina. In this case succinate was oxidized, but plant mitochondria were also present (Akkermans et al., 1983). A combination of malate, glutamate, and NAD also stimulates respiration. From this and other evidence a malate-aspartate shuttle has been suggested that might transfer reducing equivalents from the host cytoplasm to the microsymbiont (Akkermans et al., 1981). It should also be noted that NADH is rapidly oxidized by vesicle clusters (Huss-Danell, 1990).

There has been only a single study of the cytochromes of vesicle clusters. Cytochromes a, b, c, and o were found (Ching et al., 1983).

3.3 Nitrogen and hydrogen metabolism

(i) Nitrogenase and hydrogenase. The nitrogenase of *Frankia* appears similar to that from other organisms, both with respect to its biochemical properties (Benson et al., 1979) and its DNA sequence (Mullin and An, 1990). In contrast, there are some differences with respect to hydrogenase when frankiae and rhizobia are compared. Most notably, actinorhizal nodules nearly always possess a vigorous hydrogenase, while hydrogenase is lacking from many strains of rhizobia. Only one strain of *Frankia* has been shown to lack a hydrogenase (Huss-Danell, 1990). A second difference is that the hydrogenase of frankiae is sometimes inhibited by acetylene (Winship et al., 1987). This inhibition can persist for several days after the acetylene is withdrawn.

(ii) Ammonia assimilation. The ammonia resulting from nitrogen fixation is probably assimilated in the host cytoplasm. Evidence for this is the absence of the enzymes needed for ammonia assimilation in vesicle clusters, although this could be due to the inactivation of the enzymes during extraction (Benson and Schultz, 1990). Ammonia excreted by *Frankia* might be assimilated in the host cytoplasm either by the combined action of glutamine synthetase and glutamate synthase or by glutamate dehydrogenase. In spite of a number of studies, it is not yet clear which of these two pathways is used (Huss-Danell, 1990). After assimilation of ammonia, nitrogen is exported from actinorhizal nodules either as amino acids or as citrulline. The form of nitrogen exported depends on the genus and/or species of the host (Schubert, 1986; Huss-Danell, 1990).

3.4 Nitrogen fixation and oxygen

(i) Adaptations in vesicle wall thickness in frankiae. Unlike cultures of rhizobia which require microaerobic conditions for nitrogen fixation, cultures of frankiae fix nitrogen when grown at a wide range of pO_2 (Parsons et al., 1987). At high pO_2, the vesicle walls are very thick and contain a large number of laminated layers, as seen by electron microscopy of freeze-fractured preparations. The number of layers decreases as the pO_2 at which the cultures are grown decreases. At very low pO_2, vesicles are not formed at all (Murry et al., 1985). These results strongly suggest that the laminated layers of the vesicle provide a barrier to inward oxygen diffusion. The vesicles are the site of nitrogenase and are presumed to have a high rate of O_2 uptake which, coupled with the diffusion barrier, may protect nitrogenase from excessive oxygen.

Studies of the rate of O_2 uptake of cultures of _Frankia_ as a function of pO_2 are consistent with the existence of a diffusion barrier in vesicles. A much higher pO_2 is required to saturate respiration if vesicles are present than if they are absent (Murry et al., 1984).

The heterocysts of cyanobacteria appear to be remarkably similar to the vesicles of _Frankia_ with regard to their diffusive resistance to oxygen. Their walls contain a laminated layer very similar in appearance to that seen in vesicle walls; the layer consists of glycolipid with long hydrocarbon chains. Vesicles also contain a high concentration of a potentially similar long chain hydrocarbon (Benson and Schultz, 1990).

(ii) _The O_2 environment in nodules._ Although cultures of frankiae can fix N_2 vigorously under aerobic conditions, it appears that the host tissue sometimes restricts O_2 diffusion into the nodule, resulting in low pO_2 in the cytoplasm of the infected host cells. The best evidence is that strains of _Frankia_ isolated from _Casuarina_ nodules form vesicles when cultured under aerobic conditions, but not in nodules or when cultured at low pO_2 (Silvester et al., 1990). The nodules of _Casuarina_ also contain high concentrations of haemoglobin (see section 3.4.iii). In legume nodules, haemoglobins are thought to function in O_2 transport at low pO_2, and it would seem reasonable to assume a similar function in the nodules of _Casuarina_. Thus both the absence of vesicles and presence of haemoglobin are consistent with a low pO_2 in the nodules of _Casuarina_. A low pO_2 in the region of the infected cells of nodules of _M. gale_ has been measured with a microelectrode, an observation consistent with their high concentrations of haemoglobin (Silvester et al,. 1990).

Whether areas of low pO_2 occur in other actinorhizal nodules is uncertain, since only a few genera have been examined in detail. In _Alnus_ there are lenticels on the nodule surface, and intercellular air spaces are continuous between the soil atmosphere and small groups of infected cells containing _Frankia_. One would expect very little gradient in pO_2 along these intercellular air spaces between the soil atmosphere and the infected cells. However, there are no air spaces between adjacent infected cells and it is possible that a substantial pO_2 gradient could exist between the intercellular air spaces and the areas in the infected cells most distant from the air spaces (Tjepkema, 1988). Thus there could be limited regions of low pO_2 in the infected cells.

In the nodules of _Coriaria_ and _Datisca_, there are large intercellular air spaces between all of the infected cells, and these are continuous with the soil atmosphere (Silvester et al., 1990). This would suggest that these nodules are highly aerated, and that the pO_2 in the infected cells would be close to atmospheric values. However, Silvester and Harris (1989) have shown that oxygen enters the nodules of Coriaria through a single lenticel, which may provide a variable resistance to gas diffusion. The steady state rate of acetylene reduction was relatively unaffected by changes in pO_2 between 5 and 50 kPa. Thus oxygen diffusion into Coriaria nodules may be restricted by the lenticel and the pO_2 in the nodule might be substantially below atmospheric levels.

(iii) Haemoglobins. Nodules of Casuarina and M. gale have high concentrations of haemoglobins approaching those found in legume nodules (Table 2). The purified haemoglobins from these nodules have properties similar to those of leghaemoglobins (Fleming et al., 1987; Appleby et al., 1988; Pathirana, 1989).

TABLE 2 Concentration of total and CO-reactive haem in nodules and roots (± SE).

Species	Total haem	CO-reactive haem
	µmol g^{-1} fresh weight	
Lupinus albus (legume)	197 ± 11	174 ± 1
Myrica gale	127 ± 11	103 ± 7
Casuarina cunninghamiana	96 ± 23	80 ± 4
Comptonia peregrina	32 ± 2	17 ± 1
Alnus rubra	24 ± 1	15 ± 1
Ceanothus americanus	16 ± 3	12 ± 1
Hippophaë rhamnoides	13 ± 1	8 ± 0.6
Elaeagnus angustifiolia	17 ± 1	6 ± 0.5
Coriaria myrtifolia	11 ± 2	4 ± 0.4
Purshia tridentata	8 ± 1	2 ± 0.4
Datisca glomerata	13 ± 1	not detectable
Ceanothus americanus - roots only	4.4 ± 0.4	1.4 ± 0.2
Zea mays - roots	4.1 ± 0.3	1.7 ± 0.1

Modified from Silvester et al. (1990)

The CO-reactive haem quantified in Table 2 is based upon the absorption peak observed at 416 to 420 nm in crude extracts in the presence of carbon monoxide and dithionite. This assay is not specific for haemoglobins and could include other CO-reactive haemoproteins that absorb at these wavelengths. Thus it remains to be established whether the roots of Zea mays or the non-nodulated portion of the root system of Ceanothus americanus (see Silvester et al., 1990) contain any true haemoglobin. In any event, the concentration of CO-reactive haem is low in these roots and similarly low in the nodules of Coriaria, Purshia and Datisca. In the case of Comptonia, Alnus and Eleagnus, the concentration of CO-reactive haem is somewhat higher, and weak absorption spectra corresponding to deoxyhaemoglobin and oxyhaemoglobin have been observed in nodule segments of these genera (Tjepkema, 1983). More work is needed with such nodules to determine whether haemoglobin is in fact present and, if so, its function at such low concentrations.

(iv) Response of nodules to soil pO_2. The optimal pO_2 in the environment for nitrogen fixation by actinorhizal nodules is usually very close to atmospheric levels (Tjepkema et al., 1986). Plants such as M. gale, which grow on wet soils, frequently experience environmental pO_2 well below atmospheric levels, and in such conditions its roots have well-developed internal airspaces and produce upward-growing branch roots reaching better aerated surface zones (Tjepkema et al., 1986). Both of these mechanisms presumably increase the oxygenation of the root system.

Similarly, the upward-growing nodule roots of M. gale serve as aeration systems for the nodule (Tjepkema, 1978). The size of the nodule roots depends on the environment. When plants are grown in liquid culture with their root systems at 2 kPa O_2 the nodule roots are thick and 62 mm long on average but at 21 kPa (approximately atmospheric pO_2) they are thinner and only 14 mm long (Silvester et al., 1988a). Comparable variation in the size of nodule roots with changes in aeration also occurs when plants are grown in soil (Sprent and Scott, 1979; Schwintzer and Lancelle, 1983). In nodules of M. gale, adjustments in the length and thickness of the nodule roots, and possibly in the size of the airspaces within the nodules, allow them to adapt to subatmospheric pO_2 (down to 2 kPa) in the environment with little if any loss of nitrogenase activity (Silvester et al., 1988a). The ability of M. gale to adapt to low pO_2 has limits, however. Seedlings growing in soils with the water table only 3 cm below the surface make much less growth than seedlings growing higher above the water table (Schwintzer and Lancelle, 1983). Moreover, nodules on M. gale plants resuming growth in spring while fully flooded are able to reinitiate nitrogenase activity, but only at a much lower rate. After nine weeks of growth while flooded, these plants are clearly nitrogen deficient and have produced only 70% as much biomass as the unflooded controls (Schwintzer, 1985).

Alders are also frequently associated with wet soils, but are more sensitive to subatmospheric pO_2 than M. gale. In A. glutinosa, growth in liquid culture is severely limited at 1 kPa oxygen and both nitrogen content and dry weight are reduced about 50% at 5 kPa oxygen (McConnell, 1959). Alnus incana ssp. rugosa grows equally well at a pO_2 of 5 kPa as at atmospheric pO_2. The symbiosis of this species, which lacks nodule roots, adapts to low ambient pO_2 by increasing the size of air spaces within the nodule and by reducing the thickness of the laminated envelopes surrounding the Frankia vesicles (Silvester et al., 1988b).

3.5 Energy cost of nitrogen fixation

The energy cost for nitrogen fixation in actinorhizal nodules has been been measured by a variety of methods, all based on the ratio between nitrogen fixation and nodule respiration (Tjepkema, 1985; Tjepkema, 1988). The values found are similar to those reported for legumes, with the minimum experimental values approaching the minimum energy cost predicted by theory. The energy cost of nitrogen fixation in Alnus rubra is surprisingly constant over a range of temperatures and pO_2 that have a marked effect on nitrogenase activity. There is essentially no change in the energy cost as temperature is varied between 16 and 26°C, or pO_2 is varied between 5 and 20 kPa (Winship and Tjepkema, 1985).

3.6 The acetylene reduction assay in actinorhizal plants

In many legume nodules, the rate of acetylene reduction drops markedly after the first few minutes of exposure to acetylene (Minchin et al., 1986). Somewhat similar results are found in actinorhizal nodules, but often the decline in acetylene reduction rate begins after only a minute or two of exposure to acetylene, and is then followed by a recovery over a period of 10 to 80 minutes to 80% or more of the initial maximum rate (Fig. 3, curve A; Tjepkema et al., 1988; Silvester and Winship, 1990). In other cases there is no recovery or limited recovery (Fig. 3, curve B; Tjepkema et al., 1988; Silvester and Harris, 1989). The cause of the decline in acetylene reduction rate is not known, but is

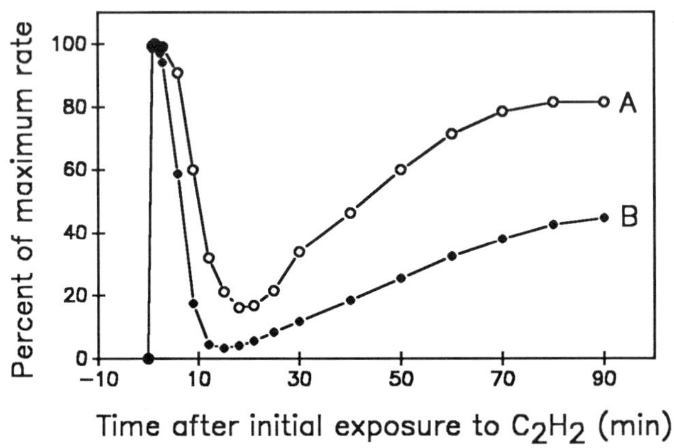

Fig. 3. Acetylene reduction in root systems of two individual Myrica gale plants grown in water culture for 11 wk after inoculation of seedlings with Frankia strain HFP161105. Unpublished data of J.D. Tjepkema and C.R. Schwintzer; methods as in Monz and Schwintzer (1989).

probably not the same as in legumes, because the acetylene-induced decline in actinorhizal plants, in contrast to legumes, is accompanied by only a small decline in nodule respiration (Monz and Schwintzer, 1989). Whatever the cause, the acetylene reduction assay should be used with caution with actinorhizal plants, and studies of activity as a function of time after addition of acetylene should be included (Winship and Tjepkema, 1990).

3.7 Spore-positive and spore-negative nodules

Nodules containing numerous *Frankia* spores are designated sp^+, and those lacking spores or containing only a few are designated sp^- (van Dijk and Merkus, 1976; van Dijk, 1978). Overall, sp- nodules are more common than sp^+ nodules, and the latter have been described in only 9 of the 25 genera of actinorhizal plants. Moreover, even in the genera in which sp^+ nodules do occur, sp^- nodules are generally more frequent (Schwintzer, 1990). *Alnus incana* ssp. *incana* is an exception and commonly bears sp^+ nodules (Weber, 1986; Daniere et al., 1986).

The presence or absence of sporulation within nodules is largely controlled by the *Frankia* strain (van Dijk, 1978; VandenBosch and Torrey, 1985; Monz and Schwintzer, 1989). Thus, pure cultures of *Frankia* strains and crushed nodule inoculum prepared from sp^- nodules induce sp^- nodules, whereas crushed nodule inoculum prepared from sp^+ nodules induces sp^+ nodules. To date, there are no strains in culture that consistently induce sp^+ nodules (Torrey, 1987), although numerous attempts have been made to isolate frankiae from sp^+ nodules (Schwintzer, 1990). The number of sporangia formed within sp^+ nodules is variable and is determined in part by the environment in which the host is growing (Schwintzer, 1990).

(i) Physiology of sp^+ and sp^- nodules. In general, sp^- nodules seem to be superior symbioses and are more effective than sp^+ nodules in supporting growth of the host plant. Overall, sp^- nodules have greater specific nitrogenase activity than sp^+ nodules, although there is considerable variation among strains belonging to the same spore type and some sp^- nodules have low specific activities (Monz and Schwintzer, 1989). In *M. gale*, specific nitrogenase activity is 32% greater on average in nodules induced by four sp^- strains than in nodules induced by five sp^+ strains (Monz and Schwintzer, 1989).

Overall, sp^- nodules also have a lower energy cost of nitrogen fixation. The bulk of the energy cost of nitrogen fixation is due to nodule respiration and sp^+ nodules on average require more respiration to support nitrogen fixation than sp^- nodules. In *M. gale* the ratios of carbon dioxide evolved in respiration to C_2H_2 reduced are 15% lower on average in nodules induced by four sp^{-s}trains than in nodules induced by five sp^+ strains (Monz and Schwintzer, 1989). The percentage biomass invested in nodules in *M. gale* is 31% lower on average in sp^- nodules than in sp^+ nodules (Monz and Schwintzer, 1989). Thus sp^+ nodules have greater energy costs than sp^- nodules due to both respiration and nodule production.

In general, host plant growth under nitrogen-limiting conditions is greater in plants with sp^- nodules than in those with sp^+ nodules (Schwintzer, 1990), so that in M. gale host plant biomass eight weeks after inoculation is 16% greater on average in plants nodulated by sp^- strains than in those with nodules induced by sp^+ strains (Monz and Schwintzer, 1989).

(ii) Ecological distribution of sp^+ and sp^- nodules. Nodule spore type distribution in the field has been studied extensively only in Finland and in Maine (U.S.A.). It appears that several factors interact to determine the distribution of sp^+ and sp^- nodules.

One factor may be host selection of Frankia strains (Weber, 1986), because some strains are able to form nitrogen-fixing nodules in all host species tested within a given genus or host-specificity group and others are not. In particular, some sp^+ strains have only a limited host range (VandenBosch and Torrey, 1983; Weber et al., 1987; Domenache et al., 1988; van Dijk et al., 1988.) Thus, A. incana ssp. incana is compatible with sp^+ strains found in Finnish and French soils and bears predominantly sp^+ nodules, whereas A. glutinosa is not compatible with these particular sp^+ strains and bears primarily sp^- nodules in those countries.

Because nodule spore type in A. incana ssp. rugosa in Maine varies with the length of time the host has occupied the site (Holman and Schwintzer, 1987), this may be a second factor. At sites that have only recently been colonized by alder, the nodules are almost exclusively sp^- and at sites that have long been occupied by alder sp^+ nodules are common. Moreover, Comptonia peregrina bears only sp^- nodules, and is found at locations where it has probably become established only recently (Schwintzer, 1989). It is likely that sp^- Frankia strains can grow saprophytically in soil whereas sp^+ strains can not (Weber, 1986; Holman and Schwintzer, 1987; Schwintzer, 1990). Thus, sp- strains presumably already occur in soils at sites that are being newly colonized by the host plant and consequently sp^- nodules would predominate. In contrast, sp^+ frankiae may be maintained at a given site primarily by growth within nodules; sp^+ nodules would then be prevalent where suitable hosts have long been present (Holman and Schwintzer, 1987).

A third factor may be soil pH. Sp^+ frankiae may be favoured at low pH by production of resistant spores within the protective environment of the nodule, and inhibition of the saprophytic growth of competing sp strains. Evidence for reduced growth by sp^- strains at low pH comes from growth of frankiae in culture and survival of infective particles in soil (Schwintzer, 1990). Consistent with this hypothesis is the fact that in Maine sp^+ nodules become increasingly common in M. gale and A. incana ssp. rugosa as the soil pH decreases (Kashanski and

Schwintzer, 1987). Thus, sp^+ nodules would be expected at sites with low pH and sp^- nodules at sites with higher pH (see Schwintzer, 1990).

4 CONCLUDING REMARKS

Actinorhizal plants are very deserving of the attention they are currently receiving. Not only are they economically and ecologically important, but they are also promising for use in the creation of new nitrogen-fixing symbioses by genetic engineering. Frankiae form symbioses with diverse hosts and can fix nitrogen at atmospheric pO_2 allowing the nodules to be structurally less complex than legume nodules because they do not need to provide a microaerobic environment for the microsymbiont. Thus it is likely that fewer genes would be required to extend this symbiosis to new hosts. Moreover, comparative studies of these symbioses and legumes may lead to important new insights into the basic requirements for successful nitrogen fixing symbioses.

Acknowledgements. We thank our colleagues for sending preprints and reprints. Our research has been supported by U.S. Department of Agriculture Research Grants No. 87-CRCR-1-2431 (JDT) and No. 87-CRCR-1-2440 (CRS).

5 REFERENCES

Akkermans, A.D.L., Huss-Danell, K. and Roelofsen, W., 1981. Enzymes of the tricarboxylic acid cycle and the malate-aspartate shuttle in the N_2-fixing endophyte of *Alnus glutinosa.* Physiol. Plant., 53: 289-294.

Akkermans, A.D.L., Roelofsen, W., Blom, J, Huss-Danell, K. and Harkink, R., 1983. Utilization of carbon and nitrogen compounds by *Frankia* in synthetic media and in root nodules of *Alnus glutinosa, Hippophae rhamnoides,* and *Datisca cannabina*. Can. J. Bot., 61: 2793-2800.

An, C.S., Riggsby, W.S., and Mullin, B.C., 1985. Relationships of *Frankia* isolates based on deoxyribonucleic acid homology studies. Int. J. System. Bacteriol., 35: 140-146.

Appleby, C.A., Bogusz, D., Dennis, E.S. and Peacock, W.J., 1988. A role for hemoglobin in all plant roots? Plant Cell, 11: 359-367.

Baker, D.D., 1987. Relationships among pure cultured strains of *Frankia* based on host specificity. Physiol. Plantarum, 70: 245-248.

Baker, D.D., 1989. Methods for the isolation, culture and characterization of the Frankiaceae: soil actinomycetes and symbionts of actinorhizal plants. In: D.P. Labeda (Editor), Isolation of Biotechnological Organisms from Nature. McGraw-Hill, New York, pp. 213-236.

Baker, D.D. and D. O'Keefe, 1984. A modified sucrose fractionation procedure for the isolation of frankiae from actinorhizal root nodules and soil samples. Plant Soil, 78: 23-28.

Baker, D.D. and Schwintzer, C.R., 1990. Introduction. In: C.R. Schwintzer and J.D. Tjepkema (Editors), The Biology of *Frankia* and Actinorhizal Plants. Academic Press, New York, pp. 1-13.

Benson, D.R., Arp, D.J. and Burris, R.H., 1979. Cell-free nitrogenase and hydrogenase from actinorhizal root nodules. Science, 205: 688-689.

Benson, D.R. and Schultz, N.A., 1990. Physiology and biochemistry of Frankia in culture. In: C.R. Schwintzer and J.D. Tjepkema (Editors), The Biology of Frankia and Actinorhizal Plants. Academic Press, New York, pp. 107-127.
Berry, A.M. and Sunell, L.A., 1990. The infection process and nodule development. In: C.R. Schwintzer and J.D. Tjepkema (Editors), The Biology of Frankia and Actinorhizal Plants. Academic Press, New York, pp. 61-81.
Bloom, R.A., Mullin, B.C., and Tate III, R.L., 1989. DNA restriction patterns and solution hybridization studies of Frankia isolates from Myrica pennsylvanica (bayberry). Appl. Environ. Microbiol., 55: 2155-2160.
Bond, G., Fletcher, W.W., and Ferguson, T.P., 1954. The development and function of root nodules of Alnus, Myrica, and Hippophaë. Plant Soil, 5: 309-323.
Bond, G., 1976. The results of the IBP survey of root-nodule formation in nonleguminous angiosperms. In: P.S. Nutman (Editor), Symbiotic Nitrogen Fixation in Plants. Cambridge Univ. Press, Cambridge, pp. 443-474.
Bousquet, J. and Lalonde, M., 1990. The genetics of actinorhizal Betulaceae. In: C.R. Schwintzer and J.D. Tjepkema (Editors), The Biology of Frankia and Actinorhizal Plants. Academic Press, New York, pp. 239-261.
Callaham, D., Del Tredici, P. and Torrey, J.G., 1978. Isolation and cultivation in vitro of the actinomycete causing root nodulation in Comptonia. Science, 199: 899-902.
Ching, T.M., Monaco, P.A. and Ching, K.K., 1983. Energy status and cytochromes in isolated endophytic vesicle clusters of red alder root nodules. Can. J. For. Res., 13: 921-928.
Daniere, C., Capellano, A. and Moiroud, A., 1986. Dynamique de l'azote dans un peuplement naturel d'Alnus incana (L.) Moench. Acta Oecol.-Oecol. Plant., 7: 165-175.
Dawson, J.O., 1990. Interactions among actinorhizal and associated plant species. In: C.R. Schwintzer and J.D. Tjepkema (Editors), The Biology of Frankia and Actinorhizal Plants. Academic Press, New York, pp. 299-316.
Diem, H.G. and Dommergues, Y.R., 1990. Current and potential uses and management of Casuarinaceae in the tropics and subtropics. In: C.R. Schwintzer and J.D. Tjepkema (Editors), The Biology of Frankia and Actinorhizal Plants. Academic Press, New York, pp. 317-342.
Dixon, R.O.D. and Wheeler, C.T., 1983. Biochemical, physiological and environmental aspects of symbiotic nitrogen fixation. In: J.C. Gordon and C.T. Wheeler (Editors), Biological Nitrogen Fixation in Forest Ecosystems: Foundations and Applications. Martinus Nijhoff/Dr W. Junk, The Hague, pp. 107-171.
Domenach, A.M., Kurdali, F., Danière, C. and Bardin, R., 1988. Determination de l'identité isotopique de l'azote fixé par le Frankia associé au genre Alnus. Can. J. Bot., 66: 1241-1247.
Fleming, A.I., Wittenberg, J.B., Wittenberg, B.A., Dudman, W.F. and Appleby, C.A., 1987. The purification, characterization and ligand-binding kinetics of hemoglobins from root nodules of the non-leguminous Casuarina glauca - Frankia symbiosis. Biochim. Biophys. Acta, 911: 209-220.
Gardner, I.C., 1976. Ultrastructural studies of non-leguminous root nodules. In: P.S. Nutman (Editor), Symbiotic Nitrogen Fixation in Plants. Cambridge University Press, Cambridge, pp. 485-495.
Gauthier, D.L., Diem, H.G. and Dommergues, Y., 1981. In vitro nitrogen fixation by two actinomycete strains isolated from Casuarina nodules. Appl. Environ. Microbiol., 41: 306-308.
Hahn, D., Lechevalier, M.P., Fischer, A. and Stackebrandt, E., 1989. Evidence for a close phylogenetic relationship between members of the genera Frankia, Geodermatophilus, and "Blastococcus" and emendation of the family Frankiaceae. Syst. Appl. Microbiol., 11: 236-242.
Hibbs, D.E. and Cromack, K., 1990. Actinorhizal plants in Pacific Northwest forests. In: C.R. Schwintzer and J.D. Tjepkema (Editors), The Biology of Frankia and Actinorhizal Plants. Academic Press, New York, pp. 343-363.

Holman, R.M. and Schwintzer, C.R., 1987. Distribution of spore-positive and spore-negative nodules of Alnus incana ssp. rugosa in Maine, USA. Plant Soil, 104: 103-111.
Huss-Danell, K., 1990. The physiology of actinorhizal nodules. In: C.R. Schwintzer and J.D. Tjepkema (Editors), The Biology of Frankia and Actinorhizal Plants. Academic Press, New York, pp. 129-156.
Kashanski, C.R. and Schwintzer, C.R., 1987. Distribution of spore-positive and spore-negative nodules of Myrica gale in Maine, USA. Plant Soil, 104: 113-120.
Kohls, S.J. and D.D. Baker, 1989. Effects of substrate nitrate concentration on symbiotic nodule formation in actinorhizal plants. Plant Soil, 118: 171-179.
Lalonde, M., Simon, L., Bousquest, J. and Seguin, A., 1988. Advances in the taxonomy of Frankia: Recognition of species alni and elaeagni and novel subspecies pommerii and vandijkii. In: H. Bothe, F.J. de Bruijn, and W.E. Newton (Editors), Nitrogen Fixation: Hundred Years After. Gustav Fischer, Stuttgart, pp. 671-680.
Lechevalier, M.P., 1983. Cataloging Frankia strains. Can. J. Bot., 61: 2964-2967.
Lechevalier, M.P. and Lechevalier, H.A., 1990. Systematics, isolation, and culture of Frankia. In: C.R. Schwintzer and J.D. Tjepkema (Editors), The Biology of Frankia and Actinorhizal Plants. Academic Press, New York, pp. 35-60.
MacConnell, J.T., 1959. The oxygen factor in the development and function of the root nodules of alder. Ann. Bot. 23: 261-268.
MacConnell, J.T. and Bond, G., 1957. A comparison of the effect of combined nitrogen on nodulation in non-legumes and legumes. Plant Soil, 8: 378-388.
Minchin, F.R., Sheehy, J.E. and Witty, J.F., 1986. Further errors in the acetylene reduction assay: effects of plant disturbance. J. Exp. Bot., 37: 1581-1591.
Monz, C.A. and Schwintzer, C.R., 1989. The physiology of spore-negative and spore-positive nodules of Myrica gale. Plant Soil, 118: 75-87.
Mullin, B.C. and An, C.S., 1990. The molecular genetics of Frankia. In: C.R. Schwintzer and J.D. Tjepkema (Editors), The Biology of Frankia and Actinorhizal Plants. Academic Press, New York, pp. 195-214.
Murry, M.A., Fontaine, M.S. and Tjepkema, J.D., 1984. Oxygen protection of nitrogenase in Frankia sp. HFPArI3. Arch. Microbiol., 139: 162-166.
Murry, M.A., Zang, Z., and Torrey, J.G., 1985. Effect of O_2 on vesicle formation, acetylene reduction and O_2-uptake kinetics in Frankia sp. HFPCcI3 isolated from Casuarina cunninghamiana. Can. J. Microbiol. 31: 804-809.
Newcomb, W. and Wood, S.M., 1987. Morphogenesis and fine structure of Frankia (Actinomycetales): the microsymbiont of nitrogen-fixing actinorhizal root nodules. Int. Rev. Cyt., 109: 1-88.
Normand, P., Simonet, P., Butour, J.L., Rosenberg, C., Moiroud, A., and Lalonde, M., 1983. Plasmids in Frankia sp. J. Bacteriol. 155: 32-35.
Parsons, R., Silvester, W.B., Harris, S., Gruijters, W.T.M. and Bullivant, S., 1987. Frankia vesicles provide inducible and absolute oxygen protection for nitrogenase. Plant Physiol., 83: 728-731.
Pathirana, M.S., 1989. Occurrence of hemoglobin in the nitrogen-fixing root nodules of Myrica gale L. M.S. Thesis, University of Maine.
Quispel, A., 1990. Discoveries, discussions and trends in research on actinorhizal root nodule symbioses before 1978. In: C.R. Schwintzer and J.D. Tjepkema (Editors), The Biology of Frankia and Actinorhizal Plants. Academic Press, New York, pp. 15-33.
Quispel, A., Svendsen, A.B., Schripsema, J., Baas, W.J., Erkelens, C., and Lugtenburg, J., 1989. Identification of dipterocarpol as isolation factor for the induction of primary isolation of Frankia from root nodules of Alnus glutinosa (L.) Gaertner. Mol. Plant-Micro. Int. 2: 107-112.

Reddy, A., Torrey, J.G., and Hirsch, A.M., 1988. Isolation of Frankia strain HFPCcI3 nodulation genes by cross-species complementation of Rhizobium meliloti. In: R. Palacios and D.P. Verma (Editors), Molecular Genetics of Plant-Microbe Interactions. Proceedings of the 4th International Symposium on Molecular Genetics of Plant-Microbe Interactions. APS Press, St. Paul, MN, pp. 213-214.
Ruvkun, G.B. and Ausubel, F.M., 1980. Interspecies homology of nitrogenase genes. Proc. Natl. Acad. Sci. USA, 77: 191-195.
Schubert, K.R., 1986. Products of biological nitrogen fixation in higher plants: synthesis, transport, and metabolism. Ann. Rev. Plant Physiol., 37: 539-574.
Schwintzer, C.R., 1979. Nitrogen fixation by Myrica gale root nodules in a Massachusetts wetland. Oecologia, 43: 283-294.
Schwintzer, C.R., 1983. Nonsymbiotic and symbiotic nitrogen fixation in a weakly minerotrophic peatland. Amer. J. Bot., 70: 1071-1078.
Schwintzer, C.R., 1985. Effect of spring flooding on endophyte differentiation, nitrogenase activity, and shoot development in Myrica gale. Plant Soil, 87: 109-124.
Schwintzer, C.R., 1989. All field-collected actinorhizae examined in Comptonia peregrina and Myrica pennsylvanica in Maine are spore negative. Can. J. Bot., 67: 1460-1464.
Schwintzer, C.R., 1990. Spore-positive and spore-negative nodules. In: C.R. Schwintzer and J.D. Tjepkema (Editors), The Biology of Frankia and Actinorhizal Plants. Academic Press, New York, pp. 177-193.
Schwintzer, C.R. and Lancelle, S.A., 1983. Effect of water-table depth on shoot growth, root growth, and nodulation of Myrica gale seedlings. J. Ecol., 71: 489-501.
Schwintzer, C.R. and Tjepkema, J.D., 1983. Seasonal patterns of energy use, respiration, and nitrogenase activity in root nodules of Myrica gale. Can. J. Bot., 61: 2937-2942.
Schwintzer, C.R., Berry, A.M., and Disney, L.D., 1982. Seasonal patterns of root nodule growth, endophyte morphology, nitrogenase activity, and shoot development in Myrica gale. Can. J. Bot., 60: 746-757.
Silvester, W.B., 1977. Dinitrogen fixation by plant associations excluding legumes. In: R.W.F. Hardy and A.H. Gibson (Editors), A Treatise on Dinitrogen Fixation, Sec. 4. Wiley, New York, pp. 141-190.
Silvester, W.B. and Harris, S.L., 1989. Nodule structure and nitrogenase activity of Coriaria arborea in response to varying pO_2. Plant Soil, 118: 97-109.
Silvester, W.B. and Winship, L.J., 1990. Transient responses of nitrogenase to acetylene and oxygen by actinorhizal nodules and cultured Frankia. Plant Physiol., 92: 480-486.
Silvester, W.B., Whitbeck, J., Silvester, J.K., and Torrey, J.G., 1988a. Growth, nodule morphology, and nitrogenase activity of Myrica gale with roots grown at various oxygen levels. Can. J. Bot., 66: 1762-1771.
Silvester, W.B., Silvester, J.K., and Torrey, J.G., 1988b. Adaptation of nitrogenase to varying oxygen tension and the role of the vesicle in root nodules of Alnus incana ssp. rugosa. Can. J. Bot., 66: 1772-1779.
Silvester, W.B., Harris, S.L. and Tjepkema, J.D., 1990. Oxygen regulation and hemoglobin. In: C.R. Schwintzer and J.D. Tjepkema (Editors), The Biology of Frankia and Actinorhizal Plants. Academic Press, New York, pp.157-176.
Simonet, P., Normand, P., Moiroud, A., and Lalonde, M., 1985. Restriction enzyme digestion patterns of Frankia plasmids. Plant Soil, 87: 49-60.
Simonet, P., Normand, P., and Bardin, R., 1988. Heterologous hybridization of Frankia DNA to Rhizobium meliloti and Klebsiella pneumoniae nif genes. FEMS Microbiol. Lett., 55: 141-146.
Simonet, P., Normand, P., Hirsch, A.M., and Akkermans, A.D.L., 1989. The genetics of the Frankia actinorhizal symbiosis. In: P.M. Gresshoff (Editor), The Molecular Biology of Symbiotic Nitrogen Fixation. CRC Press, Boca Raton, FL, pp. 77-109.

Smolander, A. and Sundman, V., 1987. Frankia in acid soils of forests devoid of actinorhizal plants. Physiol. Plantarum, 70: 297-303.
Smolander, A., van Dijk, C. and Sundman, V., 1988. Survival of Frankia strains introduced into soil. Plant Soil, 106: 65-72.
Sprent, J.I., Scott, R., and Perry, K.M., 1978. The nitrogen economy of Myrica gale in the field. J. Ecol., 66: 657-668.
Sprent, J.I. and Scott, R., 1979. The nitrogen economy of Myrica gale and its possible significance for the afforestation of peat soils. In: J.C. Gordon, C.T. Wheeler, and D.A. Perry (Editors), Symbiotic Nitrogen Fixation in the Management of Temperate Forests. Forest Research Laboratory, Oregon State University, Corvallis, pp. 234-242.
Thomas, K.A. and Berry, A.M., 1989. Effects of continuous nitrogen application and nitrogen preconditioning on nodulation and growth of Ceanothus griseus var. horizontalis. Plant Soil, 118: 181-187.
Tjepkema, J.D., 1978. The role of oxygen diffusion from the shoots and nodule roots in nitrogen fixation by root nodules of Myrica gale. Can. J. Bot., 56: 1365-1371.
Tjepkema, J.D., 1983. Hemoglobins in the nitrogen-fixing root nodules of actinorhizal plants. Can. J. Bot., 61: 2924-2929.
Tjepkema, J.D., 1985. Utilization of photosynthate for nitrogen fixation in seedlings of Myrica gale and Alnus rubra. In: P.W. Ludden and J.E. Burris (Editors), Nitrogen Fixation and CO_2 Metabolism. Elsevier, New York, pp. 183-192.
Tjepkema, J.D., 1988. Oxygen regulation and energy usage in actinorhizal symbioses. In: N.S. Subba Rao (Editor), Biological Nitrogen Fixation - Recent Developments. Oxford & IBH Publishing Co., New Delhi, pp. 311-333.
Tjepkema, J.D., Schwintzer, C.R. and Benson, D.R., 1986. Physiology of actinorhizal nodules. Ann. Rev. Plant Physiol., 37: 209-232.
Tjepkema, J.D., Schwintzer, C.R. and Monz, C.A., 1988. Time course of acetylene reduction in nodules of five actinorhizal genera. Plant Physiol., 86, 581-583.
Torrey, J.G., 1987. Endophyte sporulation in root nodules of actinorhizal plants. Physiol. Plant., 70: 279-288.
Torrey, J.G., 1990. Cross-inoculation groups within Frankia and host-endosymbiont associations. In: C.R. Schwintzer and J.D. Tjepkema (Editors), The Biology of Frankia and Actinorhizal Plants. Academic Press, New York, pp. 83-106.
Torrey, J.G. and Racette, S., 1989. Specificity among the Casuarinaceae in root nodulation by Frankia. Plant Soil, 118: 157-164.
VandenBosch, K.A. and Torrey, J.G., 1983. Host-endophyte interactions in effective and ineffective nodules induced by the endophyte of Myrica gale. Can. J. Bot., 61: 2898-29.
VandenBosch, K.A. and Torrey, J.G., 1985. Development of endophytic Frankia sporangia in field- and laboratory-grown nodules of Comptonia peregrina and Myrica gale. Amer. J. Bot., 72: 99-108.
van Dijk, C., 1978. Spore formation and endophyte diversity in root nodules of Alnus glutinosa (L.) Vill. New Phytol., 81: 601-615.
van Dijk, C., 1979. Endophyte distribution in the soil. In: J.C. Gordon, C. T. Wheeler and D.A. Perry (Editors), Symbiotic Nitrogen Fixation in the Management of Temperate Forests. Forest Res. Lab., Oregon State Univ., Corvallis, pp. 84-94.
van Dijk, C., 1984. Ecological aspects of spore formation in the Frankia-Alnus symbiosis. Ph.D. thesis. State Univ., Leiden, The Netherlands.
van Dijk, C. and Merkus, E., 1976. A microscopical study of the development of a spore-like stage in the life cycle of the root-nodule endophyte of Alnus glutinosa (L.) Gaertn. New Phytol., 77: 73-91.
van Dijk, C., Sluimer, A. and Weber, A., 1988. Host range differentiation of spore-positive and spore-negative strain types of Frankia in stands of Alnus glutinosa and Alnus incana in Finland. Physiol. Plant., 72: 349-358.

Vikman, P.-K. and Huss-Danell, K., 1987. Capacity for hexose respiration in symbiotic Frankia from Alnus incana. Physiol. Plantarum, 70: 349-354.
Weber, A., 1986. Distribution of spore-positive and spore-negative nodules in stands of Alnus glutinosa and Alnus incana in Finland. Plant Soil, 96: 205-213.
Weber, A., Nurmiaho-Lassila, E.-L., and Sundman, V., 1987. Features of the intrageneric Alnus-Frankia specificity. Physiol. Plant., 70: 289-296.
Wheeler, C.T. and Miller, I.M., 1990. Current and potential uses of actinorhizal plants in Europe. In: C.R. Schwintzer and J.D. Tjepkema (Editors), The Biology of Frankia and Actinorhizal Plants. Academic Press, New York, pp. 365-389.
Wheeler, C.T., Henson, I.E. and McLaughlin, M.E., 1979. Hormones in plants bearing actinomycete nodules. Bot. Gaz., 140 (Suppl.): S52-S57.
Winship, L.J. and Tjepkema, J.D., 1985. Nitrogen fixation and respiration by root nodules of Alnus rubra Bong.: effects of temperature and oxygen concentration. Plant Soil, 87: 91-107.
Winship, L.J. and Tjepkema, J.D., 1990. Techniques for measuring nitrogenase activity in Frankia and actinorhizal plants. In: C.R. Schwintzer and J.D. Tjepkema (Editors), The Biology of Frankia and Actinorhizal Plants. Academic Press, New York, pp. 263-280.
Winship, L.J., Martin, K.J. and Sellstedt, A., 1987. The acetylene reduction assay inactivates root nodule uptake hydrogenase in some actinorhizal plants. Physiol. Plantarum, 70: 361-366.

CHAPTER 16

CYANOBACTERIA AND THEIR SYMBIONTS

P. ROWELL and N.W. KERBY
Agricultural and Food Research Council Research Group on Cyanobacteria and Department of Biological Sciences, The University of Dundee, Dundee DD1 4HN, Scotland, UK.

1 INTRODUCTION

The cyanobacteria are phototrophic prokaryotes with a Gram-negative cell wall, chlorophyll *a* and phycobiliproteins. Cyanobacteria are a morphologically diverse group of organisms, ranging from unicellular forms to branched and unbranched filamentous forms, which may differentiate akinetes (spores) and specialized N_2-fixing cells called heterocysts. Their growth is predominantly photoautotrophic and, although some are facultative photo- or chemoheterotrophs, only a limited range of fixed carbon substrates are utilized. They perform oxygenic photosynthesis, using H_2O as the electron donor, and some can also perform anoxygenic photosynthesis using H_2S. Many species fix N_2 aerobically while simultaneously evolving O_2 photosynthetically. Cyanobacteria occupy, and are frequently abundant in, a wide range of habitats and also occur in symbiotic associations. Several detailed accounts of known N_2-fixing cyanobacteria have been published (see Rippka et al., 1979; Stewart, 1980a,b; Stewart and Rowell, 1986; Gallon and Chaplin, 1987). The classification of cyanobacteria is described by Rippka et al. (1979) and Rippka (1988).

This chapter concentrates mainly on more recent work, particularly the integration of N_2 fixation with cellular metabolism. For a broader coverage the reader is referred to reviews by Stanier and Cohen-Bazire (1977), Stewart (1980a, b), Carr and Whitton (1982), Stewart et al. (1983), Fay and Van Baalen (1987), Gallon and Chaplin (1987, 1988), Hallenbeck (1987), Packer and Glazer (1988) and Rai (1989a).

2 PROPERTIES OF NITROGENASE

The cyanobacterial enzyme has similar properties to that from other diazotrophs (Haystead and Stewart, 1972), being composed of an Fe protein and an MoFe protein, and has been purified from Anabaena cylindrica (Hallenbeck et al., 1979). The MoFe protein has a relative molecular mass (M_r) of about 220000, subunits (two of each type) of M_r 55000 and 52800 and Mo, Fe, and acid-labile S molar ratios of 2, 20 and 20, respectively. The partially purified Fe protein has two identical subunits of M_r of approximately 30000. Nitrogenase polypeptides of similar sizes were identified in extracts of isolated heterocysts of Nostoc muscorum (Anabaena 7120) (Fleming and Haselkorn, 1973) and Anabaena variabilis (Peterson and Wolk, 1978a). Nitrogenase complexes have been formed using the Fe protein and MoFe protein from different diazotrophs (Tsai and Mortenson, 1978; Hallenbeck et al., 1979; Hallenbeck, 1987). The A. cylindrica MoFe protein formed active complexes with all the Fe proteins tested, while its Fe protein showed poor activity with the MoFe proteins from other sources. Antiserum raised against the A. cylindrica MoFe protein cross-reacted with nitrogenases from the heterocystous cyanobacteria Mastigocladus laminosus, N. muscorum and A. variabilis, but not with those from the non-heterocystous, filamentous cyanobacterium Plectonema boryanum or from Azotobacter vinelandii and Rhodospirillum rubrum (Hallenbeck, 1987). Although the cyanobacterial nitrogenase proteins have not been extensively characterized, the three nif structural genes from Anabaena 7120 have been sequenced and the amino acid sequences predicted (Mevarech et al., 1980; Mazur and Chui, 1982; Lammers and Haselkorn, 1983).

Kentemich et al. (1988) recently demonstrated the occurence of an alternative nitrogenase (see Chapter 3) in A. variabilis. Cultures supplemented with V grew at similar rates to those given Mo, and showed a light-dependent formation of ethane (in addition to ethylene) from acetylene, accompanied by a higher rate of H_2 evolution. This suggests the presence of a V nitrogenase in cyanobacteria consistent with previous reports of the isolation of W-resistant mutants of N. muscorum (Singh et al., 1978).

3 SEPARATION OF N_2 FIXATION AND PHOTOSYNTHETIC O_2 EVOLUTION.

The ability of cyanobacteria to fix N_2 aerobically is usually, but not always, associated with the presence of heterocysts, which are morphologically distinct cells developing in an ordered sequence in the filaments of many cyanobacteria when cultured in the absence of combined N. Cyanobacterial nitrogenase, like that from other diazotrophs, is irreversibly destroyed by exposure to O_2. In heterocystous species N_2 fixation and photosynthetic O_2 evolution occur in different cells, whereas in many non-heterocystous species these processes can be temporally separated.

3.1 Temporal separation.

The first reports of aerobic N_2 fixation by non-heterocystous cyanobacteria were for unicellular Gloeocapsa spp. (Wyatt and Silvey, 1969; Rippka et al., 1971), subsequently reclassified as Gloeothece (Rippka and Waterbury, 1977), and an Aphanothece sp. (Singh, 1973). Stewart and Lex (1970) demonstrated nitrogenase activity in P. boryanum under microaerobic conditions. When incubated under N_2 at low pO_2 in low light, P. boryanum grows exponentially with N_2 fixation and photosynthesis apparently occurring simultaneously, whereas at high light there is a cyclic degradation and resynthesis of phycobiliprotein corresponding to periods of nitrogenase activity and photosynthetic O_2 evolution, respectively (Weare and Benemann, 1974; Giani and Krumbein, 1986). Using an antiserum against the MoFe protein, Smoker et al. (1989) showed the enzyme to be present in all cells of N_2-fixing P. boryanum, with no specific sub-cellular localization. Nitrogenase is widespread amongst non-heterocystous cyanobacteria (Rippka and Waterbury, 1977), although in most cases the enzyme is normally inactivated in vivo under aerobic conditions, due to its sensitivity to O_2. Aerobic N_2 fixation is therefore not a widespread property of non-heterocystous cyanobacteria (see Gallon and Chaplin, 1988).

When Gloeothece was subjected to alternating light/dark cycles (12 h L/12 h D), nitrogenase activity occurred almost exclusively in the dark (Mullineaux et al., 1981), even though this organism can grow and fix N_2 in continuous light. This represents a temporal separation of photosynthetic O_2 evolution and N_2 fixation. Similar diurnal changes were demonstrated in natural environments dominated by Nostoc (acetylene reduction mainly in the light) and by Gloeothece or Oscillatoria (acetylene reduction mainly in the dark) (Griffiths et al., 1987). Thus, temporal separation is physiologically important, rather than an artefact of laboratory culture. However, diurnal patterns of nitrogenase activity associated with Oscillatoria growing in microbial mats in North Sea sediments (Stal and Heyer, 1987) were not typical of aerobic laboratory cultures of either Oscillatoria (Stal and Krumbein, 1985) or heterocystous species. N_2 fixation by cyanobacteria, mainly in the dark period during growth in L/D cycles, has also been demonstrated for Synechococcus spp. (Huang and Chow, 1986; Mitsui et al., 1986), Oscillatoria spp. (Stal and Krumbein, 1985) and Microcoleus chthonoplastes (Pearson et al., 1981). In contrast, Oscillatoria (Trichodesmium) erythrea (Bryceson and Fay, 1981), like heterocystous cyanobacteria, fixes N_2 mainly during the light period. In a survey of diazotrophic cyanobacteria grown in alternating 12 h L/D cycles, (Khamees et al., 1987), the organisms were grouped according to whether nitrogenase activity occurred mainly or exclusively in the dark (9 Gloeothece strains, Aphanothece pallida and an Oscillatoria sp., ie. all the non-heterocystous species tested), mainly in the light (10 Anabaena strains, Aphanizomenon flos-aquae, Fremyella diplosiphon, 2 Nodularia strains and a Nostoc sp.) or in both the light and dark periods (10 Nostoc strains, Scytonema javanicum and Gloeotrichia echinulata).

N_2-fixing cultures of Oscillatoria (Stal and Krumbein, 1985, 1987) and a Synechococcus sp. (Mitsui et al., 1987) continue to show cycles of nitrogenase activity, which correspond with periods of net O_2 uptake, when transferred from L/D cycles to continuous light. In synchronous cultures of Synechococcus grown on L/D cycles or in continuous light, the temporal separation of N_2 fixation and photosynthetic O_2 evolution may be due to the confinement of these processes to different phases of the cell division cycle (Mitsui et al., 1986; 1987). This is not the case for Gloeothece (Gallon et al., 1988), since the diurnal pattern of N_2 fixation occurs in asynchronous cultures with doubling times in excess of the period of the diurnal changes in nitrogenase activity and synthesis, glucan synthesis and the ADP/ATP ratio. In Gloeothece, the degradation of endogenous reserves (glucan) via respiratory metabolism provides reductant and ATP for nitrogenase, in both the light and dark (Mullineaux et al., 1980; Maryan et al., 1986), as in heterocystous cyanobacteria in the dark (Ernst et al., 1984).

N_2 fixation by Trichodesmium may be associated with the formation of bundles of trichomes providing spatial separation from photosynthetic O_2 evolution (Fogg, 1974; Carpenter and Price, 1976; Bryceson and Fay, 1981). However, Ohki and Fujita (1988) showed aerobic N_2 fixation by axenic Trichodesmium independent of the formation of aggregates. Additionally, no evidence for cellular or intracellular differentiation within aggregates of Microcoleus chthonoplastes was found (Malin and Pearson, 1988) and no specific localization of nitrogenase was detected immunocytochemically within a Trichodesmium colony (Paerl et al., 1989). However, O_2-depleted microzones occur within aggregates (Paerl and Bebout, 1988), suggesting (Paerl et al., 1989) that conditions permitting Trichodesmium to aggregate may allow nitrogenase to function in the interior of such structures. N_2 fixation by Trichodesmium is of particular interest because of its contribution to N inputs in tropical and subtropical oceanic waters (Capone and Carpenter, 1982).

When grown on a L/D cycle, de novo enzyme synthesis is required for each cycle of nitrogenase activity in Oscillatoria (Stal and Krumbein, 1985) and in Gloeothece (Gallon et al., 1988). In Gleothece, nitrogenase synthesis preceded activity by about 3 h and activity correlated with glucan degradation in the dark periods (Gallon et al., 1988). Respiratory O_2 uptake is important for the protection of nitrogenase in Gloeothece (Gallon and Hamadi, 1984) and there is a rapid inactivation of the enzyme at the end of each dark period. Calcium is required for respiratory protection of nitrogenase in Gloeothece (Gallon and Hamadi, 1984) and Synechococcus RF-1 (Chen et al., 1988; 1989a), although its precise role is unknown. On transfer of Gloeothece from air to O_2 in the light (Maryan et al., 1986), nitrogenase synthesis is only transiently repressed, but the enzyme remains inactive, whereas an Oscillatoria species shows a recovery of nitrogenase activity in O_2 following transient repression (Gallon and Chaplin, 1988). Synechococcus RF-1 synthesises nif mRNA continuously when grown in continuous light, but only in the dark periods of a L/D cycle (Huang and Chow, 1990). The repression of nitrogenase synthesis in response to O_2 has been associated with a decrease in the negative supercoiling of the nif genes through inhibition of DNA gyrase (Kranz and Haselkorn, 1986). There is evidence that nif DNA in Gloeothece is more negatively supercoiled under microaerobic than under aerobic conditions (Gallon and Chaplin, 1988; Gallon et al., 1989). Changes in DNA gyrase activity in extracts of Gloeothece are consistent with transient relaxation of DNA and could explain the transient O_2 inhibition of nitrogenase synthesis (Maryan et al., 1986). NifAL genes, which function in the regulation of other nif genes in Klebsiella pneumoniae, have not been identified in cyanobacteria.

3.2 Spatial separation

Heterocystous cyanobacteria fix N_2 aerobically and anaerobically. Heterocysts are produced in a regular pattern and may be intercalary or terminal, occurring at one or both ends of a trichome. They are readily distinguished microscopically, with a thick envelope, relatively small septa connecting them to the neighbouring vegetative cells, a different arrangement of thylakoid membranes from vegetative cells and extensive polar plugs containing cyanophycin (see Wilcox et al., 1975; Haselkorn, 1978; Stewart, 1980a, b; Adams and Carr, 1981; Wolk, 1982). Carr (1983, 1988) suggests that cyanophycin may act as a dynamic N reservoir which shows a rapid turnover in heterocysts. There is now direct evidence that under aerobic conditions heterocysts contain nitrogenase activity and proteins and are the only N_2-fixing cells (Fay et al., 1968; Stewart et al., 1969; Van Gorkom and Donze, 1971; Fleming and Haselkorn, 1973; Peterson and Burris, 1976; Tel-Or and Stewart, 1976; Peterson and Wolk, 1978a; Murry et al., 1984a; Bergman et al., 1986). With the development of more suitable procedures for isolation and assay (at low pO_2, in the light and under H_2), the nitrogenase activity of isolated heterocysts rose from initial low values (Stewart et al., 1969) to levels able to account for most of the activity of intact filaments (Peterson and Wolk, 1878a).

Heterocysts isolated from _A. variabilis_ incorporate radioactively labelled amino acids into a number of proteins, including the two components of nitrogenase (Janaki and Wolk, 1982), although the pattern of labelling differed from that of whole filaments labelled with $H^{14}CO_3^-$. Protein synthesis was stimulated in the dark by O_2 and by illumination (possibly due to cyclic photophosphorylation), but not by H_2. After labelling with ^{55}Fe, the heterocysts of _A. variabilis_ were shown to contain 91% and 69%, respectively, of the total MoFe and Fe proteins found in extracts of whole filaments (Peterson and Wolk, 1978a). The maintenance of nitrogenase activity in mature heterocysts under aerobic conditions requires continuous enzyme synthesis (see Weare and Benemann, 1973a; Ohmori and Hattori, 1974).

Whether nitrogenase is synthesized in vegetative cells is controversial; it cannot be ruled out, particularly under low pO_2. A non-heterocystous strain (_Anabaena_ 7118) unable to fix N_2 aerobically and producing no morphologically discernible proheterocysts nevertheless synthesized nitrogenase anaerobically. However, differentiation in this strain may have been arrested at a very early stage, since early metabolic changes have been identified in _A. cylindrica_ prior to the formation of discernible proheterocysts (Bradley and Carr, 1977). Early proheterocysts may be the sites of nitrogenase synthesis in nostocacean strains in which the enzyme is induced anaerobically (Rippka and Stanier, 1978). Immunocytochemical evidence indicates that in _A. cylindrica_ nitrogenase is restricted to heterocysts, even under strictly anaerobic conditions (Murry et al., 1984a).

The heterocyst envelope consists of inner laminated, central homogeneous and outer fibrous layers (Lang and Fay, 1971). Mutants of _A. variabilis_ deficient in the glycolipids of the inner laminated layer showed microaerobic but not aerobic nitrogenase activity (Haury and Wolk, 1978). Strains of _Anabaena_ 7120 defective in glycolipid and outer polysaccharide layers have also been used to show that these layers restrict diffusion of O_2 (Murry and Wolk, 1989). Heterocyst differentiation under strictly anaerobic conditions proceeds only to proheterocysts, which are incapable of protecting nitrogenase against inactivation in air (Rippka and Stanier, 1978). These proheterocysts cannot form the glycolipid layer, probably because O_2 is required for hydroxylation of the glycolipids (Wolk, 1982). The ability of heterocysts to protect nitrogenase against O_2 is dependent on the presence of O_2 during differentiation (Weare and Benemann, 1973b; Rippka and Stanier, 1978; Murry et al., 1984b). Synthesis of heterocyst envelope glycolipids and appearance of nitrogenase activity in _Cylindrospermum licheniforme_ occur simultaneously (Van de Water and Simon, 1982), even though nitrogenase synthesis may have started earlier, as in some _Anabaena_ spp. (Fleming and Haselkorn, 1974; Rippka and Stanier, 1978). The low affinity, diffusion-limited component of respiratory O_2 uptake (Jensen and Cox, 1983) was lost in mutant strains deficient in the major glycolipid component or with a polysaccharide layer that is only loosely attached (Murry and Wolk, 1989).

Direct evidence that the heterocyst envelope acts as an O_2 diffusion barrier was obtained by measuring the rate of gas pressure rise required to collapse gas vesicles in heterocysts of Anabaena flos-aquae (Walsby, 1985). The permeability of heterocysts is such that it allows a sufficient rate of entry of N_2 for nitrogenase activity and a low enough rate of O_2 entry for respiration to maintain an internal O_2 concentration near zero, thus protecting nitrogenase. Walsby (1985) suggests that the main route of gas diffusion is via the envelope rather than the terminal pore, although no data are available for the permeability of the membranes at the heterocyst/vegetative cell junction or for restriction of diffusion by the polar plug. The dependence of nitrogenase activity in A. cylindrica on the external concentration of acetylene (Hallenbeck et al., 1979; Murry et al., 1984b) is also consistent with the heterocyst envelope acting as a barrier to gas diffusion.

In A. variabilis, nitrogenase activity can be correlated with the degradation of glycogen (Ernst et al., 1984; Ernst and Böger, 1985). The rate of loss of nitrogenase activity when A. variabilis is transferred from light to dark in air correlates with the glycogen content, being most rapid in older cultures with low glycogen contents (Ernst et al., 1984). This is consistent with the observation that nitrogenase becomes more O_2-sensitive as cultures of A. cylindrica age (Weare and Benemann, 1973a,b).

In some cyanobacteria nitrogenase synthesis is not repressed by O_2. In mutant strains of A. variabilis which show no nitrogenase activity aerobically, synthesis of the enzyme appears to be insensitive to O_2, since the MoFe protein can be detected immunochemically at levels similar to those in microaerobically grown cultures (Murry, 1983). This may suggest the presence of an O_2-stable precursor. Anabaena CA, Anabaena 1F and A. variabilis show a rapid, reversible inactivation of nitrogenase following their transfer from air to O_2 (Pienkos et al., 1983). Nitrogenase activity was less O_2-sensitive in extracts of Anabaena 1F prepared from cells previously exposed to 100% O_2 for several hours. Adaptation of A. cylindrica to high pO_2 involves a light-dependent induction of a superoxide dismutase and a catalase, protecting cells from photooxidation (Mackey and Smith, 1983).

An uptake hydrogenase (see section 5.3) in heterocysts (Peterson and Wolk, 1978b) is coupled to O_2 reduction (Peterson and Burris, 1978) and should protect nitrogenase, though this does not appear to be true in Anabaena CA (Smith et al.,

1988). On exposure to air, the loss of nitrogenase activity in an O_2-sensitive mutant of this strain with elevated uptake hydrogenase activity was accompanied by destruction of the MoFe protein. However, in the parent strain and a Hup^- mutant, nitrogenase activity decreased then recovered with no apparent change in the MoFe protein.

An alteration of the Fe protein may be involved in protection against O_2 in Anabaena spp. (Smith et al., 1987; Tabita et al., 1988; Reich and Böger, 1989). The Fe protein was rapidly and reversibly converted to a form with a higher M_r when N_2/CO_2-grown Anabaena CA was treated with O_2 (Smith et al., 1987; Tabita et al., 1988). Chloramphenicol did not inhibit the modification or its reversal. Since the modified (higher M_r) form was present in air-grown cultures and those which had recovered from O_2 inactivation of nitrogenase, the modification could not have inactivated the Fe protein. It is not yet known whether this modification is enzymic or whether it allows the enzyme to function in the presence of O_2.

3.3 Heterocyst development

Heterocyst differentiation (see Wilcox et al., 1975; Haselkorn, 1978; Adams and Carr, 1981; Wolk, 1982) is usually induced by transfer of ammonia-grown filaments to nitrogen-free medium. Early stages of induction appear to involve all cells. In A. cylindrica (18 h generation time), commitment of some cells to proheterocyst formation requires light for 0.5 to 1 h and 3 to 5 h for mature heterocyst formation (Bradley and Carr, 1977). Phycocyanin content and the rate of CO_2 fixation fall markedly, with activities of ribulose-1,5-bisphosphate carboxylase/oxygenase (RuBisCO) and phosphoribulokinase (PRuK) decreasing within 5 h (Bradley and Carr 1976, 1977), changes also characteristic of mature heterocysts (Fay and Walsby, 1966; Codd et al., 1980). Morphologically recognisable proheterocysts appear after only 4 to 5 h and mature heterocysts after about 14 h; they are regularly spaced and account for about 7% of the cells. The development of mature heterocysts correlates with the appearance of nitrogenase activity under aerobic conditions. Heterocyst differentiation is not reversible after a certain stage (Wilcox et al., 1975). Commitment to heterocyst differentiation was investigated using inhibitors of DNA, RNA and protein synthesis and by transferring filaments to darkness, or adding ammonia (Bradley and Carr, 1977; Adams and Carr, 1989). DNA replication may serve as a timer mechanism for heterocyst development (Adams and Carr, 1989), possibly involving the expression of two sets of genes, located near the extremes of the genome, for a short time in the absence of combined nitrogen. The long period between initiation of proheterocyst development and commitment to heterocyst maturation may allow flexibility in response to changes in nitrogen availability, with mature heterocyst production being limited to long term N-limitation.

Protein labelling with $^{35}SO_4^{2-}$ during aerobic heterocyst differentiation in N. muscorum indicated that nitrogenase proteins are synthesised de novo only in heterocysts and at an early stage of development (Fleming and Haselkorn, 1973, 1974). The synthesis of nitrogenase proteins decreased 20 to 30 h after induction and increased again when N_2 fixation began. The synthesis of some proteins was initiated in all cells, but continued only in developing proheterocysts. Nitrogenous compounds resulting from proteolysis (Wood and Haselkorn, 1978) may prevent the differentiation of neighbouring vegetative cells. Inhibitory zones that prevent vegetative cell differentiation occur around both heterocysts and proheterocysts (Wilcox et al., 1975).

Differentiated heterocysts of A. variabilis continue to synthesize RNA and protein (Lynn and Ownby, 1987). They contain mRNA transcripts complementary to 45% of the genome, about half of which (ca. 1000 genes) are heterocyst-specific and relatively abundant (Lynn et al., 1986). The appearance of nitrogenase mRNA and activity correlate with morphological changes during derepression in A. variabilis. Under aerobic conditions, nitrogenase mRNA and activity were first observed after 8.5 h when late proheterocysts were present (Wealand et al., 1989), unlike the apparently early appearance of nitrogenase proteins in N. muscorum (Fleming and Haselkorn, 1974). Under strictly anaerobic conditions, however, nif mRNA was detectable within 2 h of imposing N starvation and nitrogenase activity appeared 1 h later. There was no evidence of proheterocyst development and exposure to O_2 caused nif mRNA levels to decrease rapidly (Helber et al., 1988a).

4 GENETICS OF CYANOBACTERIAL NITROGENASE

Genetic analysis of N_2-fixing cyanobacteria has lagged behind that of other diazotrophs such as Klebsiella, mainly because suitable systems for genetic analysis were not available until recently (see Tandeau de Marsac and Houmard, 1987; Packer and Glazer, 1988).

The most detailed analysis of cyanobacterial nif gene organization has been carried out using Anabaena PCC 7120 (see Haselkorn, 1986; Haselkorn et al., 1987; Golden, 1988). Cloned fragments of K. pneumoniae nif genes (see chapter 6) were used as probes to isolate nifH, nifK, nifD and nifS from a genomic library of vegetative cell DNA (Rice et al., 1982). No other nif genes were identified in over 30 kb of DNA, although subsequent sequencing of DNA in the region of nifS revealed nifU and nifB (Mulligan et al., 1988). Unlike K. pneumoniae, in which nifHDK are contiguous, nifK in Anabaena is separated from nifHD by about 11 kb.

NifS is located close to nifDK and there is also a second nifH gene (nifH*), transcribed under the same conditions as nifH to give a nearly identical product (Haselkorn, 1986). Comparison of heterocyst and vegetative cell DNA revealed two developmentally-regulated gene rearrangements. These occur in heterocyst DNA at a late stage of heterocyst development (18 to 30 h after induction) (Golden et al., 1987) and at about the same time as nif mRNA is detected, indicating that the rearrangements are a consequence of, rather than a trigger for, heterocyst development (Golden et al., 1985). The first of these rearrangements is the excision of an 11 kb element from within the 3'-end of the nifD gene (Golden et al., 1985), resulting in the formation of a nifHDK operon, expressed from the nifH promoter. The xisA gene, located within the excision, probably encodes the site-specific recombinase involved in the excision (Golden and Wiest, 1988). The second rearrangement is the deletion of a 55 kb element of unknown function (Golden et al., 1988), which places the rbcLS (structural genes for the large and small subunits of RuBisCO) operon about 10 kb from the nifS gene. Whereas both rearrangements occur during heterocyst differentiation in air, only the 55 kb excision occurs under argon, when morphologically discernible heterocysts and nitrogenase activity do not develop. Inactivation of the xisA gene, by site directed mutagenesis (Golden and Wiest, 1988), prevented excision of only the 11 kb element and demonstrated that this rearrangement is necessary for nif gene expression but not for heterocyst differentiation or spacing. These rearrangements are thus differently regulated. Sequencing of DNA to either side of the excised elements revealed the presence of nifB and nifU and that the 55kb element separated the 5' and 3' ends of a bacterial-type ferredoxin gene, fdxN (Mulligan et al., 1988). This gene is transcribed as part of a nifB fdxN nifS nifU operon under aerobic or anaerobic conditions (Mulligan et al., 1988; Mulligan and Haselkorn, 1989). The 55 kb excision is probably required for expression of this operon, which is at a lower level than that of the nifHDK operon (Mulligan and Haselkorn, 1989). Although the function of this nifB-linked fdxN gene is not known, its location suggests a role in N_2 fixation. This gene is distinct from the fdxH gene of *Anabaena* 7120, which is located 5 kb downstream from nifK and encodes a heterocyst-specific ferredoxin. Its transcript is detectable 18 to 24 h after induction, and this ferredoxin is an electron donor to nitrogenase (Böhme and Haselkorn, 1988).

The organization of nifK, D and H genes has been examined in several other cyanobacteria. *A. variabilis* has a non-contiguous nifHDK operon, similar to *Anabaena* 7120, but the 55 kb element is absent (Herrero and Wolk, 1986; Brusca et al., 1989). The non-contiguous arrangement of nifHDK in vegetative cells was found for all heterocystous cyanobacteria examined (Kallas et al., 1983, Franche and Cohen-Bazire, 1985; Apte and Thomas, 1987) except a Fischerella sp. (Saville et al., 1987) and the major cyanobiont of Azolla caroliniana (Meeks et al., 1988). In contrast, the nifHDK genes are contiguous in the microaerobic and aerobic non-heterocystous N_2-fixing strains tested (Kallas et al., 1983; 1985; Barnum and Gendel, 1985; Apte and Thomas, 1987). Mutant strains of Nostoc and Calothrix unable to synthesize heterocysts and fix N_2 aerobically showed the non-contiguous nif gene arrangement and did not rearrange the nif genes when nitrogenase was induced anaerobically (Kallas et al., 1985). However, a het⁻ mutant of Nostoc 7906 undergoes this rearrangement on N starvation in air (Damerval et al., 1985). The continuous nifHDK gene organization in Fischerella may be related to its irregular heterocyst spacing and branching habit. The role of these rearrangements is unclear, although they may be involved in commitment to heterocyst differentiation or limiting nif gene expression to mature heterocysts (Golden, 1988).

5 INTEGRATION OF N_2 FIXATION AND METABOLISM

5.1. Sources of ATP

Heterocysts generate ATP by both cyclic photophosphorylation and oxidative phosphorylation (Tel-Or and Stewart, 1976; Almon and Böhme, 1982; Ernst et al., 1983; Hawkesford et al., 1983). In heterocystous cyanobacteria, such as A. cylindrica and Anabaenopsis circularis, photophosphorylation supports the maximal nitrogenase activities in the light, where oxidative and substrate level phosphorylation are probably unimportant (Bottomley and Stewart, 1977). ATP synthesis may be coupled to the uptake hydrogenase-dependent reduction of O_2 in heterocysts (Peterson and Burris, 1978). Isolated heterocysts had a lower energy charge than whole filaments of Anabaena 7120; addition of H_2 stimulated nitrogenase activity in the light and transiently affected the ATP/ADP ratio but not the energy charge (Privalle and Burris, 1983). Changes in irradiance and O_2 tension affected the energy charge in isolated heterocysts of A. variabilis, and a sharp increase in nitrogenase activity at energy charge values around 0.75 in the light suggested a regulatory effect of energy charge (Ernst et al., 1983). Dark N_2 fixation in heterocystous cyanobacteria can be supported by oxidative and substrate-level phosphorylation (Bottomley and Stewart, 1977). This is true also of non-heterocystous cyanobacteria such as Gloeothece where the ATP required for N_2 fixation in the light and dark is generated by oxidative phosphorylation (Maryan et al., 1986) and photosynthetically-generated ATP is important only at high irradiance. Substrate-level phosphorylation may be important in supporting nitrogenase activity in some cyanobacteria (Gallon, 1980; Stal and Heyer, 1987).

Dark anaerobic nitrogenase activity in Oscillatoria may be an important feature of its ability to fix N_2 in microbial mats (Stal and Heyer, 1987).

5.2. Sources of reductant.

Several potential electron sources for nitrogenase have been identified (see Gallon, 1980; Neuer et al., 1983; Neuer and Bothe, 1985; Stewart and Rowell, 1986); the main pathway involved is probably the oxidative pentose phosphate (OPP) pathway, although other catabolic pathways are also important. Heterocysts, which lack components of photosystem II (Thomas, 1970; Donze et al., 1972; Tel-Or and Stewart, 1977) and thus do not evolve O_2 photosynthetically, cannot reduce pyridine nucleotides or ferredoxin with water as the electron donor. Since heterocysts do not fix CO_2 photosynthetically, the reductant for nitrogenase and carbon skeletons for ammonia assimilation must be derived from carbon compounds synthesized in vegetative cells. Certain cyanobacteria can also use exogenous carbon compounds. Fructose, which supports the growth of A. variabilis in the dark (Wolk and Shaffer, 1976), stimulates growth and nitrogenase activity in the light (Haury and Spiller, 1981).

The nature and kinetics of metabolite transfer between heterocysts and vegetative cells have not been firmly established and this is clearly an area where more information is required. Transfer of glutamine from heterocysts to vegetative cells is established (see section 5.4). Experiments examining the fate of ^{14}C-labelled metabolites taken up by isolated heterocysts, or involving $^{14}CO_2$ labelling of filaments of A. cylindrica and heterocyst isolation, indicate that alanine, glutamate and sugars are the major organic compounds transported into heterocysts (Jüttner, 1983). The transfer of ^{14}C from alanine to glutamine, presumably explains the high alanine dehydrogenase activity of heterocysts (Rowell and Stewart, 1976), and indicates that alanine is metabolized to provide not only a substrate (glutamate) for glutamine synthetase (GS), but also reductant (NADH) for nitrogenase. However, alanine does not appear to be a good electron donor for nitrogenase in isolated heterocysts of A. variabilis and A. cylindrica (Neuer and Bothe, 1985). On the basis of the rates of sucrose turnover and the relatively high activities of sucrose synthase in vegetative cells and alkaline invertase in heterocysts, it was proposed (Schilling and Ehrnsperger, 1985) that this disaccharide is transported into heterocysts of A. variabilis.

The physiological electron donor for nitrogenase in heterocystous cyanobacteria is probably a specific ferredoxin. The ferredoxin antagonist metronidazole differentially inhibited photosynthesis and nitrogenase activity in A. cylindrica (Tetley and Bishop, 1979), suggesting that these processes might have different electron carriers. The ferredoxin from heterocysts of A. variabilis was a more effective electron donor to nitrogenase in vitro than that from vegetative cells (Schrautemeier and Böhme, 1985; Böhme and Schrautemeier, 1987).

Many metabolites, including sugars, sugar phosphates, pyruvate, NADH, NADPH and H_2, support acetylene reduction by isolated heterocysts (Houchins and Hind, 1982; Privalle and Burris, 1984; Neuer and Bothe, 1985; Böhme and Schrautemeier, 1987), with H_2 being the most effective. The enzymes of the OPP pathway, glycolysis and an incomplete TCA cycle are present in heterocysts (Neuer et al., 1983). Pyruvate-ferredoxin oxidoreductase was detected in heterocyst extracts, but not pyruvate dehydrogenase or pyruvate oxidase.

Electron transfer from H_2 occurs via photosystem I in the light and via the respiratory electron transport chain, pathways which share a number of common components (Eisbrenner and Bothe, 1979; Houchins and Hind, 1982). Pyruvate is metabolized via pyruvate-ferredoxin oxidoreductase (Neuer and Bothe, 1985), although its quantitative significance is not clear. Electron transfer from NADH to ferredoxin is light-dependent, since the membrane-bound dehydrogenase donates electrons to photosystem I (Houchins and Hind, 1982; Schrautemeier et al., 1984). In contrast, electron transfer from NADPH to ferredoxin may involve photosystem I in the light, but there is a direct electron transfer via ferredoxin-NADP oxidoreductase (FNR) in the dark (Apte et al., 1978; Lockau et al., 1978; Houchins and Hind, 1982; Schrautemeier et al., 1984; Neuer and Bothe, 1985). FNR has been identified as the membrane-associated NADPH dehydrogenase in A. variabilis (Scherer et al., 1988), although Houchins and Hind (1982) found no evidence of a membrane bound NADPH dehydrogenase in Anabaena 7120 heterocysts. Heterocyst FNR differs electrophoretically from that of vegetative cells (Rowell et al., 1981).

In heterocysts, the generation of NADPH may not be the main physiological function of isocitrate dehydrogenase (Karni and Tel-Or, 1983; Papen et al., 1983). NADPH is probably generated via glucose-6-phosphate dehydogenase (G6PDH) (Apte et al., 1978; Schaeffer and Stanier, 1978; Rowell et al., 1988; Rowell and Simpson, 1990) and 6-phosphogluconate dehydrogenase. These enzymes are present at higher activities in heterocysts than in vegetative cells (Winkenbach and Wolk, 1973). The regulation of G6PDH may be important in maintaining and controlling the flow of reductant to nitrogenase. The products of ammonia assimilation (Schaeffer and Stanier, 1978; Rowell and Simpson, 1990), the thioredoxin system (see Stewart et al., 1985; Böhme, 1987; Rowell et al., 1988) and reduction charge (NADPH/(NADPH+NADP)) (Apte et al., 1978; Böhme, 1987) have all been implicated in its control. Certain amino acids, particularly glutamine and alanine, promote activation of A. variabilis G6PDH, but have no apparent effect on the enzyme from Synechococcus 7942 (Rowell and Simpson, 1990), suggesting that this property of G6PDH is associated with N_2 fixation. Böhme (1987) found that heterocyst extracts maintain a high reduction charge (0.95) under N_2-fixing conditions, although this might be expected to strongly inhibit G6PDH activity (Apte et al., 1978). The thermodynamically unfavourable electron transfer from NADPH to ferredoxin does, however, require high NADPH/NADP ratios, which are not thought to be maintained in vivo. The effects of ionophores on both nitrogenase activity and the protonmotive force across the plasmalemma in A. variabilis and P. boryanum suggest a role for the membrane potential in driving or regulating electron transfer (Hawkesford et al., 1983). This role has, however, been questioned (Houchins and Hind, 1982; Schrautemeier et al., 1984, 1985).

5.3. Hydrogen metabolism

Under appropriate conditions, cultures of N_2-fixing cyanobacteria evolve H_2 from nitrogenase (Haystead et al., 1970; Benemann and Weare, 1974). Cyanobacterial H_2 metabolism is complex, involving nitrogenase, a membrane-bound uptake hydrogenase and a soluble hydrogenase, which catalyses both H_2 evolution and uptake (see Lambert and Smith, 1981; Bothe, 1982; Houchins, 1984).

Although exogenous H_2 can stimulate nitrogenase activity (Tel-Or et al., 1977), availability of H_2 in natural environments will normally be limited to that generated by nitrogenase. H_2 production by nitrogenase is unavoidable, irreversible, insensitive to CO and dependent on a supply of electrons and ATP. Rates of H_2 production will be affected by which substrates (N_2, C_2H_2, etc) or inhibitors (e.g., CO) of nitrogenase are present. In heterocystous cyanobacteria, the source of reductant for H_2 evolution is ultimately generated by photosynthesis in vegetative cells. H_2 evolved by nitrogenase is normally recycled, thereby providing ATP or electrons for biosynthesis and removing O_2 from the vicinity of nitrogenase. Rates of H_2 evolution are therefore not solely dependent on the activity of nitrogenase.

Most (Houchins, 1984) but not all (Chen et al., 1989b) heterocystous cyanobacteria have uptake hydrogenase. Some non-heterocystous forms lack this activity (Weisshaar and Böger, 1983; Van der Oost et al., 1987), as do nickel-depleted cultures of A. cylindrica (Daday et al., 1985), which may result in higher rates of H_2 production (Mitsui et al., 1983). Uptake hydrogenase has a high affinity for H_2, is insensitive to O_2 and donates electrons to the photosynthetic and respiratory electron transport chains. The existence of a reversible O_2-sensitive hydrogenase was the subject of controversy until recently (see Houchins and Burris, 1981; Bothe, 1982; Houchins, 1984).

5.4. Ammonia assimilation and the regulation of N_2 fixation

Ammonia is assimilated in cyanobacteria via glutamine synthetase/glutamate synthase (GS/GOGAT) (Stewart and Rowell, 1975; Thomas et al., 1975; Meeks, 1981; Kerby et al., 1987, 1989). GS activity is usually 2 to 3 times higher in heterocysts than in vegetative cells. The structural gene for GS, glnA, in Anabaena 7120 is transcribed from different promoters during growth on fixed N or N_2 (Tumer et al., 1983). Isolated metabolically-active heterocysts of A. cylindrica formed labelled glutamine from $^{13}N_2$, [^{13}N]ammonia and [^{14}C]glutamate, apparently via GS (Thomas et al., 1977). It was suggested that glutamine synthesized in heterocysts is exported to vegetative cells, where it is further metabolized by GOGAT; part of the glutamate formed (see also Jüttner, 1983) is returned to the heterocysts as a substrate for GS. The low activity of GOGAT in heterocyst extracts (Rai et al., 1982; Häger et al., 1983) may allow only limited glutamate synthesis.

In most cyanobacteria, addition of ammonia suppresses nitrogenase activity and heterocyst development completely, whereas nitrate may do so only partially (Stewart, 1980a, b; Meeks et al., 1983b). The opposite is true of Anabaena CA (Bottomley et al., 1979). The effects of ammonia are mediated by GS, although the mechanism is not known. Mutants of A. variabilis partially deficient in GS have derepressed nitrogenase, supporting a regulatory role for GS (Polukhina et al., 1982; Kerby et al., 1986; Spiller et al., 1986; section 7). Glutamine may regulate nitrogenase but not heterocyst differentiation in A. variabilis (Thiel and Leone, 1986). Low concentrations of carbamoyl phosphate caused a rapid decrease in A. variabilis nitrogenase mRNA indicating that it may be involved in nif regulation (Helber et al., 1988b). On treatment of heterocystous cyanobacteria with L-methionine-DL-sulphoximine (MSX), which irreversibly inactivates GS, heterocyst differentiation and nitrogenase synthesis occur even in the presence of ammonia (Stewart and Rowell, 1975). The effect of MSX can be blocked by some amino acid products of ammonia assimilation (Ownby, 1977). The effects of nitrate may also be mediated by GS (Ramos and Guerrero, 1983) or they may be related directly to nitrate metabolism (Bottomley et al., 1979). Nitrite may inhibit nitrogenase directly (Böhme, 1986). Cyanobacterial GS is not regulated by covalent modification involving adenylylation and deadenylylation (Rowell et al., 1977; Fisher et al., 1981) and there is no evidence for a functional global N control by an ntr system like that in Escherichia coli and

Klebsiella (Magasanik, 1988). There is, however, evidence for ntr-like DNA sequences linked to glnA in Anabaena 7120 (Machray and Stewart, 1985).

Uptake of ammonia is either by an active NH_4^+ transport system or by non-specific diffusion of NH_3 and ion trapping (Kerby et al., 1987, 1989). The NH_4^+ transport system does not appear to be necessary for retention of ammonia generated internally from N_2 or NO_3^- but does affect the rate of inhibition of nitrogenase on adding ammonia (Reglinski et al., 1989). Whether the well-established biphasic uptake of methylamine (used as an ammonium analogue) is indicative of two NH_4^+ transport systems has been controversial. It was argued that the second phase of uptake represents an MSX-inhibited transport system and that the effects of MSX on nitrogenase activity and heterocyst differentiation are due to inhibition of ammonia uptake rather than assimilation (Singh et al., 1983, 1985). MSX is widely used as a (specific) inhibitor of GS and interpretation of its effects on nitrogenase synthesis and heterocyst differentiation (Stewart and Rowell, 1975; Ownby, 1977) would need reassessment if their argument were correct. However, we believe that the primary effect of MSX is on ammonia assimilation rather than NH_4^+ transport (Stewart et al., 1985; Kerby et al., 1987; Reglinski et al., 1989), and that the experimental observations are easily reconcilable if NH_4^+ transport and subsequent assimilation are tightly coupled.

Some amino acid analogues, including DL-7-azatryptophan (AT) and ß-2-thienylalanine, induce heterocysts lacking nitrogenase activity without inhibiting GS (Stacey et al., 1979; Rogerson, 1979). The effects of AT are complex; it is incorporated into proteins and inhibits GOGAT (Chen et al., 1987) and anthranilate synthetase, indicating a possible role for tryptophan biosynthesis in the control of nitrogenase and heterocyst differentiation (Bottomley et al., 1980). The effects of AT and other amino acid analogues are discussed further by Wilcox et al. (1975), Adams and Carr (1981) and Wolk (1982).

Ammonia addition to exponentially-growing cultures of A. cylindrica results in a decrease in both nitrogenase activity and the MoFe protein level, consistent with repression of synthesis and dilution by growth. This decrease is more rapid in light-limited cultures (Murry et al., 1983), and may be due to ATP or reductant limitation (Ohmori and Hattori, 1978). At an external pH around 7, where ammonia uptake is mainly via active transport of NH_4^+, ammonia addition to A. variabilis (Reich et al., 1986; Kerby et al., 1987) results in a slow decline in nitrogenase activity which is blocked by MSX. However, at pH 9 or 10 there is a rapid loss of activity which is only partially prevented by MSX and is not simply due to uncoupling (Reich et al., 1986, Kerby et al., 1987). Using antiserum against the Fe protein, Reich and Böger (1989) showed that addition of ammonia to A. variabilis at pH 10 changed the electrophoretic mobility of the Fe protein. About half of the Fe protein was modified, but more slowly than the decline in nitrogenase activity. By contrast, incubation of air-grown cultures under O_2 resulted in a loss of nitrogenase activity accompanied by a slower, but complete, modification of the Fe protein. The mechanism of inactivation and the nature of the modification appear to be similar or identical to those that occur in Anabaena CA in response to O_2 (Smith et al., 1987; Tabita et al., 1988).

In a marine Anabaena sp. lacking GOGAT (Chen et al., 1987), an alternative route of ammonia assimilation via GS and carbamoyl phosphate synthetase has been suggested. This enzyme is involved in arginine biosynthesis and therefore of cyanophycin, which is found in polar plugs of heterocysts.

6 SYMBIOTIC ASSOCIATIONS

Cyanobacteria occur in associations with a wide range of organisms: algae (the marine diatom Rhizosolenia and the freshwater diatom Rhopalodia), fungi, bryophytes, pteridophytes, gymnosperms, angiosperms, some animals, bacteria and non-photosynthetic protists (see Stewart et al., 1983; Smith and Douglas, 1987; Rai, 1989a). Only aspects relating to N_2 fixation in the better characterized associations are discussed here. The cyanobacterial symbionts (cyanobionts) in these associations are mostly heterocystous, usually Nostoc, but also Anabaena, Calothrix, Fischerella and Scytonema spp.

About 8% of known lichen species contain a cyanobacterium. The symbiosis may be bipartite with a mycobiont (usually an ascomycete) or tripartite, involving a phycobiont (green alga) as well, with the cyanobiont found in cephalodia (see Stewart, et al., 1983; Millbank, 1984; Rowell et al., 1985; Rai, 1989b). In bipartite lichens, such as Peltigera canina, the cyanobiont has a heterocyst frequency and nitrogenase activity similar to that of the free-living Nostoc isolate. In contrast, in the tripartite Peltigera aphthosa, heterocyst frequency and nitrogenase activity are both several-fold higher. In both cases, the mycobiont receives fixed N from the cyanobiont, mainly as ammonia (Stewart and Rowell, 1977; Rai et al., 1980), but in P. canina the cyanobiont also provides carbon as glucose (Smith et al., 1969). The liberation of ammonia from the cyanobiont is related to the very low activities of GS and GOGAT; the low level of GS protein indicates transcriptional regulation (Stewart and Rowell, 1977; Rai et al., 1980, 1981a; Stewart et al., 1983; Bergman et al., 1986). The ammonia liberated is assimilated mainly by fungal glutamate dehydrogenase (GDH) (Stewart and Rowell, 1977; Rai et al., 1981b, 1983a). However, in P. canina about 50% of the ammonia may be assimilated via GS in the cyanobiont; glutamine is presumably transported to the mycobiont where there is an appreciable glutamine pool and no detectable GS or GOGAT activity. The reduction in GS and GOGAT activities is clearly specific to the cyanobionts, since in P. aphthosa there are appreciable activities in the phycobiont (Rai et al., 1980, 1981a). The increased heterocyst frequency and nitrogenase activity in the cyanobiont of P. aphthosa are related to the presence of the second photosynthetic partner which provides fixed carbon; this is the case for most other associations. N_2 fixation in P. aphthosa in the dark is sustained by the breakdown of glycogen accumulated in the cyanobiont; dark CO_2 fixation in the mycobiont is required for provision of carbon skeletons for ammonia assimilation (Rai et al., 1981c, 1983b). The kinetics of ^{15}N labelling in P. aphthosa suggest that alanine may be transferred from the cephalodia to the main thallus (Rai et al., 1981b; Stewart et al., 1983). Inhibition of nitrogenase by ammonia depends on active GS in the phycobiont of P. aphthosa, indicating a complex integration of N metabolism within the association.

Cycads are the only gymnosperms which form diazotrophic associations (see Stewart et al., 1983; Lindblad and Bergman, 1989). The cyanobiont (Nostoc) occurs intercellularly in coralloid roots located below the soil surface. Cyanobiont metabolism is therefore heterotrophic and, although appreciable RuBisCO and PRuK were present in the freshly isolated cyanobiont of Cycas revoluta (Lindblad et al., 1987), CO_2 fixation was undetectable. Restriction fragment length polymorphism (RFLP) analysis has shown that several cycad species contain different Nostoc strains and that a mixture of strains may associate with a single cycad species (Lindblad et al., 1989). Compared with free-living isolates, the cyanobionts of C. revoluta and Zamia skinneri show higher heterocyst frequencies, from 5 to 6% at the growing tip to 45% in the mature region (Lindblad et al., 1985; Lindblad and Bergman, 1986, 1989). However, nitrogenase specific activity is highest at the growing tip where it is comparable with that of free-living cyanobacteria. Thus, the nitrogenase activity, which is located only in the heterocysts (Bergman et al., 1986), is inversely related to heterocyst frequency. These findings contrast with a report that nitrogenase activity is several-fold higher in the cyanobiont than the free-living isolate (Kumar et al., 1986). GS and GOGAT activities in extracts from the cyanobionts of Cycas circinalis and C. revoluta are comparable to, or higher than, those of the free-living isolates (Lindblad et al., 1987), but the kinetics of methylamine uptake by the freshly isolated cyanobiont of C. revoluta (Rai et al., 1986) indicate that GS may be inhibited in vivo. It is not known what photosynthetic products the cyanobiont receives nor what nitrogenous products it liberates. However, the principal translocated solutes from coralloid roots of several cycads are citrulline and/or glutamine, together with some glutamate

(Pate et al., 1988). Labelling patterns from $^{14}CO_2$ and $^{15}N_2$ were consistent with assimilation of ammonia via GS to form glutamine, which may be translocated directly or converted to citrulline before both are translocated.

Gunnera is the only angiosperm known to form a symbiotic association with a cyanobacterium (Nostoc), which is located intracellularly in mucilage-filled glands at the bases of the petioles (see Stewart et al., 1983; Bonnet, 1989). Nitrogenase activity is highest in sub-apical nodules of the glands of Gunnera albocarpa, where the cyanobiont has a heterocyst frequency of about 30%; heterocyst frequencies can reach 65%, but are associated with low nitrogenase activity as the cells degenerate (Silvester, 1975). Although nitrogenase activity in isolated nodules of G. albocarpa is stimulated by light (Silvester, 1975), CO_2 fixation rates are low, and the cyanobiont is probably heterotrophic. A few bryophytes form associations with Nostoc spp., which are located in cavities on the underside of the gametophyte thallus (see Stewart et al., 1983; Meeks, 1989). The associations with mosses are usually epiphytic, whereas those with liverworts and hornworts are endophytic. These associations are easily maintained in the laboratory and reconstituted from the isolated partners (Rodgers and Stewart, 1977; Stewart and Rodgers, 1977; Enderlin and Meeks, 1983). Heterocyst frequencies of 30% in the cyanobiont of the liverwort Blasia pusilla and 45% in that of the hornwort Anthoceros punctatus (Rodgers and Stewart, 1977; Enderlin and Meeks, 1983) correlate with increased nitrogenase activity, when compared to the free-living isolates (Rodgers and Stewart, 1977). Ammonia assimilation in the A. punctatus-Nostoc association is via GS/GOGAT (Meeks et al., 1983a, 1985a), with much of the ammonia fixed by the cyanobiont being released and assimilated by Anthoceros (Stewart and Rodgers, 1977; Meeks et al., 1985a). This appears to be related to low cyanobiont GS activity, although GS protein levels were comparable to those of the free-living isolate, indicating post-translational control of activity (Meeks et al., 1985a; Joseph and Meeks, 1987). The cyanobiont shows a low rate of CO_2 fixation (Rodgers and Stewart, 1977), associated with a lower specific activity of RuBisCO than in the free-living isolate (Meeks, 1989) and N_2 fixation is supported by photosynthate transferred from the eukaryote (Stewart and Rodgers, 1977).

The Anabaena-Azolla association involves Anabaena azollae, located within cavities in the dorsal lobe of each leaf of the sporophyte of the water fern Azolla (Peters and Calvert, 1983; Stewart et al., 1983; Braun-Howland and Nierzwicki-Bauer, 1989; Peters and Meeks, 1989). RFLP studies indicate that the available free-living isolates of Anabaena azollae probably represent only minor symbionts (Franche and Cohen-Bazire 1985, 1987; Meeks et al., 1988), the major symbiont probably being obligately associated with the fern. Gram-positive, non-

N_2-fixing bacteria, identified as an Arthrobacter sp., occur in the leaf cavities and on the basis of studies of bacterial populations they are regarded as a component of the association (Petro and Gates, 1987). It is not known how the other symbionts are affected by these bacteria. Leaves and associated Anabaena colonies develop synchronously, with undifferentiated Anabaena at the apex and a progressive enlargement of vegetative cells and differentiation of heterocysts up to a frequency of 20 to 30% in mature cavities (Hill, 1975, 1977). There is a corresponding increase in nitrogenase activity (Kaplan et al., 1986) and a decrease in the O_2 concentration in the leaf cavities (Grilli Caiola et al., 1989). Nitrogenase protein is restricted to heterocysts (Braun-Howland et al., 1988) and, in contrast to free-living Anabaena and Nostoc spp., all cells of the cyanobiont have a contiguous nifHDK organization (Meeks et al., 1988; see section 4). There is little change in the phycobiliprotein content of the cyanobiont during heterocyst differentiation (Kaplan et al., 1986) and polar bodies are present in heterocysts before nitrogenase is detectable (Braun-Howland et al., 1988), indicating that the cyanobiont is not nitrogen-starved. Ammonia assimilation is via GS/GOGAT (Meeks et al., 1987) and GS activity, protein and mRNA (Orr and Haselkorn, 1982; Stewart et al., 1983; Peters et al., 1985; Nierzwicki-Bauer and Haselkorn, 1986; Lee et al., 1988) and GOGAT activity (Rai et al., 1981a) are markedly reduced in the cyanobiont, compared with free-living Anabaena and Nostoc spp. The low GS mRNA level indicates that GS is transcriptionally regulated (Nierzwicki-Bauer and Haselkorn, 1986). GS activity in Anabaena from mature cavities is lower (65%) than that in Anabaena from apical leaves (Uheda and Kitoh, 1989), when an increase might have been expected due to the high heterocyst frequency (see section 5.4). GS and GDH activities are higher in cavity hairs, which probably function as transfer cells, than in the cyanobiont (Peters et al., 1985; Uheda, 1986). The cyanobiont liberates ammonia which is metabolized by the fern (Meeks et al., 1985b, 1987) and fixed N is transported from the mature cavities to the apices, probably as glutamine, glutamate and ammonia (Peters et al., 1985). The cyanobiont appears to contribute to only a very limited extent to total photosynthetic CO_2 fixation in the association, although freshly isolated Anabaena fixes CO_2 at rates similar to free-living cyanobacteria (Kaplan and Peters, 1988). This, however, contrasts with the low level of RuBisCO mRNA reported for the cyanobiont (Nierzwicki-Bauer and Haselkorn, 1986). Fixed carbon is transported from the fern to the cyanobiont, probably as sucrose (Peters et al., 1985; Kaplan and Peters, 1988).

Athough there are similarities in cyanobiont development, N_2 fixation and ammonia assimilation, the regulatory mechanisms appear to vary. More information is needed on how the symbionts coordinate development and on control of metabolic interchanges. Our understanding of natural associations should benefit from and contribute to studies of novel N_2-fixing associations between cyanobacteria and higher plants (Gusev and Korzhenevskaya, 1989).

7 CYANOBACTERIA AS BIOFERTILIZERS

Free-living and symbiotic N_2-fixing cyanobacteria enhance the fertility of tropical rice paddies (see Achnitch et al., 1986; Venkataraman, 1986; Whitton and Roger, 1989; Kerby and Rowell, 1990) and may reduce the dependency of agriculture on fossil fuels. Approximately 50% of the cyanobacterial genera in rice paddies are heterocystous and certain countries encourage the use of cyanobacterial inoculants to increase soil fertility. However, data conclusively demonstrating the beneficial effects of this "algalization" are lacking and indigenous populations may well outgrow introduced strains. When rice paddies are inoculated with either free-living cyanobacteria or Azolla, the rice plant size, N content, number of tillers, ears and spikelets, and filled grains per panicle may be affected (see Roger and Kulasooriya, 1980). Field experiments suggest an average yield increase due to cyanobacterial N_2 fixation of 14% above controls with no added combined nitrogen, corresponding to about 450 kg ha^{-1} (Watanabe and Roger, 1984; Venkataraman, 1986). Grain and N yields of rice treated with cyanobacteria or Azolla are often quoted as being comparable to those obtained following an application of approximately 30 kg N fertilizer ha^{-1}. Since only limited field data are available, the relative contribution of cyanobacterial N_2 fixation cannot be fully assessed.

The selection of improved cyanobacterial strains for rice cultivation is clearly an area for further studies. One strategy is to develop strains which excrete their fixed N in the same way that symbiotic cyanobacteria do. Laboratory studies have demonstrated the photoproduction of ammonia, either using inhibitors of GS or mutant strains partially deficient in GS activity (see Kerby et al., 1989). Mutant strains of A. variabilis partially deficent in GS biosynthetic activity have been selected for resistance to ethylenediamine (Polukhina et al, 1982; Kerby et al., 1986) and MSX (Spiller et al., 1986). Such strains have low levels of GS activity due to catalytic deficiencies (Spiller et al., 1986; Kerby et al., 1990) or low glnA transcription (Hien et al., 1988), as in some symbioses (see section 6). They have derepressed nitrogenase and liberate ammonia, which may enhance the dry weight and N content of rice plants grown in N-free medium (Latorre et al., 1986).

8 REFERENCES

Achtnich, W., Moawad, A.M. and Johal, A.M., 1986. Azolla, a biofertilizer for rice. Int. J. Trop. Agric., 4: 188-211.

Adams, D.G. and Carr, N.G., 1981. The developmental biology of heterocyst and akinete formation in cyanobacteria. CRC Critical Rev. Microbiol., 9: 45-100.

Adams, D.G. and Carr, N.G., 1989. Control of heterocyst development in the cyanobacterium Anabaena cylindrica. J. Gen. Microbiol., 135: 839-849.

Almon, M. and Böhme, H., 1982. Photophosphorylation in isolated heterocysts from the blue-green alga Nostoc muscorum. Biochim. Biophys. Acta, 679: 279-286.

Apte, S.K., Rowell, P. and Stewart, W.D.P., 1978. Electron donation to ferredoxin in heterocysts of the N_2-fixing alga Anabaena cylindrica. Proc. Roy. Soc. Lond. B., 200: 1-25.

Apte, S.K. and Thomas, J., 1987. Nitrogen fixation genes (nif K,D,H) in the filamentous non-heterocystous cyanobacterium Plectonema boryanum do not rearrange. J. Genet., 66: 101-110.

Barnum, S.R and Gendel, S.M., 1985. Organisation of nitrogen fixation genes in a nonheterocystous filamentous cyanobacterium. FEMS Microbiol. Lett., 29: 339-342.

Benemann, J.R. and Weare, N.M., 1974. Nitrogen fixation by Anabaena cylindrica. III. Hydrogen-supported nitrogenase activity. Arch. Microbiol., 101: 401-408.

Bergman, B., Lindblad, P. and Rai, A.N., 1986. Nitrogenase in free-living and symbiotic cyanobacteria: immunoelectron microscopic localization. FEMS Microbiol. Lett., 35: 75-78.

Böhme, H., 1986. Inhibition of nitrogen fixation by nitrite in Anabaena variabilis. Arch. Microbiol., 146: 99-103.

Böhme, H., 1987. Regulation of electron flow to nitrogenase in a cell-free system from heterocysts of Anabaena variabilis. Biochim. Biophys. Acta, 891: 121-128.

Böhme, H. and Haselkorn, R., 1988. Molecular cloning and nucleotide sequence analysis of the gene coding for heterocyst ferredoxin from the cyanobacterium Anabaena sp. strain PCC 7120. Mol. Gen. Genet., 214: 278-285.

Böhme, H. and Schrautemeier B., 1987. Electron donation to nitrogenase in a cell-free system from heterocysts of Anabaena variabilis. Biochim. Biophys. Acta, 891: 115-120.

Bonnett, H.T., 1989. The Nostoc-Gunnera association. In: A.N. Rai (Editor), CRC Handbook of Symbiotic Cyanobacteria. CRC Press, Boca Raton. pp. 161-171.

Bothe, H., 1982. Hydrogen production by algae. Experientia, 38: 59-64.

Bottomley, P.J., Grillo, J.F., Van Baalen, C. and Tabita, F.R., 1979. Synthesis of nitrogenase and heterocysts by Anabaena sp. CA in the presence of high levels of ammonia. J. Bacteriol., 140: 938-943.

Bottomley, P.J. and Stewart, W.D.P., 1977. ATP and nitrogenase activity in nitrogen fixing heterocystous blue-green algae. New Phytol., 79: 625-638.

Bottomley, P.J., Van Baalen, C. and Tabita, F.R., 1980. Heterocyst differentiation and tryptophan metabolism in the cyanobacterium Anabaena sp. CA. Arch. Biochem. Biophys., 203: 204-213.

Bradley, S. and Carr, N.G., 1976. Heterocyst and nitrogenase development in Anabaena cylindrica. J. Gen. Microbiol., 96: 175-184.

Bradley, S. and Carr, N.G., 1977. Heterocyst development in Anabaena cylindrica: the necessity for light as an initial trigger and sequential stages of commitment. J. Gen. Microbiol., 101: 291-297.

Braun-Howland, E.B., Lindblad, P., Nierzwicki-Bauer, S.A. and Bergman, B., 1988. Dinitrogenase reductase (Fe-protein) of nitrogenase in the cyanobacterial symbionts of three Azolla species: localization and sequence of appearance during heterocyst differentiation. Planta, 176: 319-332.
Braun-Howland, E.B., and Nierzwicki-Bauer, S.A., 1989. Azolla-Anabaena symbiosis: biochemistry, physiology, ultrastructure, and molecular biology. In: A.N. Rai (Editor), CRC Handbook of Symbiotic Cyanobacteria. CRC Press, Boca Raton. pp. 65-117.
Brusca, J.S., Hale, M.A., Carrasco, C.D. and Golden, J.W., 1989. Excision of an 11-kilobase-pair DNA element from within the nif D gene in Anabaena variabilis heterocysts. J. Bacteriol., 171: 4138-4145.
Bryceson, I. and Fay, P., 1981. Nitrogen fixation in Oscillatoria (Trichodesmium) erythrea in relation to bundle formation and trichome differentiation. Mar. Biol., 61: 159-166.
Capone, D.G. and Carpenter, E.J., 1982. Nitrogen fixation in the marine environment. Science, 217: 1140-1142.
Carpenter, E.J. and Price, C.C., 1976. Marine Oscillatoria (Trichodesmium): explanation for aerobic nitrogen fixation without heterocysts. Science, 191: 1278-1280.
Carr, N.G., 1983. Biochemical aspects of heterocyst differentiation and function. In: G.C. Papageorgiou and L. Packer (Editors), Photosynthetic Prokaryotes: Cell Differentiation and Function. Elsevier, New York. pp. 265-280.
Carr, N.G., 1988. Nitrogen reserves and dynamic reservoirs in cyanobacteria. In: L.J. Rogers and J.R. Gallon (Editors), Biochemistry of the Algae and Cyanobacteria, (Proc. Phytochem. Soc. Eur., 28). Oxford Univ. Press, Oxford. pp. 13-21.
Carr, N.G. and Whitton, B.A., (Editors) 1982. The Biology of Cyanobacteria. Blackwell Scientific Publications, Oxford.
Chen, C., Van Baalen, C. and Tabita, F.R., 1987. DL-7-azatryptophan and citrulline metabolism in the cyanobacterium Anabaena sp. strain 1F. J. Bacteriol., 169: 1114-1119.
Chen, P-C., Almon, H. and Böger, P., 1989b. Physiological factors determining hydrogenase activity in nitrogen-fixing heterocystous cyanobacteria. Plant Physiol., 89: 1035-1038.
Chen, T-H., Huang, T-C. and Chow, T-J., 1988. Calcium requirement in nitrogen fixation in the cyanobacterium Synechococcus RF-1. Planta, 173: 253-256.
Chen, T-H., Huang, T-C. and Chow, T-J., 1989a. Calcium is required for the increase of dark respiration during diurnal nitrogen fixation by Synechococcus RF-1. Plant Science, 60: 195-198.
Codd, G.A., Okabe, K. and Stewart, W.D.P., 1980. Cellular compartmentation of photosynthetic and photorespiratory enzymes in the heterocystous cyanobacterium Anabaena cylindrica. Arch. Microbiol., 124: 149-154.
Daday, A., Mackerras, A.H. and Smith, G.D., 1985. The effect of nickel on hydrogen metabolism and nitrogen fixation in the cyanobacterium Anabaena cylindrica. J. Gen. Microbiol., 131: 231-238.
Damerval, T., Franche, C., Rippka, R. and Cohen-Bazire, G., 1985. Rearrangement of nif structural genes in Nostoc PCC 7906. In: H.J. Evans, P.J. Bottomley and W.E. Newton (Editors), Nitrogen Fixation Research Progress. Martinus Nijhoff Publishers, Dordrecht. p. 517.
Donze, M., Haveman, J. and Schiereck, P., 1972. Absence of photosystem II in heterocysts of the blue-green alga Anabaena. Biochim. Biophys. Acta, 256: 157-161.
Eisbrenner, G. and Bothe, H., 1979. Modes of electron transfer from molecular hydrogen in Anabaena cylindrica. Arch. Microbiol., 123: 37-45.
Enderlin, C.S. and Meeks, J.C., 1983. Pure culture and reconstitution of the Anthoceros-Nostoc symbiotic association. Planta, 158: 157-165.

Ernst, A., Böhme, H. and Böger, P., 1983. Phosphorylation and nitrogenase activity in isolated heterocysts from Anabaena variabilis (ATCC 29413). Biochim. Biophys. Acta, 723: 83-90.
Ernst, A. and Böger, P., 1985. Glycogen accumulation and the induction of nitrogenase activity in the heterocyst-forming cyanobacterium Anabaena variabilis. J. Gen. Microbiol., 131: 3147-3153.
Ernst, A., Kirschenlohr, M., Diez, J. and Böger, P., 1984. Glycogen content and nitrogenase activity in Anabaena variabilis. Arch. Microbiol., 140: 120-125.
Fay, P., Stewart, W.D.P., Walsby, A.E. and Fogg, G.E., 1968. Is the heterocyst the site of nitrogen fixation in blue-green algae? Nature, 220: 810-812.
Fay, P. and Van Baalen, C. (Editors), 1987. The Cyanobacteria. Elsevier, Amsterdam.
Fay, P. and Walsby, A.E., 1966. Metabolic activities of isolated heterocysts of the blue-green alga Anabaena cylindrica. Nature, 209: 94-95.
Fisher, R., Tuli, R. and Haselkorn, R., 1981. A cloned cyanobacterial gene for glutamine synthetase functions in Eschericia coli but the enzyme is not adenylylated. Proc. Natl. Acad. Sci. USA, 78: 3393-3397.
Fleming, H. and Haselkorn, R., 1973. Differentiation in Nostoc muscorum: nitrogenase is synthesized in heterocysts. Proc. Natl. Acad. Sci. USA, 70: 2727-2731.
Fleming, H. and Haselkorn, R., 1974. The program of protein synthesis during heterocyst differentiation in nitrogen fixing blue-green algae. Cell, 3, 159-170.
Fogg, G.E., 1974. Growth and heterocyst production in Anabaena cylindrica Lemm. II. In relation to carbon and nitrogen metabolism. Ann. Bot., 13: 241-259.
Franche, C. and Cohen-Bazire, G., 1985. The structural nif genes of four symbiotic Anabaena azollae show a highly conserved physical arrangement. Plant Sci., 39: 125-131.
Franche, C. and Cohen-Bazire, G., 1987. Evolutionary divergence in the nif H.D.K. gene region among nine symbiotic Anabaena azollae and between free-living heterocystous cyanobacteria. Symbiosis, 3: 159-178.
Gallon, J.R., 1980. Nitrogen fixation by photoautotrophs. In: W.D.P. Stewart and J.R. Gallon (Editors), Nitrogen Fixation (Proc. Phytochem. Soc. Eur., 18). Academic Press, London. pp. 197-238.
Gallon, J.R. and Chaplin, A.E., 1987. An Introduction to Nitrogen Fixation, Cassell Educational Ltd., London.
Gallon, J.R. and Chaplin, A.E., 1988. Nitrogen fixation. In: L.J. Rogers and J.R. Gallon (Editors), Biochemistry of the Algae and Cyanobacteria. (Proc. Phytochem. Soc. Eur., 28), Oxford Univ. Press, Oxford. pp. 147-173.
Gallon, J.R., Dingley, K.H., Le Noury, P.J. and Hipkin, C.R., 1989. Topoisomerase activity in a diazotrophic cyanobacterium. Biochem. Soc. Trans., 17: 926-927.
Gallon, J.R. and Hamadi, A.F., 1984. Studies on the effects of oxygen on acetylene reduction (nitrogen fixation) in Gloeothece sp. ATCC 27152. J. Gen. Microbiol., 130: 495-503.
Gallon, J.R., Perry, S.M., Rajab, T.M.A., Flayeh, K.A.M., Yunes, J.S. and Chaplin, A.E., 1988. Metabolic changes associated with the diurnal pattern of N_2 fixation in Gloeothece. J. Gen. Microbiol., 134: 3079-3087.
Giani, D. and Krumbein, W.E., 1986. Growth characteristics of non-heterocystous cyanobacterium Plectonema boryanum with N_2 as nitrogen source. Arch. Microbiol., 145: 259-265.
Golden, J.W., 1988. Genome rearrangements during Anabaena heterocyst differentiation. Can. J. Bot., 66: 2098-2102.
Golden, J.W., Carrasco, C.D., Mulligan, M.E., Schneider, G.J. and Haselkorn, R., 1988. Deletion of a 55-kilobase-pair DNA element from the chromosome during heterocyst differentiation of Anabaena sp. strain PCC 7120. J. Bacteriol., 170: 5034-5041.

Golden, J.W., Mulligan, M.E. and Haselkorn, R., 1987. Different recombination site specificity of two developmentally regulated genome rearrangements. Nature, 327: 526-529.
Golden, J.W., Robinson, S.J. and Haselkorn, R., 1985. Rearrangement of nitrogen fixation genes during heterocyst differentiation in the cyanobacterium. Anabaena. Nature, 314: 419-423.
Golden, J.W, and Wiest, D.R., 1988. Genome rearrangement and nitrogen fixation in Anabaena blocked by inactivation of xis A gene. Science, 242: 1421-1423.
Griffiths, M.S.H., Gallon, J.R. and Chaplin, A.E., 1987. The diurnal pattern of dinitrogen fixation by cyanobacteria in situ. New Phytol., 107: 649-657.
Grilli Caiola, M., Canini, A. and Moscone, D., 1989. Oxygen concentration, nitrogenase activity and heterocyst frequency in leaf cavities of Azolla filiculoides Lam. FEMS Microbiol. Lett., 59: 283-288.
Gusev, M.V. and Korzhenevskaya, T.G., 1989. Artificial associations. In: A.N. Rai (Editor), CRC Handbook of Symbiotic Cyanobacteria. CRC Press, Boca Raton. pp. 173-230.
Häger, K-P., Danneberg, G. and Bothe, H., 1983. The glutamate synthase in heterocysts of Nostoc muscorum. FEMS Microbiol. Lett., 17: 179-183.
Hallenbeck, P.C., 1987. Molecular aspects of nitrogen fixation by photosynthetic prokaryotes. CRC Critical Rev. Microbiol., 14: 1-48.
Hallenbeck, P.C., Kostel, P.J. and Benemann, J.R., 1979. Purification and properties of nitrogenase from the cyanobacterium Anabaena cylindrica. Eur. J. Biochem., 98: 275-284.
Haselkorn, R., 1978. Heterocysts. Ann. Rev. Plant Physiol., 29: 319-344.
Haselkorn, R., 1986. Organisation of the genes for nitrogen fixation in photosynthetic bacteria and cyanobacteria. Ann. Rev. Microbiol., 40: 525-547.
Haselkorn, R., Golden, J.W., Lammers, P.J. and Mulligan, M.E., 1987. Rearrangements of nif genes during cyanobacterial heterocyst differentiation. Phil. Trans. Roy. Soc. Lond. B, 317: 173-181.
Haury, J.F. and Spiller, H., 1981. Fructose uptake and influence on growth and nitrogen fixation by Anabaena variabilis. J. Bacteriol., 147: 227-235.
Haury, J.F. and Wolk, C.P., 1978. Classes of Anabaena variabilis mutants with oxygen-sensitive nitrogenase activity. J. Bacteriol., 136: 688-692.
Hawkesford, M.J., Rowell, P, and Stewart, W.D.P., 1983. Energy transduction in cyanobacteria. In: G.C. Papageorgiou and L. Packer (Editors), Photosynthetic Prokaryotes: Cell Differentiation and Function. Elsevier, New York. pp. 199-218.
Haystead, A., Robinson, R. and Stewart, W.D.P., 1970. Nitrogenase activity in extracts of heterocystous and non-heterocystous blue-green algae. Arch. Mikrobiol., 74: 235-243
Haystead, A. and Stewart, W.D.P., 1972. Characteristics of the nitrogenase system of the blue-green alga Anabaena cylindrica. Arch. Mikrobiol., 82: 325-336.
Helber, J.T., Johnston, T.R., Yarbrough, L.R. and Hirschberg, R., 1988a. Regulation of nitrogenase gene expression in anaerobic cultures of Anabaena variabilis. J. Bacteriol., 170: 552-557.
Helber, J.T., Johnston, T.R., Yarbrough, L.R. and Hirschberg, R., 1988b. Effect of nitrogenous compounds on nitrogenase gene expression in anaerobic cultures of Anabaena variabilis. J. Bacteriol., 170: 558-563.
Herrero, A. and Wolk, C.P., 1986. Genetic mapping of the chromosome of the cyanobacterium, Anabaena variabilis. J. Biol. Chem., 261: 7748-7754.
Hien, N.T., Kerby, N.W., Machray, G.C., Rowell, P. and Stewart, W.D.P., 1988. Expression of glutamine synthetase in mutant strains of the cyanobacterium Anabaena variabilis which liberate ammonia. FEMS Microbiol. Lett., 56: 337-342.
Hill, D.J., 1975. The pattern of development of Anabaena in the Azolla-Anabaena symbiosis. Planta, 122: 179-184.

Hill, D.J., 1977. The role of Anabaena in the Azolla-Anabaena symbiosis. New Phytol., 78: 611-616.
Houchins, J.P., 1984. The physiology and biochemistry of hydrogen metabolism in cyanobacteria. Biochim. Biophys. Acta, 768: 227-255.
Houchins, J.P. and Burris, R.H., 1981. Occurrence and localization of two distinct hydrogenases in the heterocystous cyanobacterium Anabaena sp. strain 7120. J. Bacteriol., 146: 209-214
Houchins, J.P. and Hind, G., 1982. Pyridine nucleotides and H_2 as electron donors to the respiratory and photosynthetic electron-transfer chains and to nitrogenase in Anabaena heterocysts. Biochim. Biophys. Acta, 682: 86-96.
Huang, T-C. and Chow, T-J., 1986. New type of N_2-fixing unicellular cyanobacterium (blue-green alga). FEMS Microbiol. Lett., 36: 109-110.
Huang, T-C. and Chow, T-J., 1990. Characterization of the rhythmic nitrogen-fixing activity of Synechococcus sp. RF-1 at the transcription level. Curr. Microbiol., 20: 23-26.
Janaki, S. and Wolk, C.P., 1982. Synthesis of nitrogenase by isolated heterocysts. Biochim. Biophys. Acta, 696: 187-192.
Jensen, B.B. and Cox, R.P., 1983. Effect of oxygen concentration on dark nitrogen fixation and respiration in cyanobacteria. Arch. Microbiol., 135: 287-292.
Joseph, C.M. and Meeks, J.C., 1987. Regulation of expression of glutamine synthetase in a symbiotic Nostoc strain associated with Anthoceros punctatus. J. Bacteriol., 169: 2471-2475.
Jüttner, F., 1983. ^{14}C-labelled metabolites in heterocysts and vegetative cells of Anabaena cylindrica filaments and their presumptive function as transport vehicles of organic carbon and nitrogen. J. Bacteriol., 155: 628-633.
Kallas, T., Coursin, T. and Rippka, R., 1985. Different organisation of nif genes in non-heterocystous and heterocystous cyanobacteria. Plant Mol. Biol., 5: 321-329.
Kallas, T., Rebiere, M-C., Rippka, R. and Tandeau de Marsac, N., 1983. The structural nif genes of the cyanobacteria Gloeothece sp. and Calothrix sp. share homology with those of Anabaena sp., but the Gloeothece genes have a different arrangement. J. Bacteriol., 155: 427-431.
Kaplan, D., Calvert, H.E. and Peters, G.A., 1986. The Azolla-Anabaena azollae relationship. XII. Nitrogenase activity and phycobiliproteins of the endophyte as a function of leaf age and cell type. Plant Physiol., 80: 884-890.
Kaplan, D. and Peters, G.A., 1988. Interaction of carbon metabolism in the Azolla-Anabaena symbiosis. Symbiosis, 6: 53-68.
Karni, L. and Tel-Or, E., 1983. Isocitrate dehydrogenase as a potential electron donor to nitrogenase of Nostoc muscorum. In: G.C. Papageorgiou and L. Packer (Editors), Photosynthetic Prokaryotes: Cell Differentiation and Function. Elsevier, New York. pp. 257-264.
Kentemich, T., Danneberg, G., Hundeshagen, B. and Bothe, H., 1988. Evidence for the occurrence of the alternative vanadium-containing nitrogenase in the cyanobacterium, Anabaena variabilis. FEMS Microbiol. Lett., 51: 19-24.
Kerby, N.W., Musgrave, S.C., Shestakov, S.V., Rowell, P. and Stewart, W.D.P., 1986. Photoproduction of ammonium by immobilized mutant strains of Anabaena variabilis. Appl. Microbiol. Biotechnol., 24: 42-46.
Kerby, N.W. and Rowell, P., 1990. Potential and commercial applications for photosynthetic prokaryotes. In: N.G. Carr and N.H. Mann (Editors), Biotechnology Handbook - Photosynthetic Prokaryotes. Plenum Press, London, (in press).
Kerby, N.W., Rowell, P. and Reglinski, A., 1990. Characterization of ammonia analogue resistant mutants of the cyanobacterium Anabaena variabilis, In: W. Ullrich, C. Rigano and A. Fuggi (Editors), Inorganic Nitrogen Metabolism, volume II, Springer Verlag, Berlin, (in press).

Kerby, N.W., Rowell, P. and Stewart, W.D.P, 1987. Cyanobacterial ammonium transport, ammonium assimilation, and nitrogenase regulation. NZ J. Mar. Freshwater Res., 21: 447-455.
Kerby, N.W., Rowell, P. and Stewart, W.D.P, 1989. The transport, assimilation and production of nitrogenous compounds by cyanobacteria and microalgae. In: R.C. Cresswell, T.A.V. Rees, and N. Shah (Editors), Algal and Cyanobacterial Biotechnology. Longman Scientific and Technical, Harlow, pp. 50-90.
Khamees, H.S., Gallon, J.R. and Chaplin, A.E., 1987. The pattern of acetylene reduction by cyanobacteria grown under alternating light and darkness. Br. Phycol. J., 22: 55-60.
Kranz, R.G. and Haselkorn, R., 1986. Anaerobic regulation of nitrogen-fixation genes in Rhodopseudomonas capsulata. Proc. Natl. Acad. Sci. USA, 83: 6805-6809.
Kumar, A.P., Perraju, B.T.V.V. and Singh, H.N, 1986. Carbon nutrition and the regulation of uptake hydrogenase activity in free-living and symbiotic Anabaena cycadeae. New Phytol., 104: 115-120.
Lambert, G.R. and Smith G.D., 1981. The hydrogen metabolism of cyanobacteria. Biol. Rev., 56: 589-660.
Lammers, P.J and Haselkorn, R., 1983. Sequence of the nifD gene coding for the α-subunit of dinitrogenase from the cyanobacterium Anabaena. Proc. Natl. Acad. Sci. USA, 80: 4723-4727.
Lang, N.J. and Fay, P., 1971. The heterocysts of blue-green algae. II. Details of ultrastructure. Proc. Roy. Soc. Lond. B., 178: 192-203.
Latorre, C., Lee, J.H., Spiller, H. and Shanmugam, K.T., 1986. Ammonium ion excreting cyanobacterial mutant as a source of nitrogen for the growth of rice: a feasibility study. Biotechnol. Lett., 8: 507-512.
Lee, K.Y., Joseph. C.M. and Meeks, J.C., 1988. Glutamine synthetase specific activity and protein concentration in symbiotic Anabaena associated with Azolla caroliniana. Antonie van Leeuwenhoek, 54: 345-355.
Lindblad, P. and Bergman, B., 1986. Glutamine synthetase: activity and localization in cyanobacteria of the cycads Cycas revoluta and Zamia skinneri. Planta, 169: 1-7.
Lindblad, P. and Bergman, B., 1989. The cycad-cyanobacterial symbiosis. In: A.N. Rai (Editor), CRC Handbook of Symbiotic Cyanobacteria. CRC Press, Boca Raton. pp. 137-159.
Lindblad, P., Hallbom, L. and Bergman, B., 1985. The cyanobacterium-Zamia symbiosis: C_2H_2-reduction and heterocyst frequency. Symbiosis, 1: 19-28.
Lindblad, P., Haselkorn, R., Bergman, B. and Nierzwicki-Bauer, S.A., 1989. Comparison of DNA restriction length polymorphisms of Nostoc strains in and from cycads. Arch. Microbiol., 152: 20-24.
Lindblad, P., Rai, A.N. and Bergman, B., 1987. The Cycas revoluta-Nostoc symbiosis: enzyme activities of nitrogen and carbon metabolism in the cyanobiont. J. Gen. Microbiol., 133: 1695-1699.
Lockau, W., Peterson, R.B., Wolk, C.P. and Burris, R.H., 1978. Modes of reduction of nitrogenase in heterocysts isolated from Anabaena species. Biochim. Biophys. Acta, 502: 298-308.
Lynn, M.E., Bantle, J.A. and Ownby, J.D. 1986. Estimation of gene expression in heterocysts of Anabaena variabilis by using DNA-RNA hybridization. J. Bacteriol., 167: 940-946.
Lynn, M.E. and Ownby, J.D., 1987. Transcriptional activity of heterocysts isolated from Anabaena variabilis. Arch. Microbiol., 148: 115-120.
Machray, G.C. and Stewart, W.D.P., 1985. Genetics of plant-microbe nitrogen-fixing symbiosis. Proc. Roy. Soc. (Edinburgh), 85B: 239-252.
Mackey, E.J. and Smith G.D., 1983. Adaptation of the cyanobacterium Anabaena cylindrica to high oxygen tensions. FEBS Lett., 156: 108-112.
Magasanik, B., 1988. Reversible phosphorylation of an enhancer binding protein regulates the transcription of bacterial nitrogen utilization genes. Trends Biochem. Sci., 13: 475-479.

Malin, G. and Pearson, H.W., 1988. Aerobic nitrogen fixation in aggregate-forming cultures of the non-heterocystous cyanobacterium Microcoleus chthonoplastes. J. Gen. Microbiol., 134: 1755-1763.
Maryan, P.S., Eady, R.R., Chaplin, A.E. and Gallon, J.R., 1986. Nitrogen fixation by Gloeothece sp. PCC 6909: respiration and not photosynthesis supports nitrogenase activity in the light. J. Gen. Microbiol., 132: 789-796.
Mazur, B.J. and Chui, C.F., 1982. Sequence of the gene coding for the ß-subunit of dinitrogenase from the blue-green alga Anabaena. Proc. Natl. Acad. Sci. USA, 79: 6782-6786.
Meeks, J.C., 1981. Radiotracer studies on nitrogen metabolism in cyanobacteria. In: J.W. Root and K.A. Krohn (Editors), Short-lived Radionuclides in Chemistry and Biology. American Chemical Society. pp. 269-294.
Meeks, J.C., 1989. Cyanobacterial-bryophyte associations. In: A.N. Rai (Editor), CRC Handbook of Symbiotic Cyanobacteria. CRC Press, Boca Raton. pp. 9-14.
Meeks, J.C., Enderlin, C.S., Joseph, C.M., Chapman, J.S. and Lollar, M.W.L., 1985a. Fixation of $[^{13}N]N_2$ and transfer of fixed nitrogen in the Anthoceros-Nostoc symbiotic association. Planta, 164: 406-414.
Meeks, J.C., Enderlin, C.S., Wycoff, K.L., Chapman, J.S. and Joseph, C.M., 1983a. Assimilation of $^{13}NH_4^+$ by Anthoceros grown with and without symbiotic Nostoc. Planta, 158: 384-391.
Meeks, J.C., Joseph, C.M. and Haselkorn, R., 1988. Organization of nif genes in the cyanobacteria in symbiotic association with Azolla and Anthoceros. Arch. Microbiol., 150: 61-71.
Meeks, J.C., Steinberg, C.M., Enderlin, C.S., Joseph, C.M. and Peters, G.A., 1987. Azolla-Anabaena relationship XIII. Fixation of $[^{13}N]N_2$. Plant Physiol., 84: 883-886.
Meeks, J.C., Steinberg, C.M., Joseph, C.M., Enderlin, C.S., Jorgensen, P.A. and Peters, G.A., 1985b. Assimilation of exogenous and dinitrogen-derived $^{13}NH_4^+$ by Anabaena azollae separated from Azolla caroliniana Willd. Arch. Microbiol., 142: 229-233.
Meeks, J.C., Wycoff, K.L., Chapman, J.S. and Enderlin, C.S., 1983b. Regulation and expression of nitrate and dinitrogen assimilation by Anabaena species. Appl. Environ. Microbiol., 45: 1351-1359.
Mevarech, M., Rice, D. and Haselkorn, R., 1980. Nucleotide sequence of a cyanobacterial nifH gene coding for nitrogenase reductase. Proc. Natl. Acad. Sci. USA., 77: 6476-6480.
Millbank, J.W., 1984. Nitrogen fixation by lichens. In: N.S. Subba Rao (Editor), Current Developments in Biological Nitrogen Fixation. Edward Arnold, London. pp. 197-218.
Mitsui, A., Cao, S., Takahashi, A., and Arai, T., 1987. Growth synchrony and cellular parameters of the unicellular nitrogen-fixing marine cyanobacterium, Synechococcus sp. strain Miami BG 043511 under continuous illumination. Physiol. Plant., 69: 1-8.
Mitsui, A., Kumazawa, S., Takahashi, A., Ikemoto, H., Cao, S. and Arai, T., 1986. Strategy by which nitrogen-fixing unicellular cyanobacteria grow phototautotrophically. Nature, 323: 720-722.
Mitsui, A., Phlips, E.J., Kumazawa, S., Reddy, J.K, Ramchandran, S., Matsunga, T., Haynes, L. and Ikemoto, H., 1983. Progress in research toward out- door biological hydrogen production using solar energy, sea water, and marine photosynthetic microorganisms. Ann. New York Acad. Sci., 413: 515-530.
Mulligan, M.E., Buikema, W.J. and Haselkorn, R., 1988. Bacterial-type ferredoxin genes in the nitrogen fixation regions of the cyanobacterium Anabaena sp. strain PCC 7120 and Rhizobium meliloti. J. Bacteriol., 170: 4406-4410.
Mulligan, M.E. and Haselkorn, R., 1989. Nitrogen fixation (nif) genes of the cyanobacterium Anabaena species strain PCC 7120. J. Biol. Chem., 264: 19200-19207.

Mullineaux, P.M., Chaplin, A.E. and Gallon, J.R., 1980. Effects of a light to dark transition on carbon reserves, nitrogen fixation and ATP concentrations in cultures of Gloeocapsa (Gloeothece) sp. 1430/3. J. Gen. Microbiol., 120: 227-232.
Mullineaux, P.M., Gallon, J.R. and Chaplin, A.E., 1981. Acetylene reduction (nitrogen fixation) by cyanobacteria grown under alternating light-dark cycles. FEMS Microbiol. Lett., 10: 245-247.
Murry, M.A., 1983. Evidence for the lack of oxygen repression of Fe-Mo protein in mutants of Anabaena variabilis. Curr. Microbiol., 8: 159-163.
Murry, M.A., Hallenbeck, P.C. and Benemann, J.R., 1984a. Immunochemical evidence that nitrogenase is restricted to the heterocysts in Anabaena cylindrica. Arch. Microbiol., 137: 194-199.
Murry, M.A., Horne, A.J. and Benemann, J.R., 1984b. Physiological studies of oxygen protection mechanisms in heterocysts of Anabaena cylindrica. Appl. Environ. Microbiol., 47: 449-454.
Murry, M.A., Jensen, D.B. and Benemann, J.R., 1983. Role of ammonia in regulation of nitrogenase synthesis and activity in Anabaena cylindrica. Biochim. Biophys. Acta, 756: 13-19.
Murry, M.A. and Wolk, C.P., 1989. Evidence that the barrier to the penetration of oxygen into heterocysts depends upon two layers of the cell envelope. Arch. Microbiol., 151: 469-474.
Neuer, G. and Bothe, H., 1985. Electron donation to nitrogenase in heterocysts of cyanobacteria. Arch. Microbiol., 143: 185-191.
Neuer, G., Papen, H. and Bothe, H., 1983. Heterocyst biochemistry and differentiation. In: G.C. Papageorgiou and L. Packer (Editors), Photosynthetic Prokaryotes: Cell Differentiation and Function. Elsevier, New York. pp. 219-242.
Nierzwicki-Bauer, S.A. and Haselkorn, R., 1986. Differences in mRNA levels in Anabaena living either freely or in symbiotic association with Azolla. EMBO J., 5: 29-35.
Ohki, K. and Fujita, Y., 1988. Aerobic nitrogenase activity measured as acetylene reduction in the marine non-heterocystous cyanobacterium Trichodesmium spp. grown under artificial conditions. Mar. Biol., 98: 111-114.
Ohmori, M. and Hattori, A., 1974. Effect of ammonia on nitrogen fixation by the blue-green alga Anabaena cylindrica. Plant Cell Physiol., 15: 131-142.
Ohmori, M. and Hattori, A., 1978. Transient changes in the ATP pool of Anabaena cylindrica associated with ammonia assimilation. Arch. Microbiol., 117: 17-20.
Orr, J. and Haselkorn, R., 1982. Regulation of glutamine synthetase activity and synthesis in free-living and symbiotic Anabaena spp. J. Bacteriol., 152: 625-635.
Ownby, J.D., 1977. Effects of amino acids on methionine-sulfoximine-induced heterocyst formation in Anabaena. Planta, 136: 277-279.
Packer, L. and Glazer, A.N. (Editors), 1988. Cyanobacteria, (Methods in Enzymology, Vol. 167), Academic Press, San Diego.
Paerl, H.W. and Bebout, B.M., 1988. Direct measurements of O_2-depleted microzones in marine Oscillatoria (Trichodesmium): relation to N_2-fixing capabilities. Science, 241: 441-445.
Paerl, H.W., Priscu, J.C. and Brawner, D.L., 1989. Immunocytochemical localization of nitrogenase in marine Trichodesmium aggregates: relationship to N_2 fixation potential. Appl. Environ. Microbiol., 55: 2965-2975.
Papen, H., Neuer, G., Refaian, M. and Bothe, H., 1983. The isocitrate dehydrogenase from cyanobacteria. Arch Microbiol., 134: 73-79.
Pate, J.S., Lindblad, P. and Atkins, C.A., 1988. Pathways of assimilation and transfer of fixed nitrogen in coralloid roots of cycad-Nostoc symbioses. Planta, 176: 461-471.

Pearson, H.W., Malin, G. and Howsley, R., 1981. Physiological studies on in vivo nitrogenase activity by axenic cultures of the blue-green alga Microcoleus chthonoplastes. Br. Phycol. J., 16: 139.
Peters, G.A. and Calvert, H.E., 1983. The Azolla-Anabaena azollae symbiosis. In: L.J. Goff (Editor), Algal Symbiosis. Cambridge University Press, Cambridge. pp. 109-145.
Peters, G.A., Kaplan, D., Meeks, J.C., Buzby, K.M., Marsh, B.H. and Corbin J.L., 1985. Aspects of nitrogen and carbon interchange in the Azolla-Anabaena symbiosis. In: P.W. Ludden and J.E. Burris (Editors), Nitrogen Fixation and CO_2 Metabolism. Elsevier, New York. pp. 213-222.
Peters, G.A. and Meeks, J.C., 1989. The Azolla-Anabaena symbiosis: basic biology. Ann. Rev. Plant Physiol. Plant Mol. Biol., 40: 193-210.
Peterson, R.B. and Burris, R.H., 1976. Properties of heterocysts isolated with colloidal silica. Arch. Microbiol., 108: 35-40.
Peterson, R.B. and Burris, R.H., 1978. Hydrogen metabolism in isolated heterocysts of Anabaena 7120. Arch. Microbiol., 116: 125-132.
Peterson, R.B. and Wolk, C.P., 1978a. High recovery of nitrogenase activity and of ^{55}Fe-labeled nitrogenase in heterocysts isolated from Anabaena variabilis. Proc. Natl. Acad. Sci. USA, 75: 6271-6275.
Peterson, R.B. and Wolk, C.P., 1978b. Localization of an uptake hydrogenase in Anabaena. Plant Physiol., 61: 688-691.
Petro, M.J. and Gates, J.E., 1987. Distribution of Arthrobacter sp. in leaf cavities of four species of the N-fixing Azolla fern. Symbiosis, 3: 41-48.
Pienkos, P.T., Bodmer, S. and Tabita, F.R., 1983. Oxygen inactivation and recovery of nitrogenase activity in cyanobacteria. J. Bacteriol., 153: 182-190.
Polukhina, L.E., Sakhurieva, G.N. and Shestakov, S.V., 1982. Ethylenediamine-resistant Anabaena variabilis mutants with derepressed nitrogen-fixing system. Microbiology, 51: 90-95.
Privalle, L.S. and Burris, R.H., 1983. Adenine nucleotide levels in and nitrogen fixation by the cyanobacterium Anabaena sp. strain 7120. J. Bacteriol. 154: 351-355.
Privalle, L.S. and Burris, R.H., 1984. D-Erythrose supports nitrogenase activity in isolated Anabaena sp. strain 7120 heterocysts. J. Bacteriol., 157: 350-356.
Rai, A.N., (Editor), 1989a. CRC Handbook of Symbiotic Cyanobacteria. CRC Press, Boca Raton.
Rai, A.N., 1989b. Cyanobacterial-fungal symbioses: the cyanolichens. In: A.N. Rai (Editor), CRC Handbook of Symbiotic Cyanobacteria. CRC Press, Boca Raton. pp. 9-41.
Rai, A.N., Lindblad, P. and Bergman, B., 1986. Absence of the glutamine-synthetase-linked methylammonium (ammonium)-transport system in the cyanobiont of the Cycas-cyanobacterial symbiosis. Planta, 169: 379-381.
Rai, A.N., Rowell, P. and Stewart, W.D.P., 1980. NH_4^+ assimilation and nitrogenase regulation in the lichen Peltigera aphthosa Willd. New Phytol., 85: 545-555.
Rai, A.N., Rowell, P. and Stewart, W.D.P., 1981a. Glutamate synthase activity in symbiotic cyanobacteria. J. Gen. Microbiol., 126: 515-518.
Rai, A.N., Rowell, P. and Stewart, W.D.P., 1981b. $^{15}N_2$ incorporation and metabolism in the lichen Peltigera aphthosa Willd. Planta, 152: 544-552.
Rai, A.N., Rowell, P. and Stewart, W.D.P., 1981c. Nitrogenase activity and dark CO_2 fixation in the lichen Peltigera aphthosa Willd. Planta, 151: 256-264.
Rai, A.N., Rowell, P. and Stewart, W.D.P., 1982. Glutamate synthase activity of heterocysts and vegetative cells of the cyanobacterium Anabaena variabilis Kütz. J. Gen. Microbiol., 128: 2203-2205.
Rai, A.N., Rowell, P. and Stewart, W.D.P., 1983a. Interactions between cyanobacterium and fungus during $^{15}N_2$-incorporation and metabolism in the lichen Peltigera canina. Arch. Microbiol., 134: 136-142.

Rai, A.N., Rowell, P. and Stewart, W.D.P., 1983b. Mycobiont-cyanobiont interactions during dark nitrogen fixation by the lichen Peltigera aphthosa. Physiol. Plant., 57: 285-290.
Ramos, J.L. and Guererro, M.G., 1983. Involvement of ammonium metabolism in the nitrate inhibition of nitrogen fixation in Anabaena sp. strain ATCC 33047. Arch. Microbiol., 136: 81-83.
Reglinski, A., Rowell, P., Kerby, N.W. and Stewart, W.D.P., 1989. Characterization of methlyammonium/ammonium transport in mutant strains of Anabaena variabilis resistant to ammonium analogues. J. Gen. Microbiol., 135: 1441-1451.
Reich, S., Almon, H., and Böger, P., 1986. Short-term effect of ammonia on nitrogenase activity of Anabaena variabilis (ATCC29413). FEMS Microbiol. Lett., 34: 53-56.
Reich, S. and Böger, P., 1989. Regulation of nitrogenase activity in Anabaena variabilis by modification of the Fe protein. FEMS Microbiol. Lett., 58: 81-86.
Rice, D., Mazur, B.J. and Haselkorn, R., 1982. Isolation and physical mapping of nitrogen fixation genes from the cyanobacterium Anabaena 7120. J. Biol. Chem., 257: 13157-13163.
Rippka, R., 1988. Recognition and identification of cyanobacteria. In: L. Packer and A.N. Glazer (Editors), The Cyanobacteria, (Methods in Enzymology, Vol. 167). Academic Press, San Diego, pp. 28-67.
Rippka, R., Deruelles, J., Waterbury, J.B., Herdman, M. and Stanier, R.Y., 1979. Generic assignments, strain histories and properties of pure cultures of cyanobacteria. J. Gen. Microbiol., 111: 1-61.
Rippka, R., Neilson, A., Kunisawa, R. and Cohen-Bazire, G., 1971. Nitrogen fixation by unicellular blue-green algae. Arch. Mikrobiol., 76: 341-348.
Rippka, R. and Stanier, R.Y., 1978. The effects of anaerobiosis on nitrogenase synthesis and heterocyst development by Nostocacean cyanobacteria. J. Gen. Microbiol., 105: 83-94.
Rippka, R. and Waterbury, J.B., 1977. The synthesis of nitrogenase by non-heterocystous cyanobacteria. FEMS Microbiol. Lett., 2: 83-86.
Rodgers, G.A. and Stewart, W.D.P., 1977. The cyanophyte-hepatic symbiosis. I. Morphology and physiology. New Phytol., 78: 441-458.
Roger, P.A. and Kulasooriya, S.A., 1980. Blue-Green Algae and Rice. IRRI, Manila.
Rogerson, A.C., 1979. Modifiers of heterocyst repression and spacing and formation of heterocysts without nitrogenase in the cyanobacterium Anabaena variabilis. J. Bacteriol., 140: 213-219.
Rowell, P., Darling, A.J., Amla, D.V. and Stewart, W.D.P., 1988. Thioredoxin and enzyme regulation. In: L.J. Rogers and J.R. Gallon (Editors), Biochemistry of the Algae and Cyanobacteria, (Proc. Phytochem. Soc. Eur., 28). Oxford University Press, Oxford, pp. 201-216.
Rowell, P., Diez, J., Apte, S.K. and Stewart, W.D.P., 1981. Molecular heterogeneity of ferredoxin: $NADP^+$ oxidoreductase from the cyanobacterium Anabaena cylindrica. Biochim. Biophys. Acta, 657: 507-516.
Rowell, P., Enticott, S. and Stewart, W.D.P., 1977. Glutamine synthetase and nitrogenase activity in the blue-green alga Anabaena cylindrica. New Phytol., 79: 41-54.
Rowell, P., Rai, A.N. and Stewart W.D.P., 1985. Studies on the nitrogen metabolism of the lichens Peltigera aphthosa and Peltigera canina. In: D.H. Brown (Editor), Lichen Physiology and Cell Biology. Plenum, New York. pp. 145-160.
Rowell, P. and Simpson, G.C., 1990. The role of amino acids in the regulation of cyanobacterial glucose-6-phosphate dehydrogenase. In: W. Ullrich, C. Rigano and A. Fuggi (Editors), Inorganic Nitrogen Metabolism, volume II, Springer Verlag, Berlin, (in press).

Rowell, P. and Stewart, W.D.P., 1976. Alanine dehydrogenase of the N_2-fixing blue-green alga Anabaena cylindrica. Arch. Microbiol., 107: 115-124.
Saville, B., Straus, N. and Coleman, J.R., 1987. Contiguous organization of nitrogenase genes in a heterocystous cyanobacterium. Plant Physiol., 85: 26-29.
Schaeffer, F. and Stanier, R.Y., 1978. Glucose-6-phosphate dehydrogenase of Anabaena sp. Kinetic and molecular properties. Arch. Microbiol., 116: 9-19.
Scherer, S., Alpes, I., Sadowski, H. and Böger, P., 1988. Ferredoxin-$NADP^+$ oxidoreductase is the respiratory NADPH dehydrogenase of the cyanobacterium Anabaena variabilis. Arch. Biochem. Biophys., 267: 228-235.
Schilling, N. and Ehrnsperger, K., 1985. Cellular differentiation of sucrose metabolism in Anabaena variabilis. Z. Naturforsch., 40c: 776-779.
Schrautemeier, B. and Böhme, H., 1985. A distinct ferredoxin for nitrogen fixation isolated from heterocysts of the cyanobacterium Anabaena variabilis. FEBS Lett., 184: 304-308.
Schrautemeier, B., Böhme, H. and Böger, P., 1984. In vitro studies on pathways and regulation of electron transport to nitrogenase with a cell-free extract from heterocysts of Anabaena variabilis. Arch. Microbiol., 137: 14-20.
Schrautemeier, B., Böhme, H. and Böger, P., 1985. Reconstitution of a light-dependent nitrogen-fixing and transhydrogenase system with heterocyst thylakoids. Biochim. Biophys. Acta, 807: 147-154.
Silvester, W.B., 1975. Endophyte adaptation in Gunnera-Nostoc symbiosis. In: P.S. Nutman (Editor), Symbiotic Nitrogen Fixation in Plants. Cambridge University Press, Cambridge. pp. 521-538.
Singh, D.T., Rai, A.N. and Singh, H.N., 1985. Methylammonium (ammonium) uptake in a glutamine auxotroph of the cyanobacterium Anabaena cycadeae. FEBS Lett., 186: 51-53.
Singh, H.N., Rai, U.N., Rao, V.V. and Bagchi, S.N., 1983. Evidence for ammonia as an inhibitor of heterocyst and nitrogenase formation in the cyanobacterium Anabaena cycadeae. Biochem. Biophys. Res. Commun., 111: 180-187.
Singh,NH, Vaishampayan, A. and Singh, R.K., 1978. Evidence for the involvement of a genetic determinant controlling functional specificity of group VI B elements in the metabolism of N_2 and NO_3^- in the blue-green alga Nostoc muscorum. Biochem. Biophys. Res. Commun., 81: 67-74.
Singh, P.K., 1973. Nitrogen fixation by the unicellular blue-green alga Aphanothece. Arch. Mikrobiol., 92: 59-62.
Smith, D.C. and Douglas, A.E., 1987. The Biology of Symbiosis. Edward Arnold, London.
Smith, D.C., Muscatine, L. and Lewis, D.H., 1969. Carbohydrate movement from autotrophs to heterotrophs in parasitic and mutualistic symbiosis. Biol. Rev., 44: 17-90.
Smith, R.L., Van Baalen, C. and Tabita, F.R., 1987. Alteration of the Fe protein of nitrogenase by oxygen in the cyanobacterium Anabaena sp. strain CA. J. Bacteriol., 169: 2537-2542.
Smith, R.L., Van Baalen, C. and Tabita, F.R., 1988. Hydrogen uptake in Anabaena sp. strain CA does not protect nitrogenase from inactivation by hyperbaric oxygen. FEMS Microbiol. Lett., 52: 183-188.
Smoker, J.A., Owen, H.A., Lehnen, L.P. and Barnum, S.R., 1989. Ultrastructure of the nitrogen-fixing, filamentous, non-heterocystous cyanobacterium, Plectonema boryanum. Protoplasma, 152: 130-135.
Spiller, H., Latorre, C., Hassan, M.E. and Shanmugam, K.T., 1986. Isolation and characterisation of nitrogenase-derepressed mutant strains of the cyanobacterium Anabaena variabilis. J. Bacteriol., 132: 596-603.
Stacey, G., Bottomley, P.J., Van Baalen, C. and Tabita, F.R., 1979. Control of heterocyst and nitrogenase synthesis in cyanobacteria. J. Bacteriol., 137: 321-326.

Stal, L.J. and Krumbein, W.E., 1985. Nitrogenase activity in the non heterocystous cyanobacterium, Oscillatoria sp. grown under alternating light-dark cycles. Arch. Microbiol., 143: 67-71.
Stal, L.J. and Krumbein, W.E., 1987. Temporal separation of nitrogen fixation and photosynthesis in the filamentous, non-heterocystous cyanobacterium Oscillatoria sp. Arch. Microbiol. 149: 76-80.
Stal, L.J. and Heyer, H., 1987. Dark anaerobic nitrogen fixation (acetylene reduction) in the cyanobacterium Oscillatoria sp. FEMS Microbiol. Ecol., 45: 227-232.
Stanier, R.Y. and Cohen-Bazire, G., 1977. Phototrophic prokaryotes: the cyanobacteria. Ann. Rev. Microbiol., 31: 185-244.
Stewart, W.D.P., 1980a. Systems involving blue-green algae (Cyanobacteria). In: F.J. Bergersen (Editor), Methods for Evaluating Biological Nitrogen Fixation. Wiley, New York, pp. 583-635.
Stewart, W.D.P., 1980b. Some aspects of structure and function in N_2-fixing cyanobacteria. Ann. Rev. Microbiol., 34: 497-536.
Stewart, W.D.P., Haystead, A. and Pearson, H.W., 1969. Nitrogenase activity in heterocysts of filamentous blue-green algae. Nature, 224: 226-228.
Stewart, W.D.P. and Lex, M., 1970. Nitrogenase activity in the blue-green alga Plectonema boryanum strain 594. Arch. Mikrobiol., 73: 250-260.
Stewart, W.D.P. and Rodgers, G.A., 1977. The cyanophyte-hepatic symbiosis. II. Nitrogen fixation and the interchange of nitrogen and carbon. New Phytol., 78: 459-471.
Stewart, W.D.P. and Rowell, P., 1975. Effects of L-methionine-D,L-sulphoximine on the assimilation of newly fixed NH_3, acetylene reduction and heterocyst production in Anabaena cylindrica. Biochem. Biophys. Res. Commun., 65: 846-857.
Stewart, W.D.P. and Rowell, P., 1977. Modifications of nitrogen-fixing algae in lichen symbioses. Nature, 265: 371-372.
Stewart, W.D.P. and Rowell, P., 1986. Biochemistry and physiology of nitrogen fixation with particular emphasis on nitrogen-fixing phototrophs. Plant Soil, 90: 167-191.
Stewart, W.D.P., Rowell, P., Cossar, J.D. and Kerby, N.W., 1985. Physiological studies on N_2-fixing cyanobacteria. In: P.W. Ludden and J.E. Burris (Editors), Nitrogen Fixation and CO_2 Metabolism. Elsevier, New York. pp. 269-279.
Stewart, W.D.P., Rowell, P. and Rai, A.N., 1983. Cyanobacteria - eukaryotic plant symbioses. Ann. Microbiol. (Inst. Pasteur), 134 B: 205-228.
Tabita, F.R., Van Baalen, C., Smith, R.L. and Kumar, D., 1988. In vivo protection and recovery of nitrogenase from oxygen in heterocystous cyanobacteria. In: H. Bothe, F.J. de Bruijn and W.E. Newton (Editors), Nitrogen fixation: Hundred Years After. Gustav Fischer, Stuttgart. pp. 189-194.
Tandeau de Marsac, N. and Houmard, J., 1987. Advances in cyanobacterial molecular genetics. In: P. Fay and C. Van Baalen (Editors), The Cyanobacteria. Elsevier, Amsterdam. pp. 253-302.
Tel-Or, E., Luijk, L.W. and Packer, L., 1977. An inducible hydrogenase in cyanobacteria enhances N_2-fixation. FEBS Lett., 78: 49-52.
Tel-Or, E. and Stewart, W.D.P., 1976. Photosynthetic electron transport, ATP synthesis and nitrogenase activity in isolated heterocysts of Anabaena cylindrica. Biochim. Biophys. Acta., 423: 189-195.
Tel-Or, E. and Stewart, W.D.P., 1977. Photosynthetic components and activities of nitrogen-fixing isolated heterocysts of Anabaena cylindrica. Proc. Roy. Soc. Lond. B., 198: 61-86.
Tetley, R.M. and Bishop, N.I., 1979. The differential action of metronidazole on nitrogen fixation, hydrogen metabolism, photosynthesis and respiration in Anabaena and Scenedesmus. Biochim. Biophys. Acta, 546: 43-53.

Thiel, T. and Leone, M., 1986. Effect of glutamine on growth and heterocyst differentiation in the cyanobacterium Anabaena variabilis. J. Bacteriol., 168: 769-774.
Thomas, J., 1970. Absence of the pigments of photosystem II of photosynthesis in heterocysts of a blue-green alga. Nature, 228: 181-183.
Thomas, J., Meeks, J.C., Wolk, C.P., Shaffer, P.W., Austin, S.M and Chien, W-S., 1977. Formation of glutamine from [^{13}N]ammonia, [^{13}N]dinitrogen, and [^{14}C]glutamate by heterocysts isolated from Anabaena cylindrica. J. Bacteriol., 129: 1545-1555.
Thomas, J., Wolk, C.P., Shaffer, P.W., Austin, S.M. and Galonsky, A., 1975. The initial organic products of fixation of ^{13}N-labeled nitrogen gas by the blue-green alga Anabaena cylindrica. Biochem. Biophys. Res. Commun., 67: 501-507.
Tsai, L.B. and Mortenson, L.E., 1978. Interaction of the nitrogenase components of Anabaena cylindrica with those of Clostridium pasteurianum. Biochem. Biophys. Res. Commun., 81: 280-287.
Tumer, N.E., Robinson, S.J. and Haselkorn, R., 1983. Different promoters for the Anabaena glutamine synthetase gene during growth using molecular or fixed nitrogen. Nature, 306: 337-342.
Uheda, E., 1986. Isolation of hair cells from Azolla filiculoides var. japonicum leaves. Plant Cell Physiol., 27: 1255-1261.
Uheda, E., and Kitoh, S., 1989. Preparation of cyanobionts from the apical and mature regions of Azolla. J. Gen. Microbiol. Meth., 9: 91-98.
Van de Water, S.D. and Simon, R.D., 1982. Induction and differentiation of heterocysts in the filamentous cyanobacterium Cylindrospermum licheniforme. J. Gen. Microbiol., 128: 917-925.
Van der Oost, J., Kanneworff, W.A., Krab, K. and Kraayenhof, R., 1987. Hydrogen metabolism of three unicellular nitrogen-fixing cyanobacteria. FEMS Microbiol. Lett., 48: 41-45.
Van Gorkom, H.J. and Donze, M., 1971. Localization of nitrogen fixation in Anabaena. Nature, 234: 231-232.
Venkataraman, L.V., 1986. Blue-green algae as biofertilizer. In: A. Richmond (Editor), CRC Handbook of Microalgal Mass Culture. CRC Press Inc., Boca Raton. pp. 455-471.
Walsby, A.E., 1985. The permeability of heterocysts to the gases nitrogen and oxygen. Proc. Roy. Soc. Lond. B, 226: 345-366.
Watanabe, I. and Roger, P.A., 1984. Nitrogen fixation in wetland rice. In: N.S. Subba Rao (Editor), Current Developments in Biological Nitrogen Fixation. Edward Arnold, London. pp. 237-276.
Wealand, J.L., Myers, J.A. and Hirschberg, R., 1989. Changes in gene expression during nitrogen starvation in Anabaena variabilis ATCC 29413. J. Bacteriol., 171: 1309-1313.
Weare, N.M. and Benemann, J.R., 1973a. Nitrogen fixation by Anabaena cylindrica. I. Localization of nitrogen fixation in the heterocysts. Arch. Mikrobiol., 90: 323-332.
Weare, N.M. and Benemann, J.R., 1973b. Nitrogen fixation by Anabaena cylindrica. II. Nitrogenase activity during induction and aging of batch cultures. Arch. Mikrobiol., 93: 101-112.
Weare, N.M. and Benemann, J.R., 1974. Nitrogenase activity and photosynthesis in Plectonema boryanum. J. Bacteriol., 119: 258-265.
Weisshaar, H. and Böger, P., 1983. Nitrogenase activity of the non-heterocystous cyanobacterium Phormidium foveolarum. Arch. Microbiol., 136: 270-274.
Whitton, B.A. and Roger, P.A., 1989. Use of blue-green algae and Azolla in rice culture. In: R. Campbell and R.M. Macdonald (Editors), Microbial Inoculation of Crop Plants. IRL Press, Oxford. pp. 89-100.

Wilcox, M., Mitchison, G.J. and Smith, R.J., 1975. Spatial control of differentiation in the blue-green alga Anabaena. In: D. Schlessinger (Editor), Microbiology 1975, American Society for Microbiology, Washington, D.C., pp. 453-463.
Winkenbach, F. and Wolk, C.P., 1973. Activities of enzymes of the oxidative and reductive pentose phosphate pathways in heterocysts of a blue-green alga. Plant Physiol., 52: 480-483.
Wolk, C.P., 1982. Heterocysts. In: N.G. Carr and B.A. Whitton (Editors), The Biology of Cyanobacteria, Blackwell, Oxford. pp. 359-386.
Wolk, C.P. and Shaffer, P.W., 1976. Heterotrophic micro- and macrocultures of a nitrogen-fixing cyanobacterium. Arch. Microbiol., 110: 145-147.
Wood, N.B. and Haselkorn, R., 1978. Proteinase activity during heterocyst differentiation in nitrogen-fixing cyanobacteria. In: G.N. Cohen and H. Holzer (Editors), Limited Proteolysis in Microorganisms. National Institute of Health, Bethesda. pp. 159-166
Wyatt, J.T. and Silvey, J.H.G., 1969. Nitrogen fixation by Gloeocapsa. Science, 165: 908-909.

CHAPTER 17

GENETIC ENHANCEMENT OF NITROGEN FIXATION

DONALD A. PHILLIPS
Department of Agronomy and Range Science, University of California, Davis, CA 95616, U.S.A.

1 INTRODUCTION

Early interdisciplinary discussions on biological N_2 fixation recognized the potential for using genetic techniques to enhance N_2 inputs into agricultural systems (Lie and Mulder, 1971; Phillips et al., 1971). Periodic summaries since that time have indicated a continuing interest in this topic (Bond, 1975; Cannon et al., 1988). The purpose of this review is to outline our knowledge of genetic traits that may increase N_2 fixation and to supply examples in which genetic changes have enhanced that process. Although many genetically controlled traits in plants and bacteria influence N_2 fixation (e.g., photosynthesis, respiration), only those genotypes related directly to N_2 fixation will be discussed. Benefits that might be derived from genetic changes in non-rhizobial systems will be mentioned briefly, but Rhizobium- and Bradyrhizobium-legume symbioses will be emphasized because of their agricultural importance.

2 NON-RHIZOBIAL SYSTEMS

2.1 Cyanobacteria

Cyanobacteria are seldom regarded as major contributors of N_2-derived N in agricultural systems, but long-term studies at the Broadbalk plots suggest that they probably make a sustained contribution (Witty et al., 1976). Because water is a primary limitation to algal growth in most agronomic environments, studies of cyanobacteria have centred on their role in specialized situations, such as paddy rice, where Anabaena azollae can fix significant amounts of N_2 in symbiosis with the aquatic fern Azolla (Peters and Meeks, 1989). As yet, genetic techniques have not been used in either Anabaena or Azolla to increase N_2 fixation. Methods for transferring genes in Anabaena are in their infancy (Wolk et al., 1984) and genetic advances in Azolla have been hindered by the unpredictable nature of sporulation in most Azolla lines (Peters and Meeks, 1989). Increasing N_2 fixation by the genetic modification of either organism is a reasonable objective, but dedicated effort will be required to achieve that goal.

2.2 Free-living diazotrophs

Given that plant residues incorporated into agricultural soils have C/N ratios ranging from 20:1 to 30:1, while comparable values for typical soil microbes are between 4:1 and 9:1 (Brady, 1974), selection pressures should favour organisms capable of heterotrophic N_2 fixation. Although N_2-fixing heterotrophs such as *Azotobacter*, *Klebsiella* and *Clostridium* exist in agricultural soils, they are seldom the dominant microbes, and there is little evidence that they contribute significant amounts of N_2-derived N to soils (Döbereiner, 1974; Mulder and Brontonegoro, 1974). Presumably major ecological factors in addition to N availability determine which microbes prevail. Recent work has centred on finding efficient, naturally occurring N_2-fixing organisms in the soil and characterizing the environment in which they function (Gibson et al., 1988). Inoculation of cereal residues with appropriate combinations of a cellulolytic fungus and a N_2-fixing heterotroph might increase N_2 fixation markedly (Veal and Lynch, 1984), and subsequent decomposition could make the N available to plants. Alternatively, current genetic techniques could introduce *nif* genes into cellulolytic bacteria. Ideally, organisms transformed with *nif* genes would also carry some type of conditional lethal trait that would facilitate the management, registration and acceptance of these potentially desirable microbes.

2.3 Rhizosphere diazotrophs

Rhizosphere diazotrophs offer a significant opportunity for increasing N_2 fixation in non-leguminous crops. Because a wide range of N_2-fixing capabilities has been reported for these organisms (Döbereiner, 1974), it seems doubtful that their activity has been optimized in most cropping systems. The amount of N_2 reduced in any particular rhizosphere depends on the N_2-fixing capacity of the diazotroph present, the quantity of carbon released from the plant root, and the environmental conditions. The first two factors are amenable to genetic manipulation, but few direct studies have been reported.

Many different microbes with diverse N_2-fixing capacities have been isolated from rhizospheres of crop plants (Döbereiner, 1989). If naturally occurring variants that supply a major portion of the N requirements in a crop cannot be isolated, it should be possible to introduce *nif* genes into other highly competitive rhizosphere organisms, possibly pseudomonads. Modifying the recipient to excrete fixed N may be a more complicated genetic change, but such mutants could directly benefit a host plant. Living microbes that release N_2-derived N into the soil could obviously promote the growth of weeds, so it may be desirable to tailor the microbe to use a particular carbon source released by the desired host plant. This latter genetic change could be the first step toward developing a novel plant-microbe association. Although nif-derepressed mutants of Klebsiella pneumoniae that fix N_2 and excrete NH_4^+ under laboratory conditions have been described (Andersen and Shanmugam, 1977), there are no published data on how those microbes function in a rhizosphere.

Several reports indicate that genotypic variation among plants can affect N_2 fixation by rhizosphere organisms. Direct selection for acetylene-reduction activity (ARA) by roots of pearl millet (Pennisetum americanum) inoculated with Azospirillum brasilense identified plant genotypes that maintained the selected ARA phenotype in a subsequent generation (Bouton et al., 1985). Crosses among plants with high ARA produced progenies that had higher ARA and released more ^{14}C-photosynthate than those produced from plants with low ARA. Estimates of associative N_2 fixation by the ^{15}N-dilution method in 11 ecotypes of Panicum maximum inoculated with mixtures of Azospirillum species, showed two ecotypes with significantly different ^{15}N concentrations (Miranda and Boddey, 1987), and sugar cane cultivar CB 47-89 apparently had higher levels of associative N_2 fixation than others (Lima et al., 1987). Azospirillum apparently promotes plant growth through processes other than N_2 fixation (Okon et al., 1988), and plant genotypes that enhance N_2 fixation by Azospirillum probably increase its total population and therefore benefit from whatever other effects this microbe has on plants. Because it seems likely that Pennisetum or Panicum genotypes promote Azospirillum activities by supplying more carbon to the rhizosphere, it should be established whether those plants have more carbon available for exudation or possess more permeable root membranes. If their roots exude an increased proportion of recent photosynthate, then it could be important to identify genotypes with reduced dark respiration (Wilson, 1982) which might compensate for that loss. Presumably Pennisetum and Panicum are not the only agronomic plants with different rhizosphere phenotypes. Searches for rhizosphere variants in rice, maize and wheat could yield major agricultural benefits.

2.4 Frankia symbioses

The value of N_2 fixation in Frankia symbioses has been recognized for many years (Bond, 1974). The long generation time of trees and the absence of proven methods for transferring genes in Frankia, however, have prevented scientists from using genetic techniques to increase N_2 fixation in these associations. Nevertheless, the demonstration that classical microbiological techniques can be used to isolate and culture Frankia in vitro (Callaham et al., 1978) has led to the selection of more effective strains from natural variants. The development of Frankia-based forest inoculation technologies comparable to the rhizobial industry that serves agricultural cropping systems may allow a significant extension and enhancement of biological N_2 fixation (Lalonde and Simon, 1985). Techniques for directed genetic improvement of _Frankia_ _in_ _vitro_ must await the development of appropriate methods for transferring genes between _Frankia_ isolates.

2.5 Transferring _nif_ genes to cereals

Although very large increases in agricultural N_2 fixation would be realized if monocotyledonous crops with N_2-fixing organelles or nodules were developed, there have been few advances toward that challenging goal. Incorporating _nif_ genes into existing organelles is an intriguing, but daunting, possibility. The well-known O_2 lability of the nitrogenase enzyme complex suggests that mitochondria may be more suitable recipients of _nif_ genes than chloroplasts. However, the peculiar genetic organization in mitochondria (Levings and Brown, 1989) compared with the less atypical genome structure in chloroplasts (Gruissem, 1989) suggests that the latter offer more suitable prospects for genetic engineering. Perhaps, non-photosynthetic root plastids, which do not evolve O_2, would be hospitable hosts for _nif_ genes.

A recent patent application described how conjugation among known species of rhizobia produced four new bacterial strains that formed effective root nodules on sorghum, rice, wheat and barley, respectively (Nielsen and Soerensen, 1987). Data indicating the effect of these strains on N accumulation were quite striking, but there were no controls for possible N in the inoculum carrier. It would be helpful to see data published with appropriate controls including, perhaps, photographs of the nodulated roots attached to the monocot shoots. Obviously, such unexpected results must be reproduced by other workers. Although reproducible rhizobial nodulation of monocots may be achieved in the future, it is the opinion of this reviewer than significant increases in N_2 inputs to cereal cropping systems will be derived first through optimization of N_2 fixation by free-living diazotrophs in cereal rhizospheres.

3 RHIZOBIAL-LEGUME SYMBIOSES

Because rhizobial-legume symbioses are so widely used in agriculture, a slight increase in the effectiveness of N_2 reduction by a major crop such as soybean or alfalfa would probably enhance global N_2 fixation measurably and would certainly increase the efficiency of production for that cropping system. Thus many investigators have searched for genetic traits that are associated with increased N_2 fixation in rhizobia and legumes.

One can debate whether genetic changes of the bacterial or the legume partner in the symbiosis will produce a greater practical benefit for agriculture. Arguments in favour of modifying the microbe include the following points: 1) Rhizobial cells are the site of N_2 fixation, and small changes in bacterial physiology could have major effects on nitrogenase activity. 2) Existing systems for producing and applying rhizobial inoculants could be used with genetically modified strains. 3) Methods for modifying bacterial genes are more advanced than those for altering plant genes and thus benefits could be obtained sooner by working with rhizobia. Factors opposing genetic modification of rhizobia include 1) Indigenous rhizobia already in the soil may prevent the successful introduction of new strains superior in N_2 fixation. 2) Regulatory problems associated with the release of genetically altered bacteria may prevent the rapid use of new strains. 3) Legume growers often lack the experience required to handle inoculants successfully.

Arguments in favour of alteration of plant genes include: 1) Existing plant varieties will not prevent the introduction of new cultivars. 2) Regulatory problems associated with the release of genetically altered plants may be solved more easily than those related to new bacterial genotypes. 3) Legume growers may be better able to manage new plants than new bacteria. Factors against choosing genetic modification of plants include: 1) Increases in N_2 fixation will probably be small because many environmental factors interact with the plant and can limit the effect of physiological changes. 2) Incorporating favorable genetic changes for N_2 fixation into agronomically desirable plant lines takes longer than modifying bacterial traits.

The problem of indigenous rhizobia, which argues against improving the bacterial symbiont, may be overcome through careful use of plant genetic traits. That possibility is mentioned below relative to work highlighted in Table 4, but a more complete review of the complex topic of legume x Rhizobium interactions can be found elsewhere (Phillips and Teuber, 1990).

3.1 Genetic changes in rhizobia

Genetic changes in rhizobia have yielded significant and reproducible increases in N_2 fixation on their normal host plants. Extensive work has been done with B. japonicum, but limited data from other rhizobia indicate that they too are easily modified to increase N_2 reduction (Table 1).

Random mutagenesis followed by selection for increased N_2 fixation has produced a number of potentially interesting strains. The first report of mutants with increased N_2 fixation involved two B. japonicum strains tested on soybeans (Glycine max (L.) Merr.) grown under controlled environmental conditions (Maier and Brill, 1978). Those strains were not characterized genetically, but the demonstration that random mutagenesis could double N_2 fixation under controlled conditions was a milestone. Although the effect of the mutants on soybean growth under field conditions was never reported, one can predict that the absence of combined N under the controlled conditions (Maier and Brill, 1978) maximized the positive effect of increased N_2 fixation in the mutants and that field responses would be considerably smaller.

TABLE 1

Genetically altered rhizobial strains with increased nitrogen fixation

Rhizobial strain	Host	Genetic change	Observed effect	Test site[a]	Reference
B. japonicum (2 mutants)	Soybean	NTG[b] mutation	+70-100% ARA[c] +80-115% total N	GC GC	Maier and Brill, 1978
B. japonicum C33	Soybean	NTG mutation	+100% ARA +20-40% ARA +12% seed and N yield	PC GC F	Scott et al., 1979 Williams and Phillips, 1983
B. japonicum (5 mutants)	Soybean	NTG mutation	+10-70% ARA +2-15% seed yield	GC F	Paau, 1989
B. japonicum CJ1(pRK290:4:46)	Soybean	+dct sequences	+30-94% ARA	PC	Birkenhead et al.,1988
B. japonicum PJ1-1	Soybean	Hup⁺ revertant	+26% N_2 fixed -30-0% N_2 fixed +13-23% N_2 fixed	GH GH GH	Evans et al., 1985 Drevon et al., 1987 Hungria et al., 1989
R. leguminosarum bv. phaseoli CFN4205	Bean	NTG mutation-cytochrome o	+0-300% ARA	GH	Soberon et al., 1989
bv. trifolii	White clover	+nodD604	+60-300% ARA	GC	Spaink et al., 1989
bv. viciae	Pea	+pJI1008	+31-128% N_2 fixed	GC	DeJong et al., 1982
bv. viciae RBL5561	Vetch	+nodD604	+0-130% ARA	GC	Spaink et al., 1989
bv. viciae (6 mutants)	Pea	Tn5 Hup⁻	-23-+21% N_2 fixed	GC	Cunningham et al.,1985

[a]: F = field; GC = growth chamber; GH = glasshouse; PC = pure culture
[b]: NTG = N-methyl-N'-nitro-N-nitrosoguanidine
[c]: ARA = acetylene reducing activity

As a first step toward quantifying the field benefits of increased N_2 fixation in a mutant of B. japonicum USDA 110, Williams and Phillips (1983) compared the effect of the mutant strain C33 on soybeans grown in controlled and field environments. Field tests were possible because Californian soils lack indigenous soybean rhizobia. Although previous tests with C33 in pure culture had shown that it had twice the ARA but the same amount of nitrogenase protein as the wild type (Scott et al., 1979), plants nodulated with the mutant in a growth chamber had only 20 to 40% higher ARA. The latter effect was undoubtedly minimized by the availability of 2 mM nitrate, which was supplied to make the growth chamber and field environments more similar. Field tests over 2 years showed that the mutant increased soybean seed yield and N content about 12% ($p < 0.05$) relative to the wild type. At least 80% of the nodules on plants inoculated with C33 were formed by that strain. It seems probable that the 12% benefit of C33 under field conditions would not be detectable in soils containing indigenous rhizobia because the mutant would have been unable to form 80% of the nodules. Nevertheless, the fact that a mutant increasing soybean productivity could be derived from a strain like USDA 110, which was previously identified as the best of 87 USDA strains tested under field conditions (Abel and Erdman, 1964), showed the potential for this approach to increasing N_2 fixation. Subsequent studies with additional, genetically uncharacterized mutants of B. japonicum under growth chamber and field conditions have confirmed the approach used with C33 (Paau, 1989). Since random mutagenesis of R. leguminosarum bv. phaseoli also produced isolates with large increases in N_2 fixation under glasshouse conditions (Soberon et al., 1989), the technique is applicable to at least one other rhizobial species.

Efforts to increase N_2 fixation by modifying rhizobia in a more defined genetic manner have also produced positive results in controlled environments. Some, but not all, experiments comparing strains with or without uptake hydrogenase activity (Hup^+ and Hup^-) (Cunningham et al., 1985; Evans et al., 1985; Drevon et al., 1987; Hungria et al., 1989) suggest that H_2 uptake can increase N_2 fixation. However, the magnitude of the effect (+26% maximum in the absence of combined N) is so small under controlled conditions that it seems unlikely there would be any significant benefit in the field. One example of increasing N_2 fixation by R. leguminosarum bv. viciae as much as 128% through the introduction of plasmid pJI1008 (DeJong et al., 1982) supports the concept that genes capable of enhancing N_2 fixation are available, but the desirable genetic loci on that plasmid need to be identified.

The most intriguing observations in this area are those in which individual, defined genetic changes have increased ARA markedly. In one case, dct sequences from R. meliloti promoted dicarboxylic acid uptake and increased ARA in pure cultures of B. japonicum (Birkenhead et al., 1988). In another, introduction of a flavonoid-independent, hybrid nodD gene into R. leguminosarum bv. trifolii and R. leguminosarum bv. viciae increased ARA in root nodules (Spaink et al., 1989). One must hope that both observations will be extended beyond measurements of apparent N_2 fixation to include actual N assimilation before deciding whether field tests are warranted.

For those seeking to increase N_2 fixation through genetic changes in rhizobia, studies of the molecular aspects of energy production, use and conservation in rhizobia could be fruitful (O'Brian and Maier, 1989). As more mutants with altered respiratory characteristics or modified capacities to synthesize haem for leghaemoglobin are developed (O'Brian et al, 1987a, b), it may be possible to construct strains that overproduce factors currently limiting the transport and reduction of O_2 in root nodules. Other mutants in which nitrogenase has an altered electron allocation coefficient may fix more N_2 and evolve less H_2 for a given amount of reductant. As various molecular limitations to N_2 reduction are identified and removed, it may become possible to combine a series of positive genetic effects on N_2 fixation in one organism for a maximum enhancement of the process under field conditions.

3.2 Genetic changes in legumes

Two types of genetic changes favouring increased N_2 fixation have been selected in legumes: 1) simple, reasonably well defined traits (Table 2), and 2) complex characteristics with no clearly defined genetic basis (Table 3). Both categories of phenotypes have properties that may be agriculturally useful.

TABLE 2

Genetic traits in legumes that may increase symbiotic N_2 fixation

Species	Gene	Phenotype	Reference
Phaseolus vulgaris	ntsn	Nod^{++} ± nitrate	Park and Buttery, 1988
Medicago sativa	nc[a]	more N_2 fixed with GS-r[b] in nodules inhibited	Knight and Langston-Unkefer, 1988
Glycine max	nts	Nod^{++} ± nitrate	Carroll et al., 1985a,b
	nc	Nod^{++} ± nitrate	Gremaud and Harper, 1988
Pisum sativum	nod_3	Nod^{++} ± nitrate	Jacobsen, 1984
	nc	Nod^{++} ± nitrate	Duc and Messager, 1989

[a]: nc = not characterized or assigned a symbol
[b]: GS-r = root-specific glutamine synthetase

One phenotype that may increase N_2 fixation and/or agronomic yields is the nitrate-tolerant-symbiosis (Nts) trait, which allows a legume to produce abundant nodules in the presence or absence of nitrate. The phenotype was first described in pea, where the controlling gene was designated as *nod*$_3$ (Jacobsen, 1984), but other investigators doing parallel work with soybean quickly reported a similar phenotype and use the term "supernodulation" to describe it (Carroll et al., 1985a, b). Similar selection programmes have identified additional mutants with Nts phenotypes in common bean (*Phaseolus vulgaris* L.)(Park and Buttery, 1988), soybean (Gremaud and Harper, 1989), and pea (*Pisum sativum* L.) (Duc and Messager, 1989). Data from root-shoot grafting studies show that the *nts* gene in soybean (Delves et al., 1986) and an unassigned locus in pea (Duc and Messager, 1989) control Nts phenotypes through the shoot. In pea, however, the *nod*$_3$ gene controls the Nts phenotype through the root (Postma et al., 1988). Whether the

TABLE 3

Changes in plant genotype that increased symbiotic nitrogen fixation in legumes

Species	Selected trait	Effect on plant[a]	Tested in[b]	Reference
Alfalfa	ARA (2 cycles)	+51% forage mass +66% ARA +46% N_2 fixed +22% nodule mass	C	Viands et al., 1981
	ARA (3 cycles)	equal forage mass equal N_2 fixed	F	Barnes et al., 1984
	Forage mass plus %N	+22-53% N_2 fixed +16-55% forage mass	C	Phillips et al., 1985 Teuber and Phillips, 1988
		+7% N concentration +17-21% N_2 fixed equal forage mass	F	D.A. Phillips and L.R. Teuber, unpublished data
French bean	Seed yield plus ARA	+565% N_2 fixed +32% total seed N (increased to mid-parent value)	F	St. Clair et al., 1988
Red clover	Forage mass	+21-62% yield[c] +89-196% total N[c] +42-53% ARA[c]	C	Nutman, 1984

[a]: significant at $p < 0.05$ unless noted
[b]: no analysis of statistical significance
[c]: C = controlled environment; F = field conditions

unassigned recessive locus responsible for the Nts phenotype in soybean (Gremaud and Harper, 1989) corresponds to the nts gene in soybean or to the nod_3 locus in pea is not known.

The physiology of Nts plants is poorly understood. The elevated ARA values and increased nodule number and/or mass (Jacobsen, 1984; Carroll et al., 1985a; Duc and Messager, 1989; Gremaud and Harper, 1989) suggest that Nts plants might fix more N_2 than wild-type parents. However, their dry weights (Gremaud and Harper, 1989) and fresh weights (Carroll et al., 1985a) are lower than the wild-type parent. Whether some Nts mutants produce pleiotropic effects or contain other deleterious mutations is still not clear. Direct measures of N accumulation under N_2-dependent growth conditions showed that nts1116 fixed as much N_2 as the parent Bragg, while nts1007 fixed only half as much (Hansen et al., 1989). Rolfe and Gresshoff (1988) present a convincing argument that the Nts phenotype in soybean is associated with a loss of the autoregulatory processes that limit nodulation. Detailed characterization of these mutants may explain how they can apparently produce higher seed yields in some field environments (Carroll et al., 1988). Until that time, further field tests are justified to determine the true agronomic potential of the Nts phenotype.

Another potentially important observation is the finding that interfering with NH_3 assimilation in alfalfa (Medicago sativa L.) root nodules can double seedling growth and N_2 fixation (Knight and Langston-Unkefer, 1988). A bacterial toxin, tabtoxinine-β-lactam, which inactivates the root-specific, but not the nodule-specific, glutamine synthetase decreased glutamine synthetase activity of the nodule plant fraction by about 50% and apparently altered normal control mechanisms that limit N_2 reduction. The toxin was supplied through infection by a tox^+ strain of Pseudomonas syringae pv. tabaci, while several spontaneous tox^- mutants served as controls and had no effect on nitrogenase or glutamine synthetase activities. If the toxin increases N_2 fixation in other legumes, a widespread application to agricultural systems may result. If inactivation of root-specific glutamine synthetase by toxins or chemical agents is agronomically useful, the longer term application probably depends on the availability of a genetically altered plant with lower root-specific glutamine synthetase activity.

Quantitative genetic methods have been applied to the study of traits related to increased N_2 fixation in at least thirteen species of legumes (Phillips and Teuber, 1990). Four groups have selected successfully for complex phenotypes associated with agronomically desirable traits and increased N_2 fixation (Table 3). In three cases involving the outcrossing species alfalfa and red clover, the frequency of alleles favouring N_2 fixation was increased by phenotypic selection. In the other case, unknown alleles enhancing N_2 fixation were transferred from one bean cultivar to another.

P.S. Nutman's selection of superior red clover (Trifolium pratense L.) plants was apparently the first successful attempt to use genetic methods to enhance N_2 fixation in a legume (Nutman et al., 1971; Nutman and Riley, 1981; Nutman, 1984). By selecting plants with high top dry weight under N_2-dependent growth conditions in test tubes, populations of plants were produced with 62% higher yield and 196% more total N. Presumably, these increases were statistically significant, but whether similar improvements in plant growth occurred under less rigidly controlled conditions was not reported.

Heritable genotypic variation for ARA in alfalfa was first reported by Seetin and Barnes (1977). Duhigg et al. (1978) established a correlation between ARA and actual N_2 reduction under similar conditions. Those facts were exploited by using phenotypic recurrent selection to develop alfalfa populations that fixed significantly more N_2 under controlled conditions (Viands et al., 1981). In that study, two cycles of selection for either increased ARA, shoot dry weight, nodule mass or fibrous roots all produced plants with more shoot N from N_2 reduction than the original populations, MnNC-4 and MnPL-6. When plants were selected through three cycles for high ARA, the resulting populations (MnNC-7 and MnPL-10) showed no significant change in forage yield or N_2 fixation under field conditions with or without 100 kg N ha^{-1} (Barnes et al., 1984). Although a population produced by crossing MnNC-7 and MnPL-10 fixed more N_2 under field conditions than either MnNC-4 or MnPL-6, no data were reported for the appropriate control population MnNC-4 x MnPL-6. Thus one cannot assess whether the strain-cross technique (e.g. MnNC-7 x MnPL-10) had any effect on efforts to select for increased N_2 fixation. 'Nitro' alfalfa, which was released after being bred for enhanced root N accumulation by the strain-cross method (Heichel et al., 1989), likewise cannot be evaluated for genetic gains because of a similar lack of appropriate controls for the strain-cross method.

A second research effort designed to increase forage yield and quality in alfalfa involves selection for integrative growth and N assimilation traits. The goal is to produce alfalfa populations with enhanced N_2 fixation and to use those materials to simultaneously identify individual traits affecting N_2 fixation and potential agronomic benefits. Initial tests identified genotypic variation in forage N concentration when plants were tested under N_2-dependent conditions with a single Rhizobium strain or on 8 mM NH_4NO_3 (Phillips et al., 1982). Alfalfa populations with significant increases in forage dry weight and N concentration were then developed from three different starting materials and tested under controlled conditions (Teuber et al., 1984; Teuber and Phillips, 1988). When one improved population (HP32) was tested with other strains of R. meliloti and various concentrations of combined N, it still accumulated more total carbon and N than the original HP starting material (Phillips et al., 1985). Forage quality in plants grown under controlled conditions was not impaired by selection for yield (Demment et al., 1986), and the improved plant populations had increased numbers of root nodules and stems (Kapulnik et al., 1986). Such phenotypic changes may be related to a decreased autoregulatory control of root nodule initiation as postulated for the Nts phenotype in soybean (Rolfe and Gresshoff, 1988), or with a limitation in production of flavonoid nodulation signals by the plant (Kapulnik et al., 1987). Recent studies have identified flavonoid nod-gene inducers released from Moapa 69 alfalfa seed and root exudates (Maxwell et al., 1989; Hartwig et al., 1990), but attempts to relate flavonoid differences with nodulation patterns in Moapa 69 and the improved Moapa 69-33 have not been successful (U.A. Hartwig and C.A. Maxwell, unpublished). Data from ^{15}N-dilution and N-difference tests at one location indicate that increases in forage N concentration and total forage N (Table 3) resulted from enhanced N_2 fixation. If these promising results are maintained throughout the 5-year field test, the improved alfalfa populations will be released to alfalfa growers and breeders.

In common bean, the inbred-backcross-line method successfully enhanced N_2 fixation in the low N_2-fixing cultivar Sanilac, but not in the moderate N_2-fixing cultivar Porrillo Sintetico (Attewell and Bliss, 1985; St. Clair et al., 1988). That method transferred desirable ARA traits from a donor parent, Puebla 152, into a Sanilac-like genetic background by using Sanilac as a recurrent parent and selecting progenies from each backcross for high ARA and seed yield. Lines were produced that fixed more N_2 than Sanilac during 1 year of field tests, and the increases were not apparently directly related to differences in growth habit or maturity of the two parents. The 5-6-fold increase in N_2 fixation (Table 3) was calculated for line 24-21 (306 mg N_2 fixed/plant, bush habit, 93 days maturity) relative to Sanilac (46 mg N_2 fixed/plant, bush habit, 96 days maturity), but was not significantly different from the midparent value (360 mg N_2 fixed/plant) calculated as the intermediate point between Sanilac and Puebla 152, the high parent donor (674 mg N_2 fixed/plant, vine habit, 114 days maturity). Thus the experiments support the concept that traits favouring N_2 fixation can be transferred fromn one plant to another and, most importantly, that the improvement is large enough to detect under field conditions.

3.3 Controlling competition among rhizobia

Indigenous rhizobia in many agricultural soils are a major impediment to the introduction of superior strains (Chapters 10, 14). However, genetic engineering using plant and rhizobial genes may help to overcome this problem (Table 4).

Several examples of specific rhizobial genes that affect nodulation by other strains are known (Table 4). First, in R. leguminosarum bv. viciae, a gene that is responsible for the competitive-nodulation-blocking phenotype (Cnb^+) has been cloned (Dowling et al., 1987). Strains carrying the gene can prevent R. leguminosarum bv. viciae TOM from nodulating Afghanistan pea, but the mechanism is unknown. Second, in R. leguminosarum bv. trifolii T24, a gene responsible for producing the antibiotic trifolitoxin has been isolated on the cosmid clone pTFX1 (Triplett, 1988). Strain T24 is Nod^+Fix^- and inhibits growth of several rhizobial species through the production of trifolitoxin (Schwinghamer and Belkengren, 1968). Clone pTFX1, which apparently carries all the genes required for both trifolitoxin synthesis and resistance, can be transferred to several species of Nod^+Fix^+ rhizobia and does not necessarily impair ARA (Triplett, 1988). In a third example, the competitiveness of Rhizobium fredii USDA 257 was decreased by insertion of Tn5 at two separate loci (McLoughlin et al., 1987). Cloning the wild-type genes for competitiveness in that strain should present few difficulties. Various combination of these and other genes may increase the competitiveness of rhizobia with genetically-enhanced N_2 fixation.

Legume genes affecting nodulation by rhizobia have been known for many years (Table 4). Extreme cases, in which a plant prevents nodulation by all rhizobia, allow for measurement of soil N available to the legume, but such phenotypes have little agronomic utility. Less rigorous Nod^- phenotypes that can be nodulated by a few rhizobial strains, however, offer a mechanism for preventing infection by indigenous organisms. Plant genotypes conferring strain-dependent Nod^- phenotypes have been reported in soybean and pea (Table 4), and they probably exist in other legume species.

There are many examples of plant genotypes that favour one rhizobial strain over another without necessarily excluding the less favoured strain. In clover, such effects of host species (Robinson, 1969; Masterson and Sherwood, 1974) and host cultivars (Gibson, 1964; Hardarson and Jones, 1979) are known, but no mechanisms have been described. In alfalfa, populations selected for increased N_2 fixation also can show an altered preference for certain rhizobial strains (Hardarson et al., 1982; C.M. Joseph, L.R. Teuber and D.A. Phillips, unpublished).

One of the best defined cases in which a specific plant gene affects rhizobial infection involves the introduction of the pea lectin gene, psl, into clover (Diaz et al., 1989). The transgenic white clover roots could be infected by pea rhizobia. Whether the absence of a pea lectin is responsible for any of the numerous Nod^- pea mutants listed in a recent review (Phillips and Teuber, 1990) is not known.

TABLE 4

Genetic traits potentially useful for controlling competition among rhizobia

Species	Gene	Phenotype	Reference
BACTERIA			
R. leguminosarum			
bv. trifolii	nc[a]	Trifolitoxin^{+}	Triplett, 1988
bv. viciae	nc	Cnb^{+b}	Dowling et al., 1987
LEGUMES			
Soybean	rj_1	Nod^{-}, strain-dependent	Williams & Lynch, 1954; Devine & Weber, 1977; Carroll et al., 1986
	nc	Nod^{+}Fix^{+} with African cowpea rhizobia	Kueneman et al., 1984
	nc	Recessive, Fix^{+} with Rf[c] and Fix^{-} with Bj USDA123	Devine, 1984
	nc	Host cv. restricts nodulation by Bj	Cregan and Keyser, 1986; Cregan et al., 1989a, b
Alfalfa	nc	Host cv. selects Rm	Hardarson et al., 1982
Pea	sym2	Nod^{-}, strain-dependent	Holl,1975; Kneen et al.,1984
	sym4	Nod^{-}, strain-dependent	Lie, 1984
	sym6	Nod^{-}, strain-dependent	Lie and Timmermans, 1983
	sym18	Nod^{-}, strain-dependent	T.A. LaRue, pers. comm.
	psl	Clover Nod^{+}Fix^{+} with Rlv	Diaz et al., 1989
Red clover	nc	Host spp. selects Rlt	Robinson, 1969
Subterranean clover	nc	Host spp. selects Rlt	Robinson, 1969; Masterson and Sherwood, 1974
	nc	Host cv. selects Rlt	Gibson, 1964
White clover	nc	Host spp. selects Rlt	Masterson and Sherwood, 1974
	nc	Host cv. selects Rlt	Hardarson and Jones, 1979

[a]: nc = not characterized or assigned a symbol
[b]: Cnb = competitive nodulation blocking
[c]: Bj = B. japonicum; Rf = R. fredii; Rlt = R. leguminosarum bv. trifolii; Rlv = R. leguminosarum bv. viciae; Rm = R. meliloti

Soybean germplasm offers special opportunities for excluding indigenous rhizobia. The rj_1 allele produces Nod^{-} plants (Williams and Lynch, 1954), but is Nod^{+} with some chlorosis-inducing rhizobia (Devine and Weber, 1977). The specific mechanism by which those strains overcome the rj_1 nodulation block is not known. A potentially more useful characteristic occurs in several wild-type soybean introductions that restrict, but do not prevent, effective nodulation by B. japonicum strains in the 123 serocluster (Cregan and Keyser, 1986; Cregan et al., 1989a, b). This trait could allow soybeans growing in a soil with indigenous strains from the mediocre 123 serocluster to form most of their nodules with a superior inoculant. A possibly related, but more extreme, host plant effect has been observed in the primitive G. max cv. Peking (Devine, 1984) and the wild soybean (G. soja Sieb. and Zucc.)(Cregan and Keyser, 1988) which produce Fix^- nodules with B. japonicum USDA 123 and a Fix^+ phenotype with R. fredii. Other "promiscuous" G. max lines form Fix^+ nodules with both B. japonicum and Bradyrhizobium spp. indigenous to African soils (Kueneman et al., 1984). Genetic relationships among these phenotypes have not been reported. It may be possible, however, to use the nodulation-restricting alleles to control indigenous rhizobia while allowing nodulation by more desirable and effective strains.

Another possible, but potentially less desirable, mechanism for controlling which strains of B. japonicum nodulate a soybean involves the dominant soybean genes Rj_2 (Caldwell, 1966), Rj_3 (Vest, 1970) and Rj_4 (Vest and Caldwell, 1972). These alleles allow the same plant to produce ineffective root nodules with certain rhizobia and effective ones with others. The potential for controlling rhizobia in this way was demonstrated by Materon and Vincent (1980), who showed that the Rj_2 gene could decrease symbiotic associations with incompatible strains, but no data on N_2 fixation and plant growth were reported. The concern is that Fix^- nodules may be a respiratory burden that decreases plant productivity.

Using soybean genes to control rhizobial symbionts will be most important in those areas where previous inoculation practices have produced a complex mixture of mediocre organisms, such as in the midwestern U.S. As methods for controlling existing organisms become available, and new genetically altered strains are introduced, it will be agronomically and ethically appropriate to include genetic traits that decrease persistence of the new strains without impairing their symbiotic performance. If that practice is followed, future generations will be able to introduce newly improved rhizobia far more easily than is possible today.

4 CONSEQUENCES OF ENHANCING NITROGEN FIXATION

The consequences of enhancing N_2 fixation cannot be assessed in isolation because it is only one of several primary components of cropping systems which are under the control of human managers. Other factors, such as the amount of fertilizer, water supply and the phenotypes and numbers of plants and bacteria present, can modify the effects of increased N_2 fixation. One of the greatest dangers associated with increasing N_2 fixation is the possibility of nitrate pollution. Although there is little credible evidence than N_2 fixation results directly in nitrate pollution, the possibility exists in poorly managed systems, and careful agronomic studies will be required to assess optimum fertilizer N and water inputs to any system where N_2 fixation is enhanced.

A primary distinction must be made between increasing N_2 fixation by free-living diazotrophs and enhancing N_2 reduction in symbiotic systems. In the former, N_2-derived N will be released from microbial biomass and can be lost as nitrate before it is assimilated by companion plants. In symbiotic systems, N_2-derived N is controlled and used internally by the plant. Although N_2-derived N from legumes has been detected in adjacent monocots by many workers, there is no evidence that such N is released directly from living, as opposed to dying, roots. Thus N_2 fixed by symbiotic systems is directly integrated into the crop plant and offers a lower risk for undesirable side effects than N_2-derived N inputs from free-living diazotrophs. With appropriate techniques, however, major N inputs from free-living diazotrophs can probably be managed without nitrate pollution.

5 CONCLUSIONS

Evidence now available shows that positive genetic effects on N_2 fixation by rhizobial-legume symbioses can be measured at the field level. More extensive studies of natural and induced genetic variation in free-living diazotrophs may also yield significant agricultural benefits. Indeed, future analyses of how both free-living and symbiotic N_2-fixing microbes interact with plants may define the molecular events controlling both beneficial and detrimental processes in the rhizosphere.

Acknowledgements

The author's research has been partially supported by both the U.S. National Science Foundation and the U.S. Department of Agriculture Competitive Grants Office.

6 REFERENCES

Abel, G.H. and Erdman, L.W., 1964. Response of Lee soybeans to different strains of Rhizobium japonicum. Agron. J., 56: 423-424.

Andersen, K. and Shanmugam, K.T., 1977. Energetics of biological nitrogen fixation: determination of the ratio of formation of H_2 to NH_4^+ catalyzed by nitrogenase of Klebsiella pneumoniae in vivo. J. Gen. Microbiol., 103: 107-122.

Attewell, J. and Bliss, F.A., 1985. Host plant characteristics of common bean lines selected using indirect measures of N_2 fixation. In: H.J. Evans, P.J. Bottomley and W.E. Newton (Editors), Nitrogen Fixation Research Progress. Martinus Nijhoff, Dordrecht, pp. 3-9.

Barnes, D.K., Heichel, G.H., Vance, C.P. and Ellis, W.R., 1984. A multiple-trait breeding program for improving the symbiosis of N_2 fixation between Medicago sativa L. and Rhizobium meliloti. Plant Soil, 82: 303-314.

Birkenhead, K., Manian, S.S. and O'Gara, F., 1988. Dicarboxylic acid transport in Bradyrhizobium japonicum: use of Rhizobium meliloti gene(s) to enhance nitrogen fixation. J. Bacteriol., 170: 184-189.

Bond, G., 1974. Root nodule symbioses with actinomycete-like organisms. In: A. Quispel (Editor), The Biology of Nitrogen Fixation. North-Holland, Amsterdam, pp. 342-378.

Bond, G., 1975. The possibility of extending the capacity for nitrogen fixation to other plant species. In: P.S. Nutman (Editor), Symbiotic Nitrogen Fixation in Plants. Cambridge University Press, Cambridge, pp. 567-572.

Bouton, J.H., Albrecht, S.L. and Zuberer, D.A., 1985. Screening and selection of pearl millet for root associated bacterial nitrogen fixation. Field Crops Res., 11: 131-140.

Brady, N.C., 1974. The Nature and Property of Soils, 8th edition. Macmillan, New York, pp. 151-154.

Caldwell, B.E., 1966. Inheritance of a strain-specific ineffective nodulation in soybeans. Crop Sci., 6: 427-428.

Callaham, D., Torrey, J.G. and Del Tredici, P., 1978. Isolation and cultivation in vitro of the actinomycete causing root nodulation in Comptonia. Science, 199: 899-902.

Cannon, F.C., Beynon, J., Hankinson, T., Kwiatkowski, R., Legocki, R.P., Ratcliffe, H., Ronson, C., Szeto, W. and Williams, M., 1988. Increasing biological nitrogen fixation by genetic manipulation. In: H. Bothe, F.J. de Bruijn and W.E. Newton (Editors), Nitrogen Fixation: Hundred Years After. Gustav Fischer, Stuttgart, pp. 735-740.

Carroll, B.J., Hartley, R., Ashley, D.A., Boerma, H.R., Lawn, R., Byth, D. and Gresshoff, P., 1988. Field evaluation of supernodulating soybean mutants. Loc cit., p. 808.

Carroll, B.J., McNeil, D.L. and Gresshoff, P., 1985a. A supernodulation and nitrate-tolerant symbiotic (nts) soybean mutant. Plant Physiol., 78: 34-40.

Carroll, B.J., McNeil, D.L. and Gresshoff, P., 1985b. Isolation and properties of soybean (Glycine max (L.) Merr.) mutants that nodulate in the presence of high nitrate concentrations. Proc. Natl. Acad. Sci. USA, 82: 4162-4166.

Carroll, B.J., McNeil, D.L. and Gresshoff, P., 1986. Mutagenesis of soybean (Glycine max (L.) Merr.) and the isolation of non-nodulating mutants. Plant Sci., 47: 109-119.

Cregan, P.B. and Keyser, H.H., 1986. Host restriction of nodulation by Bradyrhizobium japonicum strain USDA 123 in soybean. Crop Sci., 26: 911-916.

Cregan, P.B. and Keyser, H.H., 1988. Influence of Glycine spp. on competitiveness of Bradyrhizobium japonicum and Rhizobium fredii. Appl. Environ. Microbiol., 54: 803-808.

Cregan, P.B., Keyser, H.H. and Sadowsky, M.J., 1989a. Host plant effects on nodulation and competitiveness of the Bradyrhizobium japonicum serotype strains constituting serocluster 123. Appl. Environ. Microbiol., 55: 2532-2536.

Cregan, P.B., Keyser, H.H. and Sadowsky, M.J., 1989b. Soybean genotype restricting nodulation of a previously unrestricted serocluster 123 Bradyrhizobia. Crop Sci., 29: 307-312.

Cunningham, S.D., Kapulnik, Y., Brewin, N.J. and Phillips, D.A., 1985. Uptake hydrogenase activity determined by plasmid pRL6JI in Rhizobium leguminosarum does not increase symbiotic nitrogen fixation. Appl. Environ. Microbiol., 50: 791-794.

DeJong, T.M., Brewin, N.J., Johnston, A.W.B. and Phillips, D.A., 1982. Improvement of symbiotic properties in Rhizobium leguminosarum by plasmid transfer. J. Gen. Microbiol., 128: 1829-1838.

Delves, A.C., Mathews, A., Day, D.A., Carter, A.S., Carroll, B.J. and Gresshoff, P.M., 1986. Regulation of the soybean-Rhizobium nodule symbiosis by shoot and root factors. Plant Physiol., 82: 588-590.
Demment, M.W., Teuber, L.R., Borque, D.P. and Phillips, D.A., 1986. Changes in forage quality of improved alfalfa populations. Crop Sci., 26: 1137-1143.
Devine, T.E., 1984. Inheritance of soybean nodulation response with a fast-growing strain of Rhizobium. J. Hered., 75: 359-361.
Devine, T.E. and Weber, D.F., 1977. Genetic specificity of nodulation. Euphytica, 26: 527-535.
Diaz, C.L., Melchers, L.S., Hooykaas, P.J.J., Lugtenberg, B.J.J. and Kijne, J.W., 1989. Root lectin as a determinant of host-plant specificity in the Rhizobium-legume symbiosis. Nature, 338: 579-581.
Döbereiner, J., 1974. Nitrogen-fixing bacteria in the rhizosphere. In: A. Quispel (Editor), The Biology of Nitrogen Fixation. North-Holland, Amsterdam, pp. 86-120.
Döbereiner, J., 1989. Isolation and identification of root associated diazotrophs. In: F.A. Skinner, et al. (Editors), Nitrogen Fixation with Non-Legumes. Kluwer Academic Publ., Thc Hague, pp. 103-108.
Dowling, D.N., Samrey, U., Stanley, J. and Broughton, W.J., 1987. Cloning of Rhizobium leguminosarum genes for competitive nodulation blocking on peas. J. Bacteriol., 169: 1345-1348.
Drevon, J.J., Kalia, V.C., Heckmann, M.O. and Salsac, L., 1987. Influence of the Bradyrhizobium japonicum hydrogenase on the growth of Glycine and Vigna species. Appl. Environ. Microbiol., 53: 610-612.
Duc, G. and Messager, A., 1989. Mutagenesis of pea (Pisum sativum L.) and the isolation of mutants for nodulation and nitrogen fixation. Plant Sci., 60: 207-213.
Duhigg, P. Melton, B. and Baltensperger, A., 1978. Selection for acetylene reduction rates in 'Mesilla' alfalfa. Crop Sci., 18: 813-816.
Evans, H.J., Hanus, F.J., Haugland, R.A., Cantrell, M.A., Xu, L.S., Russell, S.A., Lambert, G.R. and Harker, A.R., 1985. Hydrogen recycling in nodules affects nitrogen fixation and growth of soybeans. In: R. Shibles (Editor), World Soybean Research Conference III: Proceedings. Westview Press, Boulder, pp. 935-942.
Gibson, A.H., 1964. Genetic control of strain-specific ineffective nodulation in Trifolium subterraneum L. Aust. J. Agric. Res., 15: 37-49.
Gibson, A.H., Halsall, S.M. and Roper, M.M., 1988. Nitrogen fixation associated with residue breakdown. In: H. Bothe, F.J. de Bruijn and W.E. Newton (Editors), Nitrogen Fixation: Hundred Years After. Gustav Fischer, Stuttgart, pp. 753-758.
Gremaud, M.F. and Harper, .E., 1989. Selection and initial characterization of partially nitrate tolerant mutants of soybean. Plant Physiol., 89: 169-173.
Gruissem, W., 1989. Chloroplast gene expression: how plants turn their plastids on. Cell, 56: 161-170.
Hansen, A.P., Peoples, M.B., Gresshoff, P.M., Atkins, C.A., Pate, J.S. and Carroll, B.J., 1989. Symbiotic performance of supernodulating soybean (Glycine max (L.) Merrill) mutants during development on different nitrogen regimes. J. Expt. Bot., 40: 715-724.
Hardarson, G., Heichel, G.H., Barnes, D.K. and Vance, C.P., 1982. Rhizobial strain preference of alfalfa populations selected for characteristics associated with N_2 fixation. Crop Sci., 22: 55-58.
Hardarson, G. and Jones, D.G., 1979. The inheritance of preference for strains of Rhizobium trifolii by white clover (Trifolium repens). Ann. Appl. Biol., 92: 329-333.
Hartwig, U.A., Maxwell, C.A., Joseph, C.M. and Phillips, D.A., 1990. Chrysoeriol and luteolin released from alfalfa seeds induce nod genes in Rhizobium meliloti. Plant Physiol., 92: 116-122.

Heichel, G.H., Barnes, D.K., Vance, C.P. and Sheaffer, C.C., 1989. Dinitrogen fixation technologies for alfalfa improvement. J. Prod. Agric., 2: 24-32.
Holl, F.B., 1975. Host plant control of the inheritance of dinitrogen fixation in the Pisum-Rhizobium symbiosis. Euphytica, 24: 767-770.
Hungria, M., Neves, M.C.P. and Döbereiner, J., 1989. Relative efficiency, ureide transport and harvest index in soybeans inoculated with isogenic Hup mutants of Bradyrhizobium japonicum. Biol. Fertil. Soils, 7: 325-329.
Jacobsen, E., 1984. Modification of symbiotic interaction of pea (Pisum sativum L.) and Rhizobium leguminosarum by induced mutations. Plant Soil, 82: 427-438.
Kapulnik, Y., Joseph, C.M. and Phillips, D.A., 1987. Flavone limitations to root nodulation and symbiotic nitrogen fixation in alfalfa. Plant Physiol., 84: 1193-1196.
Kapulnik, Y., Teuber, L.R. and Phillips, D.A., 1986. Seedling development as a component of increased yield in an improved alfalfa population. Crop Sci., 26: 770-775.
Kneen, B.E., LaRue, T.A. and Weedin, N., 1984. Genes reported to affect symbiotic nitrogen fixation by peas. Pisum News Lett., 16: 31-34.
Knight, T.J. and Langston-Unkefer, P.J., 1988. Enhancement of symbiotic dinitrogen fixation by a toxin-releasing plant pathogen. Science, 241: 951-954.
Kueneman, E.A., Root, W.R., Dashiell, K.E. and Hohenberg, J., 1984. Breeding soybeans for the tropics capable of nodulating effectively with indigenous Rhizobium spp. Plant Soil, 82: 387-396.
Lalonde, M. and Simon, L., 1985. Current research strategies for use of actinorhizal symbioses. In: H.J. Evans, P.J. Bottomley and W.E. Newton (Editors), Nitrogen Fixation Research Progress. Martinus Nijhoff, Dordrecht, pp. 667-674.
Levings, C.S. and Brown, G.G., 1989. Molecular biology of plant mitochondria. Cell, 56: 171-179.
Lie, T.A., 1984. Host genes in Pisum sativum L. conferring resistance to European Rhizobium leguminosarum strains. Plant Soil, 82: 415-425.
Lie, T.A. and Mulder, E.G. (Editors), 1971. Biological Nitrogen Fixation in Natural and Agricultural Habitats. Plant and Soil, Special Volume. Martinus Nijhoff, The Hague, 590p.
Lie, T.A. and Timmermans, P.C.J.M., 1983. Host-genetic control of nitrogen fixation in the legume-Rhizobium symbiosis: complication in the genetic analysis due to maternal effects. Plant Soil, 75: 449-453.
Lima, E., Boddey, R.M. and Döbereiner, J., 1987. Quantification of biological nitrogen fixation associated with sugar cane using a ^{15}N aided nitrogen balance. Soil Biol. Biochem., 19: 165-170.
Maier, R.J. and Brill, W.J., 1978. Mutant strains of Rhizobium japonicum with increased ability to fix nitrogen for soybean. Science, 201: 448-450.
Masterson, C.L. and Sherwood, M.T., 1974. Selection of Rhizobium trifolii strains by white and subterranean clovers. Ir. J. Agric. Res., 13: 91-99.
Materon, L.A. and Vincent, J.M., 1980. Host specificity and interstrain competition with soybean rhizobia. Field Crops Res., 3: 215-224.
Maxwell, C.A., Hartwig, U.A., Joseph, C.M. and Phillips, D.A., 1989. A chalcone and two related flavonoids released from alfalfa roots induce nod genes of Rhizobium meliloti. Plant Physiol., 91: 842-847.
McLoughlin, T.J., Merlo, A.O., Satola, S.W. and Johansen, E., 1987. Isolation of competition-defective mutants of Rhizobium fredii. J. Bacteriol., 169: 410-413.
Miranda, C.H.B. and Boddey, R.M., 1987. Estimation of biological nitrogen fixation associated with 11 ecotypes of Panicum maximum grown in nitrogen-15-labelled soil. Agron. J., 79: 558-563.

Mulder, E.G. and Brontonegoro, S., 1974. Free-living heterotrophic nitrogen-fixing bacteria. In: A. Quispel (Editor), The Biology of Nitrogen Fixation. North-Holland, Amsterdam, pp. 37-85.
Nielsen, S.E. and Soerensen, G.M., 1987. Rhizobial transformants which symbiotically fix nitrogen in nonlegumes, a material for treating seeds of a nonlegume plant, nonlegume seeds, a nonlegume plant and a method for producing rhizobia transconjugants. Chem. Abstr., 107: 235523q.
Nutman, P.S., 1984. Improving nitrogen fixation in legumes by plant breeding: the relevance of host selection experiments in red clover (Trifolium pratense L.) and subterranean clover (T. subterraneum L.). Plant Soil, 82: 285-301.
Nutman, P.S. and Riley, J., 1981. Breeding of nodulated red clover (Trifolium pratense L.) for high yield. Ann. Appl. Biol., 98: 319-331.
Nutman, P.S., Mareckova, H. and Raicheva, L., 1971. Selection for increased nitrogen fixation in red clover. Plant Soil, Special Volume: 27-31.
O'Brian, M.R. and Maier, R.J., 1989. Molecular aspects of the energetics of nitrogen fixation in Rhizobium-legume symbioses. Biochim. Biophys. Acta, 974: 229-246.
O'Brian, M.R., Kirshbom, P.M. and Maier, R.J., 1987a. Tn5-induced cytochrome mutants of Bradyrhizobium japonicum: effects of the mutations on cells grown symbiotically and in culture. J. Bacteriol., 169: 1089-1094.
O'Brian, M.R., Kirshbom, P.M. and Maier, R.J., 1987b. Bacterial heme synthesis is required for expression of the leghemoglobin holoprotein but not the apoprotein in soybean root nodules. Proc. Natl. Acad. Sci. USA, 84: 8390-8393.
Okon, Y., Fallik, E., Sarig, S., Yahalom, E. and Tal, S., 1988. Plant growth promoting effects of Azospirillum. In: H. Bothe, F.J. de Bruijn and W.E. Newton (Editors), Nitrogen Fixation: Hundred Years After. Gustav Fischer, Stuttgart, pp. 741-746.
Paau, A.S., 1989. Improvement of Rhizobium inoculants. Appl. Environ. Microbiol., 55: 862-865.
Park, S.J. and Buttery, B.R., 1988. Nodulation mutqants of white bean (Phaseolus vulgaris L.) induced by ethyl-methane sulphonate. Canad. J. Plant Sci., 68: 199-202.
Peters, G.A. and Meeks, J.C., 1989. The Azolla-Anabaena symbiosis: basic biology. Ann. Rev. Plant Mol. Biol., 40: 193-210.
Phillips, D.A. and Teuber, L.R., 1990. Plant genetics of symbiotic nitrogen fixation. In: R.H. Burris, H.J. Evans and G. Stacey (Editors), Biological Nitrogen Fixation. Chapman and Hall, New York, in press.
Phillips, D.A., Cunningham, S.D., Bedmar, E.J., Sweeney, T.C. and Teuber, L.R., 1985. Nitrogen assimilation in an improved alfalfa population. Crop Sci., 25: 1011-1015.
Phillips, D.A., Teuber, L.R. and Jue, S.S., 1982. Variation among alfalfa genotypes for reduced nitrogen concentration. Crop Sci., 22: 606-610.
Phillips, D.A., Torrey, J.G. and Burris, R.H., 1971. Extending symbiotic nitrogen fixation to increase man's food supply. Science, 174: 169-171.
Postma, J.G., Jacobsen, E. and Feenstra, W.J., 1988. Three pea mutants with an altered nodulation: studies by genetic analysis and grafting. J. Plant Physiol., 132: 424-430.
Robinson, A.C., 1969. Host selection for effective Rhizobium trifolii by red clover and subterranean clover in the field. Aust. J. Agric. Res., 20: 1053-1060.
Rolfe, B.G. and Gresshoff, P.M., 1988. Genetic analysis of legume nodule initiation. Ann. Rev. Plant Physiol. Plant Mol. Biol., 39: 297-319.
Schwinghamer, E.A. and Belkengren, R.P., 1968. Inhibition of rhizobia by a strain of Rhizobium trifolii: some properties of the antibiotic and of the strain. Arch. Microbiol., 64: 130-145.

Scott, D.B., Hennecke, H. and Lim, S.T., 1979. The biosynthesis of nitrogenase MoFe protein polypeptides in free-living cultures of Rhizobium japonicum. Biochim. Biophys. Acta, 565: 365-378.
Seetin, M.W. and Barnes, D.K., 1977. Variation among alfalfa genotypes for rate of acetylene reduction. Crop Sci., 17: 783-787.
Soberon, M., Williams, H.D., Poole, R.K. and Escamilla, E., 1989. Isolation of a Rhizobium phaseoli cytochrome mutant with enhanced respiration and symbiotic nitrogen fixation. J. Bacteriol., 171: 465-472.
Spaink, H.P., Okker, R.J.H., Wijffelman, C.A., Tak, T., Goosen-De Roo,L., Pees, E., Van Brussel, A.A.N. and Lugtenberg, B.J.J., 1989. Symbiotic properties of rhizobia containing a flavonoid-independent hybrid nodD product. J. Bacteriol., 171, 4045-4053.
St. Clair, D.A., Wolyn, D.J., DuBois, J., Burris, R.H. and Bliss, F.A., 1988. Field comparison of dinitrogen fixation determined with nitrogen-15-depleted and nitrogen-15-enriched ammonium sulfate in selected inbred backcross lines of common bean. Crop Sci., 28: 773-778.
Teuber, L.R. and Phillips, D.A., 1988. Influences of selection method and nitrogen environment on breeding alfalfa for increased forage yield and quality. Crop Sci., 28: 599-604.
Teuber, L.R., Levin, R.P., Sweeney, T.C. and Phillips, D.A., 1984. Selection for N concentration and forage yield in alfalfa. Crop Sci., 24: 553-558.
Triplett, E.W., 1988. Isolation of genes involved in nodulation competitiveness from Rhizobium leguminosarum bv. trifolii T24. Proc. Natl. Acad. Sci. USA, 85: 3810-3814.
Veal, D.A. and Lynch, J.M., 1984. Associative cellulolysis and dinitrogen fixation by co-cultures of Trichoderma harzianum and Clostridium butyricum. Nature, 310: 695-697.
Vest, 1970. Rj_3 - a gene conditioning ineffective nodulation in soybean. Crop Sci., 10: 34-35.
Vest, G. and Caldwell, B.E., 1982. Rj_4 - a gene conditioning ineffective nodulation in soybean. Crop Sci., 12: 692-693.
Viands, D.R., Barnes, D.K. and Heichel, G.H., 1981. Nitrogen fixation in alfalfa, responses to bidirectional selection for associated characteristics. USDA Tech. Bull. No. 1643.
Williams, L.F. and Lynch, D.L., 1954. Inheritance of a non-nodulating character in the soybean. Agron. J., 46: 28-29.
Williams, L.E. and Phillips, D.A., 1983. Increased soybean productivity with a Rhizobium japonicum mutant. Crop. Sci., 23: 246-250.
Wilson, D., 1982. Responses to selection for dark respiration rate of mature leaves in Lolium perenne and its effects on growth of young plants and simulated swards. Ann. Bot., 49: 303-312.
Witty, J.F., Day, J.M. and Dart, P.J.,1976. The nitrogen economy of the Broadbalk experiments. II. Biological nitrogen fixation. Rothamsted Report for 1976, part 2: 111-118.
Wolk, C.P., Vonshak, A., Kehoe, P. and Elhai, J., 1984. Construction of shuttle vectors capable of conjugative transfer from Escherichia coli to nitrogen-fixing filamentous cyanobacteria. Proc. Natl. Acad. Sci. USA, 81: 1561-1564.

INDEX